River Pollution

Causes, Consequences and Remediation

River Pollution

Causes, Consequences and Remediation

Edited by

Ningombam Linthoingambi Devi
Department of Environmental Science,
Central University of South Bihar,
Gaya, Bihar, India

Amrita Srivastava
Department of Life Science,
Central University of South Bihar,
Gaya, Bihar, India

ELSEVIER

Elsevier
Radarweg 29, PO Box 211, 1000 AE Amsterdam, Netherlands
125 London Wall, London EC2Y 5AS, United Kingdom
50 Hampshire Street, 5th Floor, Cambridge, MA 02139, United States

ISBN: 978-0-443-45024-2

For information on all Elsevier publications visit our website at https://www.elsevier.com/books-and-journals

Publisher: Peter Llewellyn
Acquisitions Editor: Maria Elekidou
Editorial Project Manager: Naman Singh
Production Project Manager: Fahmida Sultana
Cover Designer: Raman Kumar

Typeset by Integra Software Services

Contents

Section II
Impact of pollution

10 Effect on aquatic flora and fauna

Kanhaiya Kumar, Madhusmita Kalita, and Kasi Marimuthu

11 Histopathological changes in fishes

H.A.C.C. Perera

14 Effect of river pollution on marine ecosystem

Rantim Bhattacharjee, Aanchal Chauhan, Pratika Singh, and Abhilash K. Tilak

15 Human health ailments due to polluted waters

Somvir Singh, Rahul Kumar, Arti Thakur, Vishav Kiran, Frank Zakeyo Wandira, Arti Sharma, Mitanshi Sharma, and Nitesh Kumar

17 Socioeconomic impacts of river pollution

Amit Kumar Singh and Ekta Verma

Section III
Assessment of pollution and remedial measures

18 Assessing water quality with advanced instrumental methods: Concepts, methods, and new approaches

Ashutosh Kumar, Niranjan Kumar, Prakash Kumar, and Santosh Dixit

Contributors

Sahana Savappla Anand Department of Environmental Sciences, School of Life Sciences, JSS Academy of Higher Education and Research, Mysuru, India

Akriti Ashesh Department of Environmental Science, Central University of South Bihar, Gaya, Bihar, India

Sk Md Asif Department of Agriculture, Brainware University, Kolkata, West Bengal, India

Soham Bachaspati Department of Agriculture, Brainware University, Kolkata, West Bengal, India

Rantim Bhattacharjee National Centre for Polar and Ocean Research (NCPOR), Ministry of Earth Sciences, Vasco-da-Gama; Department of Microbiology, School of Biological Sciences and Biotechnology (SBSB), Goa University, Panjim, Goa, India

Aanchal Chauhan National Centre for Polar and Ocean Research (NCPOR), Ministry of Earth Sciences, Vasco-da-Gama; Department of Microbiology, School of Biological Sciences and Biotechnology (SBSB), Goa University, Panjim, Goa, India

Mansi Chundawat Sophisticated Environmental Analytical Facility, CSIR-National Environmental Engineering Research Institute (CSIR-NEERI), Nagpur, India

Parijat De Department of Agriculture, Brainware University, Kolkata, West Bengal, India

Ningombam Linthoingambi Devi Department of Environmental Science, Central University of South Bihar, Gaya, Bihar, India

Okram Zenita Devi Department of Chemistry, Lady Irwin College, University of Delhi, New Delhi, India

Santosh Dixit Amity School of Engineering & Technology, Amity University Patna, Patna, Bihar, India

Subhadwip Ghorai Department of Agriculture, Brainware University, Kolkata, West Bengal, India

Pabitra Kumar Ghosh Department of Agriculture, Brainware University, Kolkata, West Bengal, India

Soham Hazra Department of Agriculture, Brainware University, Kolkata, West Bengal, India

Jayalakshmi Jayashankar Department of Biotechnology, Vel Tech Rangarajan Dr. Sagunthala R&D Institute of Science and Technology, Chennai, India

Prafull Kumar Jha Department of Energy and Environmental Engineering, CSIR-Indian Institute of Chemical Technology (CSIR-IICT), Hyderabad, India

Mwalimu Raphael Jilani Department of Biotechnology, Vel Tech Rangarajan Dr. Sagunthala R&D Institute of Science and Technology, Chennai, India

Madhusmita Kalita Department of Environmental Science, Tezpur University, Tezpur, India

Surbhi Kharwar Department of Botany, University of Lucknow, Lucknow, Uttar Pradesh, India

Vishav Kiran LR Institute of Pharmacy Jabil Kyar, Solan, India

Abinaya Krishnan Department of Biotechnology, Vel Tech Rangarajan Dr. Sagunthala R&D Institute of Science and Technology, Chennai, India

Ashutosh Kumar Amity School of Engineering & Technology, Amity University Patna, Patna, Bihar, India

Kanhaiya Kumar Department of Environmental Science, Tezpur University, Tezpur, India

Niranjan Kumar Amity School of Engineering & Technology, Amity University Patna, Patna, Bihar, India

Nitesh Kumar Department of Biosciences, Himachal Pradesh University, Shimla, India

Prakash Kumar Amity School of Engineering & Technology, Amity University Patna, Patna, Bihar, India

Rahul Kumar Department of Biosciences, University Institute of Biotechnology, Chandigarh University, Mohali, India

Ravinsh Kumar Department of Life Science, Central University of South Bihar, Gaya, Bihar, India

Vinod Kumar Department of Botany, Pandit Prithi Nath College (Affiliated to CSJM University, Kanpur), Kanpur, Uttar Pradesh, India

Sonali Kumari Department of Life Science, Central University of South Bihar, Gaya, Bihar, India

Kasi Marimuthu Department of Environmental Science, Tezpur University, Tezpur, India

Sanjay Mochary Department of Agriculture, Brainware University, Kolkata, West Bengal, India

Rajdeep Mohanta Department of Agriculture, Brainware University, Kolkata, West Bengal, India

Ankur Mukhopadhyay Department of Agriculture, Brainware University, Kolkata, West Bengal, India

Azhagu Saravana Babu Packirisamy Department of Biotechnology, Vel Tech Rangarajan Dr. Sagunthala R&D Institute of Science and Technology, Chennai, India

H.A.C.C. Perera Department of Zoology and Environmental Management, Faculty of Science, University of Kelaniya, Kelaniya, Sri Lanka

Shivaraju Harikaranahalli Puttaiah Department of Environmental Sciences, School of Life Sciences, JSS Academy of Higher Education and Research, Mysuru; Centre for Water, Food and Energy, GREENS TRUST, Turuvekere, Karnataka, India

Ahanthem Rebika Devi G.B. Pant National Institute of Himalayan Environment, Itanagar, India

Sourav Roy Department of Agriculture, Brainware University, Kolkata, West Bengal, India

Thilomi Samarakoon Department of Zoology and Environmental Management, University of Kelaniya, Kelaniya, Sri Lanka

Abhishek Saxena Sangam University, Bhilwara, India
Zubair Ahamed Shaik Department of Biotechnology, Vel Tech Rangarajan Dr. Sagunthala R&D Institute of Science and Technology, Chennai, India
Arti Sharma Faculty of Agriculture, Abhilashi University, Mandi, Himachal Pradesh, India
Mitanshi Sharma Department of Forensic Science, University Institute of Allied Health Sciences, Chandigarh University, Mohali, India
Ekta Shukla Department of Botany, Sunbeam College for Women Bhagwanpur (An Autonomous College), Varanasi, Uttar Pradesh, India
Amit Kumar Singh Post Graduate Department of Botany, Magadh University, Bodh-Gaya, India
Ashutosh Singh Department of Life Science, Central University of South Bihar, Gaya, Bihar, India
Pratika Singh National Centre for Polar and Ocean Research (NCPOR), Ministry of Earth Sciences, Vasco-da-Gama, Goa, India
Somvir Singh Department of Biosciences, University Institute of Biotechnology, Chandigarh University, Mohali, India
Mamataben Soni Lips Research Foundation, Marthandam, Tamil Nadu, India
Vithyasree Srikanth Department of Biotechnology, Vel Tech Rangarajan Dr. Sagunthala R&D Institute of Science and Technology, Chennai, India
Amrita Srivastava Department of Life Science, Central University of South Bihar, Gaya, Bihar, India
Arti Thakur Department of Botany, Shoolini Institute of Life Sciences and Business Management, Solan, India
Abhilash K. Tilak Department of Mechanical and Industrial Engineering, Manipal Institute of Technology Bengaluru, Manipal Academy of Higher Education, Manipal, India
Ekta Verma Post Graduate Department of Botany, Magadh University, Bodh-Gaya, India
Nisha Verma Department of Environmental Science, Central University of South Bihar, Gaya, Bihar, India.
Frank Zakeyo Wandira Department of Agriculture Extension, University of Agriculture and Natural Resources, Lilongwe, Malawi
Ishwar Chandra Yadav Applied Science, SITASRM Engineering and Research Institute, Knowledge Park-III, Greater Noida, Uttar Pradesh; Environmental Science, Non-Collegiate Women's Education Board, University of Delhi, New Delhi, India
Mangala Yatawara Department of Zoology and Environmental Management, University of Kelaniya, Kelaniya, Sri Lanka

Preface

Water is fundamental to life, and rivers are among the most dynamic and essential components of Earth's hydrological system. Since ancient times, rivers have supported the development of human civilizations by providing water for drinking, agriculture, transportation, and cultural activities. Many major cities and communities worldwide have grown along riverbanks because of their significant ecological and economic value.

Despite their importance, rivers today face unprecedented levels of pollution caused by increasing human activities. Industrial discharges, untreated municipal wastewater, agricultural runoff rich in fertilizers and pesticides, and improper solid waste disposal have significantly degraded river water quality. These pressures have led to serious environmental consequences, including declining aquatic biodiversity, disruption of ecosystem services, and threats to public health.

The growing global concern about river pollution underscores the urgent need for improved understanding and effective management of river systems. Scientific research over the past few decades has significantly improved our knowledge of water pollution processes, monitoring techniques, and control strategies. However, the complexity of river pollution requires an integrated approach that combines environmental science, technology, policy, and public participation.

The primary motivation for writing this book was to provide a comprehensive and accessible resource that systematically brings together the various aspects of river pollution. The book aims to explain the scientific foundations of river pollution while also discussing practical approaches for monitoring, management, and restoration of polluted rivers.

This work is intended to support students and researchers in environmental science, civil engineering, hydrology, and related disciplines. At the same time, it is hoped that the book will also be useful for environmental managers, policymakers, and practitioners working in water quality management and river conservation programs.

It is our sincere hope that this book will increase awareness of river pollution and encourage collaborative efforts to protect and sustainably manage river ecosystems.

Dr. Ningombam Linthoingambi Devi
Dr. Amrita Srivastava

Acknowledgement

The completion of this book would not have been possible without the guidance, encouragement, and support of many individuals and institutions to whom we express our sincere gratitude.

We are deeply grateful to the Prof. Kameshwar Nath Singh, Vice-Chancellor of *Central University of South Bihar, Gaya, India*, for providing the academic environment and institutional support necessary to conduct research and scholarly work on environmental issues. We also extend our heartfelt thanks to the *Department of Environmental Science* and *Department of Life Science* for fostering a stimulating academic environment that encourages critical thinking, research, and discussion of pressing environmental challenges, such as river pollution.

We also acknowledge the valuable contributions of researchers and scientists whose published works, reports, and studies on water quality, river ecology, and pollution control have greatly informed and strengthened the content of this book.

We want to extend our gratitude to the publishers and editors for their professional guidance, constructive suggestions, and assistance throughout the preparation and publication of this manuscript.

It is our sincere hope that this book contributes meaningfully to the understanding of river pollution and encourages researchers, students, and policymakers to work collectively toward the protection and restoration of our vital river ecosystems.

Dr. Ningombam Linthoingambi Devi
Dr. Amrita Srivastava

About the book

Rivers are among the most valuable natural resources on Earth and play a crucial role in sustaining ecological balance, human livelihoods, and economic development. They support freshwater ecosystems, provide drinking water, irrigate agricultural lands, and facilitate industrial and domestic activities. Historically, rivers have also shaped civilizations, cultures, and settlements across the world. However, in recent decades, rapid population growth, urban expansion, industrialization, and unsustainable land-use practices have placed immense pressure on river systems, leading to severe degradation of water quality.

River pollution has emerged as one of the most critical environmental challenges of the 21st century. Discharges from untreated municipal sewage, industrial effluents, agricultural runoff, solid waste dumping, and emerging contaminants have significantly altered the physical, chemical, and biological characteristics of river water. These pollutants not only affect aquatic ecosystems but also pose serious risks to human health, biodiversity, and sustainable water resource management.

This book presents a comprehensive overview of river pollution and its associated environmental challenges and advance techniques to control the emerging pollutants. It systematically examines the major sources and types of river pollutants, including nutrients, heavy metals, organic compounds, pathogens, plastics, and emerging contaminants such as pharmaceuticals and personal care products. The book also discusses the transport, transformation, and fate of pollutants within riverine systems.

In addition to explaining the scientific basis of river pollution, the book highlights various methods for monitoring and assessing river water quality, including physicochemical analyses, biological indicators, and modern analytical techniques. It also examines the ecological consequences of pollution, including eutrophication, habitat degradation, and biodiversity loss.

A significant portion of the book focuses on pollution prevention and management strategies. Topics such as wastewater treatment technologies, nature-based solutions, constructed wetlands, policy frameworks, river basin management approaches, and community participation are discussed in detail. The roles of environmental regulations, the Sustainable Development Goals, and integrated water resource management in addressing river pollution are also emphasized.

By integrating scientific principles, environmental management practices, and real-world examples, this book aims to provide a holistic understanding of river pollution. It is intended to serve as a valuable resource for undergraduate and postgraduate students, researchers, environmental professionals, policymakers, and academicians working in the fields of environmental science, hydrology, ecology, and water resource management.

Section I

River pollutants and their sources

Chapter 1

Natural and anthropogenic sources of river pollution

Abhishek Saxena[a] **and Mamataben Soni**[b]
[a]*Sangam University, Bhilwara, India*
[b]*Lips Research Foundation, Marthandam, Tamil Nadu, India*

1 Introduction

Water pollution occurs when substances are introduced into lakes, streams, rivers, estuaries, and seas to the extent that they make the water less useful or prevent ecosystems from functioning properly. Water pollution may also involve the discharge of energy, such as heat or radioactivity, into bodies of water. This is in addition to the release of substances such as chemicals, debris, or microbes.

Water is one of the most significant natural resources sustaining life on Earth and has enormous relevance to ecological balance, economic development, agriculture, industry, and human health. Rivers are vital sources of freshwater for drinking, irrigation, fisheries, transportation, recreation, and biodiversity. But the quality of water in the rivers has been substantially degraded by growing demands from natural processes and human activities worldwide. River pollution has become a significant environmental issue, threatening aquatic ecosystems, public health, and sustainable development.

Water pollution is the introduction of harmful levels of physical, chemical, biological, or radioactive pollutants into water bodies, which impairs their quality and intended uses. This pollution may alter the natural characteristics of water and render it unfit for human consumption, agriculture, industry, or the survival of aquatic species. Rivers are subjected to contaminants from many sources, including natural environmental processes and human (anthropogenic) activities. The accumulation of these contaminants causes ecological degradation, biodiversity loss, eutrophication, oxygen depletion, and the spread of waterborne illnesses.

Soil erosion and sediment transport, volcanic eruptions, decaying vegetation, animal wastes, and natural mineral deposits are natural causes of river pollution. Torrential rains and floods sweep sediments, organic materials, and naturally occurring compounds into rivers, increasing water turbidity and

River Pollution. https://doi.org/10.1016/B978-0-443-45024-2.00020-8

altering water chemistry. Minerals and other geological formations may also contribute metals and minerals such as arsenic, iron, and fluoride to river systems. Although these natural sources have occurred throughout Earth's history, altering climate patterns and extreme weather events may intensify their effects on water quality.

However, man-made causes are regarded as the main sources of river pollution across the globe. Fast industrialization, urbanization, population explosion, and agricultural development have contributed to a rise in the discharge of contaminants into freshwater habitats. Industrial effluents often include toxic compounds, heavy metals, dyes, solvents, and dangerous substances that are likely to persist in the environment for long periods. Agricultural runoff carries fertilizers, pesticides, herbicides, and animal wastes into rivers. This results in nutrient enrichment and ecological imbalance. Similarly, untreated or poorly treated municipal wastewater is a source of organic contaminants, bacteria, pharmaceuticals, and microplastics that degrade water quality and pose serious health risks.

The effects of river pollution have wide-reaching ramifications and damage environmental and socioeconomic systems. Contaminated rivers may affect the availability of clean drinking water, increase healthcare expenditures due to waterborne infections, reduce agricultural production, and have significant impacts on the fisheries and tourism sectors. In addition, polluted rivers may disrupt aquatic food chains, imperil endangered species, and modify ecological functions that are vital to human life. In many developing nations, poor waste management infrastructure and weak environmental regulations worsen the situation.

Knowledge of the natural and anthropogenic causes of river pollution is essential for developing effective management strategies, pollution control techniques, and sustainable water resource regulations. Understanding the sources, pathways, and impacts of diverse contaminants helps academics, policymakers, and environmental managers to devise solutions to conserve river ecosystems and ensure long-term water security. Thus the comprehensive assessment and monitoring of river pollution are important for the maintenance of ecological integrity and achievement of sustainable development objectives related to clean water and environmental protection.

2 Types and sources of water pollutants

There are many things that may contaminate water bodies, such as pathogenic microbes, organic waste that can rot, fertilizers and plant nutrients, poisonous chemicals, sediments, heat, petroleum (oil), and radioactive materials. We discuss a few different kinds of water pollution in the following sections. See waste disposal and solid-waste management for a discussion of how to handle sewage and other types of trash that people produce.

There are two types of causes of water pollution: point sources and scattered sources. A point source is a conduit or canal, such as the ones that carry waste from a factory or a city's sewage system. A scattered (or nonpoint) source is a large, open region where a lot of different contaminants may get into the water, such as when water runs off farmland. Point sources of water pollution are simpler to address than scattered sources since the polluted water has been gathered and sent to one place where it may be cleaned. It is difficult to regulate pollution from many different places, and even though contemporary sewage treatment facilities have made a lot of progress, scattered sources still generate a lot of water pollution concerns.

2.1 Domestic sewage

Domestic sewage is the primary source of pathogens (microbes that cause illness) and organic matter that may decompose. Because germs are found in feces, all sewage from cities and towns likely contains some form of disease, which poses a direct hazard to public health. Putrescible organic matter is another threat to water quality. Bacteria and other microbes break down organic matter in the sewage, which reduces the amount of dissolved oxygen in the water. This endangers the health of lakes and streams, as fish and other aquatic life require high levels of oxygen to survive. In addition, active pharmaceutical components are often present in household sewage, which can harm aquatic life and contribute to bacteria developing resistance to antibiotics. Wastewater treatment techniques reduce the levels of pathogens and organic matter in sewage, but they do not eliminate them entirely (see also wastewater treatment).

Sewage from homes is another important source of plant nutrients, especially nitrates and phosphates. When there are too many nitrates and phosphates in water, algae may develop quickly and densely, a phenomenon known as an algal bloom. When algae die, the amount of oxygen in the water decreases because microorganisms consume oxygen to break down the algae (see also biochemical oxygen demand). Anaerobic organisms, which do not require oxygen to thrive, subsequently break down the organic waste, generating gases like methane and hydrogen sulfide that are harmful to aerobic (oxygen-requiring) life forms. Eutrophication is the process by which a lake transitions from being clean and clear, with a minimal amount of dissolved nutrients and a healthy aquatic ecosystem, to being full of nutrients and algae, and subsequently to being poor in oxygen and filled with waste. Eutrophication is a gradual, natural, and unavoidable process. However, when humans accelerate it and pollute the water (a process known as cultural eutrophication), it can cause a body of water to age and deteriorate too quickly.

2.2 Solid waste

Not disposing of solid waste properly is a significant cause of water contamination. Individuals, households, businesses, schools, and factories all produce solid waste, which includes garbage, rubbish, technological waste, trash, and construction and demolition debris. This issue is particularly severe in underdeveloped nations that may lack the capacity to properly dispose of solid waste or may not have the resources or regulations to prevent improper disposal. Some people deliberately throw solid waste into bodies of water. In addition, if animals, wind, or rain carries garbage or other waste to bodies of water, land pollution can also become aquatic pollution. A significant amount of solid waste contamination in inland bodies of water may eventually end up in the ocean. Solid waste contamination not only looks unsightly but also harms aquatic environments and may even directly harm animals. Plastics and electronic waste are two examples of solid waste that break down and release harmful chemicals into the water, making them a source of toxic or hazardous waste.

Plastic pollution is becoming an increasingly significant problem for marine environments. Because the ocean flows downstream from almost all land, it accumulates a substantial amount of plastic debris produced on land. Every year, millions of tons of trash end up in the seas, and much of it consists of plastic that has been improperly disposed of. Waves and UV radiation can break down plastic pollution into tiny particles called microplastics. These fragments are <5 mm (0.2 inch) long and cannot be naturally degraded. Primary microplastics, such as microbeads in personal care products and plastic fibers in synthetic textiles (like nylon), also enter the environment directly through various pathways. For example, they can originate from wastewater treatment plants, household laundry, or accidental spills during manufacturing or shipping. Alarmingly, microplastics have been detected in every aquatic organism studied in both freshwater and marine environments. These tiny plastic particles are believed to be infiltrating the food chains of marine animals, from zooplankton and small fish to large marine predators. They have also been found in seafood. Researchers have also discovered microplastics in drinking water. However, we still do not fully understand their impact on human health.

2.3 Toxic waste

Water is hazardous if it is poisonous, radioactive, explosive, carcinogenic (causing cancer), mutagenic (causing damage to chromosomes), teratogenic (causing birth abnormalities), or bioaccumulative (that is, rising in concentration at the upper ends of food chains). Toxic chemicals may come from poorly disposed wastewater from factories and chemical processing plants (lead, mercury, chromium), as well as surface runoff that contains pesticides used on farms and suburban lawns (chlordane, dieldrin, heptachlor). (For further information on dangerous substances, see poison and toxic waste.)

2.4 Sediment

Surface runoff may carry sediment, including silt, from eroded soil or construction sites into bodies of water. Sediment suspended in water prevents sunlight from penetrating and disrupts the natural balance of the ecosystem. It may also interfere with the reproductive cycles of fish and other aquatic organisms, and as it settles out of suspension, it can smother species that live on the bottom.

River sediments are an integral part of the aquatic environment and act as both a sink and a source of toxins. Sediments are particles that settle at the bottom of rivers, streams, and other water bodies. These particles include sand, silt, clay, organic debris, and mineral fragments. Sediments play a role in nutrient cycling, providing habitat for aquatic organisms and maintaining ecological balance. However, the accumulation of numerous contaminants in river sediments due to the increase in natural disturbances and human activities has become a serious environmental problem.

Sediment contamination is the adsorption of contaminants from natural or anthropogenic sources onto sediment particles and their subsequent deposition on the riverbed. Due to their large surface area and chemical characteristics, fine-grained sediments such as clay and organic-rich particles are capable of adsorbing and trapping pollutants to a significant extent. Once deposited, these contaminants may remain in the sediments for years to decades and constitute long-term storage of pollution.

Other natural processes such as rock weathering, volcanic activity, soil erosion, and floods may also contribute to sediment pollution. Rivers may also contain naturally occurring metals such as arsenic, mercury, chromium, iron, manganese, and lead from geological formations. Heavy rainfall causes surface runoff, which transports degraded soil, organic matter, and minerals into rivers, increasing the sediment load and altering the sediment composition. Extreme climate-driven phenomena such as floods and landslides may mobilize polluted soils and sediments from the surrounding areas.

The primary anthropogenic activities that contribute to river sediment pollution include industrial effluents that introduce heavy metals, petroleum hydrocarbons, dyes, solvents, and hazardous organic compounds into waterways. Industries such as mining, metal processing, electroplating, textile manufacturing, paper mills, and chemical production contribute contaminants that become attached to sediment particles. In particular, silt contamination from mining activities is significant. Mine tailings often introduce significant amounts of heavy metals that may persist in sediments for long periods.

Sediment pollution is also a major consequence of agricultural practices as fertilizers, insecticides, herbicides, and animal wastes run off farmed land into waterways. Many pesticides are hydrophobic in nature and tend to adsorb onto sediment particles rather than remain in solution in water. Nitrogen and phosphorus nutrients bind to the sediments and accumulate in riverbeds. When they are discharged back into the water column, they lead to eutrophication and hazardous algal blooms.

Stormwater runoff from metropolitan areas exacerbates sediment pollution. Roads, construction sites, parking lots, and residential areas create sediments containing oil and grease residues, heavy metals, microplastics, pharmaceuticals, detergents, and other contaminants. While many contaminants may be removed from water by wastewater treatment facilities, substantial amounts of pollutants can still accumulate in sewage sludge and eventually enter river sediments via effluent discharge or inadvertent discharges.

One of the most alarming aspects of sediment contamination is its ability to act as a secondary source of pollution. Environmental events such as floods, dredging, navigation operations, changes in water flow, or disturbance by aquatic creatures may resuspend contaminated sediment particles into the water column. This process releases previously sequestered contaminants, making them more bioavailable and exposing aquatic organisms to hazardous compounds. Heavy metals and persistent organic pollutants may then enter the food chain through benthic creatures and subsequently to fish, birds, and humans via biomagnifications.

Contaminated sediments have major ecological impacts. Their impacts include loss of biodiversity, changes in microbial communities, impaired fish reproduction, and harm to benthic ecosystems. In addition, contaminated sediment may induce harmful consequences such as oxidative stress, genetic alterations, endocrine disruption, and developmental abnormalities in aquatic creatures. Moreover, sediments pollution poses a major threat to human health when polluted water or infected fish are swallowed.

Hence, the processes of sediment pollution have to be understood for successful river management and restoration. The quality of the sediment should be periodically evaluated, and the sources of contamination should be recognized. Erosion control measures should be established, and industrial and municipal effluents should be treated. Sustainable farming methods play an important role in minimizing sediment pollution and safeguarding river habitats. Sediments are long-term sinks of pollutants, and their estimations provide information on historical and current patterns of pollution in river ecosystems.

2.5 Thermal pollution

Because it makes it harder for water to retain dissolved oxygen and speeds up the metabolism of fish, heat is considered a form of water pollution. Trout and other valuable game fish cannot survive in water that lacks sufficient dissolved oxygen. Power stations sometimes discharge cooling water into rivers, which may be 15°C (27°F) warmer than the existing water. This is a significant source of heat. In addition, the fact that global warming is causing water temperatures to rise is another example of thermal pollution.

Thermal pollution is the impairment of water quality by any process that increases the temperature of ambient water. It results from the discharge of heated or cooled water into rivers from industrial, agricultural, or urban sources and produces dramatic changes in the thermal regime of the aquatic environment. Thermal pollution does not add typical chemical pollutants to water, but it may have significant impacts on the physical, chemical, and biological characteristics of rivers. Thermal pollution is increasingly recognized as a major source of river pollution.

The largest source of thermal pollution from human activities is the discharge of cooling water from thermal power plants, nuclear power stations, industrial enterprises, refineries, steel factories, and chemical processing facilities. These enterprises use vast volumes of water for cooling and then discharge the hot water into neighboring rivers. Another cause of thermal pollution is urbanization, which creates stormwater runoff from roads, parking lots, rooftops, and other impervious surfaces that absorb solar radiation and add heat to runoff before it enters rivers. The removal of riparian vegetation along rivers also reduces the natural shade and allows greater solar heating of river water, thereby increasing water temperatures.

Natural factors influencing river temperature include climate change, long-term drought, reductions in flow of rivers, geothermal activity, and seasonal changes in temperature. But human activities are likely to exacerbate these consequences, causing long-term heat stress in aquatic ecosystems. Climate change is causing global warming, which causes more frequent and intense heat waves, and consequently, many rivers are more susceptible to thermal pollution.

One of the most significant effects of thermal pollution is the reduction in dissolved oxygen levels in river water. As water temperature increases, the ability of water to retain dissolved oxygen diminishes. Aquatic organisms (fish, aquatic insects, mollusks, bacteria, etc.) require adequate amounts of oxygen to thrive. Lower oxygen levels may induce physiological stress, reduced development, reduced reproductive success, and in severe circumstances mass fish mortality. This is particularly harmful in rivers already contaminated with organic materials, as oxygen is also required by the bacteria that break down organic matter.

Thermal pollution also affects the metabolism of aquatic creatures. Higher temperatures increase the rates of biochemical and physiological processes, and oxygen demand increases, while oxygen supply decreases. Such an imbalance may disrupt the normal functioning of the ecosystem and increase the susceptibility of organisms to diseases, parasites, and environmental stress. Each fish species has a temperature range for feeding, spawning, and development. Changes in these ranges might have detrimental effects on fish population dynamics and biodiversity.

Temperature change in rivers may have a profound effect on the distribution and composition of aquatic communities. Sensitive species may decline in number or shift to colder areas. Invasive species may become dominant if they are heat tolerant. Such alterations may alter food-web organization and ecological stability. Thus thermal pollution may be a factor in the loss of biodiversity and ecologically important species.

Another prominent impact of thermal pollution is the promotion of algae growth and eutrophication. Higher temperatures frequently stimulate the development of algae and cyan bacteria, particularly when nutrients such as nitrogen and phosphorus are available. Large algal blooms may block sunlight from reaching aquatic plants, disrupt their habitat, and deplete the water of dissolved oxygen as algae die and decompose. Cyanobacterial blooms may occasionally release toxins that pose risks to aquatic animals, livestock, and human health.

Thermal pollution and its impact on the behavior and transport of chemical contaminants in rivers are also significant. Higher temperatures may enhance the solubility, reactivity, and toxicity of certain contaminants, including heavy metals and organic chemicals. River sediments may also release toxins more rapidly under warmer conditions, increasing their bioavailability and environmental impact. Therefore thermal pollution typically acts synergistically with chemical pollution to augment the overall ecological harm.

Temperature variations also alter river sediments. Rising temperatures may alter sediment chemistry, increase microbial degradation of organic matter, and promote the release of nutrients and pollutants from sediments into the overlying water. These processes may produce secondary contamination and expand the impacts of toxins already present in river ecosystems.

Thermal pollution has substantial ecological and socioeconomic impacts. Poor water quality may impair fisheries, aquaculture, recreation, and drinking water sources. Economic losses may occur due to a reduction in fish productivity, increased expenses of water treatment, and deterioration of ecosystem services. Thermal pollution may also make river ecosystems less robust, meaning that they are less able to recover from other environmental disturbances.

Effective management of thermal pollution requires a combination of technical, ecological, and regulatory techniques. Cooling towers, cooling ponds, heat exchangers, and water recycling systems may be used to lower the temperature of industrial effluents before their release. Restoration of riparian vegetation provides natural shade that cools river temperatures. Environmental control of thermal discharges, continuous monitoring of river temperature, and sustainable water resource management methods are also involved in the protection of aquatic ecosystems. The health of rivers, biodiversity, and long-term water sustainability will increasingly depend on the control of thermal pollution as the impacts of climate change on freshwater habitats continue to intensify.

2.6 Petroleum (oil) pollution

When oil from highways and parking lots washes into bodies of water, it causes petroleum (oil) contamination. Accidental oil spills are another way that oil gets into the environment. For example, the Exxon Valdez tanker spilled >260,000 barrels of oil into Alaska's Prince William Sound in 1989, and the Deepwater Horizon oil rig spilled >4 million barrels of oil into the Gulf of Mexico in 2010. Oil slicks ultimately make their way to shore, where they harm aquatic life and ruin places where people may enjoy recreational activities.

2.7 Effects of water pollution on groundwater and oceans

Many people get their drinking water from aquifers, which are subterranean rock formations that hold water. For instance, roughly half of the population in the United States gets their drinking water from groundwater. Even though groundwater may seem crystal clear (since it naturally filters itself as it slowly travels through layers of soil), it may still be contaminated by dissolved chemicals, bacteria, and viruses. Chemical pollutants come from sources such as poorly designed or poorly maintained subsurface sewage-disposal systems (like septic tanks), industrial wastes that are disposed of in landfills or lagoons that are not lined properly or at all, leachates from unlined municipal refuse landfills, mining and oil production, and leaking underground storage tanks below petrol stations. As cities and industries grow, they use more groundwater from coastal regions. This may lead to saltwater intrusion, which occurs when seawater is drawn into wells as the water table decreases.

Even though estuaries and seas contain a lot of water, they cannot naturally absorb large amounts of pollution. Marine life may be harmed by sewage outfall pipes, sludge or other waste dumping, and oil spills. This is particularly true for tiny phytoplankton, which provide nourishment for larger aquatic animals. Sometimes, unsightly and hazardous trash may wash back to shore, leaving beaches littered with dangerous debris. In the early 21st century, it was estimated that all kinds of plastic pollution in the oceans amounted to between 4.8 million and 12.7 million tons (5.3 million and 14 million tons) each year. Floating plastic waste had accumulated in Earth's five subtropical gyres, which cover 40% of the world's oceans.

Another concern with ocean pollution is the seasonal creation of "dead zones" in certain coastal locations. These are places where dissolved oxygen levels drop so low that most higher forms of aquatic life disappear. The reason is the addition of nutrients from agricultural runoff and the simultaneous growth of algae. There are dead zones all over the globe. One of the largest, which may be as big as 22,730 square kilometers (8776 square miles), emerges every year in the Gulf of Mexico, starting in the Mississippi River delta.

Rivers are essential sources of freshwater that provide numerous benefits to the environment and society. They play a crucial role in sustaining life and supporting ecosystems, as well as serving as sources of irrigation, hydropower, and transportation. However, the health and well-being of rivers are under threat due to pollution, climate change, and human activities. In this chapter, we explore the significance of river water and the challenges faced by rivers today.

Importance of river water: Rivers are vital sources of freshwater that support the survival of millions of species, including humans. They play an essential role in regulating the water cycle, which involves the movement of water between the atmosphere, land, and oceans. Rivers collect rainwater and surface runoff, which are then transported to other parts of the world, replenishing groundwater reserves and sustaining wetlands, lakes, and other aquatic habitats. Moreover, rivers support the growth of vegetation and provide nutrients to terrestrial ecosystems, which, in turn, provide habitats for a variety of species. Apart from their ecological significance, rivers have immense economic and social importance. They are vital sources of irrigation water for agriculture and supply water to millions of people worldwide. Rivers also provide hydropower, which is a clean and renewable source of energy that can be used to generate electricity. In addition, they serve as transportation routes for goods and people, facilitating trade and commerce.

2.8 Societal importance of river water

Water supply: Rivers are a primary source of freshwater for human consumption, agriculture, and industries. They serve as a source of drinking water for millions of people, particularly those living near river basins. Reliable access to clean river water is essential for the well-being, sanitation, and health of communities.

Irrigation and agriculture: River water is extensively used for irrigation in agriculture, supporting crop growth and food production. Through irrigation systems, farmers can cultivate larger areas, improve crop yields, and sustain livelihoods. This plays a crucial role in food security and the economy of many regions.

Energy generation: Rivers offer immense potential for hydropower generation. By harnessing the kinetic energy of flowing water, hydroelectric power plants produce clean and renewable energy. Hydropower not only contributes to national energy grids but also helps reduce greenhouse gas emissions and combat climate change.

2.9 Challenges facing rivers

Despite their significance, rivers face numerous challenges that threaten their health and well-being. One of the biggest threats is pollution, which can come from various sources, including agricultural runoff, sewage, and industrial discharges. Polluted rivers not only harm aquatic life but also pose a risk to human health, as contaminated water can cause diseases such as cholera, typhoid, and hepatitis. Climate change is another significant threat to rivers, as it can cause changes in temperature, rainfall patterns, and river flow. These changes can lead to increased flooding, droughts, and water shortages, affecting both human and natural systems. Moreover, climate change can exacerbate existing problems such as pollution and habitat destruction, further weakening river ecosystems. Human activities such as damming, dredging, and urbanization also pose a threat to rivers. Damming can alter river flow, affecting aquatic habitats and fish populations. Dredging can disrupt river ecosystems, and urbanization can increase pollution, reduce water quality, and cause habitat destruction.

Importance of clean rivers is the first step toward preserving them for future generations.

3 The role of rivers in supporting life

3.1 Source of drinking water

Nearly **2 billion people** rely on rivers as their primary source of drinking water. Rivers are natural reservoirs that supply clean water to countless communities and cities. However, contamination from industrial waste, untreated sewage, and agricultural runoff compromises the safety of this water, posing severe health risks.

3.2 Agriculture and food security

Rivers play a critical role in irrigating farmlands, supporting 70% of global food production. From rice paddies in Asia to vineyards in Europe, freshwater from rivers ensures the growth of crops that feed the world. Polluted rivers, however, lead to degraded soil quality and reduced agricultural yields, threatening global food security.

3.3 Biodiversity hotspots

Rivers and their connected ecosystems are home to 80% of global biodiversity. Fish, amphibians, birds, and mammals depend on clean waterways for survival. When rivers are polluted, the delicate balance of aquatic ecosystems is disrupted, leading to species extinction and a loss of biodiversity.

4 The economic value of clean rivers

4.1 Tourism and recreation

Clean rivers attract millions of tourists annually, contributing significantly to local economies. Activities such as kayaking, fishing, and eco-tourism depend on the health of these waterways. Polluted rivers, on the contrary, deter visitors and diminish the economic benefits of tourism.

4.2 Hydropower and energy production

Rivers are a major source of renewable energy, with hydropower plants generating electricity for millions of households. Clean rivers ensure efficient energy production, while polluted ones lead to increased maintenance costs and reduced efficiency.

4.3 Industrial and commercial use

Industries depend on rivers for water in manufacturing processes. However, industries must also ensure their waste is treated to avoid contaminating these vital resources. Maintaining clean rivers not only supports industrial needs but also prevents long-term economic damage caused by environmental degradation.

5 Threats to river health

5.1 Pollution

Pollution is the most significant threat to river health. Industrial discharge, agricultural runoff laden with pesticides and fertilizers, and plastic waste all contribute to the contamination of rivers. This not only harms aquatic life but also poses serious risks to human health.

5.2 Climate change

Rising global temperatures and changing precipitation patterns affect river flow, leading to droughts or floods. These extreme conditions can alter the ecosystems that depend on rivers and reduce their ability to provide clean water.

5.3 Urbanization

Rapid urbanization leads to habitat destruction and increased waste entering rivers. Encroachment along riverbanks disrupts natural water filtration processes, making it harder for rivers to cleanse themselves.

6 Sources of river pollution

6.1 Natural sources

6.1.1 Sedimentation

Natural organic matter

Sediment consists of the loose sand, clay, silt, and other soil particles that settle at the bottom of a body of water. Sediment can result from soil erosion or from the decomposition of plants and animals. Wind, water, and ice help carry these particles to rivers, lakes, and streams.

Facts about sediment: The Environmental Protection Agency lists sediment as the most common pollutant in rivers, streams, lakes, and reservoirs. While natural erosion produces nearly 30% of the total sediment in the United States, accelerated erosion from human use of land accounts for the remaining 70%. The most concentrated sediment releases come from construction activities, including relatively minor home-building projects such as room additions and swimming pools.

Sediment pollution causes $16 billion in environmental damage annually. Sediment entering stormwater degrades the quality of water for drinking, wildlife, and the land surrounding streams in the following ways: Sediment fills up storm drains and catch basins that carry water away from roads and homes, which increases the potential for flooding. Water polluted with sediment becomes cloudy, preventing animals from seeing food. Murky water prevents natural vegetation from growing in the water. Sediment in stream beds disrupts the natural food chain by destroying the habitat where the smallest stream organisms live, causing massive declines in fish populations.

Sediment increases the cost of treating drinking water and can result in odor and taste problems. Sediment can clog fish gills, reducing resistance to disease, lowering growth rates, and affecting fish egg and larvae development. Nutrients transported by sediment can activate blue–green algae that release toxins and can make swimmers sick.

Sediment deposits in rivers can alter the flow of water and reduce water depth, which makes navigation and recreational use more difficult. Sweep sidewalks and driveways instead of hosing them off, as washing these areas results in sediment and other pollutants running off into streams, rivers, and lakes. Use weed-free mulch when reseeding bare spots on your lawn, and use a straw erosion control blanket if restarting or tilling a lawn.

Notify local government officials if you see sediment entering streets or streams near a construction site. Apply compost or weed-free mulch to your garden to help prevent soil from washing away. Avoid mowing within 10–25 feet of the edge of a stream or creek. This will create a safe buffer zone that helps minimize erosion and naturally filters stormwater runoff, which may contain sediment. Either wash your car at a commercial car wash or on a surface that absorbs water, such as grass or gravel.

6.2 Volcanic eruptions

Heavy metal pollution: Heavy metals may be released into the aquatic environment as a result of human activities such as the discharge of wastewater from factories, agriculture, and settlements. In addition, heavy metals from natural sources (leaching and weathering) can enter the environment, but these sources are often less significant. Untreated industrial wastewater and household sewage are the main contributors to heavy metal pollution. Wastewater from industries such as metal industries, paints, pigments, varnishes, pulp and paper, tanneries, distilleries, rubber, thermal power plants, steel plants, and mining industries can introduce heavy metals to water bodies, including zinc (Zn), arsenic (As), copper (Cu), lead (Pb), cadmium (Cd), mercury (Hg), nickel (Ni), and chromium (Cr). Table 1.1 shows the concentrations of most heavy metals present in river water, with levels sometimes exceeding the guideline values. For instance, a study of the Aji-Chay River in Iran detected high levels of As, Cr, and other heavy metals, which could be the result of the weathering of volcanic formations and industrial discharges. The concentrations of As, Cr, and Pb were 0.0008–0.046 mg/L, 0.005–0.175 mg/L, and 0.021–0.075 mg/L, respectively, exceeding the guideline values set by the WHO. The high metal concentrations can be traced back to wastewater discharged by mining activities. Drainage water and wastewater from mining regions caused a dramatic increase in heavy metal levels in the Voghji River basin in Armenia. As a result, river water quality sharply deteriorated due to elevated concentrations of heavy metals such as Cu, Cd, and Pb. These substances can leach into surface water or groundwater, be absorbed by plants, and form semipermanent bonds with soil constituents like clay or organic matter, which later impact human health. After entering waterbodies, heavy metals can be harmful to aquatic organisms and accumulate in sediment. Although the levels of heavy metals are increasing in river water due to anthropogenic activities, the concentrations of some heavy metals are above the limits set by the WHO in many case studies.

TABLE 1.1 Indian standards for drinking water specifications.

S. no.	Parameter	Desirable limit
1	Turbidity	1–5
2	pH	6.5–8.5
3	Total hardness	200–600
4	Calcium	75–200
5	Magnesium	30–100
6	Chlorides	250–1000
7	Sulfates	150–400
8	Nitrate	45 (no relaxation)
9	Fluoride	0.6–1.2
10	Iron	(Cr + 6) 0.05
11	Alkalinity	200–600
12	BOD	30–100
13	TDS	500–2000
14	Oil and grease	0.5 (no relaxation)

Adapted from IS 10500:2012 Drinking Water—Specification (Second Revision), Bureau of Indian Standards (BIS), New Delhi, India.

Mount Merapi is an active volcano on Java Island that experiences both minor and major eruptions, with pyroclastic material spreading to 10 rivers, including (1) Opak, (2) Bedog, (3) Krasak, (4) Code (or Boyong), (5) Gendol, (6) Winongo, (7) Gadjahwong, (8) Kuning, (9) Putih, and (10) Sempor (Gob et al., 2016). Despite its status as the most active volcano in Indonesia, the surrounding communities maintain special bonds that encourage them to stay. The latest eruption phase occurred on January 4, 2021, when magma supplies triggered a shallow volcanic earthquake 8 km below the Earth's surface in October 2019 (BPPTKG, 2021). This eruption caused deformation of the mountain's flanks (Gomez et al., 2022) and produced pyroclastic density currents that traveled 1–3 km from the summit (Thouret et al., 2020). The pyroclastic deposits distributed within the rivers originating at Mount Merapi transform into cold "lahar" (volcanic mudflow) during the rainy seasons (Wibowo et al., 2015; Ville et al., 2015). Although the mountain poses a high risk, people continue to reside in the area due to several ecosystem services that support their livelihoods (Chung and Kang, 2013; Dede et al., 2023).

A combination of high rainfall in tropical areas and porous material deposits guarantees the abundance of surface and groundwater around Merapi (Ikhsan et al., 2019; Sekaranom et al., 2021). Especially for surface water, changes in the ecosystem services can be observed through river and reservoir water analysis. Previous observations have shown that Merapi's volcanic material is generally unconsolidated and easily erodible, which allows rainwater flows to transport into the river networks (Kusumawardani et al., 2017; Hapsari et al., 2020). Subsequently, a matrix mixture of rock, gravel, sand, dust, and water forms sediment loads in the rivers, which combine with erosion materials and anthropogenic activities from the surrounding land use. This surrounding landscape contains primary and secondary forests, as well as settlements, wet agricultural land, sand mining areas, fields, and plantations (Yudistira et al., 2020). Population growth and increased accessibility also trigger pressure on land, thereby contributing to changes in water quantity and quality, biodiversity, and renewable energy sources.

The combination of landscape changes, the emergence of contaminants, and human intervention is a determinant for RES dynamics. RES represents the ability of natural processes and components in a water body to provide direct or indirect adequate services for life needs. Moreover, the investigation of RES has garnered significant attention since the emergence of environmental degradation and the decline in biodiversity in freshwater ecosystems. For example, in the Nenjiang River, China, it was discovered that reduced forest cover and wetlands converted for cultivation have affected RES (water regulation and supply) by 4.62%–14.34% since 1980 (Wang et al., 2015). Comprehending RES, as observed in the Emscher River, Germany, is vital for achieving optimal outcomes in environmental restoration, such as improvements in water quality, biodiversity, climate regulation, and flood retention (Böck et al., 2018; Gerner et al., 2018). Hidayat et al. (2018) stated that erosion of the riverbank due to natural and anthropogenic activities can cause an increase in sediment loads and silt, as well as a decrease in water storage capacity in the Brantas River, Indonesia. Material from the LUSI ("Lumpur Sidoarjo") mud volcano has been found to contain significantly high levels of TSS and oxygen demand. The Porong River, which is the main tributary of the mud volcano, also contains heavy metals, though these are below the recommended standard (Krisnayanti and Agustawijaya, 2014).

This study aims to assess RES in the volcanic area to understand how Merapi's eruption, landscape, and anthropogenic activities affect regulating and supporting services. The study of volcanic eruptions and RES provides a way to understand how volcanism correlates with the aquatic environment. These include particulate filtration, waste treatment, nutrient regulation, biological control, and primary productivity, respectively. RES changes occur due to landscape factors from volcanic eruptions and community behavior. A previous report has identified the river as a depository for mining materials such as sand and stone; these activities alter ecosystems surrounding rivers around Mount Merapi (de Bélizal et al., 2013; Miller, 2022). Therefore the rivers in Merapi's landscape are not only water providers but also distribution routes for cold lahar. Since RES changes affect the surrounding dynamics, evaluating its impact in volcanic areas is necessary to support adequate policy development in the context of river usage and management (Burkhard et al., 2012).

6.3 Industrial waste and its impact on river pollution

One of the biggest causes of river contamination is industrial waste. Factories and industrial companies typically dump waste straight into rivers without treating it properly or treating it poorly. Heavy metals, including lead, mercury, and cadmium, as well as organic compounds like solvents and dyes, may be found in these effluents.

These harmful chemicals in river water may have terrible consequences for ecosystems in the water. Heavy metals, for example, may build up in the tissues of fish and other creatures, which can lead to bioaccumulation and biomagnification along the food chain. This is harmful to fish and other aquatic life, and it is also detrimental to people who consume food that has been polluted. Chemical pollution is not the only way industrial waste damages rivers. Industrial activity may potentially exacerbate pollution concerns by altering the way rivers function. For instance, building dams and levees may modify the natural flow of water, which can cause sedimentation problems and changes in the structure of habitats.

Thermal pollution from factories may also elevate the temperature of the water, which can be even more stressful for aquatic creatures that are sensitive to fluctuations in temperature. The combined consequences of these variables make it clear that industrial sectors require stronger regulations and better methods for managing waste immediately.

6.4 Agricultural runoff and its contribution to river pollution

Agricultural runoff is another significant cause of river pollution that has been receiving increasing attention in recent years. When rains wash fertilizers and pesticides off agricultural fields, they may end up in neighboring rivers, where they can provide nutrients. Fertilizers that contain nitrogen and phosphorus are especially worrisome because they can cause algal blooms in bodies of water.

These blooms not only lower the amount of oxygen in the water, but also produce chemicals that may harm animals and people who come into contact with them. The effects of agricultural runoff go beyond immediate problems with water quality. Algal blooms may cause hypoxic conditions, which are areas with low oxygen levels that kill aquatic life. This has been observed in many rivers and coastal locations across the globe, such as the Dead Zone in the Gulf of Mexico, which is primarily caused by nutrients washing off farms in the Mississippi River Basin. In addition, runoff may transfer germs from farms with animals into rivers, further degrading water quality and putting the health of those who depend on these water supplies at risk.

7 Solutions to protect and restore rivers

7.1 Implementing stronger policies

Governments must enforce regulations to reduce industrial discharge, promote sustainable agricultural practices, and manage urban waste. Policies aimed at penalizing polluters and incentivizing clean technologies can significantly improve river health.

7.2 Community engagement

Grassroots efforts play a vital role in protecting rivers. Community clean-up drives, awareness campaigns, and citizen science initiatives encourage people to take responsibility for their local waterways.

7.3 Adopting green infrastructure

Constructing wetlands, planting vegetation along riverbanks, and creating natural buffers can reduce pollution and improve water quality. These solutions also enhance biodiversity and provide flood protection.

7.4 Innovations in waste management

Advancements in waste treatment and recycling can prevent harmful substances from reaching rivers. From bioremediation techniques to better plastic management, innovative solutions hold promise for reducing river pollution.

Urban development significantly influences river pollution through increased impervious surfaces such as roads, buildings, and parking lots. These surfaces prevent rainwater from soaking into the ground, leading to increased surface runoff during storms. This runoff often carries a cocktail of pollutants, including oil, heavy metals, trash, and sediments, directly into rivers without adequate treatment.

The result is a marked decline in water quality as these contaminants accumulate in river systems. Moreover, urbanization often leads to habitat destruction along riverbanks, which can destabilize ecosystems. The loss of vegetation that naturally filters pollutants from runoff exacerbates the problem by allowing more contaminants to enter waterways unimpeded.

In addition, urban areas frequently experience combined sewer overflows, where heavy rainfall overwhelms sewage systems, causing untreated sewage to be discharged into rivers. This not only poses immediate health risks but also contributes to long-term ecological damage as pathogens and nutrients enter the aquatic environment.

7.4.1 The role of industrial waste in river pollution

The role of industrial waste in river pollution cannot be overstated; it is a complex issue that intertwines with regulatory frameworks, technological advancements, and corporate responsibility. Many industries have historically prioritized profit over environmental stewardship, leading to lax waste management practices that result in significant pollution events. For example, the infamous Love Canal disaster in the 1970s highlighted the dangers of improper hazardous waste disposal, when toxic chemicals leached into groundwater and nearby waterways, affecting thousands of residents.

In response to such incidents, regulatory bodies have implemented stricter guidelines for industrial discharges; however, enforcement remains a challenge. Many industries still find loopholes or engage in practices that minimize their costs at the expense of environmental health. The introduction of technologies such as wastewater treatment plants has improved some situations; however, not all facilities are equipped with state-of-the-art systems capable of removing all harmful substances from their effluents.

Continuous monitoring and innovation are essential to ensure that industrial waste does not continue to pollute rivers.

8 Some major polluted rivers in India

The following are the major polluted rivers in India: Yamuna River, Brahmaputra River, Son River, Brahmani River, Mahanadi River, Ganga River, Cauvery River, Gomti River, Chambal River, Bandi River, Periyar River, Mithi River, Kali River, Narmada River, Damodar River, Mathabhanga River, Mula River, Musi River, and Kulti River (Panigrahi and Pattnaik, 2019).

9 Water (Prevention and Control of Pollution) Act, 1974

India's first effort to fully address environmental problems was this Act. The goal of this regulation is to prevent, regulate, and reduce water pollution, as well as to maintain or restore watercourses that are vital sources of water. This legislation also empowers established groups, such as the Central Council and the State Council, to manage the contamination of these waterways (iPleaders Blog, 2021). The people in charge are as follows:

Central Board—Central Pollution Control Board (CPCB)

State Board—State Pollution Control Board (SPCB)

According to Section 2(e) of the Act, "pollution" means any change to the physical, chemical, or biological properties of water, or the release of sewage into water that could cause problems or make the water unsafe for people or animals (iPleaders Blog, 2021).

10 Water (Prevention and Control of Pollution) Cess Act, 1977

Industrial waste is the biggest cause of water pollution, as it is often discharged into rivers, which makes the water quite dirty. This legislation allows the government to charge a fee for the water that businesses and local governments use or pollute. The goal of this law is to provide central and state authorities with more resources to prevent and regulate water contamination. The Water (Prevention and Control of Pollution) Cess Rules were established in 1978 as part of this legislation. They specify guidelines for the type and placement of meters that each water user must install (iPleaders Blog, 2021).

10.1 Government initiatives to tackle water pollution

Recently, the National Green Tribunal directed the Ministry of Jal Shakti to devise an appropriate National River Rejuvenation Mechanism for effective monitoring of steps to curb pollution and for the rejuvenation of all polluted river stretches across the country (Drishti, 2021).

National Water Policy (2012): It aims to acknowledge the existing situation, propose a framework for the creation of a system of laws and institutions, and develop a plan of action with a unified national perspective.

Established by the Ministry of Water Resources, it emphasizes the importance of water for human existence and economic development activities.

It offers frameworks for conserving water resources through optimal, economical, sustainable, and equitable means (Drishti, 2021).

National Water Mission (2010): It ensures integrated water resource management, leading to water conservation, reduced wastage, and equitable distribution, thereby forming better policies (Drishti, 2021).

National Mission for Clean Ganga: It envisages a five-tier structure at the national, state, and district levels to implement measures for the prevention, control, and abatement of environmental pollution in the river Ganga.

It aims to ensure a continuous and adequate flow of water to rejuvenate the river Ganga (Drishti, 2021).

Namami Gange Project: It integrates efforts to clean and protect the Ganga River in a comprehensive manner (Drishti, 2021).

10.2 National Directorate for the Protection of Rivers

The National Conservatory Directorate of Rivers is a subordinate body to the Ministry of Jal Shakti, which implements the protection of rivers and lakes across India under the National Rivers Conservation Plan and the National Conservatory Lakes Plan, with endorsement by respective state governments (iPleaders Blog, 2021). The objectives of these plans are to: improve the quality of rivers and recognize them as an important source of freshwater in the country; implement pollution reduction programs and labeling for the ethical use of water; assess the concentrations of pollutants in identified streams, such as plastics, metals, and chemicals; establish guidelines for prioritizing the conservation of the area, and provide an action plan for doing so; and identify, prioritize, and improve the capacity of regional institutions and organizations for long-term involvement in river conservation (iPleaders Blog, 2021).

This program covered 34 rivers, including the Tamiraparani, Vaigai, Vennar, Adyar, Betwa, Dhipu, Dhansiri, Bhadra, Cooum, Narmada, Godavari, Gomti,

Khan, Krishna, Rani Chu, Kshipra, Mahanadi, Mandovi, Mahananda, Musi, Pamba, Sabarmati, Brahmani, Cauvery, Sutlej, Subarnarekha, Pennar, Tapti, Tunga, Chambal, Damodar, Tungabhadra, Wainganga, and Yamuna (blog. iLeader, 2021).

The National River Conservation Plan is focused on cleaning up the sewage that flows into rivers, developing strategies to address this issue, and constructing sanitary facilities to prevent people from defecating on riverbanks. This initiative aims to raise awareness and knowledge about river conservation, safeguard aquatic life, mitigate human impact, improve water flow, and address the most harmful pollutants in rivers. It will contribute to the conservation of rivers and facilitate the protection of rivers in a planned manner across India (iPleaders Blog, 2021).

11 Effects of water pollution on human health

Many water-borne diseases are spreading from person to person (Halder et al., 2015). Ten percent of the population depends on food and vegetables that are grown in contaminated water (Baker, 2010). Health risks associated with polluted water include various diseases such as respiratory disease, cancer, diarrheal disease, neurological disorders, and cardiovascular disease (Ullah et al., 2014). Nitrogenous chemicals are responsible for cancer and blue baby syndrome (Krishnan et al., 2006). The mortality rate due to cancer is higher in rural areas than in urban areas because urban inhabitants use treated water for drinking, while rural people lack facilities for treated water and rely on unprocessed water. Poor people are at greater risk of disease due to improper sanitation, hygiene, and water supply (Jabeen et al., 2011). Contaminated water has significant negative effects on women who are exposed to chemicals during pregnancy; it leads to an increased rate of low birth weight, and as a result, fetal health is affected (Currie et al., 2013). Poor-quality water destroys crop production and contaminates food, which is hazardous to both aquatic life and human life (Khan et al., 2011). Pollutants disrupt the food chain (Halder et al., 2015), and heavy metals, especially iron, affect the respiratory system of fish. Iron clogs fish gills, which is lethal to fish, and when these fish are consumed by humans, it leads to major health issues (Ahmed et al., 2016). Metal-contaminated water causes hair loss, liver cirrhosis, renal failure (Salem et al., 2000), and neurological disorders (Chowdhury et al., 2016; Zimmermann et al., 2015; Haseena et al., 2017).

11.1 The global impact of clean rivers

Protecting rivers has a ripple effect on global health and sustainability. By ensuring the cleanliness of rivers, we can:

- Improve public health by reducing waterborne diseases.
- Safeguard food security by supporting sustainable agriculture.

- Combat climate change by preserving ecosystems that act as carbon sinks.
- Protect the planet's rich biodiversity for generations to come.

11.2 The impact of agricultural runoff on river pollution

Agricultural runoff presents a unique challenge in addressing river pollution due to its diffuse nature. Unlike point-source pollution from industrial discharges, agricultural runoff is often spread across vast areas and can vary significantly based on weather conditions and farming practices. This variability complicates efforts to monitor and regulate nutrient loading into rivers effectively.

For instance, during heavy rainfall events, runoff can surge dramatically, overwhelming existing mitigation measures and leading to acute pollution incidents. The long-term impact of agricultural runoff on river ecosystems is profound. Over time, excessive nutrient loading can lead to shifts in species composition within aquatic communities, as certain species thrive under nutrient-rich conditions while others decline or disappear altogether.

This shift can disrupt food webs and alter ecosystem functions, such as nutrient cycling and habitat structure. Furthermore, the economic implications for fisheries and recreational activities can be severe; communities that rely on healthy rivers for their livelihoods may find themselves facing declining fish populations and reduced tourism opportunities due to poor water quality.

12 Urban development and its influence on river pollution

Urban development continues to exert significant pressure on river systems worldwide. As cities expand, they often encroach upon natural landscapes that serve as critical buffers for waterways. Wetlands and riparian zones play essential roles in filtering pollutants and providing habitats for diverse species; however, urbanization frequently leads to their destruction or degradation.

This loss not only diminishes the natural filtration capacity but also increases vulnerability to flooding, as these areas are crucial for absorbing excess rainwater. The influence of urban development on river pollution is further compounded by socioeconomic factors. Low-income communities often bear a disproportionate burden of pollution due to inadequate infrastructure and limited access to clean water resources.

In many cases, these communities are located near industrial sites or along polluted waterways, exposing residents to health risks associated with contaminated water supplies. Addressing urban river pollution requires a multifaceted approach that includes community engagement, investment in green infrastructure such as permeable pavements and rain gardens, and policies aimed at reducing overall urban runoff.

References

Ahmed, K., et al., 2016. Impact of heavy metals on fish and human health. Environ. Toxicol. Pharmacol. 45 (2), 123–130.

Baker, L., 2010. Water contamination and food safety. Int. J. Environ. Res. Public Health 7 (12), 4173–4185.

Böck, K. et al., 2018. Ecosystem services in river landscapes. In: Riverine Ecosystem Management: Science for Governing Towards a Sustainable Future. Springer, pp. 413–433.

Burkhard, B., et al., 2012. Mapping ecosystem service supply, demand and budgets. Ecol. Indic. 21, 17–29.

Chowdhury, R., et al., 2016. Neurological disorders due to water contamination. Environ. Res. 151, 223–233.

Chung, M.G., Kang, H., 2013. Ecosystem service values and livelihood dependence around volcanic landscapes. Environ. Manag. 52 (5), 1231–1242.

Currie, J., et al., 2013. Environmental health risks and pregnancy outcomes. Am. Econ. J. Appl. Econ. 5 (4) 32–61.

de Bélizal, É., Lavigne, F., Hadmoko, D.S., Degeai, J.-P., Dipayana, G.A., Mutaqin, B.W., Marfai, M.A., Coquet, M., Le Mauff, B., Robin, A.-K., Vidal, C., Cholik, N., Aisyah, N., 2013. Rain-triggered lahars following the 2010 eruption of Merapi volcano, Indonesia: a major risk. J. Volcanol. Geotherm. Res. 261, 330–347.

Dede, O., et al., 2023. Ecosystem services and volcanic communities around Mount Merapi. Sustainability 15 (4), 2875–2891.

Drishti, I.A.S., 2021. Government initiatives for water pollution control in India. Retrieved from https://www.drishtiias.com.

Gerner, N.V., et al., 2018. River restoration and ecosystem services in the Emscher River. Sustainability 10 (3), 780–796.

Gob, F., et al., 2016. River systems affected by volcanic eruptions in Indonesia. Nat. Hazards Earth Syst. Sci. 16 (4), 1023–1038.

Gomez, C., et al., 2022. Deformation and volcanic activity of Mount Merapi. Remote Sens. 14 (6), 1357–1372.

Halder, S., et al., 2015. Water pollution and its impact on human health. J. Environ. Human 2 (1), 36–46.

Hapsari, D., et al., 2020. Sediment transport in volcanic river systems. Geomatics Nat. Hazards Risk 11 (1), 1894–1912.

Haseena, M., et al., 2017. Water pollution and human health implications. J. Pollut. Eff. Control 5 (2), 1–8.

Hidayat, R., et al., 2018. Riverbank erosion and sedimentation in Brantas River. IOP Conf. Ser.: Earth Environ. Sci. 148 (1), 012034.

Ikhsan, J., et al., 2019. Groundwater systems around Mount Merapi. Hydrogeol. J. 27 (5), 1745–1760.

iPleaders Blog, 2021. iPleaders Blog. https://blog.ipleaders.in/.

Jabeen, S., et al., 2011. Water quality and public health in developing countries. Pak. J. Med. Sci. 27 (5), 1180–1184.

Khan, M., et al., 2011. Water pollution and crop contamination. Ecotoxicol. Environ. Saf. 74 (8), 2295–2301.

Krishnan, S., et al., 2006. Nitrate contamination and human health. Environ. Monit. Assess. 120 (1–3), 413–422.

Krisnayanti, B.D., Agustawijaya, D., 2014. Heavy metal contamination in Porong River. J. Ecol. Eng. 15 (4), 56–64.

Kusumawardani, L., et al., 2017. Volcanic sediment dynamics in Merapi rivers. Arab. J. Geosci. 10 (18), 412–425.

Miller, J., 2022. Mining activities and river ecosystem degradation. Environ. Earth Sci. 81 (3), 77–91.

Panigrahi, A.K., Pattnaik, S., 2019. Major polluted rivers in India and management approaches. Int. J. Environ. Sci. 8 (2), 45–58.

Salem, H.M., et al., 2000. Heavy metals and health effects. Environ. Toxicol. 15 (3), 149–155.

Sekaranom, A., et al., 2021. Surface water availability around the Merapi volcano. Water Resour. Res. 13 (9), 1278–1291.

Thouret, J.C., et al., 2020. Pyroclastic density currents at Mount Merapi. J. Volcanol. Geotherm. Res. 401, 106966.

Ullah, R., et al., 2014. Hazards of polluted water. J. Environ. Sci. Health 49 (1), 1–10.

Ville, S., et al., 2015. Lahar dynamics and river systems around Merapi. Nat. Hazards 75 (2), 1235–1256.

Wang, X., et al, 2015. Land use change and river ecosystem services in Nenjiang River. Ecol. Eng. 83, 344–352.

Wibowo, A., et al., 2015. Cold lahar processes in Merapi rivers. Geomorphology 228, 127–138.

Yudistira, T., et al., 2020. Land use changes around Mount Merapi. Land Use Policy 99, 104831.

Zimmermann, M., et al., 2015. Neurological impacts of contaminated water exposure. NeuroToxicology 48, 152–160.

Chapter 2

Harmful algal blooms

Mangala Yatawara
Department of Zoology and Environmental Management, University of Kelaniya, Kelaniya, Sri Lanka

1 Overview

Nutrient enrichment in aquatic environments caused by anthropogenic activities, together with other favorable conditions, often results in eutrophication. Eutrophication increases cyanobacteria, a group of prokaryotes that result in harmful algal blooms (HABs) and associated toxins (Brooks et al., 2016). Eutrophication also increases green algal blooms. Cyanobacteria and green algae blooms are often confused, as both can produce significantly dense surface and water column growths. Nevertheless, unlike cyanobacteria, green algae are not generally known to produce toxins. Blooms specifically related to cyanobacteria are defined as "the increase of biomass in a water body over a relatively short period of time (between a few days and 1–2 weeks) and characterized by the predominance (>80%) of a single species or a few species within the phytoplankton community" (Humbert and Fastner, 2017; Janssen et al., 2019; Buratti et al., 2017). As stated by Yannarell and Kent (2009), cyanobacteria play a significant role in global carbon and nitrogen cycles. Cyanobacteria contribute to the mobility of essential minerals, influence soil structure and moisture retention, and are capable of thriving in both aquatic and damp terrestrial environments (Singh et al., 2020).

Cyanobacteria can be found in unicellular, colonial, and multicellular filamentous forms. The cell size of cyanobacteria varies from <1 μm in diameter (*Picocyanobacteria*) up to 100 μm (some tropical forms in the genus *Oscillatoria*) (Jasser and Callieri, 2017). The unicellular form occurs when the daughter cells separate after binary fission reproduction, and they can be ovoid, cylindrical, or spherical in shape. These cells can aggregate into irregular colonies, which are held together by a slimy matrix secreted during their growth. The filamentous form occurs when repeated cell divisions happen in a single plane at right angles to the main axis (Chorus and Bartram, 1999). Three different prevalent developmental cell types may differentiate from filamentous cyanobacterial vegetative cells, including hormogonia (motile cell type),

River Pollution. https://doi.org/10.1016/B978-0-443-45024-2.00012-9

heterocysts (nitrogen-fixing cell type), and akinetes (spore-like cells) (Singh and Montgomery, 2011). Hormogonia are short chains of cyanobacterial filaments with gliding motility (Rippka et al., 1979). The major role of hormogonia is in relocation and symbiotic colonization of hosts (Udeya et al., 2016). Heterocysts are large, round-shaped cells with a thicker cell envelope compared to vegetative cells; they are present in many multicellular cyanobacteria (Kumar et al., 2010) and are involved in fixing nitrogen. Akinetes are perennating spore-like cells that can form under unfavorable growth conditions such as cold or drought (Kumar et al., 2010). Akinetes also have a thick cell wall (Udeya et al., 2016).

Cyanobacteria contain chlorophyll-a pigments that can facilitate the light and dark phases of photosynthesis. In addition, carotene, xanthophyll, blue c phycocyanin (PC), and red c phycoerythrin (PE) are also present in cyanobacteria (Sandybayeva et al., 2023). These pigments are important for photosynthesis and play a role in the overall color of cyanobacteria. Carotenes are the most abundant carotenoids, while xanthophylls are oxygenated carotenoids. C-PC and C-PE are phycobiliproteins, which are a special class of pigments only found in cyanobacteria and red algae (Sandybayeva et al., 2023). They are key light-harvesting pigments, and PC is blue, while PE is red. Most cyanobacteria are aerobic photoautotrophs (gram negative), requiring only water, inorganic nutrients, carbon dioxide, and light for survival, but others have heterotrophic properties and can survive long periods in complete darkness (Issa et al., 2014).

Another unique property of the larger species is their ability to secrete gas bubbles within minute intracellular vesicles, which renders them positively buoyant and results in the formation of conspicuous scum on the surface (Yang et al., 2024). Cyanobacteria can form symbiotic associations with animals, plants, and fungi, supporting their growth and reproduction. They have evolved specific features that have enabled them to develop many morphological adaptations and physiological strategies, making it possible for them to survive in almost any type of environment. They are widely distributed in a variety of habitats, from aquatic systems, both freshwater and marine, to dry soils and extreme environments, including extremely dry, thermal, and polar sites (Whitton and Potts, 2000). Recent studies have found that 66% of global lakes identified by Landsat images showed an increasing trend of HAB events (Fang et al., 2022). The combination of low-head dams and run-of-the-river impoundments in regulated rivers favors this algal growth by slowing velocity, increasing water residence times, and providing improved light in shallow depths (Joo and Jeong, 2005; Kim et al., 2018).

Cyanobacteria are able to produce a wide range of bioactive compounds, some of which have benefits, such as acting as mood enhancers. Nevertheless, some cyanobacteria can produce bioactive compounds that may be harmful, called cyanotoxins (Messineo et al., 2024). Over 100 species of cyanobacteria belonging to forty genera are reported to be toxigenic (Jayatissa et al., 2006). Cyanotoxins can be produced by more than one species of cyanobacteria, and some species may produce more than one toxin at a time, resulting in blooms with different cyanotoxins. The production of toxins from HABs is a major concern as they may bioaccumulate in aquatic organisms and humans when exposed to these toxins. Cyanotoxins are usually released into the aquatic environment during cell death and lysis. Nevertheless, some species of cyanobacteria can release toxins into the water without rupturing or dying. The release of cyanotoxins into the environment can have significant effects on aquatic species, including decreased survival rates, altered development, larval mortality, reduced feeding, and death. In 2001 the Doñana National Park in Spain experienced a mass mortality event among flamingos and other waterfowl due to the ingestion of toxic cyanobacteria (Alonso-Andicoberry et al., 2003). The detrimental effects that HABs have on ecosystems include suppressing other primary producers and reducing food web resilience and biodiversity due to habitat degradation, disease, and impaired community structure (Carmichael and Boyer, 2016). Massive bloom die back and decomposition can lead to oxygen depletion, resulting in fish kills and the associated anoxic sediment release of nutrients, metals, and gases. HABs can cause degradation of aquatic habitats (Brenckman et al., 2024). HABs shade out seagrasses and corals, reducing light penetration and leading to habitat loss in marine environments. Studies have revealed that some HAB toxins are directly harmful to coral reefs and can contribute to coral bleaching and reef degradation. HABs create cascading ecological disruptions, affecting oxygen levels, food webs, habitats, and biodiversity. Even blooms of nontoxic species can cause ecological harm when they die off and bacterial decomposition consumes oxygen, leading to hypoxia and stressing both plant and animal communities. The exposure pathways of HABs in aquatic animals include skin contact (epidermis), inhalation (respiratory system), and ingestion (gastrointestinal tract). Upon exposure, HAB toxins can produce various health symptoms, such as mild skin rash, throat irritation, and gastroenteritis in more severe cases in humans. Reports of human deaths have also been documented as a result of exposure to cyanotoxins, the toxins produced by cyanobacteria. The severity of symptoms is primarily dependent on the cyanobacterial species, type of toxin, concentration level, as well as the route and medium of exposure (Carmichael and Boyer, 2016). HAB biomass and toxins have the ability to contaminate groundwater supplies, which can subsequently transport these substances into agricultural soils (Newton and Melaram, 2023). Studies have revealed that toxins produced by cyanobacteria can be taken up by higher plants and affect their growth as well (Pflugmacher et al., 1998). At the socioeconomic level, HABs can cause noxious odors and tastes, unpleasant views, and deteriorated aquatic conditions for organismal survival and ecosystem productivity. Geosmin and 2-methylisoborneol (2-MIB) are sesquiterpene and monoterpene molecules, respectively, that are responsible for the most unpleasant and widely distributed earthy-musty off-flavors in freshwater bodies. Recently, 2-MIB production has been observed in >40 species, including benthic and planktonic cyanobacteria

(Jüttner, 1995; Giglio et al., 2011). Moreover, HABs have adversely influenced tourism in some countries, reducing significant annual revenue. Thus HABs are commonly perceived as a global environmental and socioeconomic issue due to their extensive magnitude and devastating impacts on aquatic, animal, and human health (Newton and Melaram, 2023).

Understanding the factors affecting the growth, types of toxins, and consequences of HABs is crucial for addressing the management of HABs in freshwater ecosystems, especially in lotic systems such as rivers, since these systems can lead to HABs contamination in coastal and marine environments. In the following sections, the major factors associated with HABs growth, commonly found HABs, their bioactive compounds (cyanotoxins), and their significance are discussed.

2 Factors affecting the growth of HABs

Cyanotoxin production is strongly influenced by the environmental conditions that promote the growth of particular cyanobacterial species and strains. Freshwater HABs are influenced by a combination of factors, including nutrient levels, water temperature, light intensity, pH, microflora dynamics, hydrodynamic factors, and climate change.

2.1 Nutrients (nitrogen and phosphorus)

Nutrients are key environmental drivers that influence the proportion of cyanobacteria, the cyanobacterial biovolume, toxin production, and the impact that cyanobacteria may have on ecosystem function and water quality (Giblin et al., 2022). The addition of nitrogen (N) and/or phosphorus (P) to water bodies from various sources, including agricultural runoff (e.g., from fertilizers used on farms), wastewater treatment plants, stormwater runoff (e.g., from urban areas), and atmospheric deposition, influences the development of cyanobacterial blooms and may be related to cyanotoxin production. Recent surveys of cyanobacterial and algal productivity in response to nutrient pollution across geographically diverse aquatic ecosystems, and in different experimental enclosures of varying sizes, demonstrate that greater stimulation is routinely observed in response to both N and P additions (Anderson et al., 2002; Villatte et al., 2002). Some studies have revealed that certain cyanobacterial communities have the ability to fix nitrogen (Cook et al., 2019). Nevertheless, many species are incapable of N fixation. N fixation in capable species will only occur when dissolved N is insufficient for growth (Burford et al., 2006). Studies have also focused on the positive correlation between P and cyanobacterial biomass. According to Lewis et al. (2011), more attention has been given to P than to N because (1) P is more easily removed from anthropogenic sources than N, (2) N fixation by cyanobacteria has been assumed to make N control ineffective, and (3) the correlation between chlorophyll-a and total P among lakes is stronger than the correlation between chlorophyll-a and total N. Lewis et al. (2011) further stated that the stronger correlation is unsurprising, given that P is a more conserved element than N, an element strongly regulated by fluxes between water and atmosphere phases.

2.2 Temperature

Studies suggest that an increase in temperature may influence cyanobacterial dominance in the phytoplankton community. Cyanobacteria may benefit more from warming than other phytoplankton groups due to their higher optimum growth temperatures. Warming may also increase nutrient concentrations by enhancing mineralization through temperature or anoxia-mediated sediment phosphorus release (Malmaeus et al., 2006). Thus, temperature may indirectly increase cyanobacterial biomass through its effect on nutrient concentrations. At elevated temperatures, cyanobacteria routinely out-compete eukaryotic algae (Paerl, 2014). As the growth rates of eukaryotic taxa decline in response to warming, cyanobacterial growth rates reach their optima. In addition, vertical thermal stratification during dry seasons in the tropics and in late summer and early fall in temperate regions influences the spread of HABs. The increase in water column stability associated with higher temperatures may therefore favor cyanobacterial production and possibly the prevalence of cyanotoxins. In general, subtropical and tropical regions can experience cyanobacterial blooms year-round due to their ability to proliferate at high temperatures (Rangel et al., 2016). Overall, rising temperatures intensify HABs by promoting algal proliferation, extending bloom periods, altering species distributions, and potentially affecting toxicity profiles (Brenckman et al., 2024).

2.3 Light intensity

Availability of sunlight and turbidity has a strong influence on the cyanobacteria species that predominate, as well as the depth at which they occur. The literature stated that a greater proportion of the total phytoplankton biovolume is attributable to cyanobacteria in water bodies with high rates of light absorption. Nevertheless, the relationship between light intensity and toxin production in blooms is somewhat unclear. Some scientists have found evidence that toxin production increases with high light intensity (Watanabe and Oishi, 1985), while others have found little variation in toxicity at different levels of light intensity (Codd, 1995).

2.4 pH

HABs have been shown to intensify and persist at pH levels between 6 and 9. Studies have revealed that the percentage of cyanobacteria is highly correlated with pH, with an increased proportion of cyanobacteria at higher pH (Zepernick et al., 2021). Cyanobacteria have a competitive advantage over other phytoplankton species as they are efficient users of carbon dioxide in water. This characteristic is especially advantageous for cyanobacteria under conditions of higher pH, when the concentration of carbon dioxide in the water column is diminished due to photosynthetic activity.

2.5 Climate change

Several climate change-driven factors can also influence the occurrence and severity of HABs, including rising water temperatures, changes in salinity, elevated carbon dioxide (CO_2) levels, and altered rainfall patterns with respect to rivers. Elevated temperatures enable cyanobacteria to migrate within the water column, drawing nutrients from cooler, deeper layers and using them to fuel growth in the sunlight, warmer surface layers due to the presence or enhancement of stratification (Brenckman et al., 2024). In addition, Gobler (2020) stated that algal blooms absorb sunlight and release heat, which in turn raises surface water temperatures even further, creating a feedback loop that further favors their proliferation. Climate change is also projected to alter salinity levels (Brenckman et al., 2024). In some geographical areas, decreased freshwater runoff caused by intensified droughts and increased evaporation will lead to higher salinity in water bodies. Increased salt concentrations may lead to the leakage of cellular contents by accelerating the excretion processes. Salinity also alters the osmotic pressure of the surrounding environment, which can cause contraction of the internal membrane structure and disrupt membrane transport functions. On the contrary, rainfall patterns, already being disrupted by climate change, are expected to become more extreme, with heavy rainfall events and prolonged droughts. Intense rainfall increases nutrient-rich runoff from land into aquatic systems, thereby increasing HABs. Gobler (2020) stated that higher atmospheric and aquatic CO_2 concentrations, another consequence of climate change, can accelerate algal growth by providing more carbon for photosynthesis. At the same time, increased CO_2 levels acidify water bodies, altering species competition among algae and impacting organisms that graze on them.

2.6 Hydrodynamic factors

Hydrodynamic conditions either directly or indirectly affect the growth and distribution of HABs. Factors such as flow velocity, water turbulence, and shear stress can be considered direct effects, whereas indirect effects involve water retention, stratification, nutrient distribution, and light availability (Matson et al., 2020). Studies have shown that moderate water turbulence enhances algal growth by improving nutrient uptake and energy metabolism (Zhu et al., 2019). Nevertheless, excessive turbulence damages algal cells, disrupts colony formation, and inhibits growth. Wu et al. (2019) stated that small-scale turbulence at the water surface has been found to significantly contribute to the formation of surface scums. Studies have reported that under low-flow conditions, cyanobacteria remain in the Uruguay River, whereas when flow increases during the summer months, they reach the Río de la Plata estuary, impacting its beaches (Ferrari, 2020). Stratification of the water column, by reducing vertical mixing and impeding material exchange, allows cyanobacteria to concentrate near the surface where light is more abundant, thus supporting their proliferation (Song, 2023). Gao et al. (2018) also stated that stratified waters, often associated with extended residence times and low flow velocities, further enhance conditions favorable to bloom formation.

3 Microflora dynamics

Dynamics of microflora competition as blooms develop and collapse can also impact cyanotoxin concentrations in surface waters. It is evident that microbes play a crucial role in nutrient cycling within aquatic ecosystems. They can influence the availability of nutrients like nitrogen and phosphorus, which are essential for algal growth. Some microbial populations may compete with HAB-forming algae for these nutrients, limiting their growth. In addition, some bacteria and fungi can directly inhibit algal growth, either through the production of toxins (algicides) or by competing for resources like nutrients. For example, biofilms can be used to reduce HAB-forming species. Biofilms, which are microbial communities attached to surfaces, can interact with HAB-forming algae in various ways. Some biofilms can inhibit the growth of HAB species, while others may contribute to their development by altering nutrient availability or providing a suitable habitat (Zhang et al., 2021).

4 Harmful cyanobacterial species

The cyanobacteria are classified as prokaryotic organisms within the kingdom Bacteria. They are specifically part of the phylum Cyanobacteriota (formerly Cyanobacteria) and the class Cyanophyceae. This is further divided into several orders, including Chroococcales, Nostocales, Oscillatoriales, Pleurocapsales, Spirulinales, and Synechococcales. The classification, morphology, specific characteristics of commonly available harmful cyanobacterial species, and their significance are described in the following.

4.1 *Microcystis* sp.

The genus *Microcystis* is classified within the class Cyanophyceae, the order Chroococcales, and the family Microcystaceae. The genus is commonly found in temperate and subtropical lakes, reservoirs, and rivers, with blooms occurring in at least 108 countries across six continents, except Antarctica. *Microcystis* is a single-celled freshwater cyanobacterium that forms colonies surrounded by mucilage. The colonies of *Microcystis* species comprise numerous cells forming irregular spherical or perforated reticular structures, enclosed within a colorless, uniform gelatinous matrix containing granules and gas vesicles (Fig. 2.1A). Globally, more than fifty *Microcystis* species have been described. Although elevated temperatures and high light conditions are favorable for the growth of *Microcystis*, laboratory experiments have shown that nitrogen addition promotes higher growth. Similar results have been observed from Nanwan Reservoir at Xinyang, China. The nitrogen levels, particularly NO_3^-N were the primary factors influencing *Microcystis* density. Some studies have observed a decrease in the toxicity of *Microcystis* sp. after the removal of N or inorganic carbon, but no changes were observed when P was removed from cyanobacteria culture media. Nevertheless, at high pH (>9), *Microcystis* exhibits buoyancy mechanisms to form surface blooms, enhancing CO_2 absorption from the air-water interface. The colony morphology plays an important role in Microcystis ecology; i.e., positive buoyancy, particularly of large colonies, reduces zooplankton grazing pressure, enables them to overcome water turbulence, and makes them less prone to heavy metals (Wu et al., 2007).

4.2 *Dolichospermum* sp. (planktonic Anabaena)

According to the recent classification, the genus *Dolichospermum* belongs to the class Cyanophyceae, order Nostocales, and the new family Aphanizomenonaceae. Species of the filamentous genus *Dolichospermum* form slimy blooms on the surface of eutrophic water bodies. The planktonic Anabaena species are distinguished by the presence of gas vesicles in cells that allow regulation of buoyancy and vertical migration in the water (Fig. 2.1B). A further advantageous trait of *Dolichospermum* is the capacity for cell differentiation into heterocysts and akinetes. Heterocysts have the ability to fix molecular nitrogen, while akinetes sustain long-term dormancy and enable *Dolichospermum* populations to survive under harsh environmental conditions such as overwintering, dry seasons, and flooding. Unlike *Microcystis* sp., high growth rates of two *Dolichospermum* isolates have been observed with increasing phosphorus concentrations. Severe *Dolichospermum circinale* blooms were observed in the Murray and Edward rivers in 2010 (Williamson et al., 2018). *Dolichospermum* blooms have been dispersed all over the world by natural means (via migrating fishes, birds, winds, and river basins) and by human activities (ship ballast waters, aquaculture, etc.) (O'Neil et al., 2012). Under favorable conditions, *Dolichospermum* blooms form potential threats to the environment and human health due to their production of various kinds of potent toxins (Li et al., 2016).

4.3 *Planktothrix* sp.

Earlier, it was classified under the genus *Oscillatoria* as it grows in solitary trichomes without sheaths, heterocysts, and akinetes. It has only been since

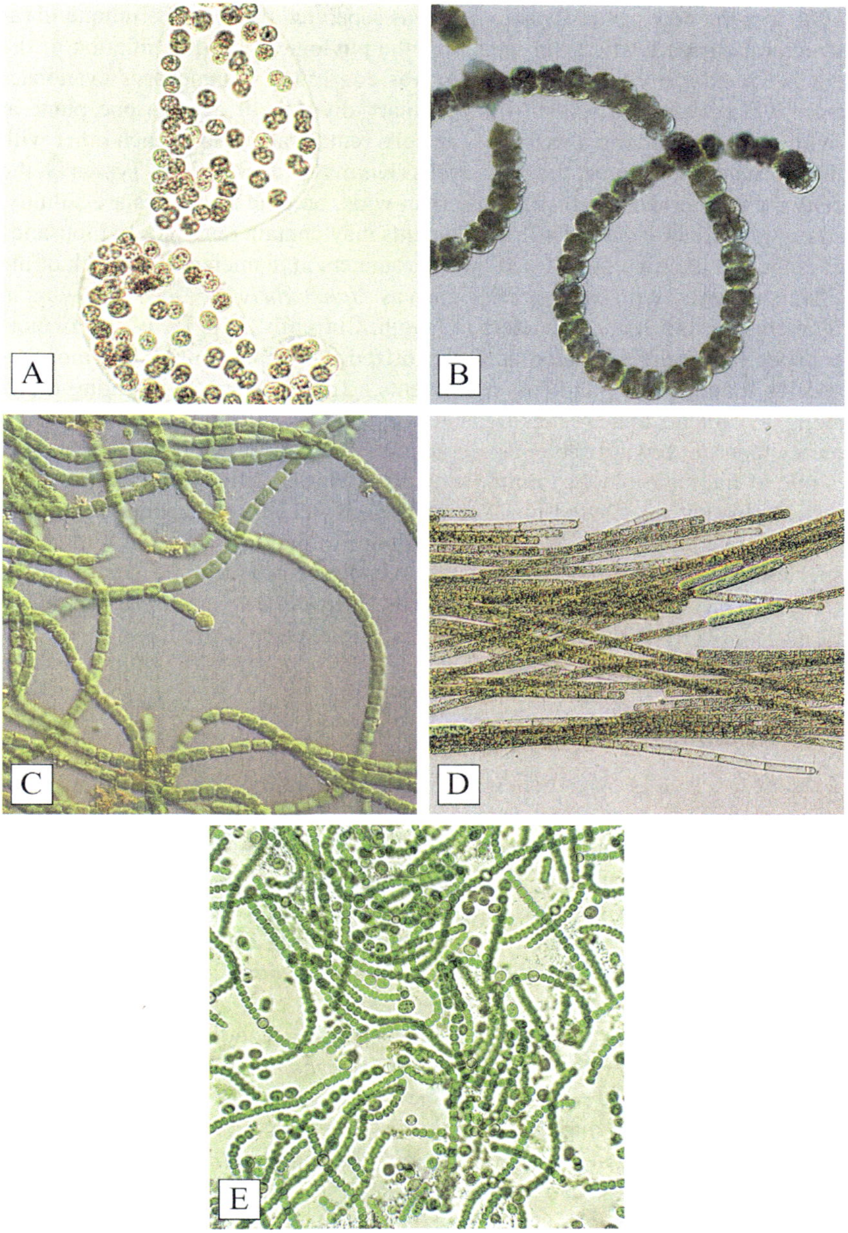

FIG. 2.1 Images showing harmful cyanobacterial species. (A) *Microcystis* sp. (B) *Dolichospermum* sp. (C) *Cylindrospermopsis* sp. (D) *Aphanizomenon* sp. (E) *Nostoc* sp.

1988 that the new genus *Planktothrix* was separated based on its unique ultrastructural characteristics. Subsequently, the phylogenetic differentiation of the genus *Planktothrix* from *Oscillatoria* was confirmed. Filamentous cyanobacteria *Planktothrix* are formed by the binary division of cells in one plane at right angles to the long axis, while the cells remain attached to each other with little or no constriction at the cross-walls (Kurmayer et al., 2016). Typically, the cells are cylindrical, slightly shorter than wide, and the trichomes are solitary, straight, or slightly curved. These filaments may contain hundreds to thousands of cells, and the trichomes are a few micrometers in diameter. The length of the filaments varies, with some species, such as *Planktothrix rubescens*, growing in filaments up to a few millimeters in length. Currently, nine *Planktothrix* morphospecies are differentiated and categorized into three groups: (1) morphospecies forming wavy and coiled filaments, (2) morphospecies forming rather straight filaments that are not attenuated and lack a cap, and (3) attenuated filaments tapering toward the ends, sometimes showing a calyptra. *Planktothrix* is one of the main bloom-forming cyanobacteria identified in temperate areas and is globally distributed in lakes and reservoirs. Recent genomic analyses revealed that all planktic *Planktothrix* belongs to the same species. Within this species, two central ecotypes exist: a red-pigmented ecotype (*Planktothrix rubescens*) and a green-pigmented ecotype (*Planktothrix agardhii*), containing PE or PC, respectively, as the principal constituent of phycobilisomes (i.e., the light-harvesting antennae of cyanobacteria).

4.4 *Cylindrospermopsis* sp.

Cylindrospermopsis sp. belongs to the phylum Cyanobacteriota, class Cyanophyceae, order Nostocales, and the family Aphanizomenonaceae. *Cylindrospermopsis* are filamentous and free-floating, solitary, bent, straight, or screw-like coiled organisms (Fig. 2.1C). Several species of *Cylindrospermopsis* are narrowed toward the ends and thus lack sheaths. Cells are cylindrical or barrel-shaped, usually distinctly longer than wide, and are pale blue–green, yellowish, or olive green. Cells divide crosswise by asymmetric division of the end cells. The cell structures contain pigments such as chlorophyll-a, carotenoids, PC, and PE. They grow well under low-light-intensity conditions and are distributed vertically in the water column. *Cylindrospermopsis* generally does not produce dense blooms and tends to be deeper in the water than most cyanobacteria. It has the ability to fix nitrogen from the air. As a result, it can continue to grow after dissolved nitrogen is depleted. Among the genus *Cylindrospermopsis*, the most widely distributed and well-known species is the pantropical type species *Cylindrospermopsis raciborskii*. The expansion of *C. raciborskii* has continued in recent times, with Antarctica being the only continent without records of its occurrence. The reasons for their wide distribution could be linked to akinete formation and a tolerance for low phosphorus and nitrogen availability in the environment. In 1979 the people of Palm Island in Queensland, Australia, were affected by a hepatoenteritis-like illness that correlated with the copper sulfate treatment of a *Cylindrospermopsis* bloom in the drinking water reservoir, Solomon Dam Reservoir (Griffiths and Saker, 2003). The symptoms of the cases included vomiting, headache, fever, and bloody diarrhea. Both hepatomegaly and renal damage were determined through urine analysis.

4.5 *Aphanizomenon* sp.

Aphanizomenon sp. belongs to the phylum Cyanobacteriota, class Cyanophyceae, order Nostocales, and the family Aphanizomenonaceae. *Aphanizomenon* sp. comprises a group of filamentous nitrogen-fixing cyanobacteria, of which several members are able to develop blooms and produce toxic metabolites (cyanotoxins), including hepatotoxins [microcystins (MCs)], neurotoxins [anatoxins and saxitoxins (STXs)], and cytotoxins [cylindrospermopsin (CYN)] (Cirés and Ballot, 2016). *Aphanizomenon flos-aquae, Aphanizomenon gracile, Cuspidothrix issatschenkoi, Sphaerospermopsis aphanizomenoides,* and *Chrysosporum ovalisporum* are the well-described species in this genus. The vegetative cells of *Aphanizomenon* often contain gas vesicles, providing buoyancy under certain environmental conditions. Apical cells in the trichomes are not (or are only slightly) narrowed toward the ends, with elongated and partially or totally hyaline apical cells (Cirés and Ballot, 2016). These species are able to differentiate into nitrogen-fixing cells, known as heterocysts (Zhang et al., 2017). These cells are generally pale yellow with various shapes (spherical to cylindrical) and are usually found in the intercalary or terminal position of the filaments (Fig. 2.1D). They contribute fixed nitrogen to vegetative cells in the filament (Singh et al., 2020). In addition, they can differentiate into other specialized spore-like cells, called akinetes, which allow these bacteria to survive in harsh conditions (Kaplan-Levy et al., 2010). The akinetes are larger, possess a thicker cell wall than vegetative and heterocyst cells, and accumulate a large amount of food reserves and DNA (Kaplan-Levy et al., 2010; Singh et al., 2020).

4.6 *Nostoc* sp.

Nostoc sp. belongs to the phylum Cyanobacteriota, class Cyanophyceae, order Nostocales, and the family Nostocaceae, which is a filamentous nitrogen-fixing cyanobacterium that forms macroscopic or microscopic colonies and is common in both terrestrial and aquatic environments (Dodds et al., 1995). Cells of *Nostoc* sp. are spherical, barrel-shaped, or oval, forming unbranched filaments. The filaments (trichomes) may contain both heterocysts and akinetes. *Nostoc* cells form freshwater colonies derived from linear sequences of cells, forming series like a necklace (Fig. 2.1E). Macroscopic *Nostoc* species have many uses, such as edible foods and medicines. Johnson et al. (2008) stated that *Nostoc commune*, used as food in the Peruvian highlands, produces the neurotoxic amino acid β methylamino-L-alanine (BMAA). In addition, *Nostoc insulare* has also been shown to produce cytotoxic metabolites, and *Nostoc spongiforme* produces the toxic compound nostocine A. The toxicity can vary depending on the specific strain and environmental conditions. Acute exposure typically occurs by drinking or swimming in water contaminated by a toxic algal bloom. These toxins are released into the water when the cyanobacterial cells rupture (Fig. 2.1).

5 Cyanotoxins and their significance

Cyanotoxins may be classified as hepatotoxins, cytotoxins, and neurotoxins. MC, nodularin, and CYN are hepatotoxins that may cause internal hemorrhage or death in cases of acute exposure to high doses, and carcinogenesis in cases of long-term exposure. Neurotoxins are substances that can damage or disrupt the function of the nervous system. They can be natural or synthetic and can cause a wide range of effects, from mild impairment to severe paralysis or death. Neurotoxins can affect the nervous system in various ways, including interfering with neurotransmitter release or binding to receptors on nerve cells. Anatoxins and STXs are neurotoxins that may cause paralysis, respiratory failure, or death. Occurrence data suggest that hepatotoxins are the most prevalent of the cyanotoxins. Cytotoxins are substances that have a toxic effect on cells, potentially leading to cell damage or death. They can interfere with essential cellular functions like DNA replication, protein synthesis, or membrane integrity.

These bioactive cyanotoxin compounds fall into several groups: cyclic peptides (MC), alkaloids [anatoxin-a (ATX-a), anatoxin-a(S) (ATX(s)), STX, and CYN], and lipopolysaccharides (Pantelić et al., 2013).

6 Hepatotoxins

6.1 Microcystins

Among the cyanotoxins produced by cyanobacterial blooms, MCs are the most ubiquitous toxins detected worldwide. These are monocyclic heptapeptides that contain seven amino acids joined end-to-end and then head-to-tail to form cyclic compounds (Table 2.1). Since they are comparatively large, their molecular weight ranges between 800 and 1100 Da. MC congeners are primarily classified based on their amino acid composition. The most common MC variant is microcystin-LR (MC-LR), which contains leucine (L) and arginine (R). Other commonly available variants include MC-RR and MC-YR (Díez-Quijada et al., 2019). The cyanotoxins produced by *Microcystis* are extremely stable and resist common chemical breakdown in most natural water bodies. They can even persist after boiling, indicating that cooking is not sufficient to destroy the toxins (Bullerman and Bianchini, 2007). Nevertheless, they are susceptible to degradation by aquatic bacteria found naturally in rivers and reservoirs. In aquatic environments, the toxin tends to remain contained within the cyanobacterial

TABLE 2.1 The structures of cyanotoxins produced by cyanobacteria.

Cyanotoxin	Structure	Reference
Microcystin	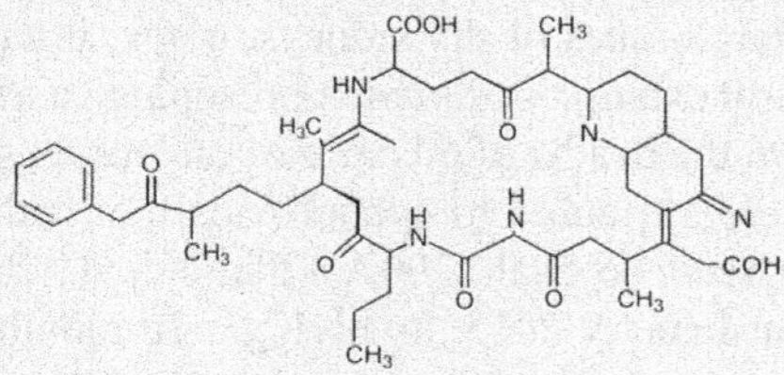	Krishnan and Mou (2020)
Anatoxin-a	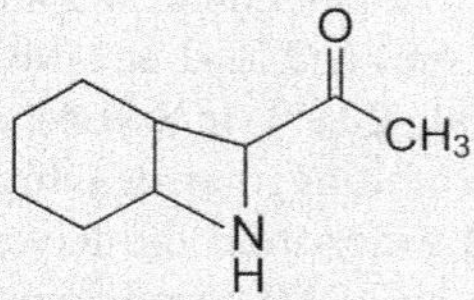	Gupta (2014)
Anatoxin-a(S)	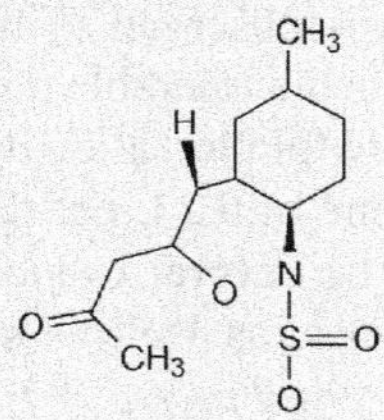	Villatte et al. (2002)
Saxitoxins	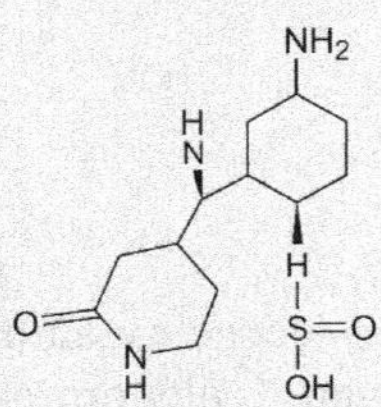	Gupta (2014)
Cylindrospermopsin	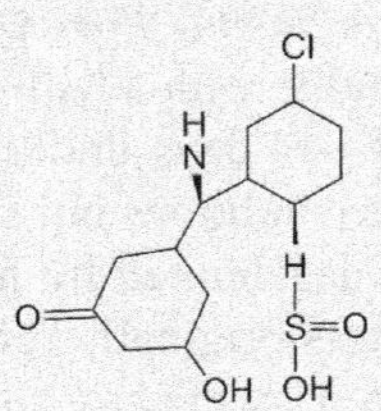	Mazmouz et al. (2010)

cell and is released in substantial amounts only upon cell lysis. In addition, MCs may adsorb onto naturally suspended solids and dried crusts of cyanobacteria, precipitate out of the water column, and reside in sediments for months. At the cellular level, MCs cause specific damage to liver cells by inhibiting protein phosphatases (PP1 and PP2A), resulting in the hyperphosphorylation of cellular proteins. Exposure to MCs has caused significant effects and deaths at the trophic level, including small planktonic invertebrates, fish, and vertebrates. Humans are exposed through contaminated drinking water. Although the liver is the major target organ, tissues of the kidneys, colon, brain, lungs, and heart are also affected. Several examples of human poisoning cases due to MCs have been described. In 1996 the deaths of 60 patients undergoing dialysis treatment in Caruaru, Brazil, were attributed to cyanotoxins that had contaminated the clinic's water supply (Azevedo et al., 2002; Villatte et al., 2002). Another case occurred in Uruguay in January 2015, involving a 20-month-old child and her family during recreational activities. These victims were admitted to the hospital with diarrhea, vomiting, fatigue, and jaundice, and analysis confirmed the presence of MC-LR (2.4 $ng \cdot g^{-1}$ tissue) and [D-Leu1]MC-LR (75.4 $ng \cdot g^{-1}$ tissue) in the explanted liver (Vidal et al., 2017). In Sweden, about 121 people experienced abdominal pain, nausea, vomiting, diarrhea, fever, headaches, and muscle pain after consuming untreated water from the River Kavlingean (Annadotter et al., 2001). In this case, a bloom of MC-producing cyanobacteria, such as *P. agardhii* and *Microcystis* spp., was observed in the river. The World Health Organization (WHO) has established a tolerable daily intake (TDI) in drinking water of 1.0 μg/L for MC-LR, while for recreational activities, a level of 20 μg/L MC-LR (based on an intake volume of 0.2 L per day) has been set (Villalobos et al., 2025). MCs are produced by several cyanobacterial species, including members of the *Microcystis* species, as well as species of *Dolichospermum (Anabaena)*, *Nodularia*, *Nostoc*, *Oscillatoria*, *Fischerella*, *Planktothrix*, and *Gloeotrichia echinulata*.

7 Nurotoxins

7.1 Anatoxin-a

Anatoxin-a (ATX-a) is a bicyclic secondary amine alkaloid (Table 2.1) with a molecular weight of 165 Da (Van Apeldoorn et al., 2007). It was first discovered in *Dolichospermum* (*Anabaena*) *flos-aquae* isolated from Canada. Although this toxin degrades rapidly, with a half-life of 1–2 days in the presence of high light intensity and ≈14 days under normal light conditions at basic pH, it is stable under sterile conditions but susceptible to microbial biodegradation. Breakdown is also accelerated by high temperature, UV light, and alkaline conditions. ATX-a was originally referred to as "very fast death factor" due to its rapid absorption from the gastrointestinal tract, as evidenced by the rapid onset of clinical effects following oral exposure. ATX-a is a potent nerve toxin that acts postsynaptically. Clinical effects of poisoning may appear within minutes to hours after exposure and may include loss of muscle coordination, muscle tremors and fasciculations, convulsions, and respiratory distress. Organs such as the liver, kidney, lung, thymus, spleen, adrenal glands, intestinal tract, immune system, and heart can be affected following toxin exposure. The principal lethal effect is respiratory failure due to loss of control over respiratory muscles. The literature highlighted only one formally documented human poisoning case involving ATX-a: the consumption of ascidians of the *Microcosmus* genus in the south of France, which intoxicated 26 people. The symptoms observed, primarily neurotoxic (blurred vision, ataxia, dizziness, asthenia, headache, muscle cramps, paresthesia), were entirely consistent with exposure to ATX-a. The study identified ATX-a at concentrations of 193.7–1240.2 $\mu g\ kg^{-1}$ in the seafood consumed (Bire et al., 2020). Guideline values for ATX-a have been proposed in some countries to address potential health risks. Specifically, California has proposed a value of 5 μg/g in fish flesh and 40 and 100 μg/L in drinking water for livestock and dogs, respectively.

ATX-a is produced by *Anabaena, Aphanizomenon, Microcystis, Planktothrix, Raphidiopsis, Arthrospira, Cylindrospermum, Phormidium, Nostoc,* and *Oscillatoria.*

7.2 Anatoxin-a(S)

ATX(s) is different both in structure and toxicity mechanism from ATX-a (Du et al., 2019). ATX(s) is a phosphate ester of a cyclic N-hydroxyguanine (Table 2.1) with a molecular weight of 252 Da. The "(s)" in the name "anatoxin-a(s)" refers to salivation, which is a characteristic sign of poisoning observed in laboratory rodents after exposure. ATX(s) was first reported in a culture of *Anabaena flos-aquae* NRC 525–17, which was isolated from a lake in Saskatchewan, Canada (Carmichael and Gorham, 2017). The toxin caused poisonings in dogs (Mahmood et al., 1988), and birds and swine (Cook et al., 1989). Poisonings related to *Anabaena flos-aquae* blooms were suspected on the basis of toxicological properties and chromatographic behavior. However, Onodera et al. (1997) confirmed that this was due to ATX(s) through other pharmacological studies. This toxin decomposes rapidly in basic solutions but is more stable under neutral and acidic conditions. It is a potent acetylcholinesterase inhibitor at the nerve synapse, which results in paralysis, asphyxiation, decreased movement, exaggerated abdominal breathing, cyanosis, convulsions, and ultimately death. Clinical signs are indistinguishable from organophosphate poisoning. ATX(s) has been detected only in *Dolichospermum* species, including *Dolichospermum flos-aquae, Dolichospermum lemmermannii,* and *Dolichospermum spiroides.*

7.3 Saxitoxins

Saxitoxins (STXs) are best known as paralytic shellfish poisons associated with their bioaccumulation in shellfish such as *Saxidomus giganteus*, from which the first identification was made. STXs are tricyclic purine derivatives (Table 2.1) with a molecular weight of around 299 Da. They are tasteless and odorless and are not destroyed by cooking. The half-life of STX depends on pH and temperature. They are stable at 25°C and neutral pH and can be detected in water intended for irrigation and in rivers for a range of ~9–28 days postbloom. They exhibit high chemical stability at acidic pH (~2–4) and high degradation at extremely alkaline pH. So far, 27 variants of STXs have been identified. STX is the most potent cyanobacterial toxin. Within the *Dolichospermum* genus, STX production has only been reported in *Dolichospermum circinalis* (*Anabaena circinalis*) blooms in Australian freshwater systems. General symptoms of STX poisoning include neurologic and respiratory paralysis. Neurologic symptoms are the most pronounced effects of STX poisoning. A tingling and burning sensation, initially occurring around the mouth and lips, is usually the first symptom. The numbness may extend to the hands and spread over the chest and abdomen, and these symptoms may progress, causing difficulty walking and arm and leg weakness. Respiratory distress from muscular paralysis may occur up to 12 h after intoxication. Respiratory paralysis from significant STX exposure may lead to death. The identification of STX analogs is regulated by international standards only in products intended for marine-origin consumption, but not in freshwater vectors. Nevertheless, Australia, Brazil, and New Zealand have established a maximum permissible level of 3.0 μg/L for drinking water.

8 Cytotoxins

8.1 Cylindrospermopsin

Cylindrospermopsin (CYN) is a small alkaloid. It possesses a tricyclic guanidine moiety combined with hydroxymethyluracil (Table 2.1) and has a molecular weight of 415 Da. CYN was chemically characterized from the species *Cylindrospermopsis raciborskii,* and two derivatives, 7-epi-cylindrospermopsin and deoxy-cylindrospermopsin, have been identified. The toxin is a protein synthesis inhibitor. The kidney has been found to be the organ most sensitive to CYN toxicity. CYN has been associated with increased risks of chronic liver disease and cancer. It was first implicated in human "Palm Island liver disease," which affected 138 children and ten adults in Palm Island, Australia (Griffiths and Saker, 2003). An ecological study of the reservoir covering a 4-year period, 1981–1984, showed that the water body suffered periodic cyanobacterial blooms, the most prominent components of which were two varieties of *Anabaena circinalis* and *Cylindrospermopsis raciborskii* (Hawkins, 1985). There is evidence that CYN may also be genotoxic. Other problems from CYN poisoning may include damage to the stomach, small intestine, lungs, adrenal glands, heart, thymus, and spleen, as well as various hemorrhages. The toxic levels of CYN have been reported in surface waters at concentrations of ~173 μg/L. This species can coexist with MCs. These high environmental levels have suggested that this group of cyanotoxins possesses the capacity to disrupt the antioxidant system and/or cause oxidative stress in a wide variety of aquatic animals. It is produced by members of cyanobacterial genera, including *Cylindrospermopsis, Aphanizomenon, Umezakia, Anabaena (now Dolichospermum), Lyngbya,* and *Raphidiopsis* (Table 2.1).

9 Summary

As anthropogenic activities continue to increase, the occurrence and intensity of HABs are expected to rise in subtropical and temperate aquatic environments. HABs represent a growing and notable environmental challenge with many ecological, economic, and human health consequences. These HABs disrupt aquatic ecosystems by depleting oxygen levels, releasing harmful toxins, leading to mass mortality of aquatic organisms, and reducing esthetic appeal. In addition, they impose economic burdens on tourism, fisheries, and water purification facilities. Moreover, HABs pose serious health risks to humans and animals through contaminated water and aquatic food. Addressing HABs requires a multifaceted approach that integrates scientific research, policymaking, and public involvement with frequent monitoring. Continuous monitoring may help minimize their impact. Nevertheless, for long-term prevention, regulatory policies targeted at reducing nutrient runoff from agricultural, industrial, and/or urban sources are essential. Public awareness of sustainable agricultural practices, responsible waste disposal practices, and active participation in conservation efforts may provide sustainable solutions for the mitigation of blooms in aquatic environments. Biotechnology and the use of artificial intelligence may also offer viable solutions for predicting HAB outbreaks with greater accuracy. Prioritizing sustainability and proactive management would minimize the adverse effects of HABs and ensure the long-term health of aquatic ecosystems.

References

Alonso-Andicoberry, C., Garcia-Villada, L., López-Rodas, V., Costas, E., 2003. Catastrophic mortality of flamingos in a Spanish National Park caused by a cyanobacteria. Vet. Rec. 151, 706–707. doi:10.1136/vr.151.23.706.

Anderson, D.M., Glibert, P.M., Burkholder, J.M., 2002. Harmful algal blooms and eutrophication: nutrient sources, composition, and consequences. Estuaries 25, 704–726. doi:10.1007/BF02804901.

Annadotter, H., Cronberg, G., Lawton, L., Hansson, H.B., Göthe, U., Skulberg, O.M., 2001. An extensive outbreak of gastroenteritis associated with the toxic cyanobacterium *Planktothrix agardhii* (*Oscillatoriales, Cyanophyceae*) in Scania, South Sweden. In: Cyanotoxins. Springer, Berlin, Germany, pp. 200–208.

Azevedo, S.M., Carmichael, W.W., Jochimsen, E.M., Rinehart, K.L., Lau, S., Shaw, G.R., Eaglesham, G.K., 2002. Human intoxication by microcystins during renal dialysis treatment in Caruaru-Brazil. Toxicology 181-182, 441–446 27.

Bire, R., Berlin, T., Dom, I., Hort, V., Schmitt, C., Diogene, J., Lemee, R., De Haro, L., Nicolas, M., 2020. First evidence of the presence of anatoxin-A in sea figs associated with human food poisoning in France. Mar. Drugs 18 (6), E285.

Brenckman, C.M., Parameswarappa Jayalakshmamma, M., Pennock, W.H., Ashraf, F., Borgaonkar, A.D., 2024. A review of harmful algal blooms: causes, effects, monitoring, and prevention methods. Water 17 (13), 1980. doi:10.3390/w17131980.

Brooks, B.W., Lazorchak, J.M., Howard, M.D., Johnson, M.V.V., Morton, S.L., Perkins, D.A., Reavie, E.D., Scott, G.I., Smith, S.A., Steevens, J.A., 2016. Are harmful algal blooms becoming the greatest inland water quality threat to public health and aquatic ecosystems? Environ. Toxicol. Chem. 35 (1), 6–13. doi:10.1002/etc.3220.

Bullerman, L.B., Bianchini, A., 2007. Stability of mycotoxins during food processing. Int. J. Food Microbiol. 0168-1605. 119 (1-2), 140–146. doi:10.1016/j.ijfoodmicro.2007.07.035.

Buratti, F.M., Manganelli, M., Vichi, S., Stefanelli, M., Scardala, S., Testai, E., Funari, E., 2017. Cyanotoxins: producing organisms, occurrence, toxicity, mechanism of action and human health toxicological risk evaluation. Arch. Toxicol 91, 1049–1130.

Burford, M.A., McNeale, K.L., McKenzie-Smith, F.J., 2006. The role of nitrogen in promoting the toxic cyanophyte *Cylindrospermopsis raciborskii* in a subtropical water reservoir. Freshw. Biol. 51, 2143–2153.

Carmichael, W.W., Boyer, G.L., 2016. Health impacts from cyanobacteria harmful algae blooms: implications for the North American Great Lakes. Harmful Algae. 1878-1470. 54, 194–212. doi:10.1016/j.hal.2016.02.002.

Carmichael, W.W., Gorham, P.R., 2017. Anatoxins from clones of *Anabaena* flos-aquae isolated from lakes of western Canada. SIL Commun., 1953-1996. 0538-4680. 21 (1), 285–295. doi: 10.1080/05384680.1978.11903972.

Chorus, I., Bartram, J., 1999. Toxic Cyanobacteria in Water: A Guide to Their Public Health Consequences, Monitoring and Management. CRC Press, London. doi:10.1201/9781482295061.

Cirés, S., Ballot, A., 2016. A review of the phylogeny, ecology and toxin production of bloom-forming *Aphanizomenon spp.* and related species within the Nostocales (cyanobacteria). Harmful Algae. 1878-1470. 54, 21–43. doi:10.1016/j.hal.2015.09.007.

Codd, G.A., 1995. Cyanobacterial toxins: occurrence, properties and biological significance. Water Sci. Technol. 0273-1223. 32 (4), 149–156. doi:10.2166/wst.1995.0177.

Cook, K.V., Li, C., Cai, H., Krumholz, L.R., Hambright, K.D., Paerl, H.W., Steffen, M.M., Wilson, A.E., Burford, M.A., Grossart, P., Hamilton, D.P., Jiang, H., Sukenik, A., Latour, D., Meyer, E.I., Padisák, J., Qin, B., Zamor, R.M., Zhu, G., 2019. The global *Microcystis* interactome. Limnol. Oceanogr. 65, S194–S207. doi:10.1002/lno.11361.

Cook, W.O., Beasley, V.R., Lovell, R.A., Dahlem, A.M., Hooser, S.B., Mahmood, N.A., Carmichael, W.W., 1989. Consistent inhibition of peripheral cholinesterases by neurotoxins from the freshwater cyanobacterium *Anabaena* flos-aquae: studies of ducks, swine, mice and a steer. Environ. Toxicol. Chem. 0730-7268. 8 (10), 915–922. doi:10.1002/etc.5620081010.

Díez-Quijada, L., Puerto, M., Gutiérrez-Praena, D., Llana-Ruiz-Cabello, M., Jos, A., Cameán, A.M., 2019. Microcystin-RR: occurrence, content in water and food and toxicological studies. A review. Environ. Res. 168, 467–489.

Dodds, W.K., Gudder, D.A., Mollenhauer, D., 1995. The ecology of NOSTOC. J. Phycol. 31 (1), 2–18. doi:10.1111/j.0022-3646.1995.00002.x.

Du, X., Liu, H., Yuan, L., Wang, Y., Ma, Y., Wang, R., Chen, X., Losiewicz, M., Guo, H., Zhang, H., 2019. The diversity of cyanobacterial toxins on structural characterization, distribution and identification: a systematic review. Toxins. 2072-6651. 11 (9), 530. doi:10.3390/toxins11090530.

Fang, C., Song, K., Paerl, H.W., Pierre-Andre, J., Wen, Z., Liu, G., et al, 2022. Global divergent trends of algal blooms detected by satellite during 1982–2018. Glob. Change Biol 28, 2327–2340. doi:10.1111/gcb.16077.

Ferrari, G., 2020. Water flow and temperature as main factors that regulate phytoplankton and cyanobacterial blooms in a large subtropical river. INNOTEC 20, 30–66.

Gao, X., Kong, B., Vigil, R.D., 2018. Multiphysics simulation of algal growth in an airlift photobioreactor: effects of fluid mixing and shear stress. Bioresour. Technol 251, 75–83.

Giblin, S.M., Larson, J.H., King, J.D., 2022. Environmental drivers of cyanobacterial abundance and cyanotoxin production in backwaters of the Upper Mississippi River. River Res. Appl. 38 (6), 1115–1128. doi:10.1002/rra.3987.

Giglio, S., Chou, W.K.W., Ikeda, H., Cane, D.E., Monis, P.T., 2011. Biosynthesis of 2-methylisoborneol in cyanobacteria. Environ. Sci. Technol. 45, 992–998.

Gobler, C.J., 2020. Climate change and harmful algal blooms: insights and perspective. Harmful Algae. 1568-9883. 91, 101731. doi:10.1016/j.hal.2019.101731.

Griffiths, D.J., Saker, M.L., 2003. The Palm Island mystery disease 20 years on: a review of research on the cyanotoxin cylindrospermopsin. Environ. Toxicol. 1522-7278. 18 (2), 78–93. doi:10.1002/tox.10103.

Gupta, R.C., 2014. Biomarkers in Toxicology. Academic Press, Elsevier Inc.

Hawkins, P.R., Runnegar, M.T.C., Jackson, A.R.B., Falconer, I.R., 1985. Severe hepatotoxicity caused by the tropical cyanobacterium *Cylindrospermopsis raciborskii* (Woloszynska) Seenaya and Subba Raju isolated from a domestic water supply reservoir. Appl. Environ. Microbiol. 50, 1292–1295.

Humbert, J.F., Fastner, J., 2017. Ecology of cyanobacteria. In: Handbook of Cyanobacterial Monitoring and Cyanotoxin Analysis. Wiley, France, 11–18. Available from: https://www.wiley.com/en-gb/Handbook+of+Cyanobacterial+Monitoring+and+Cyanotoxin+Analysis-p-9781119068686. doi:10.1002/9781119068761.ch2.

Issa, A.A., Abd-Alla, M.H., Ohyam, T., 2014. Nitrogen Fixing Cyanobacteria: Future Prospect. InTech. doi:10.5772/56995.

Janssen, A.B.G., Janse, J.H., Beusen, A.H.W., Chang, M., Harrison, J.A., Huttunen, I., Kong, X., Rost, J., Teurlincx, S.A., Troost, T., 2019. How to model algal blooms in any lake on earth. Curr. Opin. Environ. Sustain. 3, 1–10.

Jasser, I., Callieri, C., 2017. Picocyanobacteria: the smallest cell-size cyanobacteria. In: Handbook of Cyanobacterial Monitoring and Cyanotoxin Analysis. Wiley, Poland, 19–27. Available from: https://www.wiley.com/en-gb/Handbook+of+Cyanobacterial +Monitoring+and+Cyanotoxin+Analysis-p-9781119068686. doi:10.1002/9781119068761.ch3.

Jayatissa, L.P., Silva, E.I.L., Mcelhiney, J., Lawton, L.A., 2006. Occurrence of toxigenic cyanobacterial blooms in freshwaters in Sri Lanka. Syst. Appl. Microbiol. [Online] 29 (2), 156–164. doi:10.1016/j.syapm.2005.07.007.

Johnson, H.E., King, S.R., Banack, S.A., Webster, C., Callanaupa, W.J., Cox, P.A., 2008. Cyanobacteria (Nostoc commune) used as a dietary item in the Peruvian highlands produce the neurotoxic amino acid BMAA. J. Ethnopharmacol. 118 (1), 159–165. 19.

Joo, G.J., Jeong, K.S., 2005. Modelling community changes of cyanobacteria in a flow regulated river (the lower Nakdong River, S. Korea) by means of a self-organizing map (SOM). In: Modelling Community Structure in Freshwater Ecosystems. Springer Berlin Heidelberg, South Korea, 273–287. Available from: http://www.springerlink.com/openurl.asp?genre=book&isbn=978-3-540-23940-6. doi:10.1007/3-540-26894-4_24.

Jüttner, F., 1995. Physiology and biochemistry of odorous compounds from freshwater cyanobacteria and algae. Water Sci. Technol. 0273-1223. 31 (11), 69–78. doi:10.2166/wst.1995.0405.

Kaplan-Levy, R.N., Hadas, O., Summers, M.L., Rücker, J., Sukenik, A., 2010. Akinetes: dormant cells of cyanobacteria. Top. Curr. Genet. 1610-6970. 21, 5–27. doi:10.1007/978-3-642-12422-8_2.

Kim, J.S., Seo, I.W., Baek, D., 2018. Modeling spatial variability of harmful algal bloom in regulated rivers using a depth-averaged 2D numerical model. Hydro-Environ. Res 20, 63–76.

Krishnan, A., Mou, X., 2020. A brief review of the structure, cytotoxicity, synthesis, and biodegradation of microcystins. Water 13 (16), 2147. doi:10.3390/w13162147.

Kumar, K., Mella-Herrera, R.A., Golden, J.W., 2010. Cyanobacterial heterocysts. Cold Spring Harb. Perspect. Biol 2, 1–20.

Kurmayer, R., Deng, L., Entfellner, E., 2016. Role of toxic and bioactive secondary metabolites in colonization and bloom formation by filamentous cyanobacteria Planktothrix. Harmful Algae 54, 69–86.

Lewis, W.M., Wurtsbaugh, W.A., Paerl, H.W., 2011. Rationale for control of anthropogenic nitrogen and phosphorus to reduce eutrophication of inland waters. Environ. Sci. Technol. 45, 10300–10305.

Li, X., Dreher, T.W., Li, R., 2016. An overview of diversity, occurrence, genetics and toxin production of bloom-forming *Dolichospermum* (*Anabaena*) species. Harmful Algae 54, 54–68. doi:10.1016/j.hal.2015.10.015.

Mahmood, N.A., Carmichael, W.W., Pfahler, D., 1988. Anticholinesterase poisonings in dogs from a cyanobacterial (blue-green algae) bloom dominated by *Anabaena flos-aquae*. Am. J. Vet. Res. 49, 500–503.

Malmaeus, J.M., Blenckner, T., Markensten, H., Persson, I., 2006. Lake phosphorus dynamics and climate warming: a mechanistic model approach. Ecol. Model. 190 (1), 1–14.

Matson, P.G., Boyer, G.L., Bridgeman, T.B., Bullerjahn, G.S., Kane, D.D., McKay, R.M.L., McKindles, K.M., Raymond, H.A., Snyder, B.K., Stumpf, R.P., et al, 2020. Physical drivers facilitating a toxigenic cyanobacterial bloom in a major great lakes tributary. Limnol. Oceanogr. 65, 2866–2882.

Mazmouz, R., Chapuis-Hugon, F., Mann, S., Pichon, V., Méjean, A., Ploux, O., 2010. Biosynthesis of cylindrospermopsin and 7-epicylindrospermopsin in *Oscillatoria sp.* strain PCC 6506: identification of the cyr gene cluster and toxin analysis. Appl. Environ. Microbiol. 76 (15), 4943–4949. doi:10.1128/AEM.00717-10.

Messineo, V., Bruno, M., De Pace, R., 2024. The role of cyano-HAB (Cyanobacteria Harmful Algal Blooms) in the one health approach to global health. Hydrobiology 3 (3), 238–262. doi:10.3390/hydrobiology3030016.

Newton, A.R., Melaram, R., 2023. Harmful algalblooms in agricultural irrigation: risks, benefits,and management. Front. Water 5, 1325300.

O'Neil, J.M., Davis, T.W., Burford, M.A., Gobler, C.J., 2012. The rise of harmful cyanobacteria blooms: the potential roles of eutrophication and climate change. Harmful Algae. 1568-9883. 14, 313–334. doi:10.1016/j.hal.2011.10.027.

Onodera, H., Oshima, Y., Henriksen, P., Yasumoto, T., 1997. Confirmation of anatoxin-a(s), in the cyanobacterium Anabaena lemmermannii, as the cause of bird kills in Danish lakes. Toxicon. 0041-0101. 35 (11), 1645–1648. doi:10.1016/S0041-0101(97)00038-X.

Paerl, H.W., 2014. Mitigating harmful cyanobacterial blooms in a human- and climatically-impacted world. Life. 2075-1729. 4 (4), 988–1012. Available from: http://www.mdpi.com/2075-1729/4/4/988/pdf. doi:10.3390/life4040988.

Pantelić, D., Svirčev, Z., Simeunović, J., Vidović, M., Trajković, I., 2013. Cyanotoxins: characteristics, production and degradation routes in drinking water treatment with reference to the situation in Serbia. Chemosphere. 1879-1298. 91 (4), 421–441. Available from: https://www.elsevier.com/locate/chemosphere. doi:10.1016/j.chemosphere.2013.01.003.

Pflugmacher, S., Wiegand, C., Beattie, K.A., Codd, G.A., Steinberg, C.E.W., 1998. Uptake of the cyanobacterial hepatotoxin Microcystin-LR by aquatic macrophytes. J. Appl. Bot. 72, 228–232.

Rangel, L.M., Soares, M.C.S., Paiva, R., Silva, L.H.S., 2016. Morphology-based functional groups as effective indicators of phytoplankton dynamics in a tropical cyanobacteria-dominated transitional river–reservoir system. Ecol. Indic 64, 217–227.

Rippka, R., Deruelles, J., Waterbury, J.B., Herdman, M., R.Y, S., 1979. Generic assignments, strain histories and properties of pure cultures of cyanobacteria. J. Gen. Microbiol 111, 1–61.

Sandybayeva, S.K., Kossalbayev, B.D., Zayadan, B.K., Kopecký, J., Kakimova, A.B., Bolatkhan, K., Allakhverdiev, S.I., 2023. Isolation, identification and pigment analysis of novel cyanobacterial strains from thermal springs. Plants 13 (21), 2951.

Singh, S.P., Montgomery, B.L., 2011. Determining cell shape: adaptive regulation of cyanobacterial cellular differentiation and morphology. Trends Microbiol. 1878-4380. 19 (6), 278–285. doi:10.1016/j.tim.2011.03.001.

Singh, V.K., Singh, S.K., Singh, P.K., Verma, H., Pandey, K.D., Singh, P.K., Kumar, A., 2020. Chapter 10-impact of pesticides applications on the growth and function of Cyanobacteria. In: Singh, P.K., Kumar, A., Singh, V.K., Shrivastava, A.K., Eds., Advances in Cyanobacterial Biology. Academic Press, Cambridge, MA, USA, pp. 151–161 978-0-12-819311-2.

Song, Y., 2023. Hydrodynamic impacts on algal blooms in reservoirs and bloom mitigation using reservoir operation strategies: a review. J. Hydrol. 0022-1694. 620, 129375. doi:10.1016/j.jhydrol.2023.129375.

Udeya, J.C., Harmon, L.J., Blank, C.E., 2016. A comprehensive study of cyanobacterial morphological and ecological evolutionary dynamics through deep geologic time. PLoS One 11 (9), e0162539.

Van Apeldoorn, M.E., Van Egmond, H.P., Speijers, G.J.A., Bakker, G.J.I., 2007. Toxins of cyanobacteria. Mol. Nutr. Food Res. 1613-4133. 51 (1), 7–60. doi:10.1002/mnfr.200600185.

Vidal, F., Sedan, D., D'Agostino, D., Cavalieri, M.L., Mullen, E., Parot Varela, M.M., Flores, C., Caixach, J., Andrinolo, D., 2017. Recreational exposure during algal bloom in Carrasco Beach, Uruguay: a liver failure case report. Toxins 9, 267.

Villalobos, T., Suárez-Isla, B., Garcia, C., 2025. Health and environmental impacts of cyanobacteria and cyanotoxins from freshwater to seawater. Toxins. 2072-6651. 17 (3), 126. doi:10.3390/toxins17030126.

Villatte, F., Schulze, H., Schmid, R., Bachmann, T., 2002. A disposable acetylcholinesterase-based electrode biosensor to detect anatoxin-a(s) in water. Anal. Bioanal. Chem. 1618-2642. 372 (2), 322–326. doi:10.1007/s00216-001-1127-4.

Watanabe, M.F., Oishi, S., 1985. Effects of environmental factors on toxicity of a cyanobacterium (*Microcystis aeruginosa*) under culture conditions. Appl. Environ. Microbiol. 0099-2240. 49 (5), 1342–1344. doi:10.1128/aem.49.5.1342-1344.1985.

Whitton, B., Potts, M., 2000. Introduction to the cyanobacteria. In: Whitton, B.A., Potts, M. (Eds.), The Ecology of *Cyanobacteria*. Springer, Dordrecht, pp. 1–11. doi:10.1007/0-306-46855-7-1.

Williamson, N., Kobayashi, T., Outhet, D., Bowling, L.C., 2018. Survival of cyanobacteria in rivers following their release in water from large headwater reservoirs. Harmful Algae 75, 1–15.

Wu, X., Noss, C., Liu, L., Lorke, A., 2019. Effects of small-scale turbulence at the air-water interface on *Microcystis* surface scum formation. Water Res 167, 115091.

Wu, Z.X., Gan, N.Q., Huang, Q., Song, L.R., 2007. Response of Microcystis to copper stress—Do phenotypes of Microcystis make a difference in stress tolerance? Environ. Pollut. 0269-7491. 147 (2), 324–330. doi:10.1016/j.envpol.2006.05.022.

Yang, T., Pan, J., Wu, H., Tian, C., Wang, C., Xiao, B., Pan, M., Wu, X., 2024. Rapid flotation of *Microcystis wesenbergii* mediated by high light exposure: implications for surface scum formation and cyanobacterial species succession. Front. Plant Sci. 15, 1367680. doi:10.3389/fpls.2024.1367680.

Yannarell, A.C., Kent, A.D., 2009. Bacteria, distribution and community structure. In: Encyclopedia of Inland Waters. Elsevier Inc., United States, 201–210. Available from: http://www.sciencedirect.com/science/book/9780123706263. doi:10.1016/B978-012370626-3.00123-X.

Zepernick, B.N., Gann, E.R., Martin, R.M., Pound, H.L., Krausfeldt, L.E., Chaffin, J.D., Wilhelm, S.W., 2021. Elevated pH conditions associated with *Microcystis spp.* blooms decrease viability of the cultured diatom *Fragilaria crotonensis* and natural diatoms in Lake Erie. Front. Microbiol. 12, 598736.

Zhang, X., Jia, X., Yan, L., Wang, J., Kang, X., Cui, L., 2017. Cyanobacterial nitrogen fixation influences the nitrogen removal efficiency in a constructed Wetland. Water. 2073-4441. 9 (11), 865. doi:10.3390/w9110865.

Zhang, Y., Zheng, L., Wang, S., Zhao, Y., Xu, X., Han, B., Hu, T., 2021. Quorum sensing bacteria in the phycosphere of HAB Microalgae and their ecological functions related to cross-kingdom interactions. Int. J. Environ. Res. Public Health 19 (1), 163. doi:10.3390/ijerph19010163.

Zhu, C., Chi, Z., Bi, C., Zhao, Y., Cai, H., 2019. Hydrodynamic performance of floating photobioreactors driven by wave energy. Biotechnol. Biofuels 12, 54.

Chapter 3

River deterioration due to indiscriminate use of pesticides, pharmaceuticals, and personal care products

Zubair Ahamed Shaik, Jayalakshmi Jayashankar, Mwalimu Raphael Jilani, Abinaya Krishnan, Vithyasree Srikanth, and Azhagu Saravana Babu Packirisamy
Department of Biotechnology, Vel Tech Rangarajan Dr. Sagunthala R&D Institute of Science and Technology, Chennai, India

1 Introduction

River ecosystems are complex and dynamic systems that play a crucial role in maintaining biodiversity and ecological processes. These ecosystems are characterized by a shifting habitat mosaic, where the distribution of habitat patches changes over time due to factors such as flooding, channel avulsion, and sediment deposition. River ecological dynamics involve the study of ecosystem structure, function, and change, including species composition, biodiversity, and ecological processes (Chen, 2024; Stanford et al., 2017). However, river ecosystems face numerous threats, with biological invasions being a significant concern (Guareschi and Wood, 2022). These invasions can have wide-ranging impacts on local economies, biodiversity, and ecosystem stability. Assessing river ecosystem health is essential for sustainable management, with factors such as anthropogenic activities, land use, water quality, and climate change being key considerations. The current state of river deterioration requires developing comprehensive assessment methods and addressing challenges in river ecosystem conservation and restoration (Chen, 2024; Ranhao et al., 2020).

River water quality is crucial for human health, ecosystem survival, and economic activities. Key parameters for assessing river water quality include physical, chemical, and biological attributes, with fecal contamination considered particularly important by stakeholders (Okumah et al., 2020). Water quality is affected by both point-source and diffuse pollution, with the latter becoming

River Pollution. https://doi.org/10.1016/B978-0-443-45024-2.10003-X

increasingly dominant due to improved wastewater treatment (Davies-Colley, 2013). Major diffuse pollution sources include fine sediment, nutrients, and fecal microbes, primarily from agricultural intensification (Davies-Colley, 2013). Monitoring river water quality is essential for effective environmental management (Li and Liu, 2019). While some improvements have been observed in certain catchments, challenges remain in managing diffuse pollution and balancing agricultural intensification with water quality protection (Davies-Colley, 2013). Integrating stakeholder perceptions into water management policies is important, ensuring that it complements scientific measurements rather than being a replacement (Okumah et al., 2020).

Rising concerns over river pollution worldwide are well-documented, with various factors contributing to the degradation of water quality. Domestic, industrial, and agricultural wastes, including hazardous materials, are directly or indirectly discharged into rivers. This pollution affects not only water but also sediments and aquatic life, impacting clean water availability for various uses and depleting food sources from rivers. Climate change exacerbates these issues by altering temperatures, rainfall patterns, and biological communities. The pollution situation is particularly severe in developing countries, where medium-scale industries form the socioeconomic backbone, complicating the implementation of strict abatement regulations. Global scenarios suggest severe health effects from persistent pollutants in the coming decades. Addressing river pollution requires public awareness, continuous monitoring, and advanced water treatment technologies (Saad et al., 2022).

PPCPs are an essential part of our daily lives. They encompass a wide range of substances with various consumption patterns. This includes pharmaceutical compounds such as antiinflammatories, antimicrobials, and lipid-regulating agents used in both human and veterinary medicine. In addition, PPCPs cover personal hygiene items, which include fragrances, disinfectants, and other similar products (Xiang et al., 2021).

PPCPs play a crucial role in improving the quality of life. After they are metabolized, the resulting metabolites and conjugates are primarily excreted through feces and urine. These substances then enter sewage treatment systems, where they can either become part of sewage sludge or gradually disperse into surface water bodies over time. As a result, humans and other organisms are unintentionally exposed to trace amounts of these PPCPs from the environment by drinking contaminated water or consuming plants and animals that have also been exposed. This exposure can persist throughout a person's life, leading to the accumulation of these metabolites in body tissues at levels that are often many times lower than those used therapeutically (Carlsson et al., 2006; Daughton, 2008).

2 Sources of contamination

2.1 Agricultural runoff: The role of pesticides

Pesticides are a class of chemicals used globally as pest, weed, and disease control agents in crops, as well as for human and animal health, functioning as nematicides, insecticides, herbicides, rodenticides, fungicides, and plant growth regulators. The benefits of using pesticides include increased food and crop productivity, as well as a sharp decline in vector-borne illnesses. However, excessive fertilizer use poses a significant risk to surface and groundwater. The term "agricultural runoff" refers to surface water that exits farmed fields when it receives more water than the soil can absorb. Water can become contaminated by pesticides through leaching or surface runoff. The earth's hydrologic cycle is connected to these two essential activities (Kaur and Sinha, 2019). Concern over the pervasive issue of pesticides in aquatic environments is growing globally, as these substances can be deposited in aquatic organisms and soil sediment, endangering human health. The extensive use of pesticides in agricultural and nonagricultural sectors has led to the introduction of several pesticide formulations and the rapid rise of the global pesticide market. Pesticides are frequently found in water bodies due to industrial and agricultural runoff (Syafrudin et al., 2021). The use of pesticides and their effects on many environmental matrices have grown in recent years. To determine the precise situation of pesticide toxicity in aquatic environments, risk assessment and monitoring are crucial (Shah and Parveen, 2021) (Fig. 3.1).

Studies indicate that water pollution often results from human activities. In particular, modern agricultural and livestock practices contribute significantly to this issue. The indiscriminate use of pesticides, fertilizers, and hormones leads to the presence of nitrates in water. These nitrates are linked to an increased risk of congenital anomalies, including heart defects and neural tube defects (Blaisdell et al., 2019).

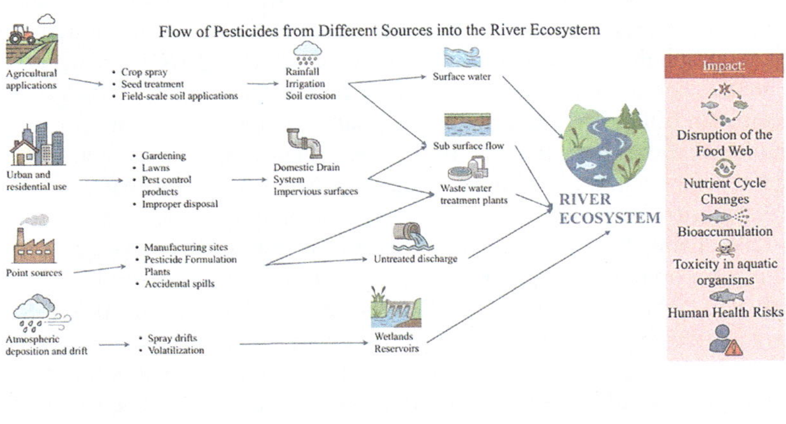

FIG. 3.1 Flow of pesticides from the source to the river ecosystem. This figure illustrates the flow of pesticides from agricultural, urban, industrial, and atmospheric sources to the river ecosystem.

2.2 Pharmaceutical disposal and wastewater effluents

Pharmaceuticals are a necessity for human health. Pharmaceutical businesses add new drugs to the market every year, in addition to the vast array of existing pharmaceutical products, due to the increasing demand for pharmaceuticals. As a result of regular usage, damage, and expiration, pharmaceutical waste is released into the environment in low concentrations by homes, farms, pharmaceutical industries, and healthcare facilities. It is unknown how active pharmaceutical ingredients (APIs) affect environmental nontarget species. Pharmaceuticals like ibuprofen and diclofenac have been found in trace concentrations in ground and surface water, as well as municipal water systems, over the years. To minimize the hazards to the public and the environment, unwanted medications should be disposed of properly and affordably. Inadequate understanding of how to properly dispose of unwanted medications causes unintended poisoning, increased medical expenses, scavenging and theft in landfills, contamination of water supplies, antimicrobial resistance, and even death. Pharmacists should be encouraged to educate the public about safe disposal procedures to lessen these impacts (Nyaga, 2020).

Significant environmental problems arise from the contamination of water by wastewater treatment plants, particularly when undesired outputs, such as residual contaminants, are not sufficiently managed. By including these unwanted outputs in the evaluation, researchers perform an efficiency analysis of wastewater treatment plants. This analysis assesses the performance of different facilities using models like data envelopment analysis, highlighting both areas that require improvement and those that are operating well. By encouraging cleaner operations and well-informed decision-making in wastewater treatment processes, the findings support sustainable water management (D'Inverno et al., 2018). Due to the excessive use and insufficient removal in traditional wastewater treatment methods, pharmaceutical pollutants such as antiinflammatories, antibiotics, and endocrine-disrupting chemicals (EDCs) are being found in aquatic habitats more frequently. These pollutants originate from agricultural runoff, landfill leachates, hospital and pharmaceutical industrial effluents, and municipal wastewater. Because of their persistence, there are serious threats to human health and the environment, such as the emergence of bacteria resistant to antibiotics, hormonal imbalances in aquatic organisms, and bioaccumulation in food chains. In addition to nature-based alternatives like artificial wetlands, mitigation options incorporate cutting-edge treatment technologies, including activated carbon, membrane filtration, bioremediation, and ozonation. Stricter laws, better pharmaceutical disposal procedures, and increased public awareness are also necessary for effective control to reduce their infiltration into water systems and protect human and environmental health (Ilham et al., 2023).

Antibiotic pollution is a significant issue among various PPCPs because it plays a major role in the horizontal transfer of antibiotic resistance genes (ARGs) across the bacterial spectrum. This process is particularly alarming since people can unknowingly ingest these resistance genes through different channels, including recreational activities in environments contaminated by these pollutants. The presence of low concentrations of antibiotics in the environment can infiltrate human systems, accumulating over time and leading to health issues ranging from reproductive disorders to muscle weakness (Kraemer et al., 2019; Manaia et al., 2024) (Fig. 3.2).

2.3 Personal care products in everyday use

Personal care products (PCPs) are a broad category of everyday household items used for wellness, cleaning, and cosmetics. These include UV filters, scents, insect repellents, preservatives, and disinfectants. Because of their abundance and detrimental effects on aquatic ecosystems, particularly those linked to endocrine disruption and reproductive disorders, some of them are regarded as chemicals of rising concern. Due to their ineffective or partial removal, such chemicals primarily enter aquatic bodies through sewage effluents (Montes-Grajales et al., 2017). Every year, millions of tons of PCPs are produced and used, and there are thousands of chemical components that make up cosmetic formulas. It is crucial to determine the dangers of the continuous discharge of large-scale quantities of these chemicals into the water. It is mostly uncertain how these products will fare in the environment; due to untreated discharge in some cases, they remain in the environment at unexpected levels, undergo bioaccumulation, and even react with other pollutants to create new forms of unpredictable contaminants. In other cases, they may be eliminated in wastewater treatment plants. It is not feasible to ban the goods that cause these issues, unless there are specific exceptions (for example, several Mexican coastal Eco

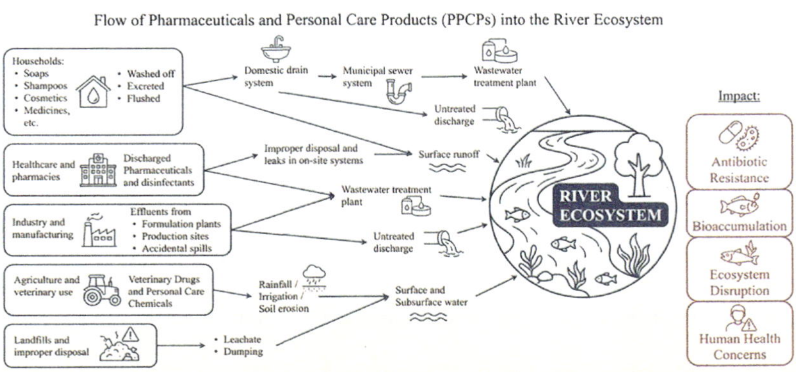

FIG. 3.2 Flow of PPCPs from source to river ecosystem. Schematic representation of the transport pathways of pharmaceuticals and personal care products from various sources to the river ecosystem.

parks forbid the use of sunscreen). Realistically addressing this problem calls for many methods and techniques. The sophisticated technologies of analytical chemistry have contributed to our heightened understanding of these compounds' potential for contamination. Thus the creation of better extraction and analytical techniques would enable a more thorough and precise assessment of environmental contaminants in intricate matrices. However, further research on the acute and long-term toxicity of these pollutants is needed to provide a more accurate estimate of the true ecological and health risks they pose. Lastly, people may be persuaded to use cosmetics more responsibly and knowledgeably if they are aware of the environmental impact of their packaging (Juliano and Magrini, 2017).

3 Chemical composition and persistence

3.1 Types of harmful chemicals in contaminated rivers

While some of the harmful compounds that are currently entering freshwaters were present decades ago, others have only recently become substantial. Unidirectional transmission and dispersion of toxicants in flowing fluids alter their effects and provide the possibility of some "self-purification." The chemical composition of incoming water impacts toxicity as well. In addition, biological variables influence the final impact of contaminants. The relevance of some contaminants that may be present in water, despite extremely low ambient concentrations, is increased by the possibility of hazardous chemicals building up within tissues. After pollutants enter flowing rivers catastrophically, drift from undamaged areas upstream can help recover the biota (Hellawell, 1988). The presence of emerging contaminants (ECs) in the aquatic environment, including pesticides, medicines, and PCPs, has sparked serious concerns about the harm they could do to both humans and aquatic life. PCPs and pesticides are more likely than other contaminants to harm water systems due to their toxicity and bioactivity, which can result in a number of negative consequences such as bioaccumulation, antibiotic resistance, biomagnification, and reproductive impairment (Okoye et al., 2022). The reproductive system may be at risk from anthropogenic pollutants in water. It is well known that the majority of these substances are EDCs. EDCs can affect the endocrine system, which in turn can impact human and nonhuman animal development and fertility. There are several different sources of chemical contamination in water, including pharmaceuticals discharged into sewage, byproducts of water treatment processes, and releases from industry and livestock (Gonsioroski et al., 2020).

3.2 Persistence of contaminants in water bodies

The concentrations and types of contaminants are determined by the quality of the recharge water and the geological materials that the groundwater passes through. The impact of these contaminants varies according to their concentrations and types, since groundwater passing through soils and sedimentary rocks may absorb a variety of substances, including arsenate, calcium, chloride, fluoride, iron, magnesium, and nitrate. By evaluating the color, turbidity, odor, and taste of the water, certain contaminants can be quickly recognized. To determine the contamination in water, testing is necessary because the majority are difficult to detect. Hence, the impurities may cause unpleasant taste or smell, discoloration, and negative health impacts (Sharma & Bhattacharya, 2016). Due to their dual roles as a sink and a possible source of pollutants, sediments are essential for maintaining pollution levels. According to numerous studies, suspended sediment is crucial for the movement of heavy metals, fertilizers, and other hydrophobic pollutants through river systems, which affects the quality of the water. Sediments are an active component in the dynamics of aquatic pollution, though, as these contaminants can be released by disturbances or chemical changes. More than 90% of the flux through a river system for most trace elements occurs in conjunction with suspended sediment. Therefore the transportation, remobilization, and deposition of suspended sediment, and consequently the sediment conveyance system, regulate the transfer, distribution, and fate of sediment-associated nutrients and pollutants in river systems (Meharg, 2003).

Organic pollutants in rivers can be reduced by photochemical transformation, which breaks them down when exposed to sunlight. This can happen directly when the pollutant absorbs sunlight or indirectly through reactive species like hydroxyl radicals. Pollutants undergo these processes to become more biodegradable. The process is influenced by various factors, including temperature, chemical composition, turbidity, pH, water depth, and solar intensity, which is affected by location, cloud cover, and time of day. This approach shows potential for breaking down pharmaceuticals and new toxins in river systems, despite the fact that it is difficult to generalize for all pollutants. Methods such as UV/H_2O_2 and photocatalysis improve removal in contaminated water bodies (Wilkinson et al., 2017).

The classification of organic contaminants is extensive, with the most significant being the class of pollutants known as persistent organic pollutants (POPs).

These pollutants pose considerable risks to human health and the environment. Examples of POPs include personal care products, per- and polyfluoroalkyl substances (PFAS), pharmaceutical compounds, dyes, pesticides, phenolic compounds, hormones, surfactants, and sweeteners, among others. The detection of these substances is crucial for evaluating their detrimental effects. Many of these contaminants originate from industrial processes and are utilized in various applications such as coolants, flame retardants, and cement production. Their presence significantly contributes to water ecotoxicity, particularly in regions such as Ecuador, Argentina, and Mexico, adversely affecting the integrity of the ecosystems and the species that inhabit them (Alava et al., 2020; Oregel-Zamudio et al., 2021).

3.3 Long-term environmental impact

When discharged into aquatic habitats, contaminants like pesticides, heavy metals, and petroleum hydrocarbons can have direct harmful impacts. Lethality may decimate sensitive species, whereas sublethal effects may damage them. This ecological change may trigger a trophic cascade or a relief from competition, which, in turn, may trigger responses in tolerant species. Foundation species and keystone facilitators are susceptible to direct impacts from contaminants, and ecosystem function may be modified by alterations in nutrient and oxygen dynamics brought on by contaminants. Pollutant exposure can, therefore, have a direct or indirect impact on natural communities and populations. Although toxicants typically have a negative direct impact on organism abundance, indirect impacts can result in either a rise or fall in abundance (Fleeger et al., 2003).

Both land and marine ecosystems are seriously endangered by contaminated sites. They are significant polluters and can have ecotoxicological impacts on aquatic, groundwater, and terrestrial ecosystems. Acute impacts do occur in highly contaminated areas, but long-term chronic effects are the main issue. At every level of biological organization, from the molecular to the ecosystem level, ecotoxicological impacts can be observed. The function and structure of terrestrial and aquatic ecosystems as a whole may be impacted, in addition to specific organisms. Critical characteristics, including chronic toxicity and/or high acute toxicity, high environmental persistence, frequent high mobility that contaminates groundwater, and high lipophilicity that causes bioaccumulation in food webs, are frequently shared by contaminants at major contaminated sites (Fent, 2004).

The global annual consumption of antibiotics surpasses 100,000 tons, raising escalating concerns about the environmental impact of these compounds. Antibiotics are prevalent in various ecosystems, with considerable concentrations identified in freshwater bodies (Danner et al., 2019). The presence of antibiotic contamination in riverine environments has been shown to shrink the overall diversity of microorganisms, including those vital for primary productivity and carbon processing (Kraemer et al., 2019).

4 Impact on aquatic life

Pharmaceuticals are essential in various sectors, including human health, veterinary medicine, agriculture, and aquaculture, primarily for the treatment and prevention of diseases. The annual consumption of these drugs has been increasing consistently. Nevertheless, the interaction of APIs with other biological compounds may result in unexpected environmental impacts (Boxall, 2004). Therapeutic compounds enter aquatic systems from multiple sources, leading to the contamination of surface waters with pharmaceutical substances and their metabolites. To safeguard a healthy ecosystem and the biodiversity of aquatic organisms, it is essential to uphold the quality of surface water.

Analytical methods are critical for the identification and evaluation of emerging pollutants, allowing us to implement proactive strategies to mitigate their effects (Al-Odaini et al., 2011; Jones et al., 2001; Daughton, 2008).

4.1 Bioaccumulation

Recent studies highlight the widespread occurrence and potential risks of pharmaceuticals, personal care products (PPCPs), and pesticides in aquatic environments. Trophic magnification has been observed for some organic UV filters, while certain pharmaceuticals undergo trophic dilution (Liu et al., 2020). Species-specific differences in bioaccumulation factors (BAFs) have been reported, possibly due to variations in toxicokinetics (Nozaki et al., 2023). Microplastics can act as carriers for PPCPs, potentially influencing their bioavailability and toxicity (Zhou et al., 2020). The presence of these contaminants in water systems poses risks to aquatic ecosystems and human health, with some compounds showing potential for adverse effects on wild fish (Nozaki et al., 2023). The situation is particularly concerning in African water systems, where limited data on remediation and control strategies necessitate urgent intervention (Okoye et al., 2022).

The accumulation of various organic compounds, including polychlorinated biphenyls (PCBs) and polybrominated diphenyl ethers (PBDEs), has been documented in significant aquatic ecosystems, such as Lake Chapala in Mexico. This was determined through the analysis of samples collected from water, fish, and sediments across two distinct seasonal periods. The fish species examined in this study included Cyprinus carpio, Oreochromis aureus, and *Chirostoma* spp. The findings indicate that these chemical contaminants can enter the lake as a result of industrial activities, strong winds, and the inflow of the Lerma River in Mexico (Oregel-Zamudio et al., 2021).

The improper disposal of pharmaceutical compounds, along with insufficient wastewater treatment practices, can lead to the environmental accumulation of these substances. Such accumulation has the potential to negatively impact nontarget organisms and exert stress on the ecosystem (Fernandes et al., 2021). The establishment of effective management practices is crucial for the proper control of emerging pollutants (Khan and Barros, 2023; Liu et al., 2020).

4.2 Biomagnification

The widespread use of antibiotics in both medical and agricultural sectors has led to their emergence as significant environmental pollutants. Effluents released from pharmaceutical manufacturing processes are major contributors to the contamination of aquatic ecosystems with these antibiotics. These chemical compounds infiltrate various environmental compartments, including groundwater, surface water, soil, and wastewater treatment plants, with concentrations varying from nanograms to grams per liter (Eapen et al., 2024). PPCPs can affect the behavior and reproductive systems of aquatic organisms, with potential cascading effects on entire ecosystems (Khan and Barros, 2023; Srain et al., 2021). Studies have identified specific compounds, like estrone, estriol, and triclosan, as high-risk contaminants in Chinese rivers. The impacts of these pollutants vary across species, with fish showing particular sensitivity to certain EDCs. The presence of PPCPs in water bodies has raised concerns about antibiotic resistance and other long-term ecological effects, highlighting the need for improved monitoring, management, and remediation strategies (Morrison et al., 2025).

4.3 Endocrine disruptions

According to existing data, over 10 million women in the United States utilize pharmaceuticals for birth control. Research has shown that 17-α-ethinylestradiol, a type of estrogen prevalent in contraceptive formulations, can induce feminization in male fish at concentrations as low as 5–6 ng/L. This observation underscores the potential influence of steroid hormones on aquatic organisms and their ecosystems. Furthermore, the presence of these hormones can lead to diminished fertility in aquatic species, thereby highlighting the critical importance of managing and reducing their release into the environment (Selwe et al., 2022).

4.4 Nutrient cycle changes

In aquatic ecosystems, critical processes such as denitrification, nitrogen fixation, and the decomposition of organic material are predominantly mediated by specific microbial communities. The introduction of antibiotics into these environments can disrupt these processes by inhibiting or modifying the functions of the relevant bacterial groups. This disruption may result in significant disturbances in nitrogen cycling, the breakdown of organic matter, and other vital ecological functions. Consequently, the overall functionality and stability of the aquatic ecosystem may be jeopardized, with potential repercussions for water quality, nutrient dynamics, and the health of aquatic organisms. In addition, the discharge of sewage effluent containing pharmaceutical residues into streams and other water bodies poses direct risks to both human and animal health due to possible exposure to these contaminants (Boxall, 2004).

5 Human health implications

The critical importance of high-quality water for the survival of humans, as well as for plant and animal life, cannot be overstated. The aquatic ecosystem is currently confronting substantial challenges as a result of industrialization and urbanization. Anthropogenic pollutants permeate aquatic environments through various pathways, including effluents from industrial and domestic sources, pharmaceutical residues, and organic wastes. These substances tend to bioaccumulate within aquatic systems. Research conducted on the surface waters of the River Taff and River Ely indicates that pharmaceutical products are released at an average daily load of approximately 6 kg, underscoring the magnitude of contamination present (Kasprzyk-Hordern et al., 2008).

5.1 Antibiotic resistance

The occurrence of antibiotics in wastewater contributes to the development of antibiotic-resistant bacteria (ARB). This issue presents considerable threats to both human health and ecological systems. The presence of antibiotics and ARGs can produce widespread ramifications, potentially undermining the efficacy of medical treatments and disturbing the equilibrium of microbial communities within the environment (Chen et al., 2016).

The 2014 Review on Antimicrobial Resistance (AMR) underscores that antibiotic resistance is responsible for a considerable number of annual fatalities. In recognition of the projected increase in antibiotic resistance, the World Health Organization (WHO) identified it as a critical global health issue in 2014, highlighting its profound implications for public health. Addressing the emergence and dissemination of mobile resistance elements presents a significant challenge due to the inadequate understanding of how environmental factors contribute to the development of resistance and the specific conditions that facilitate this process. This limited knowledge in the field obstructs efforts to effectively manage and mitigate the evolution and transmission of resistance elements (Bengtsson-Palme et al., 2018).

The increasing threat posed by antimicrobial resistance (AMR) significantly endangers global public health, presently resulting in approximately 700,000 deaths annually due to multidrug-resistant (MDR) bacterial infections. Projections indicate that this figure could rise to a staggering 10 million fatalities by the year 2050, with associated economic repercussions estimated to reach 3.8% of global GDP, potentially thrusting millions into extreme poverty. The implementation of fundamental preventive measures, such as proper hand hygiene and the judicious use of antibiotics, has the potential to avert a considerable portion—up to three-quarters—of these fatalities at minimal expense. Consequently, addressing AMR is imperative for the protection of healthcare systems and the advancement of global development objectives (Serra-Burriel et al., 2020).

From October 2008 to June 2013, a study was conducted to investigate infections resistant to multiple antibiotics in hospitalized patients with cirrhosis. These infections were categorized as multi-drug resistant (MDR), extensively drug-resistant (XDR), or pandrug-resistant (PDR). Approved by the local ethical committee, the research focused on the characteristics of these infections, associated risk factors, and clinical outcomes. It also assessed the efficacy of recommended antibiotic treatments and classified infections as hospital-acquired, healthcare-associated, or community-acquired, with an emphasis on pathogens exhibiting resistance to three or more classes of antimicrobials

(MDR, XDR, PDR). The findings of this study provide critical insights into bacterial resistance patterns in patients with advanced cirrhosis and their implications for therapeutic strategies (Merli et al., 2015). The overuse of antibiotics in young children can adversely affect their gut microbiota, which play a crucial role in the digestion of complex carbohydrates, the prevention of pathogenic bacterial colonization, and the proper development of the immune system. A reduction in beneficial bacteria, typically resulting from antibiotic treatment, can disrupt these essential functions. This underscores the importance of maintaining a healthy and diverse composition of gut bacteria in infants (Lebeaux et al., 2022).

5.2 Effect on the endocrine system

The endocrine system plays a critical role in regulating various physiological functions within the human body, functions that can be adversely affected by a category of substances known as endocrine-disrupting compounds (EDCs). These compounds, which can be either naturally occurring or synthetically produced, are primarily derived from human activities. EDCs possess the ability to imitate, enhance, or obstruct the actions of endogenous hormonal substances. Furthermore, they may also play a role in tumorigenesis. A notable segment of EDCs disrupts sexual hormonal functions, which can result in anomalies in reproductive processes, embryonic development, sexual differentiation, and metabolic maturation (Flint et al., 2012; Tijani et al., 2013).

Gill et al. (1979) reported that EDCs adversely affect reproductive health by significantly diminishing sperm quality and count in males. These substances also possess estrogenic properties. Diethylstilbestrol (DES), a compound initially developed to prevent miscarriages, serves as a prominent illustration of this phenomenon. Investigations have offered detailed insights into the endocrine-disrupting effects of DES, revealing the deleterious reproductive consequences experienced by male offspring resulting from its administration.

6 Case studies on river deterioration

Water pollution poses a significant environmental challenge, adversely affecting both ecosystems and human health. The principal concern lies with inorganic and organic pollutants, which are characterized by their high toxicity and persistence, rendering them challenging to remediate using existing methods. Research indicates that there is a considerable diversity of contaminants in the water bodies across the American continent, impacting various ecological and health-related aspects. In certain instances, viable alternatives for the remediation of contaminated water have been identified (Warren-Vega et al., 2023).

A study conducted in the St. Lawrence River, located in Quebec, Canada, focused on the analysis of surface water to identify ultraviolet absorbents (UVAs) and industrial antioxidants. Detection of these compounds was achieved through gas chromatography–mass spectrometry (GC–MS), which identified various categories of UVAs, including organic UV filters such as benzophenone (BP), 2-ethylhexyl salicylate (EHS), 2-hydroxy-4-methoxybenzophenone (BP3), 3,3,5-trimethylcyclohexyl salicylate (HMS), 2-ethylhexyl 2-cyano-3,3-diphenylacrylate (OC), and ethylhexyl methoxycinnamate (EHMC). In addition, the study analyzed aromatic secondary amines like diphenylamine (DPA), benzotriazole UV stabilizers including 2-(2H-benzotriazol-2-yl)-4,6-di-tert-pentylphenol (UV238), and synthetic phenolic antioxidants such as 2,6-di-tert-butyl-4-methylphenol (BHT) and 2,6-di-tert-butyl-1,4-benzoquinone (BHTQ). The investigation also assessed field-based tissue-specific BAFs to evaluate the presence of these contaminants in the tissues of fish species, specifically lake sturgeon and northern pike. Notably, compounds that exhibited accumulation in lake sturgeon included BP3, BHT, and UV238, whereas for northern pike, accumulated compounds were BP, BP3, BHT, and BHTQ. This research establishes an environmental risk assessment concerning the potential adverse effects of these contaminants on fish populations (Castilloux et al., 2022).

In the study conducted by Ramos et al. (2021), a comprehensive water analysis was carried out on a river and its treated water over the course of a year in Minas Gerais, Brazil. The investigation utilized a single quadrupole gas chromatograph–mass spectrometer (GCMS-QP2010 SE) integrated with a flame ionization detector to detect 17 phenolic compounds. Among the samples analyzed, only 16 compounds were identified, with 3-methylphenol being the sole compound not detected. The results indicated that in raw water, 2,3,4-trichlorophenol, 2,4-dimethylphenol, and 4-nitrophenol were detected most frequently. In contrast, treated water primarily exhibited the presence of 4-nitrophenol and bisphenol A. The findings highlight the potential health risks posed to both the environment and human health due to the contamination from these phenolic compounds.

7 Conclusion

The indiscriminate application and subsequent environmental discharge of pesticides, pharmaceuticals, and personal care products (PPCPs) constitute a significant challenge to riverine ecosystems in the 21st century. Studies around the globe have shown that these pollutants' long-lasting effects, ability to affect living organisms, and complicated chemical makeup have substantial impacts on aquatic life, human health, and the stability of the ecosystem as a whole. PPCPs and pesticides differ from regular pollutants in that they can have subtle but long-lasting effects, such as disrupting hormones, reducing the effectiveness of antibiotics, harming reproduction, and accumulating in organisms while magnifying across trophic levels. These impacts are compounded by the fact that many of these compounds are resistant to degradation, evade traditional wastewater treatment processes, and interact synergistically with other pollutants, leading to an amplification of their ecological footprint. The evidence presented underscores that rivers, as dynamic and interconnected systems, are particularly vulnerable to contamination from diffuse sources such as agricultural runoff, pharmaceutical disposal, and domestic wastewater effluents. Sediments act as both sinks and secondary sources of pollutants, while seasonal variations and climate factors further influence contaminant persistence and distribution. The ecological consequences are not confined to immediate toxicity; rather, they manifest in altered species behavior, reduced biodiversity, trophic cascades, and long-term shifts in ecosystem function. Moreover, the infiltration of antibiotic residues into aquatic environments creates two major issues: (1) compromising microbial diversity essential for nutrient cycling and (2) facilitating the horizontal transfer of resistance genes, which has acute implications for global public health.

From a human perspective, the contamination of rivers by PPCPs and pesticides jeopardizes water security, food safety, and community health. Trace exposure through drinking water, irrigation, and consumption of contaminated aquatic organisms can lead to cumulative physiological effects, including hormonal imbalances, congenital anomalies, and chronic diseases. The linkage between PPCP pollution and methane production further highlights the intersection between local water quality issues and global climate change, reinforcing the urgency of integrated solutions. Despite growing recognition of these risks, current remediation strategies remain inadequate. Conventional wastewater treatment plants are not designed to fully eliminate PPCPs, and while advanced technologies such as membrane filtration, activated carbon adsorption, ozonation, and bioremediation show promise, their scalability and cost-effectiveness remain significant barriers. Equally critical is the need for improved monitoring frameworks that combine chemical analyses with ecological risk assessments, thereby enabling a more comprehensive understanding of contaminant dynamics. Public awareness and behavioral change, particularly in the disposal of pharmaceuticals and the responsible use of personal care products, are indispensable components of mitigation. Looking forward, addressing river pollution by PPCPs and pesticides requires a multi-pronged approach that integrates scientific innovation, regulatory enforcement, and societal engagement. Research must continue to refine analytical techniques for detecting contaminants at trace levels, while also exploring nature-based solutions such as constructed wetlands and phytoremediation. Policymakers must strengthen legislation governing pharmaceutical disposal, pesticide application, and industrial effluent management, ensuring compliance through transparent monitoring systems. At the same time, interdisciplinary collaboration between ecologists, chemists, public health experts, and social scientists is essential to bridge knowledge gaps and design interventions that are both environmentally sustainable and socially equitable.

In conclusion, the deterioration of rivers due to PPCPs and pesticides is emblematic of the broader challenges posed by ECs in the Anthropocene. Rivers are lifelines for biodiversity, human societies, and economic development, and their degradation threatens the ecological integrity, as well as the resilience of communities worldwide. By advancing research, fostering innovation, and cultivating collective responsibility, it is possible to mitigate the risks posed by these pollutants and safeguard river ecosystems for future generations.

References

Al-Odaini, N., Zakaria, M., Yaziz, M., Surif, S. (2011). Detecting Human Pharmaceutical Pollutants in Malaysian Aquatic Environment: a new challenge for water quality management.

Alava, J.J., Calle, P., Tirapé, A., Biedenbach, G., Cadena, O.A., Maruya, K., Lao, W., Aguirre, W., Jiménez, P.J., Domínguez, G.A., Bossart, G.D., Fair, P.A., 2020. Persistent organic pollutants and mercury in genetically identified inner estuary bottlenose dolphin (*Tursiops truncatus*) residents of the Guayaquil Gulf, Ecuador: ecotoxicological science in support of pollutant management and Cetacean conservation. Front. Mar. Sci. 2296-7745. 7. Available from: https://www.frontiersin.org/journals/marine-science#. doi:10.3389/fmars.2020.00122.

Bengtsson-Palme, J., Kristiansson, E., Larsson, D.G.J., 2018. Environmental factors influencing the development and spread of antibiotic resistance. FEMS Microbiol. Rev. 1574-6976. 42 (1), 68–80. Available from: http://femsre.oxfordjournals.org/. doi:10.1093/femsre/fux053.

Blaisdell, J., Turyk, M.E., Almberg, K.S., Jones, R.M., Stayner, L.T., 2019. Prenatal exposure to nitrate in drinking water and the risk of congenital anomalies. Environ. Res. 1096-0953. 176. Available from: http://www.elsevier.com/inca/publications/store/6/2/2/8/2/1/index.htt. doi:10.1016/j.envres.2019.108553.

Boxall, A.B.A., 2004. The environmental side effects of medication. EMBO Rep. 1469-221X. 5 (12), 1110–1116. doi:10.1038/sj.embor.7400307.

Carlsson, C., Johansson, A.K., Alvan, G., Bergman, K., Kühler, T., 2006. Are pharmaceuticals potent environmental pollutants? Part I: environmental risk assessments of selected active pharmaceutical ingredients. Sci. Total Environ. 0048-9697. 364 (1-3), 67–87. doi:10.1016/j.scitotenv.2005.06.035.

Castilloux, A.D., Houde, M., Gendron, A., De Silva, A., Soubaneh, Y.D., Lu, Z., 2022. Distribution and fate of ultraviolet absorbents and industrial antioxidants in the St. Lawrence River, Quebec, Canada. Environ. Sci. Technol. 1520-5851. 56 (8), 5009–5019. Available from: http://pubs.acs.org/journal/esthag. doi:10.1021/acs.est.1c07932.

Chen, J., Ying, G.G., Wei, X.D., Liu, Y.S., Liu, S.S., Hu, L.X., He, L.Y., Chen, Z.F., Chen, F.R., Yang, Y.Q., 2016. Removal of antibiotics and antibiotic resistance genes from domestic sewage by constructed wetlands: effect of flow configuration and plant species. Sci. Total Environ. 1879-1026. 571, 974–982. Available from: www.elsevier.com/locate/scitotenv. doi:10.1016/j.scitotenv.2016.07.085.

Chen, M., 2024. Overview of river ecological dynamics. Front. Sci. Eng. 2710-0588. 4 (7), 132–140. doi:10.54691/hwahex28.

D'Inverno, G., Carosi, L., Romano, G., Guerrini, A., 2018. Water pollution in wastewater treatment plants: an efficiency analysis with undesirable output. Eur. J. Oper. Res. 0377-2217. 269 (1), 24–34. doi:10.1016/j.ejor.2017.08.028.

Danner, M.C., Robertson, A., Behrends, V., Reiss, J., 2019. Antibiotic pollution in surface fresh waters: occurrence and effects. Sci. Total Environ. 1879-1026. 664, 793–804. Available from: https://www.elsevier.com/locate/scitotenv. doi:10.1016/j.scitotenv.2019.01.406.

Daughton, C., 2008. Pharmaceuticals as Environmental Pollutants: The Ramifications for Human Exposure. Available from: https://www.researchgate.net/publication/288227900_Pharmaceuticals_as_Environmental_Pollutants_The_Ramifications_for_Human_Exposure. doi:10.1016/B978-012373960-5.00403-2.

Davies-Colley, R.J. (2013). River water quality in New Zealand: an introduction and overview.

Eapen, J.V., Thomas, S., Antony, S., George, P., Antony, J., 2024. A review of the effects of pharmaceutical pollutants on humans and aquatic ecosystem. Explor. Drug Sci. 484–507. doi:10.37349/eds.2024.00058.

Fent, K., 2004. Ecotoxicological effects at contaminated sites. Toxicology. 0300-483X. 205 (3), 223–240. doi:10.1016/j.tox.2004.06.060.

Fernandes, J.P., Almeida, C.M.R., Salgado, M.A., Carvalho, M.F., Mucha, A.P., 2021. Pharmaceutical compounds in aquatic environments—occurrence, fate and bioremediation prospective. Toxics. 2305-6304. 9 (10). Available from: https://www.mdpi.com/2305-6304/9/10/257/pdf. doi:10.3390/toxics9100257.

Fleeger, J.W., Carman, K.R., Nisbet, R.M., 2003. Indirect effects of contaminants in aquatic ecosystems. Sci. Total Environ. 0048-9697. 317 (1-3), 207–233. Available from:https://www.elsevier.com/locate/scitotenv. doi:10.1016/S0048-9697(03)00141-4.

Flint, S., Markle, T., Thompson, S., Wallace, E., 2012. Bisphenol A exposure, effects, and policy: a wildlife perspective. J. Environ. Manag. 0301-4797. 104, 19–34. doi:10.1016/j.jenvman.2012.03.021.

Gill, W.B., Schumacher, G.F.B., Bibbo, M., Straus, F.H., Schoenberg, H.W., 1979. Association of diethylstilbestrol exposure in utero with cryptorchidism, testicular hypoplasia and semen abnormalities. J. Urol. 0022-5347. 122 (1), 36–39. doi:10.1016/s0022-5347(17)56240-0.

Gonsioroski, A., Mourikes, V.E., Flaws, J.A., 2020. Endocrine disruptors in water and their effects on the reproductive system. Int. J. Mol. Sci. 1422-0067. 21 (6). Available from: https://www.mdpi.com/1422-0067/21/6/1929/pdf. doi:10.3390/ijms21061929.

Guareschi, S., Wood, P.J., 2022. Biological invasions of river ecosystems: a flow of implications, challenges, and research opportunities. In: Imperiled: The Encyclopedia of Conservation, vols. 1-3, Elsevier, United Kingdom, pp. 485–498. Available from: https://www.sciencedirect.com/book/9780128211397. doi:10.1016/B978-0-12-821139-7.00147-1.

Hellawell, J.M., 1988. Toxic substances in rivers and streams. Environ. Pollut. 0269-7491. 50 (1-2), 61–85. doi:10.1016/0269-7491(88)90185-6.

Jones, O.A.H., Voulvoulis, N., Lester, J.N., 2010. Human pharmaceuticals in the aquatic environment a review. Environ. Technol. 0959-3330. 22 (12), 1383–1394. doi:10.1080/09593332208618186.

Juliano, C., Magrini, G.A., 2017. Cosmetic ingredients as emerging pollutants of environmental and health concern. A mini-review. Cosmetics. 2079-9284. 4 (2). Available from: https://www.mdpi.com/2079-9284/4/2/11/pdf. doi:10.3390/cosmetics4020011.

Kasprzyk-Hordern, B., Dinsdale, R.M., Guwy, A.J., 2008. The occurrence of pharmaceuticals, personal care products, endocrine disruptors and illicit drugs in surface water in South Wales, UK. Water Res. 0043-1354. 42 (13), 3498–3518. Available from: https://www.sciencedirect.com/science/journal/00431354. doi:10.1016/j.watres.2008.04.026.

Kaur, T., Sinha, A.K., 2019. Pesticides in agricultural run offs affecting water resources: a study of Punjab (India). Agric. Sci. 2156-8553. 10 (10), 1381–1395. doi:10.4236/as.2019.1010101.

Khan, A.H.A., Barros, R., 2023. Pharmaceuticals in water: risks to aquatic life and remediation strategies. Hydrobiology. 2673-9917. 2 (2), 395–409. Available from: www.mdpi.com/journal/hydrobiology. doi:10.3390/hydrobiology2020026.

Kraemer, S.A., Ramachandran, A., Perron, G.G., 2019. Antibiotic pollution in the environment: from microbial ecology to public policy. Microorganisms. 2076-2607. 7 (6). Available from: https://www.mdpi.com/2076-2607/7/6/180/pdf. doi:10.3390/microorganisms7060180.

Lebeaux, R.M., Madan, J.C., Nguyen, Q.P., Coker, M.O., Dade, E.F., Moroishi, Y., Palys, T.J., Ross, B.D., Pettigrew, M.M., Morrison, H.G., Karagas, M.R., Hoen, A.G., 2022. Impact of antibiotics on off-target infant gut microbiota and resistance genes in cohort studies. Pediatr. Res. 1530-0447. 92 (6), 1757–1766. Available from: https://www.nature.com/pr/. doi:10.1038/s41390-022-02104-w.

Li, D., Liu, S., 2019. Detection of river water quality. In: Water Quality Monitoring and Management. Elsevier, pp. 211–220. doi:10.1016/B978-0-12-811330-1.00007-7.

Liu, Y., Luo, X., Zeng, Y., Wang, Q., Tu, W., Yang, C. Mai, B., 2020. Trophic Magnification of Short- and Medium-Chain Chlorinated Paraffins in Terrestrial Food Webs and Their Bioamplification in Insects and Amphibians during MetamorphosisTrophic Magnification of Short- and Medium-Chain Chlorinated Paraffins in Terrestrial Food Webs and Their Bioamplification in Insects and Amphibians during Metamorphosis. Environ. Sci. Technol. 54 (18), 11282–11291. doi:10.1021/acs.est.0c0309610.1021/acs.est.0c03096.s001.

Manaia, C.M., Aga, D.S., Cytryn, E., Gaze, W.H., Graham, D.W., Guo, J., Leonard, A.F.C., Li, L., Murray, A.K., Nunes, O.C., Rodriguez-Mozaz, S., Topp, E., Zhang, T., 2024. The complex interplay between antibiotic resistance and pharmaceutical and personal care products in the environment. Environ. Toxicol. Chem. 1552-8618. 43 (3), 637–652. Available from: http://onlinelibrary.wiley.com/journal/10.1002/(ISSN)1552-8618. doi:10.1002/etc.5555.

Meharg, A.A., 2003. Storage of sediment associated nutrients and contaminants in river channel and floodplain systems. Appl. Geochem. 18, 195–220. doi:10.1016/S0883-2927(02)00121-X.

Merli, M., Lucidi, C., Di Gregorio, V., Falcone, M., Giannelli, V., Lattanzi, B., Giusto, M., Ceccarelli, G., Farcomeni, A., Riggio, O., Venditti, M., Friedrich, A., 2015. The spread of multi drug resistant infections is leading to an increase in the empirical antibiotic treatment failure in cirrhosis: a prospective survey. PLoS One. 1932-6203. 10 (5), e0127448. doi:10.1371/journal.pone.0127448.

Montes-Grajales, D., Fennix-Agudelo, M., Miranda-Castro, W., 2017. Occurrence of personal care products as emerging chemicals of concern in water resources: a review. Sci. Total Environ. 0048-9697. 595, 601–614. doi:10.1016/j.scitotenv.2017.03.286.

Morrison, K., Melhado, G., Bose, A.P.H., Eastment, R., Lagisz, M., Manera, J.L., Michelangeli, M., Ozeki, S., Wong, B.B.M., Yang, Y., Nakagawa, S., 2025. The impacts of pesticide exposure on fish conspecific interactions: a systematic review and meta-analysis. Environ. Pollut. 1873-6424. 376. Available from: https://www.sciencedirect.com/science/journal/02697491. doi:10.1016/j.envpol.2025.126353.

Nassri, I., Khattabi Rifi, S., Sayerh, F., Souabi, S., 2023. Occurrence, pollution sources, and mitigation prospects of Antibiotics, anti-inflammatories, and endocrine disruptors in the aquatic environment. Environmental Nanotechnology, Monitoring & Management. 2215-1532. 20, 100878. doi:10.1016/j.enmm.2023.100878.

Nozaki, K., Tanoue, R., Kunisue, T., Tue, N.M., Fujii, S., Sudo, N., Isobe, T., Nakayama, K., Sudaryanto, A., Subramanian, A., Bulbule, K.A., Parthasarathy, P., Tuyen, L.H., Viet, P.H., Kondo, M., Tanabe, S., Nomiyama, K., 2023. Pharmaceuticals and personal care products (PPCPs) in surface water and fish from three Asian countries: species-specific bioaccumulation and potential ecological risks. Sci. Total Environ. 1879-1026. 866. Available from: www.elsevier.com/locate/scitotenv. doi:10.1016/j.scitotenv.2022.161258.

Nyaga, M. (2020). Pharmaceutical waste: overview, management and impact of improper disposal.

Okoye, C.O., Okeke, E.S., Okoye, K.C., Echude, D., Andong, F.A., Chukwudozie, K.I., Okoye, H.U., Ezeonyejiaku, C.D., 2022. Occurrence and fate of pharmaceuticals, personal care products (PPCPs) and pesticides in African water systems: a need for timely intervention. Heliyon. 2405-8440. 8 (3). Available from: http://www.journals.elsevier.com/heliyon/. doi:10.1016/j.heliyon.2022.e09143.

Okumah, M., Yeboah, A.S., Bonyah, S.K., 2020. What matters most? Stakeholders' perceptions of river water quality. Land Use Policy. 0264-8377. 99. Available from: www.elsevier.com/inca/publications/store/3/0/4/5/1/. doi:10.1016/j.landusepol.2020.104824.

Oregel-Zamudio, E., Alvarez-Bernal, D., Franco-Hernandez, M.O., Buelna-Osben, H.R., Mora, M., 2021. Bioaccumulation of pcbs and pbdes in fish from a tropical lake chapala, Mexico. Toxics. 2305-6304. 9 (10). Available from: https://www.mdpi.com/2305-6304/9/10/241/pdf. doi:10.3390/toxics9100241.

Ramos, R.L., Moreira, V.R., Lebron, Y.A.R., Santos, A.V., Santos, L.V.S., Amaral, M.C.S., 2021. Phenolic compounds seasonal occurrence and risk assessment in surface and treated waters in Minas Gerais—Brazil. Environ. Pollut. 1873-6424. 268. Available from: https://www.journals.elsevier.com/environmental-pollution. doi:10.1016/j.envpol.2020.115782.

Ranhao, S., Linyuan, W., Haiping, Z., Liding, C., 2020. Current status and prospect of research on river ecosystem health—based on bibliometric research. Acta Ecol. Sin. 40 (10), 3526–3536. doi:10.5846/stxb201903250560.

Saad, A.M., Asari, F.F.A.H., Affandi, S., Zid, A., 2022. River pollution: a mini review of causes and effects. J. Tour., Hosp. Environ. Manag. 0128-178X. 7 (29), 139–151. doi:10.35631/jthem.729011.

Selwe, K.P., Thorn, J.P.R., Desrousseaux, A.O.S., Dessent, C.E.H., Sallach, J.B., 2022. Emerging contaminant exposure to aquatic systems in the Southern African development community. Environ. Toxicol. Chem. 1552-8618. 41 (2), 382–395. Available from: http://onlinelibrary.wiley.com/journal/10.1002/(ISSN)1552-8618. doi:10.1002/etc.5284.

Serra-Burriel, M., Keys, M., Campillo-Artero, C., Agodi, A., Barchitta, M., Gikas, A., Palos, C., López-Casasnovas, G., Cheungpasitporn, W., 2020. Impact of multi-drug resistant bacteria on economic and clinical outcomes of healthcare-associated infections in adults: systematic review and meta-analysis. PLoS One. 1932-6203. 15 (1), e0227139. doi:10.1371/journal.pone.0227139.

Shah, Z.U., Parveen, S., 2021. Pesticides pollution and risk assessment of river Ganga: a review. Heliyon. 2405-8440. 7 (8), e07726. doi:10.1016/j.heliyon.2021.e07726.

Sharma, S., Bhattacharya, A., 2017. Drinking water contamination and treatment techniques. Appl. Water Sci. 2190-5487. 7 (3), 1043–1067. doi:10.1007/s13201-016-0455-7.

Srain, H.S., Beazley, K.F., Walker, T.R., 2021. Pharmaceuticals and personal care products and their sublethal and lethal effects in aquatic organisms. Environ. Rev. 1181-8700. 29 (2), 142–181. Available from: https://www.nrcresearchpress.com/loi/er. doi:10.1139/er-2020-0054.

Stanford, J.A., Lorang, M.S., Hauer, F.R., 2017. The shifting habitat mosaic of river ecosystems. SIL Proceedings, 1922-2010. 0368-0770. 29 (1), 123–136. doi:10.1080/03680770.2005.11901979.

Syafrudin, M., Kristanti, R.A., Yuniarto, A., Hadibarata, T., Rhee, J., Al-Onazi, W.A., Algarni, T.S., Almarri, A.H., Al-Mohaimeed, A.M., 2021. Pesticides in drinking water-a review. Int. J. Environ. Res. Public Health. 1660-4601. 18 (2), 1–15. Available from: https://www.mdpi.com/1660-4601/18/2/468/pdf. doi:10.3390/ijerph18020468.

Tijani, J.O., Fatoba, O.O., Petrik, L.F., 2013. A review of pharmaceuticals and endocrine-disrupting compounds: sources, effects, removal, and detections. Water, Air, Soil Pollut. 1573-2932. 224 (11). Available from: http://www.kluweronline.com/issn/0049-6979/. doi:10.1007/s11270-013-1770-3.

Warren-Vega, W.M., Campos-Rodríguez, A., Zárate-Guzmán, A.I., Romero-Cano, L.A., 2023. A current review of water pollutants in American continent: trends and perspectives in detection, health risks, and treatment technologies. Int. J. Environ. Res. Public Health. 1660-4601. 20 (5). Available from: http://www.mdpi.com/journal/ijerph. doi:10.3390/ijerph20054499.

Wilkinson, J., Hooda, P.S., Barker, J., Barton, S., Swinden, J., 2017. Occurrence, fate and transformation of emerging contaminants in water: an overarching review of the field. Environ. Pollut. 1873-6424. 231, 954–970. Available from: https://www.journals.elsevier.com/environmental-pollution. doi:10.1016/j.envpol.2017.08.032.

Xiang, Y., Wu, H., Li, L., Ren, M., Qie, H., Lin, A., 2021. A review of distribution and risk of pharmaceuticals and personal care products in the aquatic environment in China. Ecotoxicol. Environ. Saf. 0147-6513. 213, 112044. doi:10.1016/j.ecoenv.2021.112044.

Zhou, R., Lu, G., Yan, Z., Jiang, R., Bao, X., Lu, P., 2020. A review of the influences of microplastics on toxicity and transgenerational effects of pharmaceutical and personal care products in aquatic environment. Sci. Total Environ. 0048-9697. 732, 139222. doi:10.1016/j.scitotenv.2020.139222.

Chapter 4

Pollution induced alteration in hydrogeological cycle

Mansi Chundawat[a] and Prafull Kumar Jha[b]
[a]*Sophisticated Environmental Analytical Facility, CSIR-National Environmental Engineering Research Institute (CSIR-NEERI), Nagpur, India*
[b]*Department of Energy and Environmental Engineering, CSIR-Indian Institute of Chemical Technology (CSIR-IICT), Hyderabad, India*

1 Introduction

River basins are integrated hydrological systems in which precipitation is partitioned into interception, infiltration, evapotranspiration, surface runoff, groundwater recharge, baseflow, and stream discharge (Gorugantula and Kambhammettu, 2025). Under natural conditions, floodplains, riparian soils, wetlands, and hyporheic zones store water temporarily and moderate discharge fluctuations, thereby sustaining dry-season flow and ecological stability (Norman et al., 2022). Pollution disrupts this balance by changing soil properties, surface connectivity, transport pathways, and the timing of water movement across the basin (Walker et al., 2019).

This chapter treats pollution not only as a contaminant load but also as a driver of hydrological change. The emphasis is on how pollution changes the natural cycle of water movement in a basin, with special attention to runoff, infiltration, groundwater exchange, seasonal variability, and extreme-event behaviour.

The hydrological cycle provides the physical background within which rivers persist as living systems and water resources. In a healthy basin, water moves through a network of linked stores and pathways that include vegetation, soils, aquifers, channels, floodplains, and the atmosphere (Miguez-Macho and Fan, 2012). When pollution is introduced, these pathways do not simply carry contaminated water, they can also become physically altered through sediment clogging, altered surface roughness, biochemical changes, and modified exchange between surface and subsurface compartments (Hartwig and Borchardt, 2015).

River Pollution. https://doi.org/10.1016/B978-0-443-45024-2.00021-X

From a management perspective, this distinction matters because it changes the question being asked. Instead of asking only how polluted the river is, one must also ask how the pollution changes the behaviour of the river basin itself. That broader framing is necessary if restoration is to address not only contaminant removal but also hydrological recovery, recharge protection, and resilience to rainfall extremes (Michael-Bitton et al., 2025).

2 Natural hydrological cycle in river basins

The hydrologic cycle involves the continuous circulation of water in the Earth-Atmosphere system. At its core, the water cycle is the motion of the water from the ground to the atmosphere and back again. Of the many processes involved in the hydrologic cycle, the most important are:

- Evaporation
- Transpiration
- Condensation
- Precipitation
- Runoff

Fig. 4.1 shows schematic representation of the natural hydrogeological (hydrological) cycle showing precipitation, infiltration, groundwater flow, surface runoff, evapotranspiration, and accumulation of water in rivers, lakes, and oceans. In a minimally disturbed basin, rainfall follows multiple pathways. A fraction is intercepted by vegetation, a fraction infiltrates through soils, some is returned to the atmosphere by evapotranspiration, and the remainder becomes runoff that enters channels and rivers (Rai et al., 2017). Floodplains and wetlands act as natural storage zones, reducing peak flows and supporting delayed release to rivers during dry periods (Ferreira et al., 2023).

The significance of this buffering is that it stabilizes the basin against short-term climatic variability and helps maintain exchange between surface water and groundwater (Pavelic et al., 2012). Rivers therefore function as part of a wider hydrological continuum rather than as isolated drainage lines (Krickov et al., 2026). Pollution weakens this continuity by degrading soils, altering channel conditions, and reducing the basin's ability to regulate water movement naturally (Nuruzzaman et al., 2025).

A natural hydrological cycle also supports gradual transformation of water quality. As water infiltrates through soil, passes through riparian zones, or exchanges with shallow aquifers, it may be filtered, buffered, or chemically transformed before returning to the river (Kudzin et al., 2026). This natural self-regulation is one of the reasons that basin health cannot be assessed only by looking at a single river reach. The river is the visible expression of a much larger system, and pollution can interfere with that system at several points simultaneously (Lin et al., 2024).

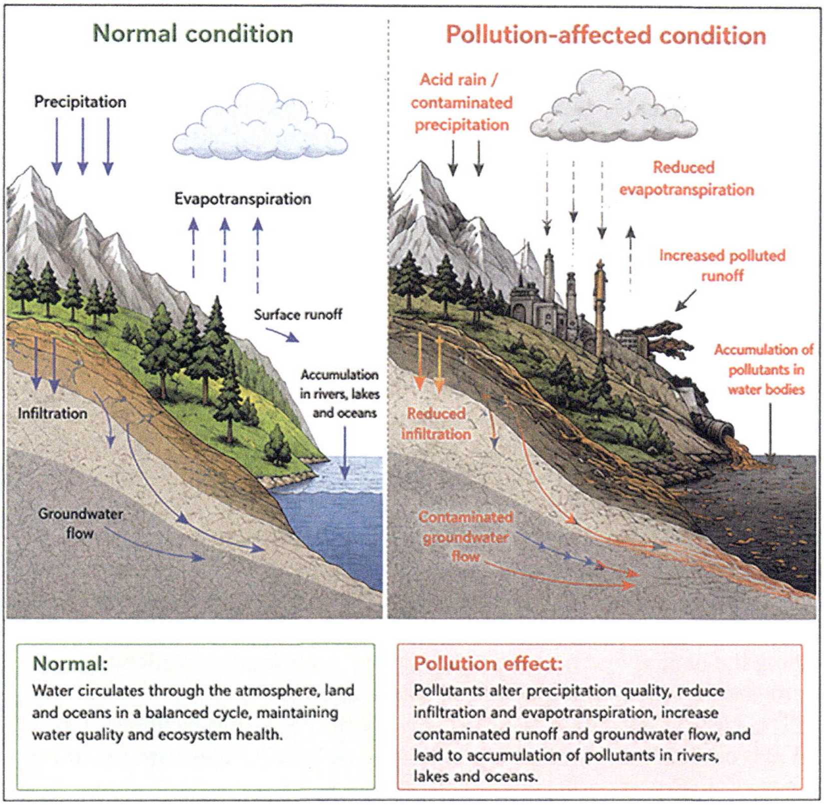

FIG. 4.1 Schematic representation of the natural hydrogeological (hydrological) cycle vs. Pollution Induced alterations in natural Hydrological cycle.

In a practical sense, this means that pollution control strategies should be designed with the hydrological cycle in mind. Actions that only treat the visible river channel but ignore runoff generation, groundwater recharge, or floodplain function may leave the underlying hydrological disturbance unresolved (Bruijnzeel et al., 2025).

3 Concept of pollution-induced alteration

Hydrological alteration refers to changes in the magnitude, timing, frequency, duration, and rate of change of flows, as well as changes in connectivity and storage across the basin (Monico et al., 2022). In the pollution context, alteration occurs when contaminant inputs and pollution-related landscape change reduce the basin's ability to absorb, store, route, and exchange water (Akhtar et al., 2021). This may happen directly, such as when sediments and wastes clog the riverbed, or indirectly, such as when urban drainage and impervious cover increase runoff and reduce recharge (Miller and Hutchins, 2017).

The concept is best understood through a simple causal chain: pollution source → transport pathway → hydrological response → basin consequence (Peeters et al., 2022). For example, untreated stormwater may enter channels quickly through drains, increase peak discharge, transport fine solids and dissolved contaminants, and reduce natural filtration time. Over time, such changes modify the river's hydrological identity rather than merely its chemical composition (Bishwakarma et al., 2025) (Fig. 4.2).

Pollution-induced alterations have significantly disrupted these freshwater reservoirs. Industrial effluents, agricultural runoff, untreated sewage, heavy metals, pharmaceuticals, microplastics, and emerging contaminants have degraded both surface water and groundwater quality (Nishmitha et al., 2025). Surface waters such as rivers and lakes are increasingly affected by eutrophication, oxygen depletion, and toxic contamination, while groundwater systems experience nitrate pollution, heavy metal leaching, and persistent organic contaminant accumulation due to infiltration and subsurface transport. Additionally, climate change-driven pollution and anthropogenic activities alter evapotranspiration, precipitation patterns, glacier melting, and groundwater recharge, thereby disturbing the natural hydrogeological cycle and reducing the availability of clean freshwater resources (Lekshmy and Achari, 2026).

The idea of "alteration" is especially useful because it captures both direct and indirect change. A river does not need to be heavily industrialized to experience hydrological modification. Even diffuse pollution from urban runoff, agricultural return flow, sewage leakage, or sediment-laden drainage can shift the balance between infiltration and runoff, degrade storage in floodplains, and alter the seasonal behavior of discharge (Janes et al., 2017). In that sense, hydrological alteration is a basin-scale phenomenon, not simply a local effect at the outfall.

This framing also aligns with the broader hydrologic literature, which has long emphasized that changes in flow regime can produce system-wide consequences. Pollution becomes particularly important because it often co-occurs

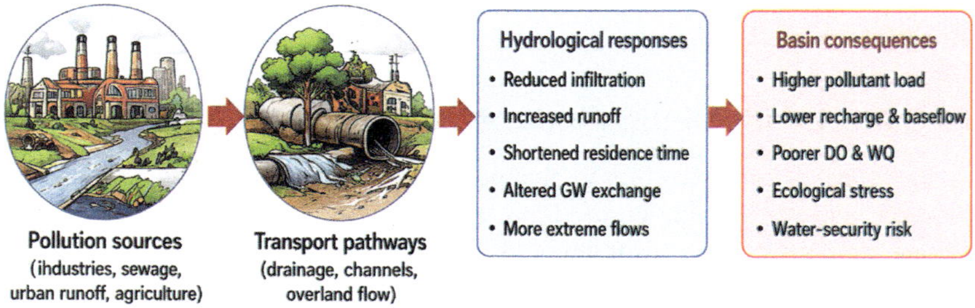

FIG. 4.2 Causal chain: from pollution source to hydrological consequences.

with land-use change, groundwater extraction, and climate variability, making the resulting hydrological signal more complex and more difficult to isolate (Sridhar and Parimalarenganayaki, 2025).

4 Mechanisms of alteration

Pollution alters the hydrological cycle through several mechanisms that reinforce one another. One major mechanism is the sealing or clogging of soils, riverbeds, and floodplain surfaces by suspended solids, organic matter, and anthropogenic residues, which reduces infiltration and increases direct runoff (Firoozi and Firoozi, 2024). Another mechanism is degradation of the riverbed–aquifer interface, where fine particles and contaminants reduce hydraulic conductivity and weaken river–groundwater exchange (Dubuis and De Cesare, 2023).

Chemical and biological processes contribute as well. Nutrient enrichment can stimulate algal growth and alter oxygen dynamics, while thermal pollution can modify evaporation behavior and dissolved oxygen patterns in shallow river reaches (Ahmad et al., 2024). In urban catchments, storm drainage networks shorten flow paths and reduce residence time, causing polluted water to reach streams more quickly during rainfall events. Pollution therefore acts not only as a chemical stressor but also as a structural modifier of hydrological pathways.

One important feature of these mechanisms is that they are cumulative. A single polluted discharge may produce a local effect, but repeated discharges, repeated storm runoff events, and repeated sediment deposition can gradually reshape the way a basin moves and stores water (Sabut et al., 2026). This cumulative character is why hydrological alteration often becomes visible only after years of incremental change. The river may still appear to flow, but its timing, stability, exchange patterns, and buffering capacity may already be significantly compromised.

A second important feature is coupling. Physical, chemical, and biological changes do not operate independently. Nutrient enrichment can promote algal biomass, which can in turn influence oxygen dynamics and suspended matter behavior, while sediment clogging can change the interaction between water quality and water movement (Astuti et al., 2026). This coupling is central to the chapter because it shows that pollution-induced hydrological alteration is not a narrow engineering issue, it is a basin-wide ecological and physical transformation (Gorugantula and Kambhammettu, 2025; Michael-Bitton et al., 2025)

The Fig. 4.3 represents the oxygen sag curve in a river system, illustrating how organic pollution alters the aquatic environment and disrupts the natural hydrological and biogeochemical processes downstream. The curve primarily reflects variations in dissolved oxygen (DO) concentration following the discharge of organic pollutants into a water body. Initially, in the clean zone, the river possesses high dissolved oxygen levels that support a diverse and healthy

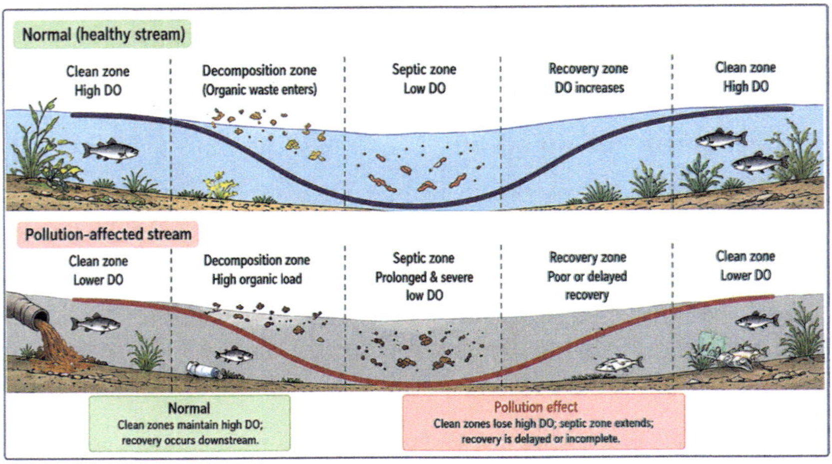

FIG. 4.3 Oxygen sag curve and aquatic life response along a normal vs. polluted stream.

aquatic ecosystem. When untreated sewage, industrial effluents, agricultural runoff, or organic wastes enter the stream, the system enters the decomposition zone, where microorganisms begin decomposing the organic matter. This microbial activity substantially increases the biochemical oxygen demand (BOD), resulting in rapid oxygen consumption and a decline in dissolved oxygen levels.

Further downstream, the river reaches the septic zone, characterized by extremely low DO concentrations. Under such oxygen-deficient conditions, only pollution-tolerant organisms survive, while sensitive aquatic species disappear (Burkholder and Glibert, 2013). Anaerobic decomposition may dominate in this zone, producing toxic gases such as methane, hydrogen sulfide, and ammonia, thereby further deteriorating water quality. As the river continues flowing, atmospheric reaeration, dilution, sedimentation, photosynthesis, and natural self-purification processes gradually restore oxygen levels in the recovery zone (Liu et al., 2016). Eventually, the river may return to a clean zone, where DO concentrations recover sufficiently to support healthy aquatic communities again.

This mechanism demonstrates how pollution-induced alterations disturb the hydrological cycle by degrading surface water quality, reducing ecosystem resilience, altering nutrient cycling, and impairing groundwater recharge through contaminated infiltration. Persistent pollution can reduce the self-purification capacity of rivers, enhance eutrophication, increase pathogen transport, and disrupt ecological balance across interconnected freshwater systems.

5 Surface runoff and infiltration

A clear signature of pollution-induced alteration is the transition from infiltration-dominated behavior to runoff-dominated behavior. In healthy landscapes, rainfall infiltrates through permeable soils and vegetated surfaces, recharging shallow groundwater and supporting delayed river discharge (Sridhar and Parimalarenganayaki, 2025). In polluted or disturbed catchments, soil compaction, impervious cover, surface sealing, and sediment deposition reduce infiltration capacity and increase the fraction of rainfall converted to rapid runoff.

This pattern is especially important in urban and peri-urban river basins in India, where intense monsoon storms interact with dense drainage networks and accumulated pollutants on roads, rooftops, and exposed land surfaces. The first flush phenomenon can transport disproportionately high contaminant loads into rivers during the early part of storm events (Perera et al., 2021). The basin therefore becomes more hydrologically "flashy," with higher peaks, lower infiltration, and stronger pollutant export than the natural system.

The runoff–infiltration shift also changes pollutant transport pathways. Instead of pollutants slowly percolating through soils and being filtered by natural interfaces, they move rapidly over the surface into channels (Nuruzzaman et al., 2025). This short-circuits the natural attenuation capacity of the basin and increases the spatial reach of pollution. In a monsoon region, this mechanism is especially important because a single intense rainfall event can move a large amount of accumulated contamination into the river network in a short time.

Infiltration refers to the downward movement of water from the surface into the soil profile. In the figure, infiltrating water passes through the unsaturated zone (soil moisture zone) and the vadose zone, eventually reaching the capillary zone and finally the saturated zone (groundwater) below the water table. Part of this infiltrated water is stored as soil moisture and later returned to the atmosphere through evapotranspiration, while the remaining water contributes to groundwater recharge and subsurface flow.

The balance between runoff and infiltration is critical for maintaining the hydrological cycle. Pollution-induced alterations significantly affect these processes. Urbanization and land sealing reuce infiltration and increase runoff, leading to flooding and reduced groundwater recharge. Agricultural chemicals, industrial wastes, sewage, heavy metals, microplastics, and emerging contaminants can infiltrate through soil layers and contaminate groundwater aquifers. Simultaneously, polluted surface runoff transports nutrients, pathogens, and toxic substances into nearby water bodies, disrupting aquatic ecosystems and overall water quality.

6 River-groundwater connectivity

River–groundwater connectivity is one of the most sensitive parts of the hydrological cycle and one of the least visible to the public. In natural settings, rivers may lose water to aquifers or gain water from them depending on hydraulic gradients, while the hyporheic zone acts as an exchange interface where water and solutes are filtered and transformed (Hester et al., 2021). Pollution can disrupt this exchange by clogging sediments, contaminating recharge zones, and altering the gradients that control water movement.

This issue is particularly relevant in India, where shallow aquifers are often hydraulically linked to rivers and are used extensively for domestic, agricultural, and industrial supply. If polluted river water infiltrates into these aquifers, contamination becomes persistent and difficult to reverse. During dry seasons, the same aquifers may feed contaminated water back into rivers, weakening baseflow quality and extending the river's pollution legacy (Irvine et al., 2024). Pollution therefore affects the entire connected surface-water–groundwater system rather than the visible channel alone.

The hydrological cycle plays a critical role in regulating water distribution, groundwater recharge, and ecosystem stability under natural conditions; however, anthropogenic pollution and land-use changes significantly disturb this balance, leading to severe hydrological stress, particularly during extreme climatic events (Elsaidy et al., 2025). In a healthy system, precipitation infiltrates into the soil, replenishes groundwater, supports evapotranspiration through vegetation, and maintains stable runoff and river flow. In contrast, pollution-affected environments experience reduced infiltration due to soil degradation, urban sealing, and vegetation loss, resulting in increased contaminated surface runoff and diminished groundwater recharge.

The hydrological cycle plays a critical role in regulating water distribution, groundwater recharge, and ecosystem stability under natural conditions; however, anthropogenic pollution and land-use changes significantly disturb this balance, leading to severe hydrological stress, particularly during extreme During extreme rainfall events, these altered conditions intensify flash floods, soil erosion, and transport of industrial, agricultural, and urban pollutants into rivers, lakes, and coastal systems. Simultaneously, contaminated groundwater flow deteriorates freshwater quality and threatens drinking water security. Under drought conditions, reduced recharge and evapotranspiration further aggravate water scarcity, ecosystem degradation, and thermal stress in aquatic environments (Pérez et al., 2026). Moreover, polluted precipitation, including acid rain, accelerates soil acidification and disrupts biogeochemical cycles. Therefore, pollution-induced modifications of the hydrological cycle not only amplify the frequency and severity of hydrological extremes but also reduce the resilience of water systems to climate change, creating long-term environmental and socio-economic challenges.

The Indian scenario makes this especially urgent because water demand is rising while the availability of renewable freshwater is spatially and temporally uneven (Rosińska et al., 2024). In such settings, a polluted river is not only a surface-water problem; it becomes a groundwater quality problem, a recharge problem, and ultimately a resilience problem. This is why river-groundwater connectivity deserves explicit treatment in a chapter about pollution-induced alteration of the natural hydrological cycle.

7 Seasonal variation in polluted rivers

Seasonal variability is one of the defining features of polluted river systems, especially in monsoon climates. In India, river water quality changes sharply between pre-monsoon, monsoon, and post-monsoon periods because rainfall, discharge, erosion, sediment transport, and dilution vary strongly through the year (Ghosh, 2024). Studies from Indian river systems have shown significant seasonal variation in turbidity, sulfate, pH, phosphate, water temperature, total alkalinity, sodium, TDS, and suspended solids, demonstrating that hydrological seasonality controls how pollution is expressed.

The hydrological meaning of this pattern is that monsoon flows may dilute some dissolved pollutants while simultaneously intensifying transport of solids, nutrients, and surface-derived contaminants. In the dry season, lower discharge reduces dilution, pollutant concentrations rise, and residence time increases, which weakens the river's capacity for self-purification (Pang and Guan, 2025). Seasonal monitoring should therefore distinguish wet-season transport dominance from dry-season concentration dominance rather than relying on annual averages alone.

The seasonal story is not simply one of "more pollution in the dry season" and "less pollution in the wet season." It is more complex. During the onset of the monsoon, accumulated surface contamination, road dust, sewage residues, and soil-bound pollutants can be rapidly mobilized by the first flush, producing high event-based loads even when the river appears visually diluted. In contrast, during prolonged dry spells, lower river volume may concentrate dissolved contaminants and heighten stress on water users and ecosystems (Joseph et al., 2026). The river thus behaves differently in each season, and pollution magnifies these differences rather than smoothing them out.

A useful way to interpret seasonal variation is to distinguish transport from concentration. During heavy rainfall, transport dominates: large volumes of water move solids and contaminants from land to river. During dry periods, concentration dominates: the same or similar pollutant inputs occur in much smaller volumes of water, producing higher ambient concentrations (Perry et al., 2024). This distinction is essential for chapter writing because it links seasonal hydrology to observed water quality and makes the chapter stronger than a generic pollution discussion.

8 Extreme events and hydrological stress

Floods, droughts, heat waves, and intense rainfall events can amplify pollution-induced hydrological alteration. Floods mobilize contaminated sediments, overwhelm sewer and drainage systems, and spread pollutants across floodplains and recharge zones. The result is not only higher pollutant exposure but also a wider spatial footprint of hydrological contamination (Talukdar et al., 2026).

Floods are especially significant because they can perform two opposing functions at once. On the one hand, they may dilute certain contaminants in the channel through high flow. On the other hand, they can spread contaminated sediments, flood polluted lowlands, and transfer river contamination into aquifers and floodplain soils (Ciszewski and Grygar, 2016). Thus, a flood is not automatically a cleansing event. In a polluted basin, it may be a redistribution event.

Droughts produce a different but equally important effect. As discharge falls, pollutants become more concentrated, low-flow dilution collapses, and the residence time of contaminants increases (Marino et al., 2024). Climate variability and climate change intensify these pressures by altering precipitation timing, event intensity, and evapotranspiration demand. Pollution and climatic stress therefore act together to reshape river responses to hydrological extremes.

Extreme heat can further worsen this picture. Higher temperatures can increase evaporation, alter dissolved oxygen dynamics, and intensify thermal stress in already contaminated waters (Motamedi et al., 2026). In low-flow reaches, this can produce a compounding effect in which warmer water holds less oxygen, residence time lengthens, and pollutant concentration rises. The combined result is a stressed hydrological system that is less able to buffer contamination or recover quickly after disturbance (Fig. 4.4).

The diurnal fluctuation of dissolved oxygen (DO) in river systems is a sensitive indicator of aquatic ecosystem health and reflects the impact of pollution-induced hydrological stress. Under normal conditions, rivers exhibit strong daily DO variation, with the lowest oxygen concentration occurring at dawn due to continuous respiration during the night, and the highest concentration at dusk as

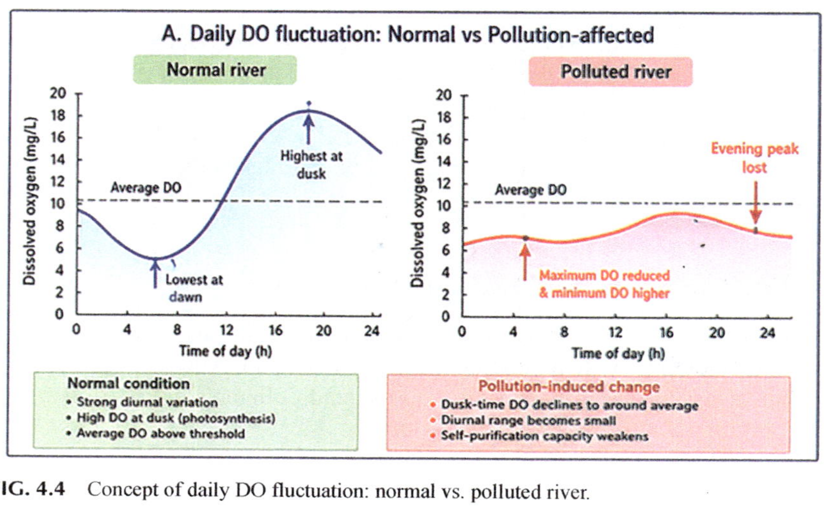

FIG. 4.4 Concept of daily DO fluctuation: normal vs. polluted river.

a result of intense daytime photosynthesis by aquatic plants and algae. This natural fluctuation maintains average DO levels above ecological threshold limits, supporting aerobic biological activity and effective self-purification processes within the river. However, in pollution-affected rivers, excessive organic loading, nutrient enrichment, industrial discharge, and reduced flow conditions alter this balance significantly. The daily DO range becomes compressed, evening oxygen peaks decline, and overall oxygen availability decreases due to elevated biochemical oxygen demand (BOD) and microbial consumption of oxygen. During extreme hydrological events such as droughts and heatwaves, low flow velocity and elevated water temperatures further intensify oxygen depletion, increasing the risk of hypoxia and aquatic ecosystem stress. Conversely, during extreme rainfall and flooding events, polluted runoff introduces additional organic matter, sediments, and contaminants into water bodies, accelerating oxygen consumption and reducing the river's natural assimilative and self-purification capacity. Thus, altered dissolved oxygen dynamics serve as a critical manifestation of hydrological stress under pollution and climate-induced extreme conditions.

9 Global and indian perspective

At the global scale, human modification of the water cycle is now widely recognized as a major environmental issue, with pollution considered part of the broader transformation of hydrological behavior (Annon, 2023). Rivers in heavily modified basins often exhibit reduced storage, higher runoff variability, lower infiltration, and impaired surface–groundwater exchange. Global monitoring frameworks increasingly address these changes through integrated water-quality and hydrological indicators.

The global perspective is also important because it shows that pollution-induced hydrological alteration is not limited to any one region or river type. Across continents, the same broad pattern appears: disturbed land surfaces increase runoff, polluted water short-circuits infiltration, and river–aquifer exchange becomes less effective as a buffer (Leader et al., 2024). What differs from basin to basin is the intensity of the effect, the seasonality of the response, and the available management capacity.

In India, these effects are often stronger because of monsoon seasonality, rapid urban expansion, mixed land use, and intensive dependence on both surface water and groundwater.Indian river basins frequently experience sharp wet–dry contrasts, which makes pollution effects highly seasonal and spatially uneven. The country also faces wide regional differences in precipitation, water availability, and water demand, which means that hydrological stress can vary dramatically from one basin to another (Zamani Sabzi et al., 2019).

India's water context makes the topic especially relevant. The country receives large annual precipitation, but rainfall is uneven in space and time, and a substantial part of the population depends on monsoon-driven systems for agriculture and domestic supply. This means that pollution-induced hydrological alteration is not merely an ecological concern; it is a water-security concern. If river systems lose their buffering capacity, the implications reach irrigation, drinking water, flood management, and groundwater sustainability.

The Indian scenario also highlights the importance of management scale. Pollution is often visible in the river, but its hydrological effects may be distributed across the broader basin through recharge zones, tributary networks, floodplains, and aquifer interactions (Anand et al., 2024) Therefore, basin-scale strategies are needed. Local clean-up without recharge protection or flow management will not fully address the underlying hydrological alteration.

10 Assessment indicators and monitoring

A reliable assessment framework should combine hydrological alteration indicators with ambient water-quality metrics. The Indicators of Hydrologic Alteration framework is especially useful because it captures changes in flow magnitude, timing, duration, frequency, and rate of change through a structured suite of hydrologic statistics (Gao et al., 2009). Representative indicators include monthly flow values, annual extremes, low-flow duration, and the timing of flow events.

Water quality should be monitored using globally recognized parameter groups. The UN-Water SDG framework emphasizes oxygen, salinity, nitrogen, phosphorus, and acidification as core ambient-water-quality categories (Ngwenya et al., 2025). This allows local monitoring to be linked with national and international reporting, while still permitting the addition of basin-specific parameters such as turbidity, suspended solids, chlorophyll-a, or temperature (Fig. 4.5).

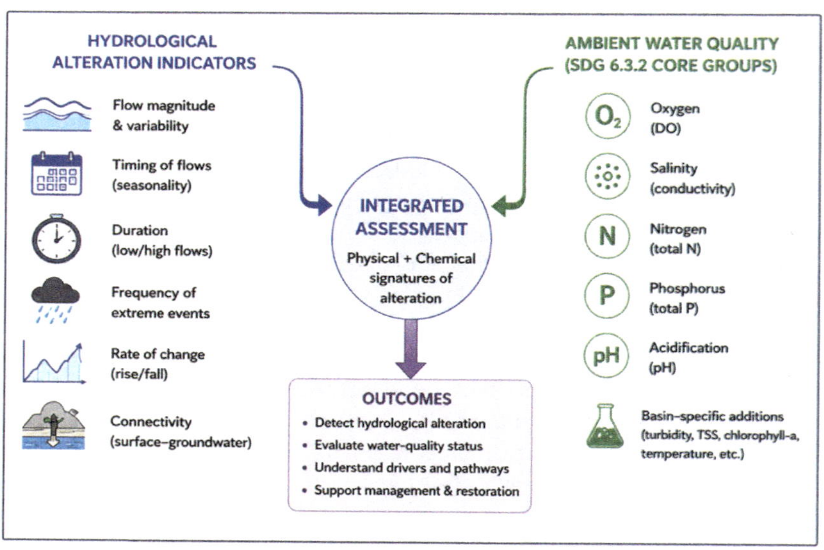

FIG. 4.5 Integrated assessment framework: linking hydrology and water quality.

This Fig. 4.7 presents an integrated assessment framework that links hydrological alterations with water quality indicators to evaluate the overall health and functioning of aquatic ecosystems. The framework emphasizes the need for combined monitoring of both hydrological behavior and physicochemical water quality parameters for effective water resource management and restoration.

The left side of the figure highlights major hydrological alteration indicators used in monitoring watershed and river basin dynamics. These include:

- Flow magnitude and variability, which assess the quantity and fluctuations of water discharge over time.
- Timing of flows (seasonality), which evaluates whether natural seasonal flow patterns are being altered due to climate change, dams, or excessive water extraction.
- Duration of low and high flows, indicating the persistence of droughts or floods.
- Frequency of extreme events, such as floods and prolonged dry periods.
- Rate of change in water levels, representing how rapidly flow conditions rise or decline.
- Surface—groundwater connectivity, which determines interactions between rivers, aquifers, wetlands, and recharge systems.

These hydrological indicators are essential for detecting disturbances in the natural hydrological cycle caused by urbanization, deforestation, industrialization, climate change, and pollution.

The right side of the figure presents ambient water quality indicators, which are monitored to assess pollution status and ecological health. Key parameters include:

- Dissolved oxygen (DO), indicating aquatic ecosystem health and biological activity.
- Salinity or conductivity, reflecting dissolved ion concentrations and possible saline intrusion or contamination.
- Nitrogen (Total N) and phosphorus (Total P), which are major nutrients responsible for eutrophication.
- pH, which indicates acidification or alkalinity changes in water systems.
- Additional basin-specific indicators such as turbidity, total suspended solids (TSS), chlorophyll-a, temperature, heavy metals, pathogens, pharmaceuticals, microplastics, and emerging contaminants.

The central "integrated assessment" component combines these hydrological and chemical signatures to identify the interactions between altered water flow and pollutant transport. Monitoring both sets of indicators together helps in understanding contaminant pathways, pollutant accumulation, nutrient cycling, and ecosystem responses.

The framework ultimately supports several important outcomes:

- Detection of hydrological and ecological alterations,
- Evaluation of water quality status,

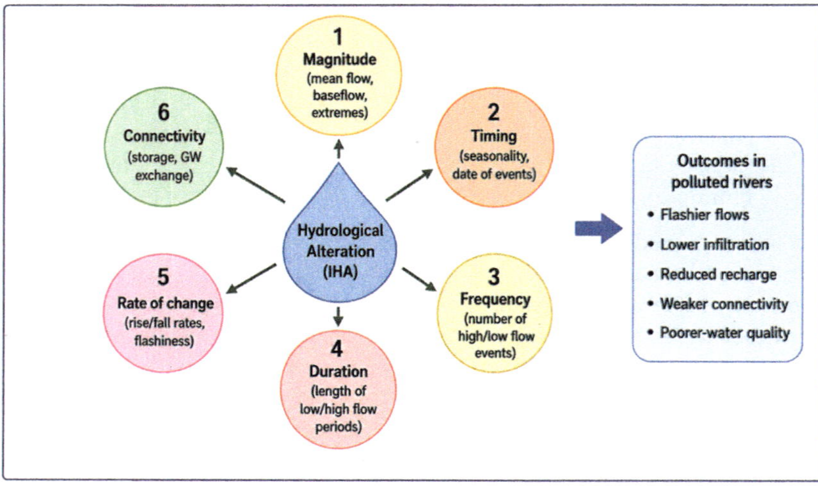

FIG. 4.6 Hydrological alteration indicators.

TABLE 4.1 Assessment indicators and monitoring.

Indicator category	Example metric	Why it matters	Suggested monitoring approach	Source
Hydrological alteration	Mean flow, baseflow, extreme flow, rise/fall rate	Quantifies changes in flow regime and timing	Continuous gauging	(Poff et al., 1997)
Seasonal variation	Monsoon/ non-monsoon concentration and load	Captures seasonal pollution behaviour	Seasonal sampling	(Kaushal et al., 2018)
River–groundwater exchange	Water table depth, seepage rate, baseflow index	Reveals connectivity loss and recharge disruption	Gauges + observation wells	(Sophocleous, 2002)
Ambient water quality	Oxygen, salinity, nitrogen, phosphorus, acidification	Internationally comparable water-quality status	*In situ* and laboratory analysis	(Schwarzenbach et al., 2010)
Suspended matter	Turbidity, TSS, sediment load	Tracks transport and clogging	Event-based sampling	(Bilotta and Brazier, 2008)
Remote sensing support	Surface temperature, turbidity proxies, bloom signals	Gives large-scale spatial coverage	Satellite monitoring	(Palmer et al., 2015)

- Identification of pollution drivers and transport pathways,
- Development of sustainable watershed management and restoration strategies.

Thus, this integrated monitoring approach is critical for assessing pollution-induced alterations in the hydrological cycle and ensuring long-term freshwater sustainability under increasing anthropogenic and climatic pressures.

The Fig. 4.6 and Table 4.1 cumulatively summarizes the major assessment indicators and monitoring approaches used to evaluate hydrological alterations and water quality changes in aquatic systems. It explains how parameters related to flow dynamics, seasonal variation, groundwater interaction, water quality, suspended matter, and remote sensing collectively help in detecting pollution impacts, ecosystem disturbances, and disruptions in the hydrological cycle.

11 Seasonal before-after perspective

A useful way to present the chapter's core message is through a before–after seasonal comparison. In a natural river basin, pre-monsoon conditions are typically characterized by low flow, stable chemistry, and limited sediment transport, while monsoon conditions produce higher discharge, recharge, and temporary dilution (Pitchaimani et al., 2026). In a polluted river basin, the same seasonal transitions produce much sharper responses because contaminants accumulated on land surfaces, in sediments, and in drainage channels are mobilized during rainfall events (Fig. 4.7).

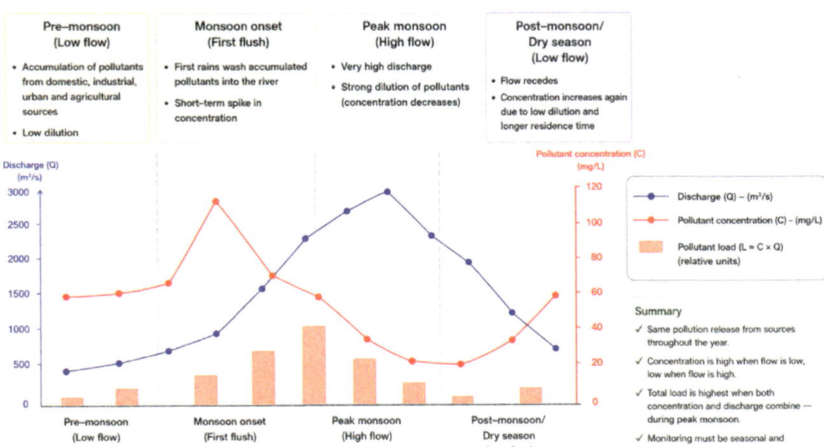

FIG. 4.7 Conceptual seasonal pollution behaviour (in case of single river system). *From Pitchaimani, V.S., Abishek, S.R., Joe, R.J.J., Promilton, A.A.A., 2026. Seasonal variations of inner shelf sediment dynamics and granulometric characteristics along the Nagapattinam coast of the Cauvery basin, Southern India. Discover Oceans 3 (1), 4. doi:10.1007/s44289-025-00109-x; Xu, G., Li, P., Lu, K., Tantai, Z., Zhang, J., Ren, Z., et al, 2019. Seasonal changes in water quality and its main influencing factors in the Dan River basin. Catena (Amst) 173, 131–140. doi:10.1016/j.catena.2018.10.014.*

TABLE 4.2 Seasonal before–after interpretation for polluted rivers.

Seasonal condition	Natural river response	Polluted river response	Interpretation	Source
Pre-monsoon	Lower flow, stable chemistry	Strong concentration of contaminants, weak dilution	Dry-season stress becomes visible	(van Vliet and Zwolsman, 2008)
Monsoon onset	Rising discharge and recharge	First flush of pollutants, sediment mobilization	Rapid transfer from land to river	(Sansalone and Buchberger, 1997)
Peak monsoon	High flow, greater dilution potential	Large pollutant transport, floodplain contamination	Transport dominates over dilution	(Paul and Meyer, 2001)
Post-monsoon	Gradual stabilization	Residual contamination persists in sediments and shallow groundwater	Legacy pollution remains after rainfall	(Vreca and Dolenec, 2005)

This difference matters because the monsoon does not simply "clean" a river. It can dilute some dissolved pollutants while simultaneously increasing total pollutant transport, sediment resuspension, floodplain contamination, and recharge-zone exposure (Xu et al., 2019). In the dry season, the pollutant burden may become even more concentrated due to reduced dilution and longer residence times. The hydrological cycle therefore acts as a seasonal amplifier of pollution impacts rather than a neutral background process.

The seasonal comparison is also useful because it provides a practical lens for readers. River pollution in a monsoon system does not have a single fixed expression. Its character changes across the year. The same basin may show relatively low concentrations during a wet period, strong transport during storm onset, sediment redistribution during peak flow, and high ambient concentrations in the dry season (Ravazzolo et al., 2026) (Table 4.2).

12 Extreme-event and basin-resilience perspective

Floods and droughts reveal the resilience or fragility of polluted river systems. Floods can redistribute contaminants across the floodplain, overwhelm drainage infrastructure, and connect polluted surface water with shallow groundwater reserves (Gupta and Arrighi, 2026). Droughts, by contrast, reduce dilution, intensify concentration effects, and extend the persistence of contaminants in channels and pools.

This matters because climate change is expected to alter the frequency and intensity of hydrological extremes in many regions, including monsoon-dominated river basins. Where pollution loads are already high, more intense storms and longer dry spells can magnify both transport and concentration effects, placing extra pressure on river health and water security (Rosińska et al., 2024). In this sense, pollution and climate stress interact to reshape basin resilience rather than acting as isolated pressures.

Extreme-event thinking is also important for management. If a basin is assessed only through average conditions, it may appear less stressed than it really is. Yet the most damaging pollution impacts often occur during a short storm, a long drought, or a sequence of extreme events (Bolan et al., 2024). That is why this chapter emphasizes not just steady-state pollution but event-driven hydrological change.

13 Monitoring strategy

A practical monitoring design should combine fixed stations, seasonal surveys, and event-based sampling. Pre-monsoon and post-monsoon sampling help identify concentration and dilution patterns, while storm-event sampling captures first flush and flood redistribution (Jiang et al., 2023). For rivers with strong groundwater interaction, paired surface-water and groundwater sampling is necessary to detect delayed contamination and baseflow feedbacks.

Remote sensing can complement ground-based measurements by providing spatially extensive information on turbidity, surface temperature, and some indicators of suspended material or biological response (Jiang et al., 2023). However, satellite methods cannot replace in situ monitoring for core SDG 6.3.2 parameters, which require direct measurement. An integrated approach is therefore best: ground stations for accuracy, remote sensing for basin-scale screening, and hydrological modelling for system interpretation.

Monitoring should also be designed with the basin's seasonal rhythm in mind. In monsoon regions, the same station may record very different pollutant behavior in pre-monsoon, onset, peak monsoon, and post-monsoon windows (Sanjeevaiah et al., 2021). If monitoring is limited to one or two campaign dates, it may miss the most important hydrological transitions. This is especially true for rivers that receive large seasonal loads from urban runoff, agricultural drainage, or floodplain washout.

A robust monitoring program should therefore use multiple scales: event-scale measurements to capture acute change, seasonal surveys to capture regime change, and long-term monitoring to identify cumulative hydrological alteration. This multi-scale strategy is essential because the chapter is about process change over time, not simply a list of pollutants.

14 Management and restoration implications

Because pollution alters both water quality and hydrological behavior, restoration must target the basin system rather than a single river segment. Riparian buffers, wetland restoration, pollution source control, stormwater detention, infiltration protection, and environmental-flow management can help recover both hydrological buffering and ecological resilience (Kasim et al., 2025) In urban catchments, reducing imperviousness and managing first-flush runoff are particularly important because they reduce both pollutant loading and abrupt flow peaks.

The policy implication is that water-quality monitoring and hydrological monitoring should not be separated. Integrated basin management should track not only pollutant concentrations but also changes in runoff response, recharge, connectivity, and low-flow persistence. This combined view is essential for achieving sustainable river restoration, especially in monsoon-dominated basins such as those in India.

Management also needs to be preventive rather than reactive. Once river–groundwater connectivity is degraded or floodplain storage is clogged, restoring natural hydrological buffering becomes much more difficult than preventing the damage in the first place (Zheng et al., 2025). That is why source control, infiltration protection, and riparian restoration are not supplementary measures; they are central hydrological interventions.

At the basin scale, restoration success should be measured not only by reduced pollutant concentrations but also by improved hydrological function. Indicators such as more stable dry-season flow, stronger infiltration, healthier recharge zones, better exchange with groundwater, and lower event-driven pollutant spikes can all be used as evidence of recovery (Zheng et al., 2025). This helps shift river restoration from a narrow cleanup model toward a more complete basin rehabilitation model.

15 Conclusion

In conclusion, pollution-induced alteration of the natural hydrological cycle is not merely a water-quality issue, but a basin-scale disruption that affects the entire functioning of riverine ecosystems. Anthropogenic activities modify flow regimes, infiltration capacity, groundwater recharge, river–groundwater interactions, seasonal hydrological patterns, and the response of watersheds to extreme events. These alterations weaken the natural resilience and self-purification capacity of aquatic systems, making rivers increasingly vulnerable to droughts, floods, contaminant accumulation, and ecological degradation. This chapter highlights that river pollution must be understood as both a chemical and hydrological disturbance. Water quality and hydrological processes are inherently interconnected, as pollution influences the movement, storage, and transport of water, while altered hydrology governs the fate, mobility, and persistence of pollutants. Loss of infiltration, recharge, storage, and hydrological connectivity reduces the capacity of river basins to sustain ecological balance and recover from environmental stress. Therefore, effective pollution management cannot be separated from hydrological restoration. Sustainable river basin management requires integrated monitoring, restoration of natural flow and recharge processes, protection of groundwater–surface water connectivity, and reduction of pollutant inputs to ensure long-term freshwater security and ecosystem sustainability under increasing anthropogenic and climatic pressures.

References

Ahmad, N.B., Jaafaru, M.S., Isa, Z., Abdulhamid, Y., Kakudi, R.A., Ugya, A.Y., et al., 2024. High pollution loads engineer oxygen dynamics, ecological niches, and pathogenicity shifts in freshwater environments. J. Hazard. Mater. Adv. 14, 100425. doi:10.1016/j.hazadv.2024.100425.

Akhtar, N., Syakir Ishak, M.I., Bhawani, S.A., Umar, K., 2021. Various Natural and Anthropogenic Factors Responsible for Water Quality Degradation: a Review. Water (Basel) 13 (19), 2660. doi:10.3390/w13192660.

Anand, J., Gosain, A.K., Khosa, R., 2024. Hydrological and hydrodynamic modelling for flood management: a case study of the Yamuna River Basin in Delhi. J. Hydrol. Reg. Stud. 56, 101960. doi:10.1016/j.ejrh.2024.101960.

Annon, 2023. Water Cycle Changes. In: Climate Change 2021 – The Physical Science Basis. Cambridge University Press, pp. 1055–1210. doi:10.1017/9781009157896.010.

Astuti, S.P., Hashmi, Z., Bilad, M.R., Payus, C.M., Affandi, M., Soegianto, A., 2026. Freshwater microalgae as bioindicators: a review of trait-based advances, ecosystem insights, and biomonitoring innovations. Environ. Chall. 22, 101428. doi:10.1016/j.envc.2026.101428.

Bilotta, G.S., Brazier, R.E., 2008. Understanding the influence of suspended solids on water quality and aquatic biota. Water Res. 42 (12), 2849–2861. doi:10.1016/j.watres.2008.03.018.

Bishwakarma, K., Wang, G., Zhang, F., Pant, R.R., Wu, G., 2025. Concentration-discharge relationships and chemical weathering in the silicate-rich Tamor River Basin: insights from weekly hydrochemical time-series. J. Hydrol. Reg. Stud. 62, 102824. doi:10.1016/j.ejrh.2025.102824.

Bolan, S., Padhye, L.P., Jasemizad, T., Govarthanan, M., Karmegam, N., Wijesekara, H., et al, 2024. Impacts of climate change on the fate of contaminants through extreme weather events. Sci. Total Environ. 909, 168388. doi:10.1016/j.scitotenv.2023.168388.

Bruijnzeel, L.A., Peña-Arancibia, J.L., Sheil, D., Ziegler, A.D., Zhang, J., Zwartendijk, B.W., et al, 2025. Potential for improved groundwater recharge and dry-season flows through forest landscape restoration on degraded lands in the tropics. For. Ecosyst. 14, 100376. doi:10.1016/j.fecs.2025.100376.

Burkholder, J.M., Glibert, P.M., 2013. Eutrophication and Oligotrophication. In: Encyclopedia of Biodiversity. Elsevier, 347–371. doi:10.1016/B978-0-12-384719-5.00047-2.

Ciszewski, D., Grygar, T.M., 2016. A Review of Flood-Related Storage and Remobilization of Heavy Metal Pollutants in River Systems. Water Air Soil Pollut. 227 (7), 239. doi:10.1007/s11270-016-2934-8.

Dubuis, R., De Cesare, G., 2023. The clogging of riverbeds: a review of the physical processes. Earth Sci. Rev. 239, 104374. doi:10.1016/j.earscirev.2023.104374.

Elsaidy, A., Yimer, E.A., Mogheir, Y., Huysmans, M., Villani, L., van Griensven, A., 2025. Groundwater drought and anthropogenic amplifiers: a review of assessment and response strategies in arid and semi-arid areas. Sci. Total Environ. 978, 179406. doi:10.1016/j.scitotenv.2025.179406.

Ferreira, C.S.S., Kašanin-Grubin, M., Solomun, M.K., Sushkova, S., Minkina, T., Zhao, W., et al, 2023. Wetlands as nature-based solutions for water management in different environments. Curr. Opin. Environ. Sci. Health 33, 100476. doi:10.1016/j.coesh.2023.100476.

Firoozi, A.A., Firoozi, A.A., 2024. Water erosion processes: mechanisms, impact, and management strategies. Results Eng. 24, 103237. doi:10.1016/j.rineng.2024.103237.

Gao, Y., Vogel, R.M., Kroll, C.N., Poff, N.L., Olden, J.D., 2009. Development of representative indicators of hydrologic alteration. J. Hydrol. (Amst) 374 (1–2), 136–147. doi:10.1016/j.jhydrol.2009.06.009.

Ghosh, B., 2024. Assessing the effects of human interventions on the morphodynamics and health of a lowland tropical river: a case study of the river Dwarkeswar, India. Geosyst. Geoenviron. 3 (1), 100234. doi:10.1016/j.geogeo.2023.100234.

Gorugantula, S.S., Kambhammettu, B.P., 2025. An integrated hydrological model to simulate terrestrial water storage in a large river basin: evaluation using GRACE data. J Hydrol Reg Stud 59. 102309. doi:10.1016/j.ejrh.2025.102309.

Gupta, K., Arrighi, C., 2026. From drivers to damage: a conceptual framework for integrated flood risk assessment of inland hydrological assets. Sci. Total Environ. 1031, 181813. doi:10.1016/j.scitotenv.2026.181813.

Hartwig, M., Borchardt, D., 2015. Alteration of key hyporheic functions through biological and physical clogging along a nutrient and fine-sediment gradient. Ecohydrology 8 (5), 961–975. doi:10.1002/eco.1571.

Hester, E.T., Santizo, K.Y., Nida, A.A., Widdowson, M.A., 2021. Hyporheic transverse mixing zones and dispersivity: laboratory and numerical experiments of hydraulic controls. J. Contam. Hydrol. 243, 103885. doi:10.1016/j.jconhyd.2021.103885.

Irvine, D.J., Singha, K., Kurylyk, B.L., Briggs, M.A., Sebastian, Y., Tait, D.R., et al, 2024. Groundwater-Surface water interactions research: past trends and future directions. J. Hydrol. (Amst) 644, 132061. doi:10.1016/j.jhydrol.2024.132061.

Janes, V.J., Grabowski, R.C., Mant, J., Allen, D., Morse, J.L., Haynes, H., 2017. The Impacts of Natural Flood Management Approaches on In-Channel Sediment Quality. River Res. Appl. 33 (1), 89–101. doi:10.1002/rra.3068.

Jiang, Y., Bao, X., Huang, Z., Chen, Y., Wu, X., Li, X., et al, 2023. Identification of pollutant delivery processes during different storm events and hydrological years in a semi-arid mountainous reservoir basin. Sci. Total Environ. 883, 163606. doi:10.1016/j.scitotenv.2023.163606.

Joseph, N., Shareef, N., Preetha, P., Liu, S., 2026. Evaluating long-term water quality variations to climate variability and land use change in California's Sacramento watershed. Sci. Total Environ. 1020, 181597. doi:10.1016/j.scitotenv.2026.181597.

Kasim, R., Lihawa, F., Buffers, K.B.D.W.R., 2025. Connectivity, and Water Quality: a Systematic Review of Land-Use Gradients in Agro-Urban Watersheds. West Sci. Interdiscip. Stud. 3 (12), 2313–2321. doi:10.58812/wsis.v3i12.2514.

Kaushal, S.S., Likens, G.E., Pace, M.L., Utz, R.M., Haq, S., Gorman, J., et al, 2018. Freshwater salinization syndrome on a continental scale. Proc. Natl. Acad. Sci. U. S. A. 115 (4). doi:10.1073/pnas.1711234115.

Krickov, I.V., Loiko, S.V., Lim, A.G., Kuzmina, D.M., Istigechev, G.I., Shirokova, L.S., et al, 2026. Hydrological continuums across climate and permafrost gradients: spatial patterns of organic carbon, greenhouse gases, and major and trace elements. Water Res. 294, 125507. doi:10.1016/j.watres.2026.125507.

Kudzin, M.H., Mrozińska, Z., Sikora, M., Żyłła, R., 2026. Ecohydrological Pathways of Water Quality Under Climate Change: nature-Based Solutions for Pollutant Flux Regulation. Water (Basel) 18 (3), 347. doi:10.3390/w18030347.

Leader, S., Kettridge, N., Hannah, D., Mendoza, C., Devito, K.J., 2024. Diverse response of shallow lake water levels to decadal weather patterns in a heterogeneous glacial Boreal Plains landscape. Hydrol. Process. 38 (4). doi:10.1002/hyp.15125.

Lekshmy, S.A.J., Achari, V.S., 2026. Climate change impact on groundwater and soil geochemistry—emerging challenges: a review. Appl. Water Sci. 16 (5), 160. doi:10.1007/s13201-026-02827-9.

Lin, F., Ren, H., Qin, J., Wang, M., Shi, M., Li, Y., et al, 2024. Analysis of pollutant dispersion patterns in rivers under different rainfall based on an integrated water-land model. J. Environ. Manage. 354, 120314. doi:10.1016/j.jenvman.2024.120314.

Liu, H., Hu, Z., Zhang, J., Ngo, H.H., Guo, W., Liang, S., et al, 2016. Optimizations on supply and distribution of dissolved oxygen in constructed wetlands: a review. Bioresour. Technol. 214, 797–805. doi:10.1016/j.biortech.2016.05.003.

Marino, A., Bertolotti, S., Macrì, M., Bona, F., Bonetta, S., Falasco, E., et al, 2024. Impact of wastewater treatment and drought in an Alpine region: a multidisciplinary case study. Heliyon 10 (15), e35290. doi:10.1016/j.heliyon.2024.e35290.

Michael-Bitton, G., Zemah-Shamir, S., Portnov, B., 2025. Managing stream restoration: framing and assessing the stream ecosystem services and biodiversity index (SESBI). J. Environ. Manage. 388, 125938. doi:10.1016/j.jenvman.2025.125938.

Miguez-Macho, G., Fan, Y., 2012. The role of groundwater in the Amazon water cycle: 1. Influence on seasonal streamflow, flooding and wetlands. J. Geophys. Res. Atmos. 117 (D15). doi:10.1029/2012JD017539.

Miller, J.D., Hutchins, M., 2017. The impacts of urbanisation and climate change on urban flooding and urban water quality: a review of the evidence concerning the United Kingdom. J. Hydrol. Reg. Stud. 12, 345–362. doi:10.1016/j.ejrh.2017.06.006.

Monico, V., Solera, A., Bergillos, R.J., Paredes-Arquiola, J., Andreu, J., 2022. Effects of environmental flows on hydrological alteration and reliability of water demands. Sci. Total Environ. 810, 151630. doi:10.1016/j.scitotenv.2021.151630.

Motamedi, M., Iranfar, S., Khudhur, F.W.K., Hassanpouryouzband, A., 2026. When heat meets water: environmental trade-offs of underground thermal energy storage. J. Energy Storage 149, 120265. doi:10.1016/j.est.2025.120265.

Ngwenya, B., Paepae, T., Bokoro, P.N., 2025. Monitoring ambient water quality using machine learning and IoT: a review and recommendations for advancing SDG indicator 6.3.2. J. Water Process Eng. 73, 107664. doi:10.1016/j.jwpe.2025.107664.

Nishmitha, P.S., Akhilghosh, K.A., Aiswriya, V.P., Ramesh, A., Muthuchamy, M., Muthukumar, A., 2025. Understanding emerging contaminants in water and wastewater: a comprehensive review on detection, impacts, and solutions. J. Hazard. Mater. Adv. 18, 100755. doi:10.1016/j.hazadv.2025.100755.

Norman, L.M., Lal, R., Wohl, E., Fairfax, E., Gellis, A.C., Pollock, M.M., 2022. Natural infrastructure in dryland streams (NIDS) can establish regenerative wetland sinks that reverse desertification and strengthen climate resilience. Sci. Total Environ. 849, 157738. doi:10.1016/j.scitotenv.2022.157738.

Nuruzzaman, M., Bahar, M.M., Naidu, R., 2025. Diffuse soil pollution from agriculture: impacts and remediation. Sci. Total Environ. 962, 178398. doi:10.1016/j.scitotenv.2025.178398.

Palmer, S.C.J., Kutser, T., Hunter, P.D., 2015. Remote sensing of inland waters: challenges, progress and future directions. Remote Sens. Environ. 157, 1–8. doi:10.1016/j.rse.2014.09.021.

Pang, X., Guan, M., 2025. Urban rivers at wet seasons exhibit predominant self-purification capacity in pollutant concentrations than dry seasons at subtropical climate. J. Hydrol. (Amst) 662, 133992. doi:10.1016/j.jhydrol.2025.133992.

Paul, M.J., Meyer, J.L., 2001. Streams in the Urban Landscape. Annu. Rev. Ecol. Syst. 32 (1), 333–365. doi:10.1146/annurev.ecolsys.32.081501.114040.

Pavelic, P., Patankar, U., Acharya, S., Jella, K., Gumma, M.K., 2012. Role of groundwater in buffering irrigation production against climate variability at the basin scale in South-West India. Agric. Water Manag. 103, 78–87. doi:10.1016/j.agwat.2011.10.019.

Peeters, L.J.M., Holland, K.L., Huddlestone-Holmes, C., Boulton, A.J., 2022. A spatial causal network approach for multi-stressor risk analysis and mapping for environmental impact assessments. Sci. Total Environ. 802, 149845. doi:10.1016/j.scitotenv.2021.149845.

Perera, T., McGree, J., Egodawatta, P., Jinadasa, K.B.S.N., Goonetilleke, A., 2021. New conceptualisation of first flush phenomena in urban catchments. J. Environ. Manage. 281, 111820. doi:10.1016/j.jenvman.2020.111820.

Pérez, P.A., Quiroz, W., Echeveste, P., 2026. Coupled pollution and water scarcity heighten ecological degradation and social vulnerability in global dryland rivers. Environ. Res. 293, 123708. doi:10.1016/j.envres.2026.123708.

Perry, W.B., Ahmadian, R., Munday, M., Jones, O., Ormerod, S.J., Durance, I., 2024. Addressing the challenges of combined sewer overflows. Environ. Pollut. 343, 123225. doi:10.1016/j.envpol.2023.123225.

Pitchaimani, V.S., Abishek, S.R., Joe, R.J.J., Promilton, A.A.A., 2026. Seasonal variations of inner shelf sediment dynamics and granulometric characteristics along the Nagapattinam coast of the Cauvery basin, Southern India. Discover Oceans 3 (1), 4. doi:10.1007/s44289-025-00109-x.

Poff, N.L., Allan, J.D., Bain, M.B., Karr, J.R., Prestegaard, K.L., Richter, B.D., et al, 1997. The Natural Flow Regime. Bioscience 47 (11), 769–784. doi:10.2307/1313099.

Rai, R.K., Singh, V.P., Upadhyay, A., 2017. Hydrologic Computations. In: Planning and Evaluation of Irrigation Projects. Elsevier, 83–229. doi:10.1016/B978-0-12-811748-4.00004-2.

Ravazzolo, D., Casasola, H., Fortesa, J., García-Comendador, J., Cuello-Llobel, F., Moragues, A., et al, 2026. Bedload and suspended sediment transport analysis in a Mediterranean intermittent river. Catena (Amst) 267, 110009. doi:10.1016/j.catena.2026.110009.

Rosińska, W., Jurasz, J., Przestrzelska, K., Wartalska, K., Kaźmierczak, B., 2024. Climate change's ripple effect on water supply systems and the water-energy nexus – a review. Water Resour. Ind. 32, 100266. doi:10.1016/j.wri.2024.100266.

Sabut, A., Bhat, A.M., Mohapatra, P.K., 2026. Stormwater control measures, design aspects, and impact of climate change – a review. Water-Energy Nexus 9, 46–74. doi:10.1016/j.wen.2025.12.001.

Sanjeevaiah, S.H., Rudrappa, K.S., Lakshminarasappa, M.T., Huggi, L., Hanumanthaiah, M.M., Venkatappa, S.D., et al, 2021. Understanding the Temporal Variability of Rainfall for Estimating Agro-Climatic Onset of Cropping Season over South Interior Karnataka, India. Agronomy 11 (6), 1135. doi:10.3390/agronomy11061135.

Sansalone, J.J., Buchberger, S.G., 1997. Partitioning and First Flush of Metals in Urban Roadway Storm Water. J. Environ. Eng. 123 (2), 134–143. doi:10.1061/(ASCE)0733-9372(1997)123:2(134).

Schwarzenbach, R.P., Egli, T., Hofstetter, T.B., von Gunten, U., Wehrli, B., 2010. Global Water Pollution and Human Health. Annu. Rev. Environ. Resour. 35 (1), 109–136. doi:10.1146/annurev-environ-100809-125342.

Sophocleous, M., 2002. Interactions between groundwater and surface water: the state of the science. Hydrogeol. J. 10 (1), 52–67. doi:10.1007/s10040-001-0170-8.

Sridhar, D., Parimalarenganayaki, S., 2025. Review on urban groundwater pollution an Indian perspective: sources, analysis, and management strategies. Desalin. Water Treat. 324, 101453. doi:10.1016/j.dwt.2025.101453.

Talukdar, M.J., Borah, P., Reang, D., Chabukdhara, M., 2026. Spatial patterns and health risks of heavy metals in groundwater along an urban–periurban gradient in Kamrup, Assam, India. J. Trace Elem. Min. 17, 100304. doi:10.1016/j.jtemin.2026.100304.

van Vliet, M.T.H., Zwolsman, J.J.G., 2008. Impact of summer droughts on the water quality of the Meuse river. J. Hydrol. (Amst) 353 (1–2), 1–17. doi:10.1016/j.jhydrol.2008.01.001.

Vreca, P., Dolenec, T., 2005. Geochemical estimation of copper contamination in the healing mud from Makirina Bay, central Adriatic. Environ. Int. 31 (1), 53–61. doi:10.1016/j.envint.2004.06.009.

Walker, D.B., Baumgartner, D.J., Gerba, C.P., Fitzsimmons, K., 2019. Surface Water Pollution. In: Environmental and Pollution Science. Elsevier, 261–292. doi:10.1016/B978-0-12-814719-1.00016-1.

Xu, G., Li, P., Lu, K., Tantai, Z., Zhang, J., Ren, Z., et al, 2019. Seasonal changes in water quality and its main influencing factors in the Dan River basin. Catena (Amst) 173, 131–140. doi:10.1016/j.catena.2018.10.014.

Zamani Sabzi, H., Moreno, H.A., Fovargue, R., Xue, X., Hong, Y., Neeson, T.M., 2019. Comparison of projected water availability and demand reveals future hotspots of water stress in the Red River basin, USA. J. Hydrol. Reg. Stud. 26, 100638. doi:10.1016/j.ejrh.2019.100638.

Zheng, H., Tetzlaff, D., Birkel, C., Chmieleski, J., Comte, J.C., Freymueller, J., et al, 2025. Hydrological connectivity and biogeochemical dynamics in the function and management of the lower Oder floodplain. J. Hydrol. (Amst) 653, 132708. doi:10.1016/j.jhydrol.2025.132708.

Chapter 5

Micro- and nanoplastic pollution

Thilomi Samarakoon
Department of Zoology and Environmental Management, University of Kelaniya, Kelaniya, Sri Lanka

1 Introduction

Since the invention of the world's first synthetic plastic, Bakelite, in 1907, plastic production worldwide has been increasing year by year due to its versatility in a wide range of applications (Tursi et al., 2022). These applications include packaging, building and construction, automotive components, electrical and electronic equipment, textiles, agriculture, healthcare, and household goods. In fact, global plastic production has increased from 2 million metric tons in 1950 to 367 million tons in 2020, and it is predicted to double within the next 20 years (Tursi et al., 2022; Walker and Fequet, 2023).

Most consumer plastics are single-use plastics, and more than half of all plastic products are reported to become waste in less than a year after production (Tursi et al., 2022). Globally, only 9% of plastic waste is recycled, 12% is incinerated, and the remaining 79% is released into the environment (Walker and Fequet, 2023), making plastic waste a significant portion of all anthropogenic waste worldwide.

As a result of massive production, unsustainable use, and disposal, combined with the low degradation rate, pollution due to plastic waste has become one of the most severe environmental issues in the Anthropocene. This issue affects all types of ecosystems, including terrestrial, aquatic, and atmospheric ecosystems, but it is particularly severe in aquatic ecosystems. Most plastic waste originates from land-based sources, and a significant proportion of this eventually ends up in oceans (Jambeck et al., 2015). The inland fluvial systems, comprised of a network of streams and rivers, hence act as transport routes for these pollutants toward the oceans, making rivers a critical conduit for plastic debris from land to sea (Jayaweera et al., 2025). It is estimated that rivers worldwide export approximately 0.5 million tons of plastic per year to the ocean (Strokal et al., 2023).

Rivers are rich in biodiversity, and they also play an important role in sustaining human health by providing drinking water, food, and water for

River Pollution. https://doi.org/10.1016/B978-0-443-45024-2.00016-6

irrigation. However, plastic pollution poses a severe threat to both riverine biodiversity and human well-being worldwide. Plastic contamination affects sensitive freshwater species at the individual level, which cascades to the population and community levels, ultimately disrupting riverine ecosystem processes by altering their community structure and functional traits (Nava et al., 2024). Since historical times, humans have maintained a strong connection with river systems, relying on them as essential life-supporting ecosystems. The health of rivers, therefore, directly reflects the health and resilience of the communities that depend on them. Emerging evidence indicates that plastic pollution in river water affects human health (Xu et al., 2022).

2 Micro- and nanoplastics and their physical and chemical properties

Plastic particles are diverse in their physical and chemical properties. The physical properties, such as size, shape, color, and specific density, as well as their chemical composition, vary greatly across different plastics, making them a unique category of pollutants in receiving environments. These characteristics greatly influence the environmental fate and the ecological and human health impacts of plastics.

Depending on their size, plastics can be classified as megaplastics, macroplastics, mesoplastics, microplastics (MPs), and nanoplastics (NPs) (Fig. 5.1). However, clear-cut size ranges have not been established for each category, and it is generally accepted that plastics larger than 50 cm are considered megaplastics; those between 5 and 50 cm are macroplastics; those between 0.5 and 5 cm are mesoplastics; and those smaller than 0.5 cm are MPs (Tursi et al., 2022). Although a lower size limit for MPs has not been strictly established for several years (Blair et al., 2017), recent researchers consider the very fine fraction within the microplastic category, specifically particles smaller than 100 nm, as NPs (Tursi et al., 2022; Gigault et al., 2016).

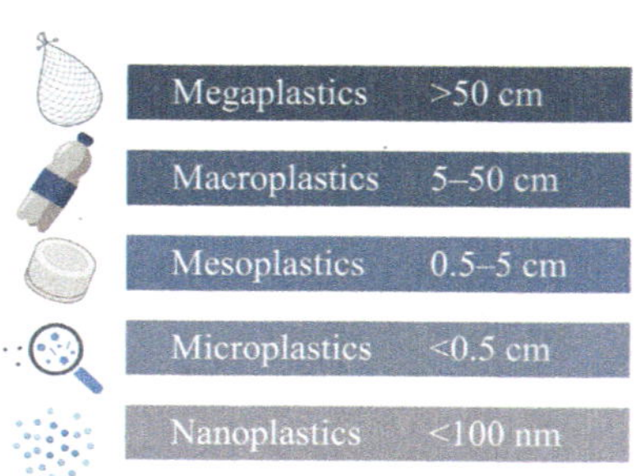

FIG. 5.1 Size categories of plastics. Categorization of plastics according to size, ranging from megaplastics (>50 cm) to nanoplastics (<100 nm).

This chapter focuses on micro- and nanoplastics (MNPs) as a single category for ease of reference, and due to the similarities, they share in terms of sources, fate, and ecological and human health impacts. However, when significant differences exist, they are discussed separately.

The shapes of MNPs also show significant diversity, which often reflects their potential sources. The most common shapes include fragments, fibers/lines, films, foams, beads/spheres, and pellets/granules (Kooi and Koelmans, 2019; Wu et al., 2018) (Fig. 5.2). Some researchers group beads/spheres together with pellets/granules (Gupta et al., 2023), while others classify them as separate categories (Hu et al., 2022). However, there are no standardized guidelines for the shape classification of MNPs, and in general, the categories are defined based on their microscopic appearance. Thick particles with irregular edges are often identified as fragments, although some researchers classify particles with uniform thickness and sharp edges as sheets (Wu et al., 2018; Hu et al., 2022). Another shape category, known as flakes, is also reported in some studies to describe relatively flat, irregularly shaped plate-like particles with rough edges (Turner et al., 2022). However, these categories are more commonly grouped together as fragments in most studies. Fibers/lines are thread-like plastic particles, while films are thin, flat pieces. Foams have a characteristic foamy appearance, and beads/spheres are round or oval-shaped particles. Pellets/granules appear similar to beads/spheres but typically have rougher edges rather than the smooth surfaces seen in beads/spheres.

Although color may seem like a simple property, in MNP research, color is often reported as a feature that can influence its ecological impacts. For example, some organisms appear to consume particles of specific colors while avoiding others (Horie et al., 2024). The color of MNPs can be inherited from

FIG. 5.2 Shapes of micro- and nanoplastics. This figure illustrates the shapes of micro- and nanoplastics: (A) Fragment, (B) fragment/sheet, (C) fragment/flake, (D) fibers/lines, (E) film, (F) foam, (G) bead/sphere, and (H) pellet/granule.

their parent plastic product, hence being useful to speculate on the potential source, or it can also result from degradation and weathering processes in the environment.

As with their physical properties, the chemical properties of MNPs are also diverse. They differ in terms of constituent polymer types, plastic additives, and chemicals adsorbed from the surrounding environment. The most commonly detected plastic polymer types include polyethylene (PE), polypropylene, polystyrene (PS), polyvinyl chloride, polyethylene terephthalate, and polyurethane (PU). A critical feature of most of these polymers is their carbon-based backbone, which is resistant to degradation. While this property is beneficial for various human applications, it also leads to their persistence and accumulation in the environment (Okoye et al., 2022). Along with this property, the density, a physical property of plastics, also varies, which in turn influences the fate of the plastic particles in the environment. A wide range of chemicals that can be categorized as plasticizers, inert/reinforcing fillers, lubricants, antioxidants, UV stabilizers, and flame retardants are added during plastic production to achieve desired plastic properties such as color and resistance to thermal, oxidative, and photodegradation (Okoye et al., 2022). In addition to these inherent and additive chemicals associated with MNPs, these plastic particles provide surfaces for hundreds of other chemical substances such as polychlorinated biphenyls, polycyclic aromatic hydrocarbons, pharmaceuticals and personal care products, per- and polyfluoroalkyl substances, and heavy metals (Rai et al., 2022; Yu et al., 2019).

3 Sources of MNPs in riverine environments

MNPs are classified as primary or secondary (Fig. 5.3) based on how they originate. Primary MNPs are intentionally manufactured at the size scale of MNPs, whereas secondary MNPs are generated in the environment due to the breakdown of larger plastic items into smaller fragments at the size scale of MNPs. Both types are present in riverine environments, although secondary MNPs are typically more dominant, according to several studies (Gao et al., 2024).

Primary MNPs are either used as raw materials in industrial processes or directly incorporated into consumer products. Plastic pellets or spheres, primarily composed of polymers such as PS and PE, are used as raw materials in several industries, including construction, furniture, packaging, electronic devices, vehicles, and medicine (Gupta et al., 2023; Gao et al., 2024). Synthetic fibers such as polyester, nylon (polyamide), acrylic, and spandex/elastane (PU) are used as raw materials in the textile industry (Acharya et al., 2021). On the contrary, MNPs are directly used in various consumer products. PE and PS microbeads are frequently incorporated into personal care products such as facial cleansers, scrubs, creams, and cosmetics as abrasive scrubbers, exfoliating agents, bulking agents, or as components designed to control the release of active ingredients and extend shelf life (Diana et al., 2025; Fang et al., 2024).

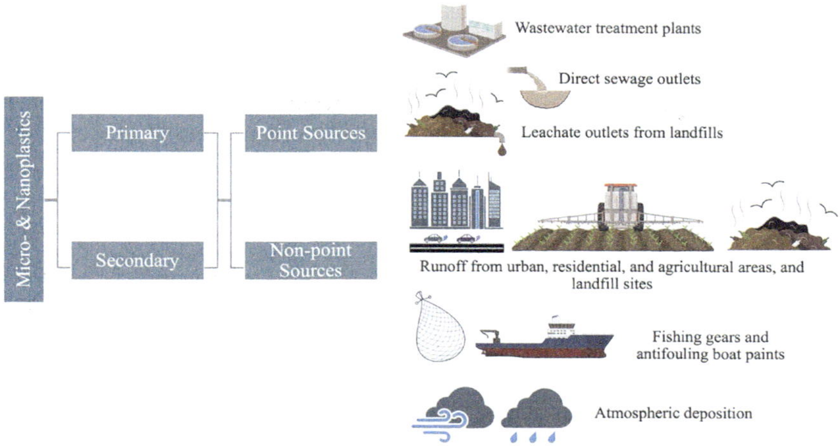

FIG. 5.3 Sources of micro- and nanoplastics. This figure illustrates the origins of micro- and nanoplastics and their sources in riverine environments.

Another direct use of MNPs is the incorporation of microbeads and microfibers into paints and coatings to enhance paint elasticity, improve resistance, thicken coating layers, and reduce the weight of the product (Diana et al., 2025; Fang et al., 2024).

On the contrary, secondary MNPs are produced in the environment due to physical, chemical, and biological degradation of large plastic items. These processes include photodegradation by sunlight, mechanical degradation caused by wind, wave action, and sand or soil friction, chemical degradation such as oxidation and hydrolysis, and biodegradation mediated by macro- and microorganisms (Tursi et al., 2022). A distinctive characteristic of secondary MNPs in riverine environments is that they may originate externally and be transported to riverine environments, or they may be generated in situ from larger plastic debris flowing into riverine environments. Therefore secondary MNPs have diverse origins, and their shapes and chemical composition exhibit a wider spectrum compared to primary MNPs (Gao et al., 2024). They typically exhibit irregular shapes, including fragments, films, foams, and flakes. Fibers may also result from the degradation of fibrous materials in the environment. However, pellets and spheres cannot be fully excluded as shapes of secondary MNPs, since environmental weathering processes may also produce these shapes (Efimova et al., 2018).

The primary and secondary MNPs enter riverine systems from both point and nonpoint sources (Fig. 5.3).

The point sources include industrial wastewater treatment plants, industrial and domestic sewage treatment plants, direct sewage outlets, and landfill leachate outlets. Even though wastewater treatment plants, including sewage treatment plants, remove MNPs to a significant extent, they are still considered a major source of MNPs in receiving environments, particularly riverine environments. According to some studies, the physical, chemical, and biological stresses in these wastewater treatment plants increase the production of MNPs in their effluent (Iskandarani et al., 2025; Rout et al., 2022). Industries, including textiles, paints, petrochemicals and manufacturing, tire and rubber, pharmaceuticals, cosmetics and personal care products, recycling and plastic waste treatment, and paper and printing, release significant amounts of MNPs in their effluents (Rout et al., 2022). A major source of MNPs in domestic sewage is synthetic fibers shed from textiles during laundering and personal care products (Iskandarani et al., 2025). A recent study estimated that >70,000 microfibers/m^3 can be released per wash in certain synthetic textiles (Yang et al., 2019). Landfill leachate, when collected through leachate collection tanks, can be considered a point source of MNPs. MNPs present in landfill leachate are characteristically heavily contaminated with other chemical pollutants, amplifying their impacts severalfold (Gao et al., 2024).

The nonpoint sources of larger plastics and MNPs in riverine environments include runoff from urban areas, agricultural areas, landfills, fishing gear, antifouling boat paints, and atmospheric deposition via wind and rainfall. The major nonpoint source of MNPs in riverine environments is runoff. Runoff across urban areas carries MNPs along with large plastic litter from roads, roofs, parking lots, and other impervious surfaces, as well as lawns and gardens belonging to residential and commercial areas. The major sources of MNPs in urban runoff include tire wear particles, road markings, building materials, city dust, and artificial grass. Moreover, the overflow of sewage pipes also contributes to MNPs in urban runoff (Birch et al., 2020; He et al., 2022). Runoff across agricultural areas is also a significant source of MNPs in rivers. Similar to other human activities, modern agricultural practices depend on various plastic products, such as containers for pesticides and fertilizers, seed packages, irrigation pipes, and plastic mulch films, all of which contribute to the origin of MNPs in agricultural areas. Moreover, organic fertilizers such as compost and biosolids, as well as domestic sewage irrigation, are significant sources of MNPs in agricultural soils (Athavuda et al., 2025; He et al., 2022). Recent studies confirm that runoff from landfills is one of the major sources of MNPs in riverine environments, especially in developing countries (Jayaweera et al., 2025). A unique nonpoint source of MNPs in aquatic environments includes lost, abandoned, or discarded fishing gear such as nets, lines, and ropes that can degrade into MNPs. Apart from that, antifouling paints applied to boat hulls can release MNPs into aquatic environments through abrasion (Nelms et al., 2021). Atmospheric transport and deposition of MNPs by wind, rainfall, or snow into riverine environments is also possible. For example, MPs have been detected in Lake Toma, Switzerland, located at a high altitude in the absence of industry, agriculture, or settlement, and the particles were speculated to have been generated by waste incineration, transported by wind, and deposited via wet or dry deposition (Gao et al., 2024). Plastic pollution in riverine environments also occurs through accidental spills and intentional dumping.

4 The fate of MNPs in rivers

The fate of MNPs in riverine environments is a complex process determined by both the intrinsic physical and chemical properties of the particles (e.g., size, shape, color, density, and chemical composition) and the interacting physical, chemical, and biological processes within the fluvial system (e.g., flow conditions, hydrology, river morphology, water chemistry, and biological activities). Studies indicate that most MNPs originating from land-based sources will ultimately end up in marine habitats, with rivers serving as the primary transport routes that carry them toward the ocean (Tursi et al., 2022). However, during this transit, MNPs undergo a range of processes that determine their fate within the riverine environments. These processes include transport, aggregation, sedimentation–resuspension, degradation, biofilm formation, burial, and biota transport (Besseling et al., 2017) (Fig. 5.4).

Once entered or generated inside riverine environments, the MNPs are subjected to transportation. The particles can undergo both horizontal and vertical transportation (Yan et al., 2021). Horizontal transportation carries MNPs downstream under the influence of hydraulic conditions, rainfall, river morphology, dams, and vegetation. A certain proportion of MNPs travel vertically in the river column. Particles with lower density float to the surface or disperse in the water column, while those with higher density tend to settle in benthic environments under the influence of gravity. Other processes, such as aggregation, may increase their density, supporting their settlement in benthic environments. Aggregation can be either homogeneous (MNP–MNP) or heterogeneous

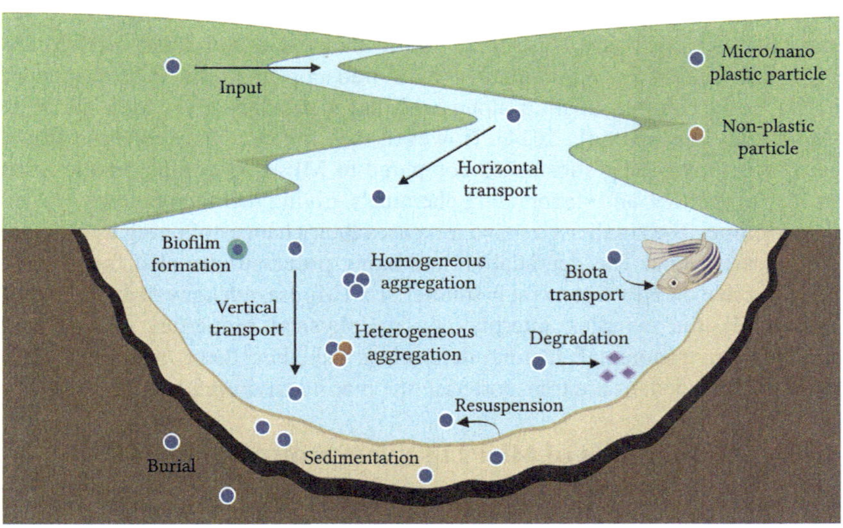

FIG. 5.4 Fate of micro- and nanoplastics in riverine environments. This figure illustrates the fate of micro- and nanoplastics in riverine environments, including horizontal and vertical transport, biota transport, biofilm formation, degradation, sedimentation, aggregation, and resuspension.

(MNP–nonplastic particles such as clay minerals and organic matter) (Besseling et al., 2017; Yan et al., 2021). Surface chemistry and size of MNPs, along with water chemistry, govern the aggregation processes. Size plays a complex role in the settling process, where particles larger than 0.5–1 mm tend to remain in the water column, while smaller particles exhibit efficient settling due to their tendency to aggregate (Li et al., 2020; Munz et al., 2024). Moreover, the colonization of microorganisms around particles (e.g., biofilm formation) may also result in an increase in density, allowing the particles to settle at the river bottom (Yan et al., 2021). Riverbeds, therefore, act as temporary or long-term sinks for MNPs. Several studies report the accumulation of MNPs in benthic sediments compared to surface waters in rivers (Jayaweera et al., 2025; Drummond et al., 2022). However, resuspension of MNPs occurs during rainfall events and turbulence (Blair et al., 2017). MNPs may also undergo degradation in riverine environments due to physical abrasion caused by water flow and collisions with riverbed sediments and rocks, as well as thermal and chemical stressors and microbial activities. This degradation may modify the color, shape, size, density, and surface chemistry of MNPs, which in turn affects their fate. Transportation of MNPs to aquatic biota is also common. Aquatic organisms may mistakenly recognize MNPs as food due to their micrometric size and variable coloration. Apart from direct ingestion from water, MNPs can be ingested by aquatic organisms such as fish, tadpoles, and invertebrates, including zooplankton, through their prey along the food chain (Tursi et al., 2022). The MNPs transferred to biota will eventually settle in sediments along with excretion or debris, which can be identified as another mechanism of their vertical transport in the water column (Yan et al., 2021).

When the fate of riverine environments is considered, notable differences exist between MPs and NPs. NPs, due to their light and small size, follow Brownian motion that limits their vertical transport and tend to be suspended in the water column, drifting along with the water flow over long distances (Mousazadehgavan et al., 2024). However, they are also more likely to aggregate with surrounding substances compared to MPs due to their smaller size, surface properties, and surrounding chemicals, giving them the potential to settle over time. The smaller size also increases their chances of being transported into aquatic biota. The degradation and aging processes are also found to be less effective on NPs due to their smaller size (Mousazadehgavan et al., 2024).

Overall, the complex interplay of the physical properties of MPs and dynamic environmental factors determines whether these particles settle, remain suspended, or are transported to the marine environment.

5 Measured levels of MNPs in rivers around the world

Compared to the marine and estuarine environments, the reported levels of MNPs in riverine systems worldwide are relatively limited. Since MNPs in riverine environments are partitioned between the water column and sediments, it is equally important to monitor their concentrations in both media to obtain a comprehensive understanding of their distribution and environmental behavior. However, most existing studies focus primarily on MNP concentrations in the water column, while data on sediment-associated MNPs remain comparatively limited or vice versa.

The concentrations of MNPs reported in river water and sediment samples from some selected rivers of the world are summarized in Table 5.1. The dataset highlights distinct spatial, temporal, and environmental patterns in MNP contamination across global river systems.

Generally, sediment compartments contain a higher MNP concentration than surface waters. For example, the Aarin River in Nigeria exhibited an extremely high sediment concentration of MNPs (30,384 ± 55.0 particles/kg) compared to water (130.5 ± 3.5 particles/L). Similar sediment dominance is demonstrated in the River Elbe in Germany, the Vistula River in Poland, the Wei River in China, and the Weras Ganga in Sri Lanka. This supports the understanding that riverine sediments act as short- or long-term sinks for MNPs. However, in some rivers, the MNP concentrations in water and sediment levels are comparable or higher, highlighting the complex interplay of physical and chemical characteristics of MNPs and environmental conditions, such as river morphology, water velocity, and water chemistry, in determining their fate and behavior. For example, samples collected from rivers of the Tibet Plateau have characteristically lower amounts of MNP levels in sediments (50–195 items/kg) compared to the levels in water (483–967 items/m^3) owing to the faster water flow in headwaters at high elevations.

Seasonal variation is another key determinant of MNP pollution. For example, a significant increase in MNP levels in both water and sediment of the Aarin River in Nigeria during the rainy season (130.5 ± 3.5 particles/L and 30,384 ± 55.0 particles/kg, respectively) compared to the dry season (113.0 ± 5.0 particles/L and 2763.5 ± 28.5 particles/kg, respectively) indicates that stormwater and surface runoff are major drivers of MNP mobilization from urban, agricultural, and landfill sources into nearby waterways. On the contrary, depending on the river morphology and other environmental factors, the sediment load of MNPs can be reduced during rainfall events due to the resuspension of MNPs, resulting in lower levels in sediments. For example, the sediment concentration of MNPs reduced from 9.4–50.3 items/kg in the dry season to 0.0–21.3 items/kg in the rainy season in the Defeng River in China.

Land-use characteristics and anthropogenic pressure strongly influence the MNP abundance in riverine environments. Consistently higher levels of MNPs have been reported in rivers flowing across urbanized or industrialized catchments, such as the Ciwalengke River in Indonesia, the Ergene River in Turkey, the Ganga River in India, the Mekong River Delta in Vietnam, the Po River tributaries in Italy, the River Elbe in Germany, and the Seine River in France. Conversely, rivers sampled from remote or less populated areas, such

TABLE 5.1 Micro- and nanoplastic concentrations reported in riverine water and sediment from selected rivers worldwide.[a]

River (location)	Characteristic feature of the sampling location	Average concentration/range in water	Average concentration/range sediments	Reference
Aarin River (Southwestern Nigeria)	Flowing through rural villages, prevailing farming, fishing, washing practices, and local industries	130.5 ± 3.5 particles/L (rainy season) 113.0 ± 5.0 particles/L (dry season)	30,384 ± 55.0 particles/kg (rainy season) 2763.5 ± 28.5 particles/kg (dry season)	(Oladeji et al., 2025)
Amazon River (Brazil)	Mixed landscapes: natural forests, pastures, urban areas	NR	0–8178 particles/kg	(Gerolin et al., 2020)
Arakawa River (Saitama and Tokyo, Japan)	Populated area, flooding history, tourism activities	2.21 ± 1.48 pieces/m^3	NR	(Wang et al., 2024)
Brisbane River (Australia)	Multiple land uses: natural, residential, commercial, industrial, and various urbanization levels	NR	10–520 items/kg	(He et al., 2020)
Ciwalengke River (Majalaya District, Indonesia)	Slum and industrial area	5.85 ± 3.28 particles/L	3.03 ± 1.59 particles/100 g	(Alam et al., 2019)
Defeng River (China)	A remote river	0.00004–0.0009 items/L (rainy season) 0.0003–0.0025 items/L (dry season)	0.0–21.3 items/kg (rainy season) 9.4–50.3 items/kg (dry season)	(Liu et al., 2021)

Ergene River (Turkey)	Industrial area	4.65 ± 2.06–6.90 ± 5.16 items/L	97.90 ± 71.72–277.76 ± 207.21 items/kg	(Akdogan et al., 2023)
Ganga River (Eastern India)	Highly populated, industrial area	NR	99.27–409.86 items/kg	(Sarkar et al., 2019)
Ganga River (India)	Lower stretches	380–684 items/1000 m^3	17–36 items/kg	(Singh et al., 2021)
Mekong River Delta (Vietnam)	Populated area with agriculture and aquaculture-related activities and industries	53.8 ± 140.7 items/m^3	6.0 ± 2.0 items/g	(Kieu-Le et al., 2023)
Po River Tributaries (Northern Italy)	Industrialized and urbanized area	0.9 ± 0.5 plastics/m^3 (lowest: Ticino) 62.9 ± 53.9 plastics/m^3 (highest: Mincio)	6.8 ± 4.5 plastics/kg (lowest: Ticino) 26.5 ± 13.3 plastics/kg (highest: Mincio)	(Sbarberi et al., 2024)
Rhine River (Near Basel, Switzerland)	–	4.48 number/m^{3a}	NR	(Erni-Cassola et al., 2024)
Rhine River (Switzerland and Germany)	Different discharge regimes: nival (Swiss Rhine catchment), pluvial (German locations)	0.04 ± 0.04–3.3 ± 2.1 particles/m^3 (Nival) 2.7 ± 0.4–6.3 ± 2.6 particles/m^3 (Pluvial)	NR	(Mani and Burkhardt-Holm, 2020)
River Elbe (Germany)	Urban river	5.57 particles/m^3	3,350,000 particles/m^3	(Scherer et al., 2020)

Continued

TABLE 5.1 Micro- and nanoplastic concentrations reported in riverine water and sediment from selected rivers worldwide—cont'd

River (location)	Characteristic feature of the sampling location	Average concentration/ range in water	Average concentration/range sediments	Reference
Rivers of the Tibet Plateau	Remote area, high elevation	483–967 items/m^3	50–195 items/kg	(Jiang et al., 2019)
Seine River (Greater Paris Area, France)	Heavily urbanized area	600 particles/m^3	NR	(Stratmann et al., 2024)
The Mississippi River System (USA)	Flash drought and nondrought periods	16–381 particles/L	NR	(Wontor et al., 2024)
Vistula River (Poland)	Urban river	1.6–2.55 items/L	190–580 items/kg	(Sekudewicz et al., 2021)
Wei River (Northwest China)	Heavily populated and agricultural area	3.67–10.7 items/L	360–1320 items/kg	(Ding et al., 2019)
Weras Ganga (Bolgoda, Sri Lanka)	Closer to a landfill site	373.43 ± 125.80 particles/m^3	4612.85 ± 1797.02 particles/kg	(Jayaweera et al., 2025)

[a]*Median.*

NR, Not reported.

as the Defeng River in China and the rivers of the Tibetan Plateau, exhibited comparatively lower contamination levels, reflecting minimal anthropogenic influence. However, it should be noted that even in these remote locations, man-made pollutants such as MNPs were still detected, even at low concentrations, highlighting their global presence. This indicates that long-range atmospheric transport, followed by wet and dry deposition processes, may play a significant role in the world-wide distribution of MNPs.

Proximity to landfills seems to be a key determinant in the elevated levels of MNPs in nearby rivers. For example, the Weras Ganga, located near an open landfill, exhibited high levels of MNPs in both water and sediments.

Even though not widely studied, different hydrological regimes also play a vital role in determining MNP load in riverine systems. For example, a significant difference in observed MNP levels in water in the Rhine River has been documented between nival (snowmelt-driven) and pluvial (rainfall-driven) discharge regimes. In the Swiss Rhine catchment (nival regime), MNP concentrations in the water ranged from 0.04 ± 0.04 to 3.3 ± 2.1 particles/m^3, which were considerably lower than those found in the German locations (pluvial regime), where concentrations ranged from 2.7 ± 0.4 to 6.3 ± 2.6 particles/m^3. Nival systems, dominated by snowmelt, experience relatively lower anthropogenic inputs and slower runoff rates, which may reduce the immediate transport of MNPs. In contrast, pluvial systems, which are strongly influenced by rainfall events, often show enhanced surface runoff and stormwater discharges that mobilize and introduce greater quantities of plastic debris from urban and industrial surfaces into the river network.

6 Ecological impacts of riverine MNPs

As primary transport routes for MNPs, while simultaneously acting as both long-term and short-term reservoirs for MNPs, riverine ecosystems are under significant pressure from MNP pollution. The ecological impacts of riverine MNPs range from toxicological effects at the cellular and individual organism levels to broader impacts on populations and communities. These effects may arise from the direct toxicity of MNPs themselves, as well as their indirect toxic effects as vectors for a wide range of other environmental pollutants.

Several studies indicate that MNPs affect a wide range of aquatic species inhabiting rivers and streams, including phytoplankton (Amaneesh et al., 2023), macrophytes (van Weert et al., 2019), benthic macroinvertebrates (Redondo-Hasselerharm et al., 2020), zooplankton (Funke et al., 2024), fish (Rahman et al., 2022), and amphibians (Fang et al., 2025).

There is evidence that MNPs get adsorbed on to the surfaces of phytoplankton and macrophytes and, in some cases, become internalized into their cells, inducing toxic responses such as growth inhibition, altered photosynthetic pigment production, oxidative stress, and other physiological and metabolic changes (Amaneesh et al., 2023; Mateos-Cárdenas et al., 2021).

The other aquatic organisms, including benthic macroinvertebrates, zooplankton, fish, and amphibians, are primarily exposed to MNPs through ingestion, either unintentionally or intentionally when plastic particles are mistaken for prey (Hu et al., 2022). This may lead to both lethal and sublethal effects, including gut obstruction, physical damage to digestive tissues, altered feeding behavior, and the crossing of the gut epithelium to transport to other organs (Strungaru et al., 2019). Laboratory and field studies show increased mortality, reduced growth, and impaired reproduction in freshwater organisms exposed to MNPs due to mechanisms such as oxidative stress, modulation of immune responses, endocrine disruption, and altered gene expression (Vo and Pham, 2021).

MNPs tend to be transferred through trophic levels in aquatic food chains, ultimately reaching organisms at higher trophic levels (Mateos-Cárdenas et al., 2022; Chae et al., 2018). This process increases exposure and the potential for bioaccumulation in top consumers, including humans. Compared to MPs, NPs are more likely to bioaccumulate within organisms due to their smaller size and greater bioavailability, thereby enhancing their transfer across trophic levels (Chae et al., 2018). The trophic transfer of MNPs can further disrupt energy flow in riverine food webs, potentially impairing ecosystem functioning at broader ecological scales (Ma et al., 2020).

Lethal and sublethal effects due to exposure to MNPs at the individual level can scale up to influence the entire population and community dynamics in riverine ecosystems. Species that are extremely sensitive to MNPs, such as zooplankton and green algae (Nam et al., 2023), may decline where MNP burdens are high, while tolerant or opportunistic taxa may increase, resulting in shifts in community composition. Such changes in community composition are often accompanied by shifts in ecosystem functional traits. For example, a reduction in zooplankton populations may increase the abundance of phytoplankton due to decreased grazing pressure, potentially leading to altered ecosystem productivity and nutrient cycling. Furthermore, MNPs disrupt photosynthetic pigment production in green algae, resulting in low photosynthetic activity, which in turn results in altered nutrient cycling (Ma et al., 2020).

The community shifts not only occurred due to the toxicological impacts of MNPs on individual organisms. Recent studies show that the MNPs provide substrates for the colonization of biofilms consisting of microbial communities composed of diverse prokaryotes, microalgae, fungi, and sometimes macroorganisms. This new artificial ecosystem on the surface of MNPs is identified as the "plastisphere" (Nava et al., 2024; Li et al., 2021). It has been demonstrated that this new artificial system is capable of modifying ecosystem productivity by causing reductions in oxygen concentrations within river water, with impacts on system metabolism and nutrient cycling (Nava et al., 2024). Moreover, it also increases the chance of the occurrence of pathogenic microbes attached to the MNPs, making it an ecological and a public health concern.

7 Exposure and human health impacts of riverine MNPs

Humans are exposed to riverine MPs primarily through drinking water and contaminated food sources, such as fish and other species.

Rivers serve as major sources of drinking water in many regions of the world. For example, the Arakawa and Toyo Rivers in Japan provide tap water for approximately 20 million people in the Tokyo Metropolitan Area (Wang et al., 2024), while the Rhine and Meuse Rivers account for about 40% of the drinking water supply in the Netherlands (Sjerps et al., 2017). Usually, raw water extracted from rivers is directed to drinking water treatment plants before distribution for human consumption. However, several studies have reported the partial removal of MNPs from the raw water during the treatment process. For instance, raw water sourced from the Úhlava River in the Czech Republic for two drinking water treatment plants contained total MP concentrations of 23 ± 2 and 1296 ± 35 particles/L, and after treatment, 14 ± 1 and 151 ± 4 particles/L were still detected in the treated water (Pivokonský et al., 2020). Moreover, some studies have observed significant increases in MP concentrations during the treatment process (Wu et al., 2022).

The majority of existing studies focus on MNP contamination in edible marine species. However, several communities around the world fulfill their protein requirements by consuming freshwater organisms that inhabit rivers. Only a few studies have quantified MNP concentrations in such freshwater organisms. For example, 44% of the tested *Oreochromis niloticus* cultured in water sourced from the Betania Dam, Magdalena River, and Bache River in Colombia, and 75% of the tested *Prochilodus magdalenae* and *Pimelodus grosskopfii* inhabiting the Magdalena River in Colombia, contained MPs in their tissues (Garcia et al., 2021). All three of these species are edible freshwater fish. In addition to fish, MPs have also been detected in the edible freshwater gastropod *Sinotaia quadrata* collected from the Arno River Basin in Central Italy (two sites: 0.8 ± 1.30 and 1 ± 0.37 items/specimen) (Pastorino et al., 2021), as well as in edible freshwater arthropods, *Caridina* sp., *Macrobrachium* sp., *Aethriamanta* sp., *Aciagrion* sp., and *Sphaerodema molestum*, caught from the Chi and Pong Rivers in Thailand (0.25–8.0 items/individual) (Thamsenanupap et al., 2022).

Although less common, unintentional ingestion of contaminated water during recreational activities, such as swimming, can also be an exposure route for MNPs. These particles cannot easily travel through the human skin, but research has found that NPs can enter through wounds, sweat glands, or hair follicles owing to their smaller size (Schneider et al., 2009; Rashed et al., 2023).

Although the effects of MNPs on animals and plants have been extensively studied, their impacts on humans remain unclear. However, the detection of MNPs in human lung tissue, stool, urine, placenta, meconium, blood, breast milk, and several other organs, including the colon, liver, spleen, and testis, indicates potential health risks associated with exposure to these particles (Hu et al., 2022; Brits et al., 2024). It has been documented that MNPs smaller than

150 μm can readily cross the gastrointestinal epithelium, those smaller than 10 μm can penetrate the placenta and the blood–brain barrier, and particles smaller than 2.5 μm can reach systemic circulation via endocytosis (Tursi et al., 2022). Therefore compared to MPs, NPs are considered more toxic due to their higher bioavailability, as they can cross biological membranes more easily because of their smaller size (Xu et al., 2022; Rashed et al., 2023).

The possible health effects of MNPs in humans arise from both the direct toxicity of the plastic particles themselves and the indirect toxicity caused by leachable additives and adsorbed contaminants, such as heavy metals and persistent organic pollutants (Chang et al., 2020). Based on the results from several in vitro, in vivo, and ex vivo studies, the direct effects of MNPs on humans have been associated with several health outcomes, including gastrointestinal toxicity, hepatotoxicity, neurotoxicity, reproductive toxicity, and carcinogenicity (Tursi et al., 2022; Xu et al., 2022; Chang et al., 2020). Studies investigating the effects of MNPs on human subjects through epidemiological research are limited; however, a recent study established a correlation between the abundance of fecal MNPs and inflammatory bowel disease (Yan et al., 2022). In addition to the direct effects, a wide range of additional toxic effects may arise depending on the plastic additives and other adsorbed chemicals on MNPs, which are beyond the scope of this chapter. Potential mechanisms of MNP toxicity in humans include induction of oxidative stress and cytotoxicity, disruption of energy and metabolism, translocation to the circulatory system and other tissues, disruption of immune function, and neurotoxicity and neurodegenerative diseases (Hu et al., 2022; Chang et al., 2020).

8 Future directions for MNP research

Even though research on MNPs is rapidly advancing at present, several key challenges remain, especially regarding their occurrence and behavior in riverine environments. These challenges encompass the areas of monitoring and quantifying MNPs, understanding their transport and fate in riverine environments, assessing ecological and human health impacts at detected levels in riverine environments, developing effective treatment and mitigation strategies, and addressing policy implications.

One prominent issue is the overall scarcity of monitoring data on MNPs in riverine systems compared to marine and estuarine environments. Furthermore, assessing MNP concentrations in different river compartments, including the water column, sediment, and biota, can provide a comprehensive understanding of their distribution and dynamics. Most existing studies primarily focus on the spatial distribution of MNPs by sampling across varying land-use types to speculate on potential sources. However, long-term and seasonal monitoring data are limited for many riverine systems, despite their importance in understanding the temporal variability and discharge patterns of MNPs. Another issue related to MNP monitoring is the lack of standard sampling, analytical, and reporting protocols, which makes interstudy comparisons challenging. Reviews have highlighted that differences in sampling mesh size, size cut-off, sample processing methods, and normalization metrics (e.g., particles per liter, particles per cubic meter, weight per liter, weight per cubic meter, etc.) contribute to significant challenges in comparing studies (Zhao et al., 2024; Birch et al., 2020).

Another critical research direction involves enhancing the understanding of transport and fate mechanisms of MNPs within riverine systems. Processes such as biofilm formation, aggregation, sedimentation, resuspension (especially during high-flow or flood events), as well as the influence of hydraulic and geomorphological features (e.g., flow velocity, water depth, bed morphology, channel slope) in determining the behavior and distribution of MNPs in riverine environments, should be investigated systematically to better predict MNP dynamics in fluvial environments in future studies (Owowenu et al., 2023).

Ecological and human health risk assessment related to MNPs in riverine environments remains underdeveloped. While numerous studies have demonstrated MNP presence in water and sediments, fewer have linked those environmental concentrations to adverse biological effects in aquatic biota under natural exposure conditions (e.g., sublethal effects, trophic transfer, chronic exposure) or evaluated the implications for human health via drinking water, fish consumption, or exposure through sediments (Xu et al., 2022). Review works call for more toxicological studies using environmentally realistic doses and particle types.

Although the treatment strategies for MNPs are beyond the scope of this chapter, there is an urgent need for research on effective treatment methods for MNPs in wastewater treatment plants and stormwater, as they are the major point and nonpoint sources of MNPs, respectively, in riverine systems. Once MNPs are released into riverine environments, remediation becomes extremely challenging due to their persistence, small size, and complex interactions with sediments and biota (Mousazadehgavan et al., 2024). Therefore preventive measures focused on source reduction, as well as improved treatment technologies, are essential to minimize their entry into aquatic systems.

Lastly, there is an urgent need for stronger integration of MNP research outcomes for policy and regulation formulation (da Costa, 2018). While environmental monitoring for MNPs has expanded, policies mandating regular monitoring, setting allowable limits for MNP discharge (from wastewater treatment plants and industrial effluents), and strengthening waste management, including banning certain plastic types, improving landfill controls, and promoting extended producer responsibility, are not yet universally applied or enforced. Cross-disciplinary work among environmental scientists, toxicologists, engineers, and policymakers will be the key to addressing this emerging pollutant and safeguarding freshwater ecosystems for future generations.

References

Acharya, S., Rumi, S.S., Hu, Y., Abidi, N., 2021. Microfibers from synthetic textiles as a major source of microplastics in the environment: a review. Text. Res. J. 17467748. 91 (17–18), 2136–2156. Available from: https://journals.sagepub.com/home/TRJ. doi:10.1177/0040517521991244.

Akdogan, Z., Guven, B., Kideys, A.E., 2023. Microplastic distribution in the surface water and sediment of the Ergene River. Environ. Res. 00139351. 234, 116500. doi:10.1016/j.envres.2023.116500.

Alam, F.C., Sembiring, E., Muntalif, B.S., Suendo, V., 2019. Microplastic distribution in surface water and sediment river around slum and industrial area (case study: ciwalengke River, Majalaya district, Indonesia). Chemosphere. 18791298. 224, 637–645. Available from: www.elsevier.com/locate/chemosphere. doi:10.1016/j.chemosphere.2019.02.188.

Amaneesh, C., Balan, S.A., Silpa, P.S., Kim, J.W., Greeshma, K., Mohan, A.A., Antony, A.R., Grossart, H.P., Kim, H.S., Ramanan, R., 2023. Gross Negligence: impacts of Microplastics and Plastic Leachates on Phytoplankton Community and Ecosystem Dynamics. Environ. Sci. Technol. 15205851. 57 (1), 5–24. Available from: http://pubs.acs.org/journal/esthag. doi:10.1021/acs.est.2c05817.

Athavuda, S., Weerasinghe, T., Pathirana, K., Dabare, P., Rathnayake, N., Samarakoon, T., Hemachandra, C.K., 2025. Occurrence and distribution of microplastics in agricultural lands in the Gampaha district of Sri Lanka: insights from selected paddy fields, vegetable plots, and coconut cultivations. Next Res. 30504759. 2 (1), 100101. doi:10.1016/j.nexres.2024.100101.

Besseling, E., Quik, J.T.K., Sun, M., Koelmans, A.A., 2017. Fate of nano- and microplastic in freshwater systems: a modeling study. Environ. Pollut. 18736424. 220, 540–548. Available from: https://www.journals.elsevier.com/environmental-pollution. doi:10.1016/j.envpol.2016.10.001.

Birch, Q.T., Potter, P.M., Pinto, P.X., Dionysiou, D.D., Al-Abed, S.R., 2020. Sources, transport, measurement and impact of nano and microplastics in urban watersheds. Rev. Environ. Sci. Biotechnol. 15729826. 19 (2), 275–336. Available from: http://www.kluweronline.com/issn/1569-1705. doi:10.1007/s11157-020-09529-x.

Blair, R.M., Waldron, S., Phoenix, V., Gauchotte-Lindsay, C., 2017. Micro- and Nanoplastic Pollution of Freshwater and Wastewater Treatment Systems. Sci. Rev. 2213-7793. 5 (1-2), 19–30. doi:10.1007/s40362-017-0044-7.

Brits, M., van Velzen, M.J.M., Öykü Sefiloglu, F., Scibetta, L., Groenewoud, Q., Garcia-Vallejo, J.J., Vethaak, A.D., Brandsma, S.H., Lamoree, M.H., 2024. Quantitation of micro and nanoplastics in human blood by pyrolysis-gas chromatography–mass spectrometry. Microplastics Nanoplastics. 2662-4966. 4 (1). doi:10.1186/s43591-024-00090-w.

Chae, Y., Kim, D., Kim, S.W., An, Y.J., 2018. Trophic transfer and individual impact of nano-sized polystyrene in a four-species freshwater food chain. Sci. Rep. 20452322. 8 (1). Available from: www.nature.com/srep/index.html. doi:10.1038/s41598-017-18849-y.

Chang, X., Xue, Y., Li, J., Zou, L., Tang, M., 2020. Potential health impact of environmental micro- and nanoplastics pollution. J. Appl. Toxicol. 10991263. 40 (1), 4–15. Available from: http://onlinelibrary.wiley.com/journal/10.1002/(ISSN)1099-1263. doi:10.1002/jat.3915.

da Costa, J.P., 2018. Micro- and nanoplastics in the environment: research and policymaking. Curr. Opin. Environ. Sci. Health. 24685844. 1, 12–16. Available from: www.journals.elsevier.com/current-opinion-in-environmental-science-and-health. doi:10.1016/j.coesh.2017.11.002.

Diana, Z.T., Chen, Y., Rochman, C.M., 2025. Paint: a ubiquitous yet disregarded piece of the microplastics puzzle. Environ. Toxicol. Chem. 15528618. 44 (1), 26–44. Available from: https://global.oup.com/academic/product/environmental-toxicology-and-chemistry-9780195117134?lang=en&cc=us#. doi:10.1093/etojnl/vgae034.

Ding, L., Mao, R.F., Guo, X., Yang, X., Zhang, Q., Yang, C., 2019. Microplastics in surface waters and sediments of the Wei River, in the northwest of China. Sci. Total Environ. 18791026. 667, 427–434. Available from: www.elsevier.com/locate/scitotenv. doi:10.1016/j.scitotenv.2019.02.332.

Drummond, J.D., Schneidewind, U., Li, A., Hoellein, T.J., Krause, S., Packman, A.I., 2022. Microplastic accumulation in riverbed sediment via hyporheic exchange from headwaters to mainstems. Sci. Adv. 23752548. 8 (2). Available from: https://www.science.org/doi/10.1126/sciadv.abi9305. doi:10.1126/sciadv.abi9305.

Efimova, I., Bagaeva, M., Bagaev, A., Kileso, A., Chubarenko, I.P., 2018. Secondary microplastics generation in the sea swash zone with coarse bottom sediments: laboratory experiments. Front. Mar. Sci. 22967745. 5. Available from: https://www.frontiersin.org/articles/10.3389/fmars.2018.00313/full. doi:10.3389/fmars.2018.00313.

Erni-Cassola, G., Dolf, R., Burkhardt-Holm, P., 2024. Microplastics in the Water Column of the Rhine River Near Basel: 22 Months of Sampling. Environ. Sci. Technol. 0013-936X. 58 (12), 5491–5499. doi:10.1021/acs.est.3c08364.

Fang, C., Zhou, W., Hu, J., Wu, C., Niu, J., Naidu, R., 2024. Paint has the potential to release microplastics, nanoplastics, inorganic nanoparticles, and hybrid materials. Environ. Sci. Eur. 21904715. 36 (1). Available from: https://enveurope.springeropen.com. doi:10.1186/s12302-024-00844-6.

Fang, P., Shuzi, Y., Luo, Z., Guo, R., Jiang, Y., Liu, L., Siwen, L., Xiao, F., 2025. Nanoplastics under the charge effects: unveiling the potential threats to amphibian (Rana nigromaculata) growth, intestinal damage and microbial ecology. Environ. Pollut. 02697491. 381, 126566. doi:10.1016/j.envpol.2025.126566.

Funke, E., Webb, L., Wolinska, J., 2024. The effect of microplastics on Daphnia fitness – systematic review and meta-analysis. Freshw. Biol. 13652427. 69 (4), 461–476. Available from: http://onlinelibrary.wiley.com/journal/10.1111/(ISSN)1365-2427. doi:10.1111/fwb.14228.

Gao, S., Orlowski, N., Bopf, F.K., Breuer, L., 2024. A review on microplastics in major European rivers. Wiley Interdiscip. Rev. Water. 20491948. 11 (3). Available from: http://wires.wiley.com/WileyCDA/WiresJournal/wisId-WAT2.html. doi:10.1002/wat2.1713.

Garcia, A.G., Suárez, D.C., Li, J., Rotchell, J.M., 2021. A comparison of microplastic contamination in freshwater fish from natural and farmed sources. Environ. Sci. Pollut. Res. 16147499. 28 (12), 14488–14497. Available from: https://link.springer.com/journal/11356. doi:10.1007/s11356-020-11605-2.

Gerolin, C.R., Pupim, F.N., Sawakuchi, A.O., Grohmann, C.H., Labuto, G., Semensatto, D., 2020. Microplastics in sediments from Amazon Rivers, Brazil. Sci. Total Environ. 18791026. 749. Available from: www.elsevier.com/locate/scitotenv. doi:10.1016/j.scitotenv.2020.141604.

Gigault, J., Pedrono, B., Maxit, B., Halle, A.T., 2016. Marine plastic litter: the unanalyzed nano-fraction. Environ. Sci. Nano. 2051-8153. 3 (2), 346–350. doi:10.1039/C6EN00008H.

Gupta, D.K., Choudhary, D., Vishwakarma, A., Mudgal, M., Srivastava, A.K., Singh, A., 2023. Microplastics in freshwater environment: occurrence, analysis, impact, control measures and challenges. Int. J. Environ. Sci. Technol. 17352630. 20 (6), 6865–6896. Available from: https://www.springer.com/journal/13762. doi:10.1007/s13762-022-04139-2.

He, B., Goonetilleke, A., Ayoko, G.A., Rintoul, L., 2020. Abundance, distribution patterns, and identification of microplastics in Brisbane River sediments, Australia. Sci. Total Environ. 18791026. 700. Available from: www.elsevier.com/locate/scitotenv. doi:10.1016/j.scitotenv.2019.134467.

He, L., Zhongwen, O., Fan, J., Zeng, B., Guan, W., 2022. Research on the non-point source pollution of microplastics. Front. Chem. 2296-2646. 10. doi:10.3389/fchem.2022.956547.

Horie, Y., Mitsunaga, K., Yamaji, K., Hirokawa, S., Uaciquete, D., Ríos, J.M., Yap, C.K., Okamura, H., 2024. Variability in microplastic color preference and intake among selected marine and freshwater fish and crustaceans. Discover Oceans. 2948-1562. 1 (1). doi:10.1007/s44289-024-00005-w.

Hu, K., Yang, Y., Zuo, J., Tian, W., Wang, Y., Duan, X., Wang, S., 2022. Emerging microplastics in the environment: properties, distributions, and impacts. Chemosphere. 00456535. 297, 134118. doi:10.1016/j.chemosphere.2022.134118.

Iskandarani, S., Manjunath, S., Busquets, R., Raeli, L., Saikaly, P.E., Campos, L.C., 2025. Characterization of nanoplastics and small-sized microplastics in sewage treatment. Sci. Rep. 20452322. 15 (1). Available from: https://www.nature.com/srep/. doi:10.1038/s41598-025-15504-9.

Jambeck, J.R., Geyer, R., Wilcox, C., Siegler, T.R., Perryman, M., Andrady, A., Narayan, R., Law, K.L., 2015. Plastic waste inputs from land into the ocean. Science. 10959203. 347 (6223), 768–771. Available from: http://www.sciencemag.org/content/347/6223/768.full.pdf. doi:10.1126/science.1260352.

Jayaweera, M., Bastians, S., Dissanayake, J., Hemachandra, C., Samarakoon, T., 2025. Microplastics in Weras Ganga, Bolgoda North Lake, Sri Lanka: a Reflection of Urban Waste Mismanagement. Results Eng. 25901230. 26. Available from: https://www.sciencedirect.com/science/journal/25901230. doi:10.1016/j.rineng.2025.105243.

Jiang, C., Yin, L., Zhiwei, L., Wen, X., Luo, X., Shuping, H., Yang, H., Long, Y., Deng, B., Huang, L., Liu, Y., 2019. Microplastic pollution in the rivers of the Tibet Plateau. Environ. Pollut. 02697491. 249, 91–98. doi:10.1016/j.envpol.2019.03.022.

Kieu-Le, T.C., Thuong, Q.T., Truong, T.N.S., Le, T.M.T., Tran, Q.V., Strady, E., 2023. Baseline concentration of microplastics in surface water and sediment of the northern branches of the Mekong River Delta, Vietnam. Mar. Pollut. Bull. 18793363. 187. Available from: www.elsevier.com/locate/marpolbul. doi:10.1016/j.marpolbul.2023.114605.

Kooi, M., Koelmans, A.A., 2019. Simplifying Microplastic via Continuous Probability Distributions for Size, Shape,and Density. Environ. Sci. Technol. Lett. 23288930. 6 (9), 551–557. Available from: http://pubs.acs.org/page/estlcu/about.html. doi:10.1021/acs.estlett.9b00379.

Li, C., Wang, L., Shuping, J., Chang, M., Wang, L., Gan, Y., Liu, J., 2021. The ecology of the plastisphere: microbial composition, function, assembly, and network in the freshwater and seawater ecosystems. Water Res. 00431354. 202, 117428. doi:10.1016/j.watres.2021.117428.

Li, J., Huang, W., Xu, Y., Jin, A., Zhang, D., Zhang, C., 2020. Microplastics in sediment cores as indicators of temporal trends in microplastic pollution in Andong salt marsh, Hangzhou Bay, China. Reg. Stud. Mar. Sci. 23524855. 35. Available from: http://www.journals.elsevier.com/regional-studies-in-marine-science/. doi:10.1016/j.rsma.2020.101149.

Liu, S., Chen, H., Wang, J., Lei, S., Wang, X., Zhu, J., Lan, W., 2021. The distribution of microplastics in water, sediment, and fish of the Dafeng River, a remote river in China. Ecotoxicol. Environ. Saf. 01476513. 228, 113009. doi:10.1016/j.ecoenv.2021.113009.

Ma, H., Pu, S., Liu, S., Bai, Y., Mandal, S., Xing, B., 2020. Microplastics in aquatic environments: toxicity to trigger ecological consequences. Environ. Pollut. 18736424. 261. Available from: https://www.journals.elsevier.com/environmental-pollution. doi:10.1016/j.envpol.2020.114089.

Mani, T., Burkhardt-Holm, P., 2020. Seasonal microplastics variation in nival and pluvial stretches of the Rhine River – from the Swiss catchment towards the North Sea. Sci. Total Environ. 707. 10.1016/j.scitotenv.2019.135579

Mateos-Cárdenas, A., Moroney, A.V.D.G., van Pelt, F.N.A.M., O'Halloran, J., Jansen, M.A.K., 2022. Trophic transfer of microplastics in a model freshwater microcosm; lack of a consumer avoidance response. Food Webs. 23522496. 31. Available from: https://www.journals.elsevier.com/food-webs. doi:10.1016/j.fooweb.2022.e00228.

Mateos-Cárdenas, A., van Pelt, F.N.A.M., O'Halloran, J., Jansen, M.A.K., 2021. Adsorption, uptake and toxicity of micro- and nanoplastics: effects on terrestrial plants and aquatic macrophytes. Environ. Pollut. 18736424. 284. Available from: https://www.journals.elsevier.com/environmental-pollution. doi:10.1016/j.envpol.2021.117183.

Mousazadehgavan, M., Khademi, S., Naeini, A.M., Yoosefdoost, I., Vashisht, V., Hashemi, M., Manouchehri, M., Hashim, K., 2024. Fate of micro- and nanoplastics in water bodies: A critical review of current challenges, the next generation of advanced treatment techniques and removal mechanisms with a special focus on stormwater. J. Water Process Eng. 67, 106159. doi:10.1016/j.jwpe.2024.106159.

Munz, M., Loui, C., Postler, D., Pittroff, M., Oswald, S.E., 2024. Transport and retention of micro-polystyrene in coarse riverbed sediments: effects of flow velocity, particle and sediment sizes. Microplastics Nanoplastics. 26624966. 4 (1). Available from: https://www.springer.com/journal/43591. doi:10.1186/s43591-023-00077-z.

Nam, S.H., Kim, S.A., Lee, T.Y., An, Y.J., 2023. Understanding hazardous concentrations of microplastics in fresh water using non-traditional toxicity data. J. Hazard. Mater. 18733336. 445. Available from: www.elsevier.com/locate/jhazmat. doi:10.1016/j.jhazmat.2022.130532.

Nava, V., Leoni, B., Arienzo, M.M., Hogan, Z.S., Gandolfi, I., Tatangelo, V., Carlson, E., Chea, S., Soum, S., Kozloski, R., Chandra, S., 2024. Plastic pollution affects ecosystem processes including community structure and functional traits in large rivers. Water Res. 18792448. 259. Available from: https://www.sciencedirect.com/science/journal/00431354. doi:10.1016/j.watres.2024.121849.

Nelms, S.E., Duncan, E.M., Patel, S., Badola, R., Bhola, S., Chakma, S., Chowdhury, G.W., Godley, B.J., Haque, A.B., Johnson, J.A., Khatoon, H., Kumar, S., Napper, I.E., Niloy, M.N.H., Akter, T., Badola, S., Dev, A., Rawat, S., Santillo, D., Sarker, S., Sharma, E., Koldewey, H., 2021. Riverine plastic pollution from fisheries: insights from the Ganges River system. Sci. Total Environ. 18791026. 756. Available from: www.elsevier.com/locate/scitotenv. doi:10.1016/j.scitotenv.2020.143305.

Okoye, C.O., Addey, C.I., Oderinde, O., Okoro, J.O., Uwamungu, J.Y., Ikechukwu, C.K., Okeke, E.S., Ejeromedoghene, O., Odii, E.C., 2022. Toxic Chemicals and Persistent Organic Pollutants Associated with Micro-and Nanoplastics Pollution. Chem. Eng. J. Adv. 26668211. 11. Available from: https://www.journals.elsevier.com/chemical-engineering-journal-advances. doi:10.1016/j.ceja.2022.100310.

Oladeji, R.D., Azeez, L., Agbaogun, B.K., Shakirat, M.L., Oladejo, A.A., Aremu, H.K., Busari, H.K., Awolola, O.S., Hammed, A., 2025. Seasonal dynamics of micro- and nanoplastics and associated pollutants in Aarin River: insights into abundance and digestion impact. Cleaner Water. 29502632. 3. Available from: https://www.sciencedirect.com/journal/cleaner-water. doi:10.1016/j.clwat.2024.100059.

Owowenu, E.K., Nnadozie, C.F., Akamagwuna, F., Noundou, X.S., Uku, J.E., Odume, O.N., 2023. A critical review of environmental factors influencing the transport dynamics of microplastics in riverine systems: implications for ecological studies. Aquat. Ecol. 15735125. 57 (2), 557–570. Available from: https://www.springer.com/journal/10452. doi:10.1007/s10452-023-10029-7.

Pastorino, P., Nocita, A., Ciccotelli, V., Zaccaroni, A., Anselmi, S., Giugliano, R., Tomasoni, M., Silvi, M., Menconi, V., Vivaldi, B., Pizzul, E., Renzi, M., Prearo, M., 2021. Health Risk Assessment of Potentially Toxic Elements, Persistence of NDL-PCB, PAHs, and Microplastics in the Translocated Edible Freshwater Sinotaia quadrata (Gasteropoda, Viviparidae): a Case Study from the Arno River Basin (Central Italy). Expo. Health. 24519685. 13 (4), 583–596. Available from: http://link.springer.com/journal/volumesAndIssues/12403. doi:10.1007/s12403-021-00404-w.

Pivokonský, M., Pivokonská, L., Novotná, K., Čermáková, L., Klimtová, M., 2020. Occurrence and fate of microplastics at two different drinking water treatment plants within a river catchment. Sci. Total Environ. 18791026. 741. Available from: www.elsevier.com/locate/scitotenv. doi:10.1016/j.scitotenv.2020.140236.

Rahman, M.M., Ferdouse, Z., Nur, N., Islam, M.N., Rouf, M.A., Arafat, S.T., Rahman, S.M., Rahman, M.M., 2022. Microplastic ingestion alters the expression of some sexually selected traits in a model fish guppy (Poecilia reticulata Peters 1859). Mar. Freshw. Behav. Physiol. 10290362. 55 (3-4), 87–106. Available from: www.tandf.co.uk/journals/titles/10236244.asp. doi:10.1080/10236244.2022.2100772.

Rai, P.K., Sonne, C., Brown, R.J.C., Younis, S.A., Kim, K.H., 2022. Adsorption of environmental contaminants on micro- and nano-scale plastic polymers and the influence of weathering processes on their adsorptive attributes. J. Hazard. Mater. 18733336. 427. Available from: www.elsevier.com/locate/jhazmat. doi:10.1016/j.jhazmat.2021.127903.

Rashed, A.H., Yesilay, G., Hazeem, L., Rashdan, S., AlMealla, R., Kilinc, Z., Ali, F., Abdulrasool, F., Kamel, A.H., 2023. Micro- and Nano-Plastics Contaminants in the Environment: sources, Fate, Toxicity, Detection, Remediation, and Sustainable Perspectives. Water. 20734441. 15 (20). Available from: http://www.mdpi.com/journal/water. doi:10.3390/w15203535.

Redondo-Hasselerharm, P.E., Gort, G., Peeters, E.T.H.M., Koelmans, A.A., 2020. Nano-and microplastics affect the composition of freshwater benthic communities in the long term. Available from: https://www.science.org.

Rout, P.R., Mohanty, A., Aastha, A.S., Miglani, M., Liu, D., Varjani, S., 2022. Micro- and nanoplastics removal mechanisms in wastewater treatment plants: a review. J. Hazard. Mater. Adv. 27724166. 6. Available from: https://www.journals.elsevier.com/journal-of-hazardous-materials-advances. doi:10.1016/j.hazadv.2022.100070.

Sarkar, D.J., Sarkar, S.D., Das, B.K., Manna, R.K., Behera, B.K., Samanta, S., 2019. Spatial distribution of meso and microplastics in the sediments of river Ganga at eastern India. Sci. Total Environ. 18791026. 694. Available from: www.elsevier.com/locate/scitotenv. doi:10.1016/j.scitotenv.2019.133712.

Sbarberi, R., Magni, S., Boggero, A., Torre, C.D., Nigro, L., Binelli, A., 2024. Comparison of plastic pollution between waters and sediments in four Po River tributaries (Northern Italy). Sci. Total Environ. 00489697. 912, 168884. doi:10.1016/j.scitotenv.2023.168884.

Scherer, C., Weber, A., Stock, F., Vurusic, S., Egerci, H., Kochleus, C., Arendt, N., Foeldi, C., Dierkes, G., Wagner, M., Brennholt, N., Reifferscheid, G., 2020. Comparative assessment of microplastics in water and sediment of a large European river. Sci. Total Environ. 00489697. 738, 139866. doi:10.1016/j.scitotenv.2020.139866.

Schneider, Stracke, Hansen, Schaefer, 2009. Nanoparticles and their interactions with the dermal barrier. Derm 1 (4), 197–206.

Sekudewicz, I., Dąbrowska, A.M., Syczewski, M.D., 2021. Microplastic pollution in surface water and sediments in the urban section of the Vistula River (Poland). Sci. Total Environ. 18791026. 762. Available from: www.elsevier.com/locate/scitotenv. doi:10.1016/j.scitotenv.2020.143111.

Singh, N., Mondal, A., Bagri, A., Tiwari, E., Khandelwal, N., Monikh, F.A., Darbha, G.K., 2021. Characteristics and spatial distribution of microplastics in the lower Ganga River water and sediment. Mar. Pollut. Bull. 18793363. 163. Available from: www.elsevier.com/locate/marpolbul. doi:10.1016/j.marpolbul.2020.111960.

Sjerps, R.M.A., ter Laak, T.L., Zwolsman, G.J.J.G., 2017. Projected impact of climate change and chemical emissions on the water quality of the European rivers Rhine and Meuse: a drinking water perspective. Sci. Total Environ. 18791026. 601-602, 1682–1694. Available from: www.elsevier.com/locate/scitotenv. doi:10.1016/j.scitotenv.2017.05.250.

Stratmann, C.N., Dris, R., Gasperi, J., Buschman, F.A., Markus, A.A., Guerin, S., Vethaak, A.D., Tassin, B., 2024. Monitoring microplastics in the Seine River in the Greater Paris area. Front. Earth Sci. 22966463. 12. Available from: https://www.frontiersin.org/journals/earth-science. doi:10.3389/feart.2024.1386547.

Strokal, M., Vriend, P., Bak, M.P., Kroeze, C., van Wijnen, J., van Emmerik, T., 2023. River export of macro- and microplastics to seas by sources worldwide. Nat. Commun. 20411723. 14 (1). Available from: https://www.nature.com/ncomms/. doi:10.1038/s41467-023-40501-9.

Strungaru, S.A., Jijie, R., Nicoara, M., Plavan, G., Faggio, C., 2019. Micro- (nano) plastics in freshwater ecosystems: abundance, toxicological impact and quantification methodology. TrAC Trends Anal. Chem. 18793142. 110, 116–128. Available from: www.elsevier.com/locate/trac. doi:10.1016/j.trac.2018.10.025.

Thamsenanupap, P., Tanee, T., Kaewsuk, J., 2022. Evidence of microplastics in the Chi River Basin, Thailand: anthropogenic influence and potential threats to edible arthropods. Limnologica. 00759511. 97. Available from: http://www.urbanfischer.de/journals/limno/limno.htm. doi:10.1016/j.limno.2022.126030.

Turner, A., Ostle, C., Wootton, M., 2022. Occurrence and chemical characteristics of microplastic paint flakes in the North Atlantic Ocean. Sci. Total Environ. 18791026. 806. Available from: www.elsevier.com/locate/scitotenv. doi:10.1016/j.scitotenv.2021.150375.

Tursi, A., Baratta, M., Easton, T., Chatzisymeon, E., Chidichimo, F., De Biase, M., De Filpo, G., 2022. Microplastics in aquatic systems, a comprehensive review: origination, accumulation, impact, and removal technologies. RSC Adv. 2046-2069. 12 (44), 28318–28340. doi:10.1039/d2ra04713f.

van Weert, S., Redondo-Hasselerharm, P.E., Diepens, N.J., Koelmans, A.A., 2019. Effects of nanoplastics and microplastics on the growth of sediment-rooted macrophytes. Sci. Total Environ. 18791026. 654, 1040–1047. Available from: www.elsevier.com/locate/scitotenv. doi:10.1016/j.scitotenv.2018.11.183.

Vo, H.C., Pham, M.H., 2021. Ecotoxicological effects of microplastics on aquatic organisms: a review. Environ. Sci. Pollut. Res. 16147499. 28 (33), 44716–44725. Available from: https://link.springer.com/journal/11356. doi:10.1007/s11356-021-14982-4.

Walker, T.R., Fequet, L., 2023. Current trends of unsustainable plastic production and micro(nano) plastic pollution. TrAC Trends Anal. Chem. 18793142. 160. Available from: www.elsevier.com/locate/trac. doi:10.1016/j.trac.2023.116984.

Wang, Q., Yamada, Y., Enyoh, C.E., Wang, W., Sankoda, K., 2024. Characterization of Microplastics and Adsorbed/Dissolved Polycyclic Aromatic Hydrocarbons in the Biggest River System in Saitama and Tokyo, Japan. Nanomaterials. 20794991. 14 (12). Available from: http://www.mdpi.com/journal/nanomaterials. doi:10.3390/nano14121030.

Wontor, K., Olubusoye, B.S., Cizdziel, J.V., 2024. Microplastics in the Mississippi River System during Flash Drought Conditions. Environments. 20763298. 11 (7). Available from: www.mdpi.com/journal/environments. doi:10.3390/environments11070141.

Wu, C., Zhang, K., Xiong, X., 2018. Microplastic pollution in inland waters focusing on Asia. Handb. Environ. Chem. 58, 85–99. Available from: http://www.springer.com/series/698. doi:10.1007/978-3-319-61615-5_5.

Wu, J., Zhang, Y., Tang, Y., 2022. Fragmentation of microplastics in the drinking water treatment process - A case study in Yangtze River region, China. Sci. Total Environ. 00489697. 806 (Pt 1), 150545. doi:10.1016/j.scitotenv.2021.150545.

Xu, J.L., Lin, X., Wang, J.J., Gowen, A.A., 2022. A review of potential human health impacts of micro- and nanoplastics exposure. Sci. Total Environ. 18791026. 851. Available from: www.elsevier.com/locate/scitotenv. doi:10.1016/j.scitotenv.2022.158111.

Yan, M., Wang, L., Dai, Y., Sun, H., Liu, C., 2021. Behavior of Microplastics in Inland Waters: aggregation, Settlement, and Transport. Bull. Environ. Contam. Toxicol. 14320800. 107 (4), 700–709. Available from: https://link.springer.com/journal/128. doi:10.1007/s00128-020-03087-2.

Yan, Z., Liu, Y., Zhang, T., Zhang, F., Ren, H., Zhang, Y., 2022. Analysis of Microplastics in Human Feces Reveals a Correlation between Fecal Microplastics and Inflammatory Bowel Disease Status. Environ. Sci. Technol. 15205851. 56 (1), 414–421. Available from: http://pubs.acs.org/journal/esthag. doi:10.1021/acs.est.1c03924.

Yang, L., Qiao, F., Lei, K., Li, H., Kang, Y., Cui, S., An, L., 2019. Microfiber release from different fabrics during washing. Environ. Pollut. 18736424. 249, 136–143. Available from: https://www.journals.elsevier.com/environmental-pollution. doi:10.1016/j.envpol.2019.03.011.

Yu, F., Yang, C., Zhu, Z., Bai, X., Jie, M., 2019. Adsorption behavior of organic pollutants and metals on micro/nanoplastics in the aquatic environment. Sci. Total Environ. 00489697. 694, 133643. doi:10.1016/j.scitotenv.2019.133643.

Zhao, B., Richardson, R.E., You, F., 2024. Microplastics monitoring in freshwater systems: a review of global efforts, knowledge gaps, and research priorities. J. Hazard. Mater. 03043894. 477, 135329. doi:10.1016/j.jhazmat.2024.135329.

Chapter 6

Atmospheric wet deposition-related river pollution: Causes, effects, assessment, and remediation

Ishwar Chandra Yadav[a,b]
[a]*Department of Applied Science, SITASRM Engineering and Research Institute (SERI), Greater Noida, Uttar Pradesh, India*
[b]*Environmental Science, Non-Collegiate Women's Education Board, University of Delhi, New Delhi, India*

1 Introduction

A river is a large natural watercourse that flows into a lake, ocean, or other body of water. Usually, the river is fed by tributaries that meet along its course. In addition to providing water, rivers are essential to human civilization since all living things depend on them (Adeloye, 2009). Rivers provide the freshwater needed for drinking, transportation, agriculture, electricity generation, and recreation (Adeloye, 2009). The greater environment and human civilization depend on these rivers, and their pollution can have a number of detrimental effects on human, animal, and plant life (Wang and He, 2022). The riverine environment, which ultimately transports both organic and inorganic contaminants to the sea, is where they are concentrated in ways that impact human health, fauna, and vegetation. Despite the fact that there is no systematic global rating for these rivers, many rivers throughout the world are seriously polluted (Huang et al., 2008; Wang and He, 2022; Sinha and Loganathan, 2015). One of the primary causes of this is the concentration of the three primary sources of pollution—industrial, agricultural, and domestic—along the rivers. Historically, villages and industries have been located near rivers because they provide mobility and have been a convenient place to dispose of waste (Sinha and Loganathan, 2015).

One of the most important ways that air pollution affects rivers is through atmospheric wet deposition, which is the process by which pollutants are removed from the atmosphere by precipitation such as rain, snow, or fog (Majewski and Capel, 1995; Liu et al., 2020). Despite providing vital nutrients

River Pollution. https://doi.org/10.1016/B978-0-443-45024-2.10006-5

to river systems, atmospheric deposition poses a significant threat to aquatic ecosystems because it can result in eutrophication, acidification, and the accumulation of harmful materials (Greaver et al., 2012; Grennfelt and Hultberg, 1986). Atmospheric wet deposition is widely recognized as one of the primary techniques for eliminating both man-made and natural contaminants (Andreae and Rosenfeld, 2008; Guo et al., 2015; Pan and Wang, 2015). It is one of the main methods for removing human and natural contaminants from the atmosphere. A growing number of studies are utilizing atmospheric wet deposition to address the problem of environmental pollutants entering river ecosystems (Duce et al., 1991; Pan and Wang, 2012, 2015).

Although atmospheric deposition allows a range of vital nutrients to reach river ecosystems, the potential for acidification, eutrophication, and the buildup of toxic materials makes this process a serious danger to these ecosystems (Greaver et al., 2012; Grennfelt and Hultberg, 1986). In addition to harming the ecosystem, atmospheric pollution may have a major negative impact on the quality of rivers and other freshwater sources, endangering human health and the lives of people who rely on them (Lepori and Keck, 2012). Depending on their composition and amounts, chemical species can have either a positive or negative effect on aquatic environments. The potential toxicity, environmental persistence, interactions with other chemicals, bioaccumulation, and alterations that take place following their deposition into river water all affect the potential toxicity of deposited chemical species (Swackhamer et al., 2004). Nonetheless, the nutritional significance of a chemical species is determined by the conditions of nutrient limitation and the requirements of the river ecosystem (Thomas and Cebrian, 2008). Fig. 6.1 makes clear that pollutants stay

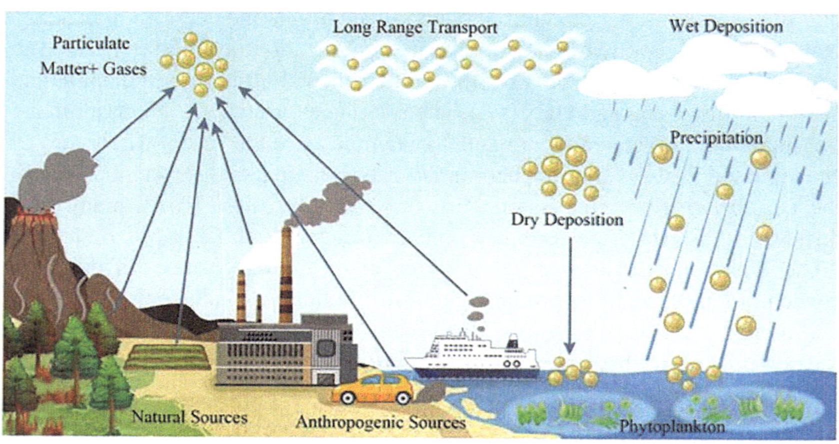

FIG. 6.1 Conceptual framework of the wet deposition process. *(Data from Sundarambal, P., Balasubramanian, R., Tkalich, P., He, J., 2010. Impact of biomass burning on ocean water quality in Southeast Asia through atmospheric deposition: field observations. Atmos. Chem. Phys. 10, 11323–11336. https://doi.org/10.5194/acp-10-11323-2010.)*

in the atmosphere after being released because of a number of conditions that might trap them and give them a lengthy atmospheric lifespan, ranging from hours to thousands of years. There are pollutants that are removed from the atmosphere and end up in river water, while some convert into new compounds and others stay in the atmosphere (Varney and Mccormac, 1971). Varney and Mccormac (1971) state that pollutants can migrate anywhere from a few yards to thousands of miles before they contribute to our river water pollution load. With an emphasis on the effects on rivers, this chapter examines the causes, consequences, and possible remediation techniques for pollution resulting from atmospheric wet deposition.

2 Types of atmospheric wet deposition

A wide variety of pollutants, both man-made and natural, can be present in wet deposition, including organic chemicals like pesticides and herbicides, as well as inorganic elements such as nitrogen, sulfur, basic cations, mercury, and various metals (Pacyna, 2008; Polkowska et al., 2011). The deposition of these pollutants can lead to a range of environmental consequences. For example, compounds containing nitrogen and sulfur may result in the leaching of base cations and soil nutrients, contribute to the eutrophication of estuaries and coastal waters, cause the decline of aquatic species, increase the acidity of freshwater bodies, and alter plant communities.

1. Acidic compounds: Acid rain, which lowers the pH of river water and damages aquatic life by making dangerous metals like mercury more soluble, can be caused by sulfuric acid (H_2SO_4) and nitric acid (HNO_3).
2. Nitrogen compounds: Wet deposition is the process by which nitrogen from the atmosphere is deposited in organic forms (dissolved organic nitrogen) and inorganic forms [ammonium (NH_4^+) and nitrate (NO_3^-)].
3. Heavy metals: Heavy metals (e.g., mercury, lead, cadmium) that are poisonous to river creatures can build up in the food chain and cause long-term ecological harm.
4. Persistent organic pollutants: Wet deposition may also carry chemicals, such as pesticides and industrial byproducts, which are known as persistent organic pollutants (POPs).

3 Source, emission, and deposition pathways

Wet deposition is one of the most important ways that air pollutants reach the surface of the Earth. Pollutants in the atmosphere are scavenged by clouds and/or liquid or solid precipitation until they are returned to the surface. The mechanisms for scavenging include in-cloud scavenging, which involves diffusive and inertial collisions between aerosol and precipitation particles; below-cloud scavenging by rain, snow, or hail particles as a result of gas absorption or inertial scavenging of aerosol particles larger than 1 μm; and nucleation, which is the process of pollutant scavenging from the air via water condensation on the surfaces of aerosol particles serving as cloud condensation nuclei (Atkinson, 1986; Boyce and Hoffman, 1984; Fowler, 1984; Dore et al., 1992; Hill et al., 1981a; Haiden, 1995). The location and mode of precipitation, surface condensation and direct deposition, weather and topography, the type and amount of precipitation or condensation, the interaction of aerosols, clouds, and gaseous pollutants, the presence of particles that act as cloud condensation nuclei, the concentration of a particular substance in the atmosphere, the solubility of a particular substance in the atmosphere, and the presence of substances that act as catalysts for the chemical transformations of pollutants absorbed in precipitation and/or condensation droplets are just a few of the numerous variables that affect wet deposition (Fowler, 1984). Fig. 6.2 lists the types of wet deposition that contribute to the transfer of pollutants from the atmosphere to the surface of the Earth.

Atmospheric wet deposition occurs when pollutants are removed from the atmosphere by precipitation. Numerous hazardous substances, including sulfur compounds, nitrogen oxides, particulate matter, and toxic heavy metals, are among these contaminants (Fernandez and Grimalt, 2003; Lintern et al., 2018). This phenomenon, where contaminants are brought to the Earth's surface by atmospheric moisture—be it rain, snow, or fog—is known as wet deposition. The process begins when air pollutants interact with water droplets, either dissolving into the water or adhering to the surfaces of particles (Pacyna, 2008). As the droplets or particles grow in size, they eventually fall to the ground, carrying the pollutants with them. Once the rainwater or snowmelt enters a river or watershed, these contaminants can enter the water directly through precipitation or indirectly via runoff. These pollutants are typically caused by human activities such as industrial emissions, transportation, and agriculture (Lepori and Keck, 2012; Vodyanitskii, 2013; de Vries et al., 2014, 2015; Jones et al., 2014; Duan et al., 2016; Rodríguez-Estival and Smits, 2016). Once these pollutants are on the surface of water bodies, they can alter the chemical balance of rivers, causing a series of negative effects on aquatic life, water quality, and the surrounding environment.

Rainout and washout are the two main mechanisms that contribute air pollutants to wet deposition. Rainout is a phenomenon that occurs in clouds, resulting in the development of water droplets on the pollutant through gas dissolution, condensation, or nucleation. Rainfall subsequently removes the contaminants from the atmosphere. The process by which precipitation scavenges airborne particles and aerosols between the ground and the cloud base is known as washout. Generally speaking, rainout is more effective than washout as a removal process. Since most storms generate convective air currents that contribute significant amounts of near-surface air to clouds above, these processes can occur continuously during wet deposition (Schroder and Hedley, 1986). Both natural and man-made contaminants can be found in atmospheric deposition.

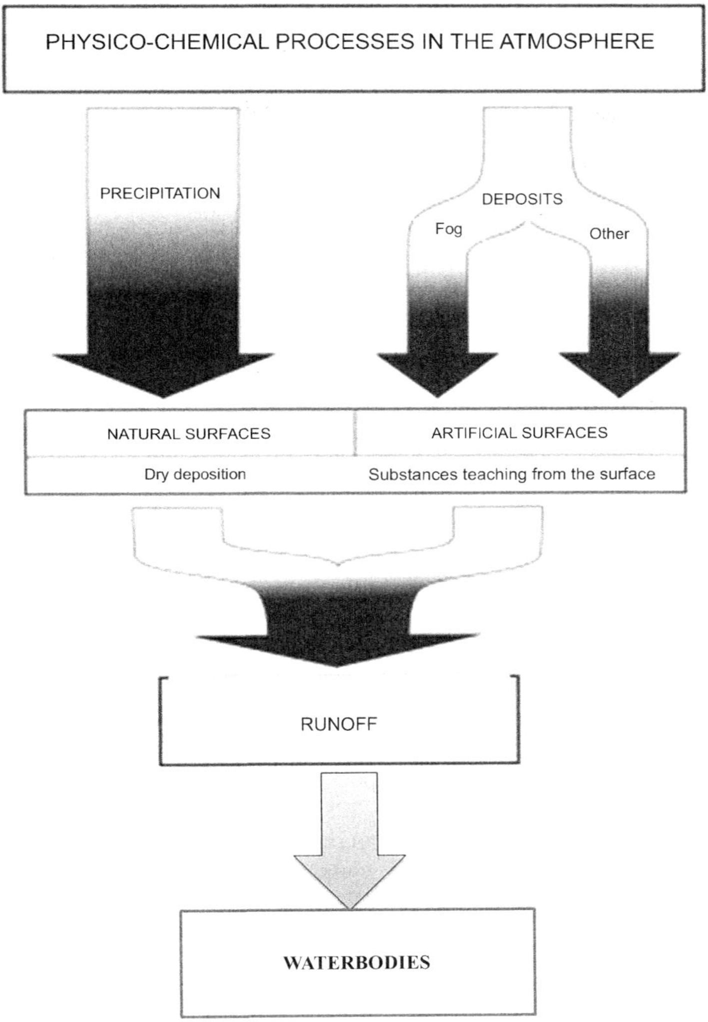

FIG. 6.2 Various kinds of wet deposition that contribute to the movement of pollutants from the atmosphere to the Earth's surface. *(Modified from Polkowska, Ż., Gorecki, T., Namieśnik, J., 2011. Determination of atmospheric pollutants in wet deposition. Environ. Rev. 19, 185–213.)*

3.1 Anthropogenic source

Anthropogenic (man-made) activities are the main causes of air pollution. Large amounts of sulfur oxides (SOx), nitrogen oxides (NOx), and particulate matter are released into the atmosphere when fossil fuels are used for transportation, energy generation, and industrial processes (United States Environmental Protection Agency, 2001). Before these contaminants are deposited by precipitation, they can travel great distances. Large amounts of sulfur dioxide (SO_2), nitrogen oxides (NO_x), and particulate matter (PM) are released when coal, oil, and gas are burned in power plants, factories, and automobiles. Winds in the atmosphere carry these emissions, which are then finally deposited into bodies of water by rain or snow. Nitrogen molecules like ammonia (NH_3) and nitrate (NO_3) are released when fertilizers and insecticides are used in agriculture. Livestock dung breaks down, and nitrogen fertilizers volatilize to create NH_3. When these chemicals undergo atmospheric reactions to form ammonium ions (NH_4^+) or nitric acid (HNO_3), they can reach the atmosphere and contribute to atmospheric deposition. Transportation generates nitrogen oxides (NO_x) and other volatile organic compounds (VOCs), especially when fossil fuels are used in automobiles. These emissions can lead to the development of acid rain and other types of wet deposition. Other significant sources that can release different gases and particulate matters (PMs) are landfills and other waste disposal facilities. Before ending up in neighboring rivers, these contaminants can travel great distances.

3.2 Natural sources

Man-made sources are the primary cause of air pollution, although natural sources can also contribute, albeit they are far less extensive (Pacyna, 2008). Volcanic eruptions emit sulfur compounds, while forest fires can discharge a range of pollutants, including PMs that contribute to air deposition. Forests, marshes, and seas naturally contain sulfur, nitrogen compounds, and particulate matter, all of which can contribute to air pollution. When vegetation burns, chemicals are released into the atmosphere, including carbon monoxide, nitrogen compounds, and ozone (Pacyna, 2008). The release of VOCs from forests or wetlands is one of the biological sources of air pollution. These pollutants may then end up in aquatic bodies as a result of wet deposition.

4 Impact of Atmospheric wet deposition (AWD) on river ecosystem

Wet deposition of atmospheric pollutants may have both measurable and incalculable ecological effects, including soil contamination, soil erosion, changes in terrain features, rock erosion, and contamination of surface and groundwater used as potable water sources (Tang et al., 2004; Nyssen et al., 2005; Tang and Davidson, 2004; D'eom et al., 2005). In addition, through exposure to and/or absorption of the toxins, it may affect the health of living things, such as humans, animals, and plants. The degradation of metals, concrete, glass, and other materials as a result of atmospheric corrosion or damage can have a substantial economic impact, as can wet deposition of air pollutants (Magaino, 1997). Runoff from wet deposition enters surface and groundwater, which are both sources of drinkable water.

Simultaneous atmospheric precipitation and fog water deposition cause substantial variations in total wet deposition, which, in turn, affects the kind and quantity of contaminants released into the environment. A few millimeters of precipitation per hour lower the water content of fog by 50%–75% and reduce the effectiveness of direct fog water deposition. The fog water content stays relatively consistent during the event (0.07–0.1 g/m^3), and fog may not entirely clear even after many hours of heavy precipitation. In the few minutes following the end of the precipitation, the amount of fog water may increase twofold to threefold, rapidly returning to preprecipitation levels (Błaś and Sobik, 2004). At or below freezing temperatures, there is a significantly lower chance of simultaneous precipitation and deposition occurrences.

The ecology of the basin and the river's water quality are greatly impacted by atmospheric moist deposition. In addition to affecting the overall amount of river water, wet deposition is an important source of nutrients, a solvent, and a transporter (Lu et al., 2017). Wet deposition is regarded as the main technique for eliminating air pollutants in regions affected by the monsoon (Akata and Asakura, 2011). River systems can experience several adverse effects when contaminants from atmospheric wet deposition enter them. These impacts may manifest in various ways, including loss of biodiversity, economic impacts, and ecological disruptions. The nature and extent of these effects are influenced by the types and quantities of contaminants present, along with the vulnerability of the river ecosystem. In addition to providing a direct impact on watersheds, wet deposition can also indirectly affect the migration of soil organic matter, leading to an increase in organic matter in water (Siddiqui et al., 2020). Atmospheric phosphorus deposition serves as a crucial pathway for phosphorus input into aquatic systems (He et al., 2007; Zhai et al., 2009), primarily occurring through windborne dust from sources such as pollen, sea spray, and resuspended dust (Pollman et al., 2002). The rise in nitrogen and phosphorus loads contributes to eutrophication, which subsequently encourages toxic algae proliferation and decreases biodiversity (Shi et al., 2021). Accumulation of heavy metals (such as cadmium, arsenic, lead, and mercury) can exceed ecological thresholds, posing risks to aquatic organisms and human health through bioaccumulation in food webs (Guo et al., 2025). Atmospheric wet deposition can influence rivers in multiple ways, which are included in the following.

4.1 Alteration of water chemistry

The most immediate effect of wet deposition is the alteration of river water chemistry. Pollutants such as nitric and sulfuric acids can reduce the pH of water, leading to acidification. Acidified waters can harm aquatic organisms by disrupting their biological functions, particularly in species with calcium carbonate structures like mollusks and crabs. The deposition of nitrogen compounds may cause eutrophication due to excessive nutrients and acidity.

As reported by Zhang et al. (2024), nitrogen deposition can lead to the acidification of lakes; specifically, nitrate nitrogen directly contributes to acidification, while ammonium nitrogen has an indirect effect by influencing nitrification and assimilation processes.

4.2 Effects on aquatic life

Wet deposition can affect aquatic organisms in rivers by introducing pollutants such as nitrogen and heavy metals, which can disrupt the food web and alter water quality, along with causing acidification from acid rain that can harm living organisms. Aquatic habitats can experience a cascading effect due to acidification and an excess of nutrients. Lower pH levels can interfere with the regular functioning of aquatic species and lead to harmful substances leaking from the soil into the water. The availability of essential minerals, such as calcium, important for the growth and development of aquatic species, may be reduced due to acidification. Moreover, increased nutrient concentrations, particularly nitrogen and phosphorus, can lead to rampant algae growth, which depletes oxygen levels in the water and creates "dead zones" where fish and other aquatic creatures cannot survive. River ecosystems may be affected by heavy metals like mercury due to wet deposition. These metals can transform into more toxic forms in water, accumulating in the food chain and potentially leading to neurological issues and reproductive problems in both humans and wildlife. In addition, microplastics can be transported by wet deposition, posing risks to aquatic organisms both chemically and physically. They can leach toxic substances into the water or be ingested, resulting in internal injuries.

4.3 Bioaccumulation and biomagnification

The entry of pollutants, such as pesticides and mercury, into a river via wet deposition leads to their absorption by living organisms (bioaccumulation), which then results in higher concentrations at elevated trophic levels (biomagnification). This chain of events contaminates the river ecosystem, affecting all organisms, from the smallest species to the largest predators. Aquatic creatures can acquire heavy metals and persistent organic contaminants, deposited by precipitation, in their tissues. These contaminants have the potential to accumulate in top predators, such as fish that people consume, as they move up the food chain over time. High quantities of pollutants can build up in top predators, causing serious health problems or even death. Toxin buildup can upset the ecosystem's equilibrium by influencing the well-being and reproductive ability of animals at all food web levels. Wildlife and human populations that depend on rivers for drinking water or fish for sustenance are seriously at risk from this bioaccumulation.

4.4 Decline in water quality and human health

Water quality can be severely deteriorated by pollutants that are introduced into rivers through wet deposition. River waters can become unsuitable for many aquatic creatures due to acid rain, which is caused by the deposition of sulfuric and nitric acids. Fish species could face extinction if pH levels drop below a critical threshold, and the reproductive cycles of other organisms may be disrupted. In addition, eutrophication, a phenomenon that depletes oxygen in aquatic environments and suffocates marine life, could be triggered by the accumulation of nitrogen compounds (Shi et al., 2021). A decline in the quality of river water also poses risks to human health. The accumulation of pollutants like heavy metals, including lead and mercury, can lead to various health issues. For example, the neurotoxic substance mercury can accumulate in fish, which serve as a staple food source for many communities. Consuming contaminated seafood can result in mercury poisoning, leading to neurological harm, particularly for vulnerable groups such as pregnant women and children.

4.5 Economic consequences

River pollution caused by atmospheric wet deposition has numerous adverse economic impacts. A decline in river water quality significantly influences industries such as agriculture, tourism, and fishing. Fish populations in rivers contaminated by fertilizers, heavy metals, or acid rain are decreasing. Communities that depend on fish as their primary protein source may face substantial economic setbacks and reduced food security due to the destruction of local fisheries. Rivers often serve as attractions for their natural beauty or for recreational activities like boating and fishing. As water quality worsens, tourist destinations are diminished, leading to a decrease in income related to tourism. Rivers are a vital source of water for agriculture. Pollutants that harm crops, such as heavy metals and acidic waters, can reduce agricultural output and put financial pressure on farmers. The food chain may become increasingly contaminated as tainted water enters irrigation systems.

4.6 Socioeconomic and policy challenges

The pollution of rivers through atmospheric moist deposition often leads to complex socioeconomic issues, both directly and indirectly. The detrimental effects of air pollution on the environment can significantly influence the local and regional economies reliant on clean water. In addition, alongside technological solutions, strong national and international policy measures are necessary to tackle these challenges. Addressing cross-border pollution requires global cooperation. Transboundary contamination is common, particularly in regions where countries share rivers or when industrial emissions cross national borders. This situation can lead to conflicts over water resources, especially when one country's pollution adversely impacts another nation's water quality.

5 Monitoring method of wet deposition to river system

Tracking air pollutants in wet deposition and assessing water quality are vital for effective environmental management, as runoff from wet deposition flows into surface and groundwater, which serve as drinking water sources. A variety of methods and tools are needed to monitor the air contaminants that cause wet deposition. It takes a lot of time and effort to monitor air pollution. It frequently requires sophisticated equipment and skilled workers. There are two main ways to monitor atmospheric wet deposition to river water: direct precipitation collection and indirect estimation using a watershed-based technique. The most thorough monitoring systems employ a variety of methods, collecting information on rainfall volume, frequency, and chemical makeup, and comparing it to the chemistry of river water.

5.1 Direct collection of precipitation

5.1.1 Wet only collector

To assess wet deposition, instruments are commonly positioned in open fields to gather rain or snow over periods ranging from several days to a month (Zhang et al., 2021). These automated samplers provide the highest accuracy for measuring wet deposition. To prevent contamination from dust and other particulates, they are equipped with a lid that slides shut over the collection bucket when it is not raining. When a moisture sensor detects precipitation, the lid opens to allow a clean sample to be collected. The formulas utilized to determine wet deposition fluxes are based on the concentrations of detected pollutants in the precipitation and the total amount of precipitation.

$$C_w = \sum_{i=1}^{n}(C_i P_i) / \sum_{i=1}^{n} Pi \tag{1}$$

where C_w is the volume-weighted mean concentration (mgL^{-1}) of pollutants computed from the n precipitation samples throughout a month or a year, and C_i is the concentration of each individual sample, weighted by the quantity of rainfall Pi for each sample.

$$D_w = P_t C_w / 100 \tag{2}$$

where D_w is the wet/bulk deposition flux (kg pollutants ha^{-1}), P_t is the total quantity of precipitation events (mm), and 100 is a unit conversion factor.

5.1.2 Bulk collector

For wet deposition in rivers, a bulk collector is a simple, open-container device that gathers both wet (rain, snow) and dry (dust, gas) atmospheric deposition into a single sample. It is a cost-effective method of monitoring pollution and air quality in large networks. These are more basic, open-topped containers that gather both dry (dust, gas) and wet (rain, snow) deposits. They can record the whole atmospheric intake, albeit they are less accurate when monitoring wet deposition alone. The most popular samplers for monitoring wet N deposition are rain gauges (or open buckets) of a certain size (Xu et al., 2015; Zhu et al., 2015). However, the fluxes produced by this approach are larger than realistic wet N deposition and should be characterized as bulk N deposition (wet-only deposition plus portions of dry-deposited gaseous and particulate N-compounds) (Dammgen et al., 2005). The ion-exchange resin approach is better than conventional methods (such as traditional rainwater collecting), which are labor-intensive and costly to execute on a wide scale, for bulk deposition or through fall deposition in the forest environment (Fenn and Poth, 2004).

5.1.3 Sampling

In order to get a sample for wet deposition analysis, automated or manual collection equipment is used to gather precipitation in a sterile container, following standard procedures to prevent cross-contamination. Samples should be gathered using event-based sampling for a full event or sequential sampling for high temporal resolution. The samples should then be moved to storage bottles, frozen if feasible, and brought to a laboratory for analysis. Keep in mind that some analytes may require particular sample management, including immediate transfer or dark storage. Samples are gathered over a certain time period (such as weekly or monthly) for both procedures (direct collection of precipitation and watershed approaches), and they are then submitted to a laboratory for chemical analysis. To test for contaminants, including heavy metals, nitrates, phosphates, and acidic chemicals, chemical sampling entails collecting water samples at several locations along the river. To monitor changes in water quality over time, this is often done on a regular basis.

5.1.4 Laboratory analysis

Samples that have been frozen are taken to a laboratory to have their chemical makeup examined. Ions, including ammonium (NH_4^+), nitrate (NO_3^-), sulfate (SO_4^{2-}), and trace metals, are among the chemical species whose concentrations are measured in laboratory analysis using methods like ion chromatography. Heavy metals, organic pollutants, nitrogen compounds, conductivity, pH, and other contaminants are measured using standard analytical methods. For some hydrometeors, such as fog and horizontal precipitation, it might be difficult to ensure representative and accurate sampling. Since the composition of wet deposition samples is strongly linked to air pollution, and because collection, sample preparation, and analysis may be made simpler in this situation, it may be an intriguing option. Analyzing condensation and precipitation in the atmosphere is still difficult. Table 6.1 summarizes the general issues found during the examination of wet deposition samples. Horizontal precipitation (fog) and surface condensation and (or) deposition are important components of the total atmospheric water balance, even though atmospheric precipitation is the primary source of water on the ground. The chemical makeup of fog is continuously monitored at many locations worldwide due to the increased interest in horizontal precipitation and its function in the deposition of air contaminants.

5.1.5 Satellite-based monitoring

Large-scale evaluations require integrating satellite observations and models with data from ground monitoring, but precisely reproducing scavenging mechanisms remains challenging. A more comprehensive picture of pollution levels across regions and even the world may be obtained through satellite-based monitoring, which can also be highly significant. Nitrogen dry and wet deposition may be estimated using the satellite remote sensing approach. Atmospheric remote sensing, such as the Infrared Atmospheric Sounding Interferometer for NH_3 (Van Damme et al., 2021) and the Ozone Monitoring Instrument for NO_2 (Miyazaki et al., 2017), can currently detect NH_3 and NO_2 in the atmosphere and has provided continuous monitoring of daily NH_3 and NO_2 at the global scale. As stated by Liu et al. (2017a), the amounts of NH_3 and NO_2 obtained through this technique represent the total molecular content (molec.cm^{-2}) of the atmospheric column concentration from the Earth's surface up to the tropopause. The wet deposition of inorganic nitrogen, as assessed through satellite remote sensing, can be expressed as the product of the erosion coefficient, precipitation, and the atmospheric concentration of inorganic nitrogen (Liu et al., 2017). Remote sensing is capable of monitoring atmospheric levels of NO_2 and NH_3 to ascertain the concentration of inorganic nitrogen, while precipitation data can be acquired from reanalysis or ground monitoring sources (Table 6.2).

5.1.6 Model-based monitoring

Hydrological models may also be used to forecast the movement and settling of contaminants in river systems, as they take into account variables, including topography, wind patterns, and precipitation trends. To anticipate the implications of atmospheric moist deposition in the future, these models must be employed. Various atmospheric chemistry transport models (ATMs), including the Goddard Earth Observing System (GEOS)-Chem model (Xu et al., 2018a; Zhao et al., 2017), the Weather Research and Forecasting model integrated with chemistry (WRF-Chem), the EMEP/MSC-W model (EMEP) (Chang et al.,

TABLE 6.1 Problems identified in the analysis of wet deposition samples.

Independent of the analytical procedure used	Problem related to analytical procedure	Problem related to data interpretation
Variety of pathways of pollutant transport from the atmosphere to other environmental compartments (precipitation and direct deposition)	Different types of samplers required	Changes in analyte concentrations during various stages of the deposition process (temporal variability)
Nonhomogeneity of the wet deposition stream (simultaneous occurrence of precipitation and direct deposition)	Small volumes of samples collected	Intercorrelations between sample components
Transformation of the deposition on contact with substances present on the surface or dry-deposited on the surface	Selection of appropriate analytical procedures	Effect of meteorological parameters on the amount and composition of precipitation
Seasonality of occurrence	Unavailability of reference materials and standards	Effect of anthropopression and terrain topography on the pollutants present in precipitation and direct deposition
Spatial variability of the amount of precipitation and direct deposition (local conditions)	Potential changes in chemical composition during sampling and/or sample storage (method, time)	Differences in the toxicity of various forms of wet deposition
	Great variety of pollutants present in precipitation and direct deposition samples	Need to use chemometric techniques during the processing of multivariate sets of measurement data
	Specific matrix composition	
	Low levels of the analytes	
	Interactions between sample components	

Data from Polkowska et al. (2011).

TABLE 6.2 Efficiency, advantages, and disadvantages of different physical/engineering-based treatment methods for river water.

Treatment techniques	Process description	Advantages	Disadvantage	References
Artificial aeration	Air flow into river water increases microbial diversity and degrades organic compounds in the water	Effectively improve water quality, simple and easy to apply, sustainable, and widely applicable	Cost-intensive during the operation and maintenance phase	Bai et al. (2020) and Zhang (2016)
Water transfer/ diversion	Mixing clean water with polluted river water and the dilution of pollution	Improve river water quality, enhance water supply, control river pollution, and promote the self-purification process	Potential destruction of the ecosystem, cost, and labor-intensive	Song et al. (2018) and Davies et al. (2010)
Mechanical algae removal	Removal of algae by mechanical process	Improve river water and sediment quality	Cost-intensive during the operation and maintenance phase	Middlebrooks et al. (1974)
Dredging river sediment	Removal of polluted sediment by a dredging machine	Improve sediment and river water environment	Potential increase in pollution, cost-intensive mechanical process	Kondratyev (2003) and Bai et al. (2020)
Building hydraulic structures	Irrigation weirs or infra-structure built on the river	Improve river water quality for irrigation purposes	Potential destruction of ecosystem health, cost-intensive	Ensink et al. (2010) and Radspinner et al. (2010)

2020), and the Community Multiscale Air Quality (CMAQ) model (Qiao et al., 2015), have been used to simulate nitrogen (N) deposition. Liu et al. (2001) describe the modeling of wet N deposition as comprising in-cloud processes, below-cloud processes, and precipitation scavenging, while the inferential technique calculates dry N deposition by multiplying the ambient concentrations of

N species and their deposition velocity (Vd). To accurately simulate the flow of atmospheric N deposition with ATMs, emission inventories of NH_3 and NOx, derived from either top-down or bottom-up methods, are generally required (Kang et al., 2016; Liu et al., 2016; Zhang et al., 2018). The quantitative results of N deposition are significantly influenced by the accuracy of these emission inventories.

5.1.7 Biological monitoring

In biological monitoring, the condition of aquatic organisms serves as a measure of water quality. This includes the examination of microbial populations, fish species, and invertebrate communities. The overall health of the river ecosystem can be reflected by the presence or absence of specific species, as well as changes in their population numbers or behaviors.

5.2 Indirect estimation via watershed-based approach

This indirect approach simulates the movement of pollutants throughout a river basin to assess the atmospheric influence on river water. It involves comparing the total chemical burden of the river with data obtained from air deposition monitoring. A watershed-based technique for monitoring wet deposition utilizes the entire drainage basin as a natural environment for gathering and assessing airborne pollutants. This method examines the total deposition of elements such as nitrogen and sulfur through the use of hydrological models and observations of streamflow, providing a comprehensive understanding of how air pollution impacts ecosystems.

5.2.1 River water sampling

The chemical makeup of river water is routinely assessed through sampling, which gauges the total concentrations of specific elements like phosphorus and nitrogen. This sampling is conducted by the long-term monitoring network of the US Environmental Protection Agency to assess the ecological health of ecosystems. The main challenge when sampling river water to measure wet deposition is distinguishing the signal of air pollution from the various other contaminants and natural substances already found in the river. Thus comparing data from precipitation samples with those from river water can help evaluate the impact of wet deposition on river ecology. To take spatial variations into account, water is gathered from multiple locations and depths in a large river and then combined into a single sample. For a sample to be representative, the US Geological Survey recommends using equal-width or equal-discharge increment techniques. By analyzing the concentrations of pollutants in both the precipitation and river water samples, researchers can determine the extent of the river's pollution burden that is due to wet deposition.

5.2.2 Watershed characteristics

River water quality is affected by wet deposition, which is the process by which contaminants are drawn from the atmosphere by rain, snow, or fog, depending on the features of the watershed. Land use, plant cover, soil type, and topography are all examined because they impact the retention or flushing of deposited air pollutants into the river. Water quality, erosion, and runoff are all greatly impacted by elements, including agriculture, forests, and urbanization. Vegetation density and type influence evapotranspiration, erosion prevention, and water absorption. The amount of water that soaks into the ground vs that flows off the surface depends on the type, depth, and infiltration capacity of the soil. Runoff and water flow direction are influenced by topography, which includes elevation, slope, and aspect.

5.2.3 Watershed modeling

Pollutant flow and transformation can be simulated and measured using a hydrological and biogeochemical model. Models can distinguish between atmospheric inputs and other sources, including runoff from cities or farms, by integrating data on measured atmospheric deposition. To simulate the flow of pollutants across a watershed, hydrological models like the Soil and Water Assessment Tool (SWAT) can be fed the modeled deposition data.

5.2.4 Wet deposition flux calculation

The quantity of a material deposited per unit area per unit time is known as the wet deposition flux. It is computed by multiplying the total amount of precipitation by the concentration of contaminants in the precipitation. The standard formula is:

$$W_{D_i} = C_i \times I \tag{3}$$

where I is the intensity of the rain, C_i is the concentration of a specific chemical species, and W_{D_i} is the flow of that particular chemical species in rainwater. This can be a total of monthly or event-based fluxes for yearly computations, where the monthly concentration (C_{ij}) is multiplied by the monthly precipitation (Pi) and added up across the months, frequently with molar mass modifications.

A passive MTX ARS 1010 automated deposition sampler (MTX Italia SPA, Modane, Italy) is used to gather the wet deposition flux of insoluble air particles. The collector consists of two buckets, each measuring 30 cm in diameter. An aluminum lid is activated by a humidity sensor to cover or reveal the buckets. Samples are taken either every week or following each precipitation occurrence. Calculating the wet deposition flux of soluble species involves examining the precipitation's chemical makeup. An automated wet-only precipitation collector, created especially for the International Network to Research Deposition and Atmospheric Composition in Africa, is used to collect rainwater samples (Yoboué et al., 2005; Akpo et al., 2015; Galy-Lacaux et al., 2009).

5.2.5 *Mass balance approach*

With this approach, scientists develop a budget of chemical inputs and outputs using a watershed. They can calculate the net effect of atmospheric deposition by summing up the amounts of a chemical that enter the watershed (via wet deposition) and those that exit it (through streamflow and other outflows). To calculate the net effect of atmospheric deposition, measurements are made of the volume and concentration of streamflow, as well as the chemical composition of incoming precipitation. This indicates a pollutant's discharge or retention within the environment. The pollutant is likely being held in the ecosystem by the soil or plants if, for instance, the amount of the chemical measured at the stream exit decreases in comparison with the atmospheric intake.

6 Remediation strategies of polluted river by wet deposition

Either pollution control at the source or in-situ water treatment can be used to rehabilitate contaminated river water. Although there are a variety of physical, chemical, biological, ecological, and engineering techniques that can be used to remediate polluted river water, sometimes a single technique is insufficient to purify highly contaminated river water (Gao et al., 2018; Bai et al., 2020). Therefore for effective treatment, hybrid techniques—which mix two or more single methods—are more commonly advised. Water transfer or diversion and dilution, mechanical algae removal, dredging river silt, hydraulic structure construction, and mechanical aeration are examples of engineering and physical treatment procedures. Employing these technical methods can effectively improve the quality of river sediment and water, resulting in the restoration of rivers. However, many of these methods carry significant initial and ongoing expenses, which may lead to adverse effects such as the destruction of natural ecosystems and financial strain (Ge et al., 2019). Thus to accelerate restoration efforts sustainably, these methods should be integrated with biological, chemical, and ecological processes.

A comprehensive approach that combines source reduction with physical, chemical, and biological techniques for river restoration is essential to address rivers impacted by atmospheric wet deposition. Often, treatments must target both the water column and the riverbed, since pollutants like heavy metals and POPs can build up in riverbed sediments. The complexity of river contamination and atmospheric wet deposition calls for a diverse cleanup approach. The aim of these strategies is to reduce pollution discharges, restore harmed river ecosystems, and prevent further degradation. The most effective remediation strategies for mitigating river pollution from deposition focus on managing river catchments, minimizing pollution sources, and employing advanced treatment technologies. The following remediation techniques may be implemented to protect the contaminated river.

6.1 Source emission control

One of the main strategies to limit pollutants entering the atmosphere in the first place is to reduce nitrogen, phosphorus, heavy metals, and organic pollutants from the air through tougher industrial regulations, cleaner agricultural practices (limited fertilizer use, manure management), and lower vehicle emissions. To stop more river pollution from rainfall, the most crucial step is to limit pollutants at their source. Strict environmental regulations for power plants and industrial facilities, such as the installation and upkeep of scrubbers and filters to absorb nitrogen oxides, sulfur dioxide, and particulate matter before they are released into the atmosphere, can aid in the management of the polluting river. Eco-friendly agricultural methods should be used to reduce ammonia volatilization from manure and fertilizers, which can cause nutrient pollution (eutrophication) and acid rain. To reduce pollutants from car emissions, catalytic converters and the shift to greener energy sources are being used.

6.2 Advanced water treatment technologies

The effective methods for treating polluted water include adsorption, ion exchange, electrocoagulation, and phytoremediation. Utilizing plants and microorganisms, biological techniques (bioremediation and phytoremediation) eliminate or immobilize contaminants, especially heavy metals and surplus nutrients. Algae and suspended solids (SS) can be eliminated from contaminated water through chemical treatment using flocculation, precipitation, oxidation, and algaecides. Although chemical methods may quickly clean up contaminated river water (Ge et al., 2019), they are only temporary and may result in secondary pollutants that pose further risks. Therefore the use of ecologically acceptable chemicals to destroy suspended particles and algae through chemical treatment should be the main focus of flocculation or precipitation operations. Poly aluminum chloride has been found by Wang et al. (2012) to be an excellent nonpolluting flocculation agent for removing algae from water. In hybrid remediation methods, pollutants are frequently removed from river water via adsorption on the surface of minerals or other materials.

Plant purification treatment, membrane bioreactor technology, microbial bioremediation, biofilms, contact oxidation, ecological ponds, ecological floating beds, and artificial wetlands are a few of the biological–ecological treatment technologies available in the literature (Bai et al., 2020). The characteristics of polluted river water include turbidity, odor, low transparency, and high levels of organic and inorganic pollutants, as well as chemical and biological oxygen demand (COD and BOD). To remove organic chemical pollutants, absorb metals (inorganic pollutants), and completely eliminate COD, BOD, odors, turbidity, and other organic and inorganic contaminants from river water, various studies have utilized microorganisms, plants, and aquatic animals to assist in bioremediation or biodegradation processes (Zheng and Wang, 2017; Cui et al., 2018; Ge et al., 2019).

Aquatic plants, including algae, hold significant promise for purifying contaminated water, wastewater discharges, and river systems (Anawar et al., 2008). By absorbing, adsorbing, accumulating, and decomposing pollutants, resilient plants positioned along riverbanks can effectively clean the water. These plants have shown a remarkable capacity to absorb nutrients such as total nitrogen (TN) and phosphorus from aquatic environments. Extensive and robust rhizosphere networks present in plant roots foster the proliferation of bacteria and various microorganisms, as well as the secretion of root exudates (Steenhoudt and Vanderleyden, 2000; Zimmels et al., 2008). These bacteria play a crucial role in eliminating both organic and inorganic pollutants from wastewater and contaminated water by breaking down organic compounds and accumulating nutrients and heavy metals. This phytoremediation technique can be implemented at sites where wastewater or stormwater is discharged or along the riverbanks. It is cost-effective and widely embraced by local communities, irrespective of the area. Commonly utilized aquatic plants for wastewater treatment include reed (*Phragmites australis*), water hyacinth (*Eichhornia crassipes*), alligator weed (*Alternanthera philoxeroides*) (Sato et al., 2008), water lettuce (*Pistia stratiotes*) (Zimmels et al., 2008), whorl-leaf watermilfoil (*Myriophyllum verticillatum*), pondweed (*Potamogeton* spp.), cattail (*Typha latifolia*), duckweed (*Lemna gibba*), and canna (*Canna indica*) (Allam et al., 2016).

6.3 Nature-based solution

To enhance the removal of pollutants while fostering biodiversity and the resilience of ecosystems, these approaches employ sustainable, ecosystem-based methods such as artificial wetlands, biofiltration systems, and the rehabilitation of natural hydrological processes. Restoring wetlands and establishing vegetative buffer zones along rivers contribute to capturing contaminants from runoff, thereby reducing the direct inflow into rivers. These natural mechanisms can absorb heavy metals and nutrients before they enter the water. As noted by Anawar et al. (2019), wetlands constructed with sediment-rooted plants exhibit significant potential for purifying polluted water in urban rivers and creeks. Constructed wetlands combine physical and biogeochemical processes to effectively eliminate water pollutants and restore the natural ecology of rivers.

As stated by Konnerup et al. (2009), these methods present several benefits over traditional water treatment techniques, including minimal operating expenses, straightforward maintenance, no secondary pollution, environmental and economic advantages, and effectiveness in rehabilitating river ecosystems and improving river water quality. However, they exhibit low hydraulic capacity, require significant space, and are sensitive to high levels of pollution (Bai et al., 2020). In addition, they can display signs of plant diseases and experience seasonal die-offs (Deng and Ni, 2013).

Aqua mats and artificial seaweed, which can promote the growth of microorganisms and circumvent the aforementioned problems, are thus recommended to be used in place of real plants (Xiao et al., 2010). In order to remediate urban river water, Bai et al. (2020) developed floating bed wetlands, horizontal subsurface flow constructed wetlands, and surface flow constructed wetlands in succession due to the poor removal rate of single constructed wetlands. These three wetland systems worked together to efficiently remove SS, TN, NH_4^+–N, COD, and total phosphorus from river water. Compared to traditional sewage treatment techniques, these multiple treatment systems were significantly less expensive. As a result, artificial wetlands are categorized into several types, including vertical flow, horizontal flow, subsurface flow, and surface flow wetlands. When organic contaminants come into contact with the bed matrix and biofilm in a subsurface flow wetland, they undergo aerobic degradation (Samal et al., 2018; Vymazal, 2007).

The biofilm reactors are made out of a biomembrane that is attached to a media substrate, such as a microcarrier or the riverbed. The biofilm reactor's microbial diversity and biomass growth are regulated by the discharge of wastewater and sediment. The formation of microbial communities in biofilm and the effectiveness of biofilm reactors with various biofillers in cleaning up contaminated river water were assessed by Xu et al. (2012). The efficacy of the treatment system in eliminating ammonia, nitrogen, and COD from water was exceptionally high. Through adsorption, degradation, and filtering processes, aeration or the addition of DO to the water aids biofilm approaches in eliminating both organic and inorganic impurities. The hydraulic loading rate, water flow velocity, temperature, and biofilm composition all affect how well pollutants are removed and how structurally stable biofilm-based systems are (Paulsen et al., 1997). Although many developed and developing nations have successfully implemented the technology (Wang et al., 2012), there are some disadvantages, including the possibility of system failure due to strong water flow and waves (Li and Chu, 2003) and the need for substantial construction work (Bai et al., 2020). Gravel contact oxidation, aerated biofilter biological fluidized bed, artificial packing contact oxidation, thin-layer flow method, underground stream purification method, and suspended carrier biofilm reactors are some of the various biofilm technologies used to remediate contaminated river water.

6.4 Monitoring, modeling, and policy integration

Keeping a constant watch on river water quality and atmospheric wet deposition is essential for assessing pollution levels and evaluating the effectiveness of restoration efforts. For effective solutions, it is important to implement integrated approaches that simultaneously tackle air and water pollution, focusing on source management, ecosystem restoration, and the involvement of the community. Advanced modeling tools, like the SWAT, can help identify high-risk areas and monitor the movement of pollutants. By combining strategies for managing water and air quality, comprehensive control of pollutants across different environmental media is achieved.

6.5 Community engagement and awareness

Technical initiatives aimed at ensuring long-term sustainability are complemented by community involvement through efforts to reduce waste, organize clean-up activities, promote sustainable farming practices, and provide education. Engaging local residents in river clean-up events and awareness campaigns helps cultivate a culture of environmental responsibility and discourages illegal dumping. To address pollutant inputs from atmospheric wet deposition and safeguard river ecosystems, the most effective remediation strategy is a multifaceted approach that integrates emission reduction, efficient catchment management, innovative and nature-based treatments, policy enforcement, and community participation.

7 Future challenge and research direction

Addressing the problem is not without its difficulties. For instance, policies and initiatives pertaining to water and air have historically been handled differently at all governmental levels. Atmospheric deposition is frequently overlooked in the institutional framework and media-specific initiatives of the majority of resource management bodies. Furthermore, unlike discharges that are dumped directly into a river, the sources of air pollution might be located far away, potentially in another state or even another country, or they may be situated adjacent to a water source.

8 Conclusion

Atmospheric wet deposition is a complex yet vital process that significantly influences water quality, ecological integrity, and public health by transferring pollutants from the atmosphere to river systems. A diverse array of chemical substances is introduced to aquatic environments through atmospheric deposition. The introduction of these chemicals alters the primary productivity of aquatic ecosystems and affects biogeochemical cycles. The influence of atmospheric wet deposition on river pollution is a multifaceted issue that requires input from diverse fields such as ecology, atmospheric science, and environmental policy. The direct input of pollutants from the atmosphere into waterways highlights the pressing need for comprehensive measures to address the long-term impacts on aquatic life, human health, and water quality. To combat pollution resulting from wet deposition, it is essential to reduce emissions at their source, monitor pollution levels, and apply remediation strategies in affected river systems. Protecting water resources for future generations calls for collaborative, interdisciplinary efforts focused on both river management and cutting down emissions. Recognizing the importance of sustainable practices in urban development, industry, and agriculture is vital to prevent further degradation of river ecosystems. By fostering international cooperation, advancing clean technology, and enforcing effective regulations, we can mitigate the adverse effects of air pollution on river systems and ensure that future generations benefit from clean, healthy rivers.

References

Adeloye, A.J., 2009. Rivers and human development. Fresh Surf. Water 3, 117.

Akata, M., Asakura, K., 2011. Atmospheric dry deposition of trace elements at a site on Asian-continent side of Japan. Atmos. Environ. 45 (5), 1075–1083.

Akpo, A.B., Galy-Lacaux, C., Laouali, D., Delon, C., Liousse, C., Adon, M., Gardrat, E., Mariscal, A., Darakpa, C., 2015. Precipitation chemistry and wet deposition in a remote wet savanna site in West Africa: djougou (Benin). Atmos. Environ. 115, 110–123.

Allam, A., Tawfik, A., El-Saadi, A., Negm, A.M., 2016. Potentials of using duckweed (Lemna gibba) for treatment of drainage water for reuse in irrigation purposes. Desalin. Water Treat. 57, 459–467.

Anawar, H.M., Ahmed, G., Strezov, V., 2019. Long-term performance and feasibility of using constructed wetlands for treatment of emerging and nanomaterial contaminants in municipal and industrial wastewater. In: Mishra, A.K., Anawar, H.M., Nadjib, D. (Eds.), Emerging and Nanomaterial Contaminants in Wastewater: Advanced Treatment Technologies, first ed. Elsevier, Cambridge, CA, USA. pp. 63–81.

Anawar, H.M., Garcia-Sanchez, A., TariKul Alam, M., Majibur Rahman, M., 2008. Phytofiltration of water polluted with arsenic and heavy metals. Int. J. Environ. Pollut. 33, 292–312.

Andreae, M.O., Rosenfeld, D., 2008. Aerosol-cloud-precipitation interactions. Part 1. The nature and sources of cloud-active aerosols. Earth Sci. Rev. 89, 13e41.

Atkinson, R. 1986. Kinetics and mechanisms of the gas-phase reactions of the hydroxyl radical with organic compounds under atmospheric conditions. Chem. Rev. 86 (1), 69–201. doi:10.1021/cr00071a004

Bai, X.Y., Zhu, X.F., Jiang, H.B., Wang, Z.Q., He, C.G., Sheng, L.X., Zhuang, J., 2020. Purification effect of sequential constructed wetland for the polluted water in urban river. Water 12, 1054.

Błaś, M., Sobik, M., 2004 The distribution of fog frequency in the Carpathians. Geographia Polonica. In: Kuchcik, M., Błażejczyk, K. (Eds.), General and Applied Climatology: Selected Aspects, vol 77. Polish Academy of Science, Institute of Geography and Spatial Organization, Warszawa, Poland, pp. 19–34.

Boyce, S.D., Hoffman, M.R., 1984. Kinetics and mechanism of the formation of hydroxymethane-sulfonic acid at low pH. J. Phys. Chem. 88 (20), 4740–4746. doi:10.1021/j150664a059.

Chang, M., Cao, J.C., Ma, M.R., Liu, Y.M., Liu, Y.Q., Chen, W.H., Fan, Q., Liao, W.H., Jia, S.G., Wang, X.M., 2020. Dry deposition of reactive nitrogen to different ecosystems across eastern China: a comparison of three community models. Sci. Total Environ. 720, 137548. doi:10.1016/j.scitotenv.2020.137548.

Cui, N.X., Chen, G.F., Liu, Y.Q., Zhou, L., Cai, M., Song, X.F., Zou, G.Y., 2018. Comparison of two different ecological floating bio-reactors for pollution control in hyper-eutrophic freshwater. Sci. Rep. 8, 14306.

D'eom, A., Gouyon, R., Berne, C., 2005. Rain erosion resistance characterizations: Link between on-ground experiments and in-flight specifications. Wear 258 (1–4), 545–551. doi:10.1016/j.wear.2004.09.053.

Dammgen, U., Erisman, J.W., Cape, J.N., Grünhage, L., Fowler, D., 2005. Practical considerations for addressing uncertainties in monitoring bulk deposition. Environ. Pollut. 134, 535–548. 10.1016/j.envpol.2004.08.013.

Davies, B.R., Thoms, M., Meador, M., 2010. An assessment of the ecological impacts of inter-basin water transfers, and their threats to river basin integrity and conservation. Aquat. Conserv. 2, 325–349.

de Vries, W., Du, E., Butterbach-Bahl, K., 2014. Short and long-term impacts of nitrogen deposition on carbon sequestration by forest ecosystems. Curr. Opin. Environ. Sustain. 9, 90–104.

de Vries, W., Hettelingh, J.P., Posch, M., 2015. Critical Loads and Dynamic Risk Assessments: Nitrogen, Acidity and Metals in Terrestrial and Aquatic Ecosystems. Springer, Dordrecht, Netherlands

Deng, Y., Ni, F.Q., 2013. Review of Ecological Floating Bed Restoration in Polluted Water. J. Water Resour. Prot. 5, 1203–1209.

Dore, A.J., Choularton, T.W., Fowler, D., 1992. An improved wet deposition map of the United Kingdom incorporating the seeder-feeder effect over mountainous terrain. Atmos. Environ. 26A, 1375–1381.

Duan, L., Yu, Q., Zhang, Q., Wang, Z., Pan, Y., Larssen, T., Tang, J., Mulder, J., 2016. Acid deposition in Asia: emissions, deposition, and ecosystem effects. Atmos. Environ. 146, 55–69.

Duce, R.A., Liss, P.S., Merrill, J.T., Atlas, E.L., Buat-Menard, P., Hicks, B.B., Miller, J.M., Prospero, J.M., Arimoto, R., Church, T.M., Ellis, W., Galloway, J.N., Hansen, L., Jickells, T.D., Knap, A.H., Reinhardt, K.H., Schneider, B., Soudine, A., Tokos, J.J., Tsunogai, S., Wollast, R., Zhou, M., 1991. The atmospheric input of trace species to the world ocean. Glob. Biogeochem. Cycles 5, 193-259.

Ensink, J.H.J., Scott, C.A., Brooker, S.J., 2010. Sewage disposal in the Musi-River, India: water quality remediation through irrigation infrastructure. Irrig. Drainage Syst. 24, 65–77.

Fenn, M.E., Poth, M.A., 2004. Monitoring nitrogen deposition in throughfall using ion exchange resin columns: a field test in the San Bernardino Mountains. J. Environ. Qual. 33, 2007–2014. doi:10.2134/jeq2004.2007.

Fernandez, P., Grimalt, J.O., 2003. On the global distribution of persistent organic pollutants. CHIMIA 57, 514. doi:https://doi.org/10.2533/000942903777679000.

Fowler, D., 1984. Transfer to terrestrial surfaces. Philos. Trans. R. Soc. Lond. B Biol. Sci. 30, 281–297.

Galy-Lacaux, C., Laouali, D., Descroix, L., Gobron, N., Liousse, C., 2009. Long-term precipitation chemistry and wet deposition in a remote dry savanna site in Africa (Niger). Atmos. Chem. Phys. 9, 1579–1595.

Gao, H., Xie, Y.B., Hashim, S., Khan, A.A., Wang, X.L., Xu, H.Y., 2018. Application of Microbial Technology Used in Bioremediation of Urban Polluted River: a Case Study of Chengnan River, China. Water 10, 643.

Ge, P.L., Chen, M., Zhang, L.C., Song, Y.J., Mo, M.H., Wang, L.Y., 2019. Study on water ecological restoration Technology of river. IOP Conf. Ser. Earth Environ. Sci. 371, 032025.

Greaver, T.L., Sullivan, T.J., Herrick, J.D., Barber, M.C., Baron, J.S., Cosby, B.J., Deerhake, M.E., Dennis, R.L., Dubois, J.J.B., Goodale, C.L., Herlihy, A.T., Lawrence, G.B., Liu, L., Lynch,

J.A., Novak, K.J., 2012. Ecological effects of nitrogen and sulfur air pollution in the US: what do we know? Front. Ecol. Environ. doi:10.1890/110049.

Grennfelt, P., Hultberg, H., 1986. Effects of nitrogen deposition on the acidification of terrestrial and aquatic ecosystems. Water Air Soil Pollut. 30, 945–963. doi:10.1007/BF00303359/METRICS.

Guo, J., Kang, S., Huang, J., Zhang, Q., Tripathee, L., Sillanpa€€a, M., 2015. Seasonal variations of trace elements in precipitation at the largest city in Tibet, Lhasa. Atmos. Res. 153, 87e97.

Guo, T., Li, M., He, S., Mo, Z., Kang, X., Pei, J., Liao, W., Chang, M., Wang, X., 2025. Characterization of atmospheric arsenic wet deposition transport pathways and potential sources areas in the Pearl River Delta region. J. Environ. Sci. 158, 372–385.

Haiden, T., 1995. Analytical solution of washout in stratiform clouds. Contrib. Atmos. Phys. 68, 3–14.

He, C.E., Liu, X., Fangmeier, A., Zhang, F., 2007. Quantifying the total airborne nitrogen input into agroecosystems in the North China Plain. Agriculture, Ecosystems & Environment 121 (4), 395–400.

Hill, F.F., Browning, K.A., and Bader, M.J. 1981a. Radar and rain-gauge observations of orographic rain over South Wales. Quart. J. Roy. Met. Soc., 107 (453), 643–670. doi:10.1256/smsqj.45311.

Huang, X., Sillanpaa, M., Duo, B., Gjessing, E.T., 2008. Water quality in the Tibetan Plateau: metal contents of four selected rivers. Environ. Pollut. 156, 270–277.

Jones, L., Provins, A., Holland, M., Mills, G., Hayes, F., Emmett, B., Hall, J., Sheppard, L., Smith, R., Sutton, M., 2014. A review and application of the evidence for nitrogen impacts on ecosystem services. Ecosyst. Serv. 7, 76–88.

Kang, Y., Liu, M., Song, Y., Huang, X., Yao, H., Cai, X., Zhang, H., Kang, L., Liu, X., Yan, X., He, H., Zhang, Q., Shao, M., Zhu, T., 2016. High-resolution ammonia emissions inventories in China from 1980 to 2012. Atmos. Chem. Phys. 16, 2043–2058. doi:10.5194/acp-16-2043-2016.

Kondratyev, S., 2003. A system for ecological and economic assessment of the use, preservation and restoration of urban water bodies: st Petersburg as a case study. Intern. Assoc. Hydrol. Sci 281, 327–333.

Konnerup, D., Koottatep, T., Brix, H., 2009. Treatment of domestic wastewater in tropical, subsurface flow constructed wetlands planted with Canna and Heliconia. Ecol. Eng. 35, 248–257.

Lepori, F., Keck, F., 2012. Effects of atmospheric nitrogen deposition on remote freshwater ecosystems. Ambio 41, 235–246.

Li, X.Y., Chu, H.P., 2003. Membrane bioreactor for the drinking water treatment of polluted surface water supplies. Water Res. 37, 4781–4791.

Lintern, A., Webb, J.A., Ryu, D., Liu, S., Bende-Michl, U., Waters, D., Leahy, P., Wilson, P., Western, A.W., 2018. Key factors influencing differences in stream water quality across space. WIREs Water 5, e1260. doi:10.1002/wat2.1260.

Liu, F., Zhang, Q., van der, A., Zheng, R.J., Tong, B., Yan, D., Zheng, L., He, K.B, Y.X., 2016. Recent reduction in NOx emissions over China: synthesis of satellite observations and emission inventories. Environ. Res. Lett. 11, 114002. doi:10.1088/1748-9326/11/11/114002.

Liu, H.Y., Jacob, D.J., Bey, I., Yantosca, R.M., 2001. Constraints from Pb-210 and Be-7 on wet deposition and transport in a global three-dimensional chemical tracer model driven by assimilated meteorological fields. J. Geophys. Res. -Atmos. 106, 12109–12128. doi:10.1088/1748-9326/11/11/114002.

Liu, L., Liu, Y., Wen, W., Liang, L., Ma, X., Jiao, J., Guo, K., 2020. Source identification of trace elements in PM2.5 at a rural site in the North China plain. Atmosphere (Basel). 11, 179. doi:10.3390/atmos11020179.

Liu, L., Zhang, X., Xu, W., Liu, X., Li, Y., Lu, X., Zhang, Y., Zhang, W., 2017a. Temporal characteristics of atmospheric ammonia and nitrogen dioxide over China based on emission data, satellite observations and atmospheric transport modelingsince 1980. Atmos. Chem. Phys. 17, 1–32. doi:10.5194/acp-17-9365-2017.

Lu, J., Zhang, S., Fang, J., Yan, H., Li, J., 2017. Nutrient fluxes in rainfall, throughfall, and stemflow in Pinus densata natural forest of Tibetan Plateau. CLEAN–Soil, Air, Water 45(7), 1600008.

Magaino, S. 1997. Corrosion rate of copper rotating-disk-electrode in simulated acid rain. Electrochim. Acta 4 (3), 377–382. doi:10. 1016/S0013-4686(96)00225-3.

Majewski, M.S., Capel, P.D., 1995. Pesticides in the Atmosphere, Distribution, Trends, and Governing factors—Pesticides in the Hydrologic System, Vol. 1. Ann Arbor Press Inc., Chelsea, Mich., pp. 3–21,77–152.

Middlebrooks, E.J., Porcella, D.B., Gearheart, R.A., Marshall, G.R., Reynolds, J.H., Grenney, W.J., 1974. Review Paper: Evaluation of Techniques for Algae Removal from Wastewater Stabilization Ponds. Available from: https://digitalcommons.usu.edu/cgi/viewcontent.cgi?referer=https://www.google.com.hk/&httpsredir=1&article=1202&context=water_rep (22 July 2019).

Miyazaki, K., Eskes, H., Sudo, K., Boersma, K.F., Bowman, K., Kanaya, Y., 2017. Decadal changes in global surface NOx emissions from multi-constituent satellite data assimilation. Atmos. Chem. Phys. 17, 807–837. doi:10.5194/acp-17- 807–837.

Nyssen, J., Vandenreyken, H., Poesen, J., Moeyersons, J., Deckers, J., Haile, M., Salles, C., Govers, G., 2005. Rainfall erosivity and variability in the northern Ethiopian Highlands. J. Hydrol. 311 (1–4), 172–187. doi:10.1016/j.jhydrol.2004.12.016.

Pacyna, J.M., 2008, Atmospheric deposition. Encycl. Ecol. 5, 275–285. doi:10.1016/B978-008045405-4.00258-5.

Pan, K., Wang, W.X., 2012. Trace metal contamination in estuarine and coastal environments in China. Sci. Total Environ. 421-422, 3-16.

Pan, Y.P., Wang, Y.S., 2015. Atmospheric wet and dry deposition of trace elements at 10 sites in Northern China. Atmos. Chem. Phys. 15, 951-972.

Paulsen, J.E., Oppen, E., Bakke, R., 1997. Biofilm morphology in porous media, a study with microsopic and image techniques. Water Sci. Technol. 36, 1–9.

Polkowska, Ż., Gorecki, T., Namieśnik, J., 2011. Determination of atmospheric pollutants in wet deposition. Environ. Rev. 19, 185–213.

Pollman, C.D., Landing, W.M., Perry, J.J., Fitzpatrick, T., 2002. .Wet deposition of phosphorus in Florida. Atmos. Environ. 36 (14), 2309–2318.

Qiao, X., Tang, Y., Hu, J.L., Zhang, S., Li, J.Y., Kota, S.H., Wu, L., Gao, H.L., Zhang, H.L., Ying, Q., 2015. Modelingdry and wet deposition of sulfate, nitrate, and ammoniumions in Jiuzhaigou National Nature Reserve, China using a source-oriented CMAQ model: Part I. Base case model results. Sci. Total Environ. 532, 831–839. doi:10.1016/j.scitotenv.2015.05.108.

Radspinner, R.R., Diplas, P., Lightbody, A.F., Sotiropoulos, F., 2010. River training and ecological enhancement potential using in-stream structures. J. Hydraul. Eng. 136, 967–980.

Rodríguez-Estival, J., Smits, J.E., 2016. as and in . Ecotoxicol. Environ. Saf. 124, 285–295. Small mammals as sentinels of oil sands related contaminants and health effects in northeastern Alberta, Canada

Samal, K., Dash, R.R., Bhunia, P., 2018. A comparative study of macrophytes influence on performance of hybrid vermifilter for wastewater treatment. J. Environ. Chem. Eng. 6, 4714–4726.

Sato, K., Sakui, H., Sakai, Y., Tanaka, S., 2008. Long-term experimental study of the aquatic plant system for polluted river water. Water Sci. Technol. 46, 217–224.

Schroder, L.J., Hedley, A.G., 1986. Variation in precipitation quality during a 40-hour snowstorm in an urban environment—Denver, Colorado. Int. J. Environ. Stud. 28, 131–138.

Shi, M., Geng, B., Zhao, T., Wang, F., 2021. Influence of atmospheric deposition on surface water quality and DBP formation potential as well as control technology of rainwater DBPs: a review. Environ. Sci. Water Res. Technol. 7 (12), 2156–2165.

Siddiqui, E., Pandey, J., Pandey, U., Mishra, V., V, S.A., 2020. Integrating atmospheric deposition-driven nutrients (N and P), microbial and biogeochemical processes in the watershed with carbon and nutrient export to the Ganga River. Biogeochemistry 147 (2), 149–178.

Sinha, R.K., Loganathan, B.G., 2015. Ganges river contamination issues: challenges and solutions. In: Ahuja, S., et al. (Eds.), Water Challenges and Solutions on the Global Scale. American Chemical Society Symposium, vol. 1206. pp.129–159.

Song, W.W., Xu, Q., Fu, X.Q., Zhang, P., Pang, Y., Song, D.H., 2018. Research on the Relationship between Water Diversion and Water Quality of Xuanwu Lake, China. Int. J. Environ. Res. Public Health 15, 1262.

Steenhoudt, O., Vanderleyden, J., 2000. Azospirillum, a free-living nitrogen-fixing bacterium closely associated with grasses: genetic, biochemical and ecological aspects. FEMS Microbiol. Rev. 24, 487–506.

Swackhamer, H., Paerl, S., Eisenreich, S., Hurley, J., Hornbuckle, K., McLachlan, ., Mount, M., Muir, D., Schindler, D., 2004. Impacts of atmospheric pollutant on aquatic ecosystems. Issues Ecol. 12.

Tang, W., Davidson, C.I., 2004. Erosion of limestone building surfaces caused by wind-driven rain: 2. Numerical modeling. Atmos. Environ. 38(33): 5601–5609. doi:10.1016/j.atmosenv.2004.06.014.

Tang, W., Davidson, C.I., Finger, S., Vance, K., 2004. Erosion of limestone building surfaces caused bywind-driven rain: 1. Field measurements. Atmos. Environ. 38 (33), 5589–5599. doi:10.1016/j.atmosenv.2004.06.030.

Thomas, S.A., Cebrian, J., 2008. Ecosystem patterns and processes. In: Encyclopedia of Ecology, Five-Volume Set. Elsevier Inc., pp. 1139–1148. doi:10.1016/B978-008045405-4.00308-6.

United States Environmental Protection Agency, 2001. Frequently Asked Questions about Atmospheric Deposition: A Handbook for Watershed Managers, USEPA, Washington, D.C..

Van Damme, M., Clarisse, L., Franco, B., Sutton, M.A., Erisman, J.W., Wichink Kruit, R., Van Zanten, M., Whitburn, S., Hadji-Lazaro, J., Hurtmans, D., Clerbaux, C., Coheur, P.F., 2021. Global, regional and national trends of atmospheric ammonia derived from a decadal (2008-2018) satellite record. Environ. Res. Lett. doi:10.1088/1748-9326/abd5e0.

Varney, R., Mccormac, B.M., 1971. Atmspheric Pollutants. In: Introduction to the Scientific Study of Atmospheric Pollution. Springer Netherlands, Dordrecht, pp. 8–52.

Vodyanitskii, Y.N., 2013. Contamination of soils with heavy metals and metalloids and its ecological hazard (analytic review). Eurasian Soil Sci. 46, 793–801.

Vymazal, J., 2007. Removal of nutrients in various types of constructed wetlands. Sci. Total Environ. 380, 48–65.

Wang, H., He, G., 2022. Rivers: linking nature, life, and civilization. River 1 (1), 25–36.

Wang, J., Liu, X.D., Lu, J., 2012. Urban River Pollution Control and Remediation. Procedia Environ. Sci. 13, 1856–1862.

Xiao, L.J., Ouyang, H., Li, H.M., Chen, M.R., Lin, Q.Q., Han, B.P., 2010. Enclosure study on phytoplankton response to stocking of silver carp (Hypophthalmichthys molitrix) in a eutrophic tropical reservoir in South China. Int. Rev. Hydrobiol. 95, 428–439.

Xu, W., Luo, X.S., Pan, Y.P., Zhang, L., Tang, A.H., Shen, J.L., Zhang, Y., Li, K.H., Wu, Q. H., Yang, D.W., Zhang, Y.Y., Xue, J., Li, W.Q., Li, Q.Q., Tang, L., Lu, S.H., Liang, T., Tong, Y.A., Liu, P., Zhang, Q., Xiong, Z.Q., Shi, X.J., Wu, L.H., Shi, W.Q., Tian, K., Zhong, X.H., Shi, K., Tang, Q.Y., Zhang, L.J., Huang, J.L., He, C.E., Kuang, F.H., Zhu, B., Liu, H., Jin, X., Xin, Y.J.,

Shi, X.K., Du, E.Z., Dore, A.J., Tang, S., Collett Jr. J.L., K., Goulding, Y.X., Sun, J., Ren, F.S., Zhang, X.J., Liu, ., 2015. Quantifying atmospheric nitrogen deposition through a nationwide monitoring network across China. Atmos. Chem. Phys. 15, 12345–12360.

Xu, W., Zhao, Y.H., Liu, X.J., Dore, A.J., Zhang, L., Liu, L., Cheng, M.M., 2018a. Atmospheric nitrogen deposition in the Yangtze River basin: Spatial pattern and source attribution. Environ. Pollut. 232, 546–555.

Xu, X.Y., Feng, L.J., Zhu, L., Xu, J., Ding, W., Qi, H.Y., 2012. Biofilm formation and microbial community analysis of the simulated river bioreactor for contaminated source water remediation. Environ. Sci. Pollut. Res. 19, 1584–1593.

Yoboué, V., Galy-Lacaux, C., Lacaux, J.P., Silué, S., 2005. Rainwater Chemistry and Wet Deposition over the Wet Savanna Ecosystem of Lamto (Côte d'Ivoire). J. Atmos. Chem. 52, 117–141.

Zhai, S., Yang, L., Hu, W., 2009. Observations of atmospheric nitrogen and phosphorus deposition during the period of algal bloom formation in Northern Lake Taihu, China. Environ. Manage. 44 (3), 542–551.

Zhang, L., Chen, Y., Zhao, Y., Henze, D.K., Zhu, L., Song, Y., Paulot, F., Liu, X., Pan, Y., Lin, Y., Huang, B., 2018. Agricultural ammonia emissions in China: reconciling bottom-up and top-down estimates. Atmos. Chem. Phys. 18, 339–355. 10.5194/acp-18-339-2018.

Zhang, L., 2016 A Study of Polluted River Remediation by Aeration. In: Qi, E. (Ed.), Proceedings of the 6th International Asia Conference on Industrial Engineering and Management Innovation. Atlantis Press, Paris, France, pp. 451–461.

Zhang, L., Wang, J., Wang, S., Wang, C., Yang, F., Li, T., 2024. Chemical characteristics of long-term acid rain and its impact on lake water chemistry: a case study in Southwest China. J. Environ. Sci. 138, 121–131.

Zhang, Q., Li, Y., Wang, M., Wang, K., Meng, F., Liu, L., Zhao, Y., Ma, L., Zhu, Q., Xu, W., Zhang, F., 2021. Atmospheric nitrogen deposition: a review of quantification methods and its spatial pattern derived from the global monitoring networks. Ecotoxicol. Environ. Saf. 216, 112180.

Zhao, Y., Zhang, L., Chen, Y.F., Liu, X.J., Xu, W., Pan, Y.P., Duan, L., 2017. Atmospheric nitrogen deposition to China: a model analysis on nitrogen budget and critical load exceedance. Atmos. Environ. 153, 32–40. doi:10.1016/j. atmosenv..01.018.

Zheng, L.G., Wang, H.P., 2017. Effect of combined ecological floating bed for eutrophic lake remediation. AIP Conf. Proc. 1839, 020058.

Zhu, J.X., He, N.P., Wang, Q.F., Yuan, G.F., Wen, D., Yu, G.R., Jia, Y.L., 2015. The composition, spatial patterns, and influencing factors of atmospheric wet nitrogen deposition in Chinese terrestrial ecosystems. Sci. Total Environ. 511, 777–785. doi:10.1016/j.scitotenv.2014.12.038.

Zimmels, Y., Kirzhner, F., Malkovskaja, A., 2008. Application and features of cascade aquatic plants system for sewage treatment. Ecol. Eng. 34, 147–161.

Further reading

Hu, B., Teng, Y.G., Zhai, Y.Z., Zuo, R., Li, J., Chen, H.Y., 2016. Riverbank filtration in China: a review and perspective. J. Hydrol. 541, 914–927.

Tyagi, S., Dobhal, R., Kimothi, P.C., Adlakha, L.K., Singh, P., Uniyal, D.P., 2013. Studies of river water quality using riverbank filtration in Uttarakhand, India. Water Qual. Expo. Health. 5, 139–148.

Chapter 7

Emerging Pollutants: Detection, assessment, and mitigation of emerging contaminants in rivers

Akriti Ashesh and Ningombam Linthoingambi Devi
Department of Environmental Science, Central University of South Bihar, Gaya, Bihar, India

1 Introduction

"Emerging contaminants (ECs)," as the name suggests, refer to the "contaminants of emerging concern." They came to light in the early 20th century along with advancements in new technologies (Khan et al., 2022). Simultaneously, the number of scientific publications related to ECs grew at a very fast rate over 20 years (Noguera-Oviedo and Aga, 2016). The realization of the ubiquitous nature of ECs emerged in 2002 when the US Geological Survey conducted a water quality survey and reported the presence of hormones and pharmaceuticals in many US streams (Kolpin et al., 2002). Although these pollutants have been present in the environment since their discovery, their emergence as pollutants was detected much later due to a lack of instrumental developments. ECs are alternatively referred to as endocrine-disrupting compounds (EDCs) because they can have significant impacts on the endocrine systems of wildlife and humans (Rodriguez-Narvaez et al., 2017). ECs constitute a wide group of chemical compounds, and their list is growing as new compounds are being identified. Researchers have added a number of chemicals to their studies, including artificial sweeteners (e.g., sucralose), fertilizers, pesticides, fire retardants, personal care products (PCPs), hormones, antibiotics, drugs, steroids, analgesics, contraceptives, sunscreens, fragrances, insect repellents, and other pharmaceutical products (Khan et al., 2022; Noguera-Oviedo and Aga, 2016; Rodriguez-Narvaez et al., 2017). Some of these synthetic products are essential in the day-to-day activities of our contemporary society. Their production across the world has risen from 1 to 5 million tons, which is equivalent to a 500-fold increase (Thomaidis et al., 2012). Research conducted in 72 countries across six continents found that 65% of the sites were contaminated with 14 antibiotics. For example, metronidazole, used to treat bacterial infections in the skin and mouth, was found at concentration levels

River Pollution. https://doi.org/10.1016/B978-0-443-45024-2.00005-1

300 times greater than safe levels at one site in Bangladesh (Evidence rising: the emerging pollutants poisoning our environment, 2019). A global study highlighted the concentration of antibiotics in the River Thames as 233 ng/L (Evidence rising: the emerging pollutants poisoning our environment, 2019). Other antibiotics, such as trimethoprim and ciprofloxacin, were also found at levels exceeding safe thresholds (Keely, 2019). These chemicals are linked to adverse health and environmental problems. The health problems could include immunological, neurological, reproductive, and developmental impacts (Houtman, 2010). Even a minute quantity of manmade ECs could cause similar responses in nontarget organisms for which they were originally and specifically synthesized (Houtman, 2010; Khan et al., 2022; Wilkinson et al., 2016). These ECs are often prevalent in sites located near waste dumps, sewage treatment plants, and effluent treatment plants. A wastewater expert from the UN Environment Program (UNEP), Birguy Lamizana, stated, "Municipal, industrial, and increasingly domestic, wastewater are primary sources of emerging pollutants in ecosystems" (Evidence rising: the emerging pollutants poisoning our environment, 2019). Agricultural fields, livestock, horticulture, poultry farming, and other sectors could be vulnerable to pollution from ECs if located near these sites. Lamizana also highlighted "the dearth of research related to possible toxicity and health impacts of EDCs present in domestic water supplies as well as wastewater" (Evidence rising: the emerging pollutants poisoning our environment, 2019). The evolution of scientific findings has made it clear that these invisible intruders could contribute to ecological imbalances, antimicrobial resistance, and unknown but significant health hazards. Their inclusion in routine monitoring and environmental regulations is extremely rare, particularly in developing countries in Asia and Africa. The menace of ECs has spread globally. Even developed nations in North America, South America, and Europe have reported concerning levels of ECs in aquatic environments. Henceforth, the cumulative aim of this chapter is to explore the categories, sources, environmental behavior, and health impacts of ECs in riverine ecosystems. In addition, this chapter includes advancements in analytical methods for EC detection, mitigation strategies, and their status in national and international regions.

2 Categories of ECs

As highlighted earlier, ECs are a huge part of our day-to-day activities. From applying sunscreen in the morning to consuming antibiotics at night, our day revolves around these ECs. Now that our technologies have advanced and we are at the peak of industrialization, ECs are successfully infiltrating every environmental matrix. The vast list of ECs is summarized into various categories (Fig. 7.1), which are described in detail here.

2.1 Pharmaceuticals

Pharmaceutical compounds are chemical substances, either natural or manmade, which have site-specific biological activity. These chemicals cannot be

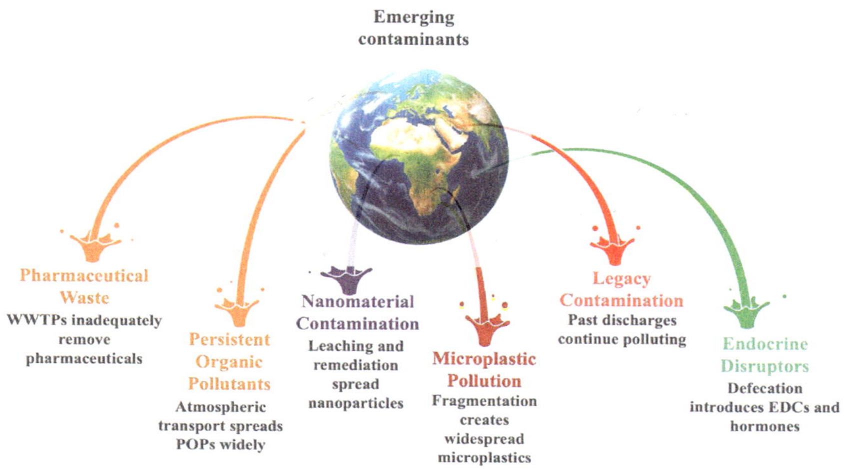

FIG. 7.1 Categories of emerging contaminants. Emerging pollutants are categorized into various subgroups, with the major ones presented in this diagram.

removed completely by sewage treatment plants. The drugs were not manufactured to pollute the environment; they are designed to perform a specific function. It took 10 years to identify diclofenac as the culprit for the dramatic loss in the vulture population, which died due to kidney failure after feeding on dead cows dosed with diclofenac (Mathew and Unnikrishnan, 2012). Subsequently, several drugs were identified in effluent samples collected from a Hyderabad suburb, Patancheru, by Swedish researchers in 2007. These two events together played an important role in ringing the warning bells about the hazards caused by the presence of pharmaceutical compounds in the environment. When these drugs enter the river, apart from their target environment, they tend to affect the degradation of complexes present in the river, kill beneficial microbes, and benefit the survival of pathogens. Drugs such as antibiotics, analgesics, antiepileptics, antiseptics, antihypertensives, contraceptives, heart drugs (β-blockers), psychotherapeutics, and hormones are reported as environmental pollutants (Shalini et al., 2010). The discharge from sewage treatment plants, sludge, sediments, soil, surface water, and ground water has been contaminated with over 100 drugs (Singare and Dhabarde, 2014). Humans/animals consume these pharmaceutical products, which undergo several transformations and are removed via excretion. These compounds are not completely metabolized inside human and animal bodies (Anjanapriya et al., 2021). After excretion, these products eventually make their way to surface waters via treatment plants and sewage water. Properties of pharmaceutical compounds, such as high polarity and low volatility, facilitate their transport to, from, and within the aquatic ecosystem. The Resource Conservation and Recovery Act lists pharmaceuticals, therapeutic agents, and other chemical products as P-listed waste, which is often categorized as acutely hazardous (Anjanapriya et al., 2021; Ashesh et al., 2022).

2.2 Pesticides

Aquatic organisms use the multixenobiotic resistance (MXR) mechanism against toxic compounds as a "first line of defense." This mechanism is mediated by the multidrug resistance (MDR) phenomenon, which is regulated by a group of proteins under the ATP-binding cassette (ABC) family. The most prominent ABC protein involved in MDR is P-glycoprotein (P-gp), which is present in the cell membrane and is responsible for the removal of xenobiotic compounds and drugs from cells (Smital et al., 2004). Pesticides such as ivermectin, thiodicarb, dacthal, sulfallate, pentachlorophenol, and chlorbenside have the potential to inhibit MXR mechanisms. Some other pesticides, including chlorpyrifos, dichlorvos, endosulfan, malathion, and pirimicarb, have shown high inhibition potential for MXR mechanisms, whereas diazinon, dimethoate, 2,4-D, fenoxycarb, furathiocarb, glyphosate, maleic hydrazide, methomyl, phosalone, and propiconazole exhibit moderate to weak inhibitory potential for MXR mechanisms (Smital et al., 2004). Pesticides from the family of persistent organic pollutants (POPs), particularly organochlorine pesticides (OCPs) such as hexachlorobenzene, are considered EPs (Morin-Crini et al., 2022). The contamination caused by POPs is not new; since they are persistent, they have been ubiquitous in the environment. However, their presence in compartments within the aquatic ecosystem has been widely studied over the past two decades (Ashesh et al., 2022). An estuary in Sydney, Australia, was found to contain one food additive (acesulfame), seven pesticide compounds, and eight pharmaceuticals. Pesticides such as 2,4-D, 3,4-dichloroaniline, carbaryl, mecoprop, diuron, and simazine were detected in the estuary water (Birch et al., 2015). Other pesticides, including metaldehyde, atrazine, and metolachlor, were detected in trace amounts in groundwater (Stuart et al., 2011).

2.3 Microplastics and nanomaterials

Plastic waste in the environment is not a recent addition. However, the occurrence of microplastics (<5 mm), microfibers, and nanomaterials (1–100 nm) in the aquatic ecosystem has raised global concern. These contaminants have the tendency to move up the trophic levels, eventually leading to health concerns for the human race. Microplastics gained popularity as a topic of concern when their presence began affecting us. They refer to plastic and fiber particles with a diameter of <5 mm or particles that can be retained when passed through a neuston net with a mesh size equivalent to 333 μm. Unlike microplastics, nanomaterials include more than one type of material. They refer to any natural or man-made material consisting of particles in an aggregate, unbound, or agglomerate state, where 50% or more of the particles have a diameter of <100 nm

(Hartl et al., 2015). Their miniature characteristics enable them to be freely ingested by many aquatic organisms such as fish, invertebrates, filter-feeding organisms, aquatic mammals, and, eventually, humans, leading to potentially fatal effects (Koigoora et al., 2019). Once these contaminants enter the aquatic ecosystem, they can move to other matrices through environmental transport from land to rivers to oceans; degradation via altered tensile strength, fragmentation of molecular structure, and thermal impact; and interactions with other materials via physical adsorption processes, chemical adsorption processes, and pore filling (Lambert and Wagner, 2018).

2.4 Endocrine-disrupting compounds

Though all the ECs are known as EDCs, some specific xenobiotics that have the potential to interfere with specific hormones are discussed here. Approximately 1000 compounds are categorized as EDCs. Many chemical compounds have the tendency to disrupt the hormonal system of our body. The EDCs act in a similar manner to natural hormones to alter metabolic activity at the cellular level. The challenge is to identify such compounds in the environment. At the beginning of the 1990s, the US-EPA developed a procedure called Toxicity Identification and Evaluation to assess the endocrine-disrupting effects of compounds present in complex environmental matrices (Petrovic et al., 2004). The priority EDCs in the riverine system are alkylphenols, phthalates, steroid sex hormones, and polybrominated diphenyl ethers (PBDEs). The presence and accumulation of EDCs in rivers pose significant impacts on wildlife and human health. Chronic diseases such as cancer, obesity, diabetes, and cardiovascular illnesses are some of the major concerns due to EDC exposure. Many aquatic animals are vulnerable to illnesses such as reproductive, behavioral, and developmental effects (Kasonga et al., 2021). EDCs are further classified into persistent or nonpersistent groups. Persistent EDCs are hydrophobic and lipophilic, with a tendency to bioaccumulate and bio-magnify in the food chain. Some compounds included in this group are chlorinated pesticides (such as DDE, heptachlor), perfluorooctane sulfonates (PFOS), metals (such as arsenic, lead), dioxins, polychlorinated biphenyls (PCBs), and PBDEs. On the contrary, nonpersistent groups are less lipophilic, have shorter half-lives, and can readily transform. These include parabens (methyl paraben, ethyl paraben, butyl paraben, and polyparaben), glyphosate, bisphenol A (BPA) and its phthalates, benzophenones, permethrins, and triclosans (Bukhari et al., 2025).

2.5 Personal care products

Personal care products (PCPs) consist of various daily-use items such as sunscreens, synthetic hormones, cosmetics, analgesics, insect repellents, toiletries, steroids, shampoos, preservatives, lipid regulators, and fragrances (Chopra and Kumar, 2018). PCPs are categorized into parabens and polycyclic musks, which prevent bacterial decay. Several industries use disinfectants such as triclosan, chloroprene, siloxanes, and benzophenones to manufacture PCPs such as toothpaste, soap, sportswear, lotions, sunscreens, air fresheners, hair products, and mouthwash (Patel et al., 2020). We encounter most of these products in our daily lives. PCPs are different from pharmaceuticals, as the latter are ingested, while most PCPs are directly applied to the body externally (Juliano and Magrini, 2017). PCPs are an important class of ECs because their discharge into the environment, particularly into aquatic systems, is continuous due to frequent showering, washing, and bathing. They are released in their unaltered form since they are applied externally and do not undergo metabolic degradation. The heavy and continuous discharge poses a significant threat to the ecological environment due to their high persistence, ability to bioaccumulate, and low volatility (Juliano and Magrini, 2017).

2.6 Food additives

A major food additive, acesulfame, is an artificial sweetener that is calorie-free and used to provide a long shelf life to food products. It has been recognized as an EC in the aquatic ecosystem in Sydney, Australia (Birch et al., 2015). Compounds such as phthalates and plasticizers are also used in food packaging (Morin-Crini et al., 2022). Per- and polyfluoroalkyl substances (PFAS) may be present in materials used for food packaging, processing equipment, or machines. PFAS exhibit properties such as resistance to degradation, bioaccumulation, and hazardous nature, which categorize them as hazardous ECs present in the environment (Eze et al., 2024). Nonstick utensils, fast food packaging, popcorn bags, and grease-resistant papers are made of PFAS, which make their way into rivers through run off from open dumping sites and intentional dumping of waste near aquatic bodies. The recent heavy floods in the tropics also introduced these contaminants into aquatic ecosystems. Bisphenols are endocrine-disrupting chemicals (EDCs) also used in food packaging, specifically to line metal cans for protection against corrosion (Trasande et al., 2018).

3 Sources of ECs

Aquatic ecosystems act as sinks for ECs (Fig. 7.2). Municipal wastewater treatment plants are not designed to treat pharmaceutical compounds. Conversely, they are designed to treat the most common and consistently occurring waste, consisting of microbes or biodegradable phosphorus, nitrogen, and carbon compounds (Gogoi et al., 2018). Thus pharmaceutical waste and PCPs, in their parent form or as their metabolites, enter rivers through waste showers or discharges from waste water treatment plants (WWTPs), effluent treatment plants (ETPs), etc. Unused or expired products and medicines are disposed of in sinks or flushed down toilets, leading to treatment plants and eventually to rivers (Tijani et al., 2013). These pollutants are detected in various riverine systems

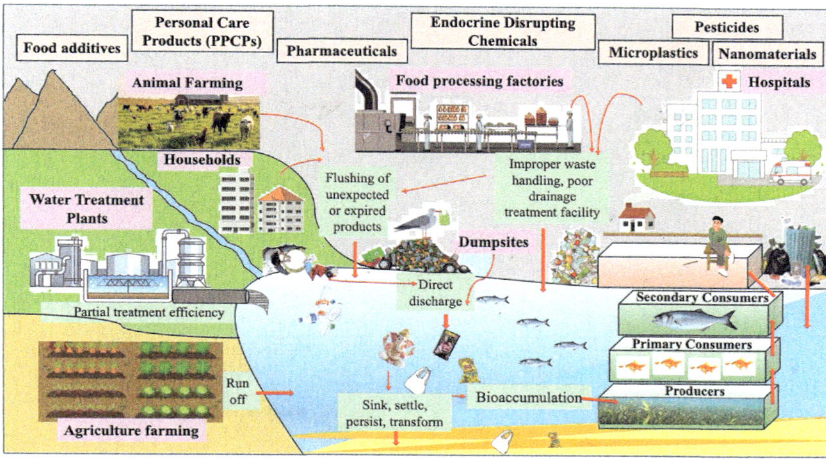

FIG. 7.2 Source and environmental behavior of ECs in riverine ecosystems. Various phenomena contribute ECs (such as food additives, personal care products (PCPs), pharmaceuticals, endocrine-disrupting chemicals, microplastics, pesticides, and nanomaterials) into the riverine ecosystem. To name a few, discharges coming from food processing factories, hospitals, households, waste water treatment plants, agricultural farming, and animal farming lead to the contamination of ECs in river bodies.

across the world (Fig. 7.3) and in India (Fig. 7.4). POPs, including OCPs, PCBs, polycyclic aromatic hydrocarbons (PAHs), and PBDEs, enter the atmosphere via primary and secondary sources. Primary sources include industries, urban localities, and agriculture, while secondary sources involve long-range atmospheric transport, even to areas where there is no trace of applications of such compounds (Morin-Crini et al., 2022). In rivers, these compounds can bioaccumulate, which is why they are detected in aquatic organisms (Yang et al., 2019; Singh et al., 2022). The hydrophobic properties of these contaminants cause them to settle in sediments and accumulate via electrostatic interactions. Other sources include discharge from animal feed, land treated with biosolids, runoff from roads and highways, accidental spills, recreational activities such as swimming and bathing, and traditional offerings to rivers (Tijani et al., 2013). In developing continents such as Asia and Africa, open dumping is very common among civilians. If not handled properly, all the open dumps eventually make their way into river systems, increasing their load and contamination levels.

In the aquatic ecosystem, the persistent or hydrophobic contaminants bioaccumulate, while the hydrophilic contaminants undergo dilution, photolysis, and biodegradation.

Nanoparticles can make their way into aquatic ecosystems through leaching from textiles, paint, and building materials treated with coatings such as stain or scratch-resistant materials, self-cleaning materials, and substances with antimicrobial properties. The remediation technologies nowadays use nanomaterials

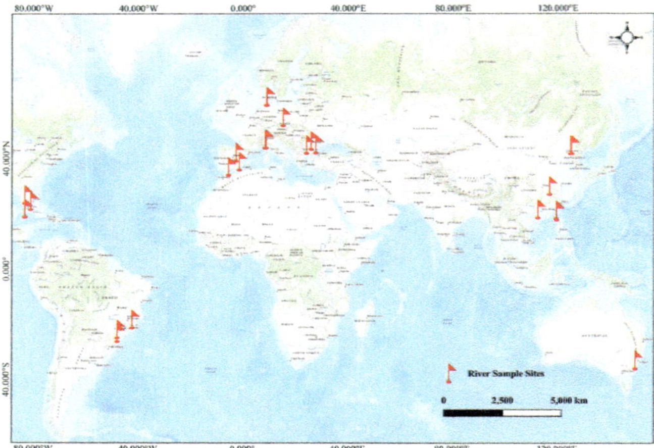

FIG. 7.3 World map representing river sample sites (red flags) studied for various emerging contaminants. The graph represents locations where studies were conducted on EC contamination in river bodies across the world.

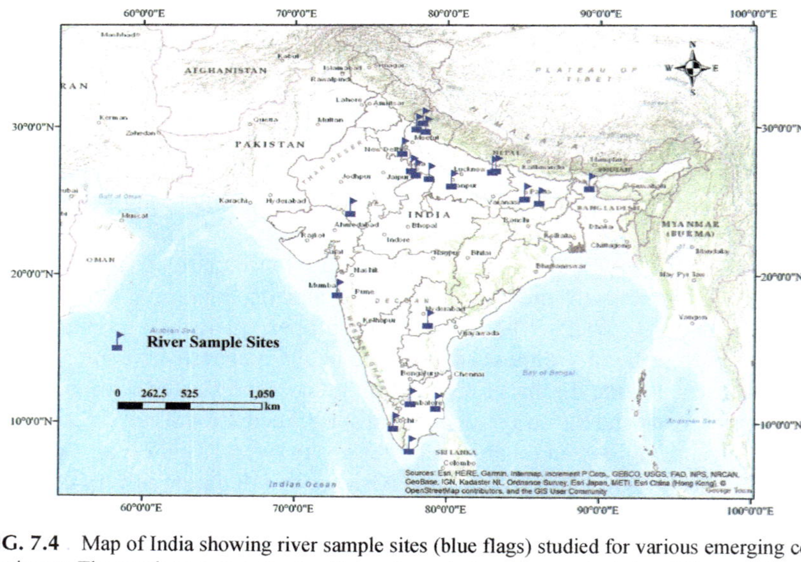

FIG. 7.4 Map of India showing river sample sites (blue flags) studied for various emerging contaminants. The graph represents the locations where studies were conducted on EC contamination in river bodies across India.

for adsorption or absorption phenomena. These nanomaterials could migrate from remediation processes to contamination in aquatic bodies. These materials include carbon, metallic, nonmetallic, and composite-based nanomaterials (Hartl et al., 2015). Microplastics are a product of plastic, which is an umbrella term covering a wide range of substances like nylon, polyethylene, polystyrene, polypropylene, and polyvinyl chloride (Koigoora et al., 2019). The primary sources of microplastics include products such as paints, cleaning materials, cosmetics, personal care items, and detergents. Secondary sources include the fragmentation of fishing nets, plastic bags, soda bottles, and large-sized plastics into smaller particles of size < 5mm (Stapleton and Hai, 2023) through friction, mechanical abrasion, thermal degradation, ultraviolet (UV) exposure, and photodegradation. The physical alteration of microplastics due to degradation mechanisms influences the rate of their sinking in river bodies. As the size of microplastics decreases, their bioaccumulation potential increases (Wagner et al., 2014). Even if we stop all discharges into rivers, the system would still be contaminated because of the heavy overload from earlier discharges. Some more specific sources of microplastics in river ecosystems include nurdle (plastic pellet) spills, textile washing, discharges containing tea bags, and crumbed rubber present in playgrounds. EDCs and steroid hormones can primarily make their way into rivers through animal and human defecation. Otherwise, these compounds reach the WWTPs and are then discharged into rivers. Unnikrishnan et al. (2024) reported the contamination of EDCs like steroids in the Periyar River in Kerala (Table 7.1). They noted the use of steroids in veterinary and human therapy and their subsequent direct or indirect discharge into the Periyar River. The contamination of ECs is a global issue. Some international river sites with contamination ranges are reported in Table 7.2.

4 Health impacts of ECs on riverine ecosystem

ECs have detrimental health impacts on the aquatic organisms present in riverine ecosystems. The circulation of ECs in food webs could pose potential threats to human health through exposure to river bodies.

4.1 Health impact on aquatic organisms

Some of the ECs, including BPA, phthalates, pesticides, synthetic contraceptives, and estrogens, could disrupt the normal functioning of hormones in aquatic ecosystems, which is a major cause of decreased fertility, immune suppression, reproductive abnormalities, genotoxicity, development of female traits, and low sperm counts in amphibians and fish (Patel et al., 2020). A controlled laboratory experiment with synthetic estrogens and fish observed their population extinction, concluding that the abnormal reproductive signals possibly resulted in population decline (Tijani et al., 2013). ECs exposure has been linked to an increased risk of thyroid, breast, and testicular cancers. Further accumulation of ECs in the food web, due to their fat-loving (lipophilic) nature, could amplify their impact on predators such as birds and mammals. In goldfish and trout, diclofenac (an analgesic) and carbamazepine (an antiepileptic) could lead to gill alterations, renal lesions, and oxidative stress (Gogoi et al., 2018). In vultures, diclofenac exposure resulted in visceral gout and renal failure, which

TABLE 7.1 Emerging contaminants study in the Indian rivers.

Locations	Rivers	Types	Contaminants	Concentration (ng/L)	References
Dehradun, Delhi, Mathura, Agra, Etawah	Yamuna River	PCPs and hormones	Caffeine, trimethoprim, ciprofloxacin, sulfamethoxazole, enrofloxacin, testosterone, progesterone, estrone, diclofenac, naproxen, ketoprofen, carbamazepine, atenolol, triclosan, gemfibrozil, 2-hydroxybenzothiazole	25.50–2187.50	Biswas and Vellanki (2021)
		Pharmaceuticals	Antibiotics, NSAID, analgesic, stimulant, antihistamine, antimicrobial, anticonvulsant, beta blocker, lipid regulator, plasticizer, vulcanization accelerators	1.15–7598.70	Biswas et al. (2024)
		Hormones	Estrone, 17β-estradiol, ethinyl estradiol, progesterone, testosterone	1.70–28,532.20	
Hyderabad	Musi River	Pharmaceuticals and PCPs	Antiinflammatory drugs, stimulants, anticonvulsants, xenobiotics, antifungal compounds	10.00–19,295.00	Konda et al. (2024)
Indo -gangetic region	Ganga River	ASWs	Acesulfame K, aspartame, cyclamate, saccharin, sucralose	0.20–102.00	Sharma et al. (2019)
		PCPs	diethyltoluamide (DEET), triclocarban, triclosan	54.70–826.00	
Kerala	Periyar River	Pharmaceutical, PCPs, and surfactants	16-hydroxyhexadecanoic acid, benzophenone, lignocaine, metoprolol, 2-dodecyl benzene sulfonic acid, mefenamic acid, β-estradiol, oxybenzone, and diisobutyl phthalate	4.30–1012.00	Khalid et al. (2018)
		Steroid hormones (EDCs)	Estradiol, estriol, and estrone	1.00–15.00	Unnikrishan et al. (2024)
		Beta blockers	Metoprolol, atenolol, timolol, and propranolol	0.30–25.00	
		Pharmaceutical and PCPs	Lidocaine, mefenamic acid, isoniazid, triamterene, caffeine	1.00–5600.00	

Maharashtra	Mithi River	PCPs	Triclosan	260.00–2960.00	Kumar et al. (2024)
Tamil Nadu	Kaveri River		Synthetic musks	1.00–198.00	Vimalkumar et al. (2021)
	Vellar River			1.00–1932.00	
	Thamiraparani River			2.00–107.00	
	Kaveri River (fish)			1.00–350.00 ng/g	
	Vellar River (fish)			1.00–8.00 ng/g	
Tamil Nadu	Kaveri, Tamiraparani, Vellar River	Pharmaceutical and PCPs	Triclosan, carbamazepine, parabens	2.67–13.00	Ramasamy et al. (2011)
Udaipur	Ahar River	Pharmaceuticals	Antibiotics, active pharmaceutical ingredients (APIs)	1.00–1887.00	Williams et al. (2019)
		EDCs	Estrone, 17β-estradiol	0.30–124.00	
		PCPs	Benzotriazole, bisphenol A, caffeine, DEET	1.00–37,476.00	

Continued

TABLE 7.1 Emerging contaminants study in the Indian rivers—cont'd

Locations	Rivers	Types	Contaminants	Concentration (ng/L)	References
Varanasi and Prayagraj	Ganga River	PAHs	Naphthalene, fluorene, phenanthrene, anthracene, fluoranthene, pyrene, chrysene, benzo(b)fluoranthene, benzo(k)fluoranthene, benzo(a)pyrene, dibenz(a,h)anthracene, benzo(ghi)perylene	45.10–1138.60	Satyanarayana et al. (2023)
		Pharmaceuticals	Aspirin, ibuprofen, nonylphenol, tramadol, fluoxetine, metoprolol, naproxen, diclofenac, pindolol, estrone, β-estradiol, 17α-ethinylestradiol, estriol.	29.00–4613.60	
		Pesticides	Propoxur, dichlorvos, phorate, α-HCH, β-HCH, γ-HCH, δ-HCH, phosphamidon, chlorpyrifos-methyl, parathion-methyl, aldrin, malathion, chlorpyrifos, pendimethalin, o,p'-DDD, α-endosulfan, butachlor, o,p'-DDE, p,p'-DDE, β-endosulfan, o,p'-DDT, p,p'-DDD, ethion, endosulfan sulfate, bifenthrin, fenpropathrin, λ-cyhalothrin, α-cypermethrin, fenvalerate, deltamethrin	35.80–600.90	
West Bengal	Torsa River	Pharmaceuticals	Triclosan and Triclocarban	41.00–184.00	Das et al. (2020)

TABLE 7.2 Emerging contaminants study in the international rivers.

Locations	Aquatic ecosystem	Types	Contaminants	Concentrations (ng/L)	References
Brazil	Doce River	Pharmaceuticals and pesticides	Acyclovir, amoxicillin, azithromycin, ciprofloxacin, enrofloxacin, fluoxetine, erythromycin, sulfadiazine, sulfamethoxazole, glyphosate, and aminomethylphosphonic acid	4600.00–14,500.00	Gomes et al. (2022)
		Pharmaceuticals	28 compounds	MDL-2060.00	Garrido et al. (2016)
		Surfactants	LAS C10, LAS C11, LAS C12, LAS C13, NP, NP1EO, NP2EO	85.00–2660.00	
		Plasticizers	DEHP, BPA	89.70-1661	
		PFAS	PFBuA, PFPeA, PFHxA, PFHpA, PFOA, PFOS	<MDL-13.10	
	Upper Iguassu Basin	PCPs	Triclosan	50.70–1070.60	Baudisch et al. (2022)
		Pharmaceuticals	Fenofibrate	40.20–1602.20	
			Ibuprofen	441.50–1466.30	
		Sterol	Ethinylestradiol	140.00–242.80	
	Piraí River	Pharmaceuticals	Caffeine (CFN), ibuprofen (IBP), naproxen (NPX), diclofenac (DCF), carbamazepine (CBZ), atenolol (ATL), propranolol (PPL), triclosan (TCS), estrone (E1), 17-β-estradiol (E2), and 17-α-ethinylestradiol (EE2)	11.1–213.6	de Sousa et al. (2014)
	Jundiaí River			993.9–19,329.7	
China	Pearl River Basin	CECs	Pharmaceuticals, antibiotics, flame retardants, chemical intermediates, and hormones	1.13–3.41 (Average)	Zhu et al. (2024)

Continued

TABLE 7.2 Emerging contaminants study in the international rivers—cont'd

Locations	Aquatic ecosystem	Types	Contaminants	Concentrations (ng/L)	References
Czech Republic	Oder River	PFAS	14 compounds	7.60–68.00	Zarębska et al. (2024)
Denmark	Aarhus River	Pharmaceuticals, fragrances, antiseptics, fire retardants, pesticides, and plasticizers.	17 compounds (diclofenac, 2-methyl-4-chlorophenoxyacetic acid (MCPA), caffeine, and tris(2-chloroethyl) phosphate (TCEP), etc.)	3.00–1475.00	Matamoros et al. (2012)
	Lyngbygaards River			4.00–382.00	
Florida	River	Pharmaceutical	Diclofenac	<MDL-11.40	Ng et al. (2021)
		EDCs	10 compounds	3.10–1980.00	
Iberia	The Ebro, the Llobregat, the Júcar, and the Guadalquivir	PhACs	73 compounds	40.00–3000.00	Castro-Catala et al. (2015)
		EDCs	39 compounds	50.00–2300.00	
		PFAS	16 compounds	Nd-800.00	
		Pesticides	50 compounds	0-600.00	
Italy	Lambro River basin	EDCs	39 compounds	1000.00–6000.00	Riva et al. (2019)
		Pharmaceuticals	Atenolol, carbamazepine, clarithromycin, diclofenac, ibuprofen, hydrochlorothiazide, naproxen, ofloxacin, salbutamol	0.20–226.70	
		PFAS	PFOS, PFOA	4.00–34.00	
		PCPs	PBSA, BP4, BP3, 4-MBC	0.60–262.90	
Northeast China	Jilin Songhua River	Pharmaceuticals, Environment estrogens, PCPs	22 compounds	Nd-37.30	He et al. (2018)
	Jilin Songhua River (Sediment)			Nd-745.90 (ng/g)	

Romania	Danube River	Pharmaceuticals	Sulfamethoxazole, erythromycin, carbamazepine, clopidogrel	Nd-41.40 (ng/g)	Bujagic et al. (2019)
		Pesticides	Carbendazim, imidacloprid, carbofuran, atrazine, propazine, linuron, malathion.	Nd-8.8 ng/g	
		Steroids	Coprostanol, cholestanol, cholesterol, cholestanone, desmosterol, stigmasterol, campesterol, β-sitosterol, sitostanol, mestranol	Nd-901.30 ng/g	
		PFAS	PFOS	Nd-1.10 ng/g	
Sydney, Australia	Estuary	Pesticides	7 compounds	1.70–55.00	Birch et al. (2015)
		Pharmaceutical	8 compounds	1.10–17.00	
		Food additives	Acesulfame	16.80	
Taiwan	Shijhong river	Pharmaceuticals	30 compounds	0.00–863.00	Jiang et al. (2015)
	Baoli River			0.00–3293.00	
	Linbian River			0.00–1623.00	

led to their extinction (Wilkinson et al., 2016). The bioaccumulation of perfluorinated compounds has caused gonadal and liver damage in fish.

In some fish and algae, β-blockers and lipid regulators reduce the viability of eggs along with growth (Gogoi et al., 2018). These compounds not only positively stimulate bodily functions, but also negatively impact our responses. For instance, triclocarban, also known as triclosan, is an antimicrobial compound that could also lead to growth retardation in algae. Parabens, originally used as preservatives, also exhibit factors leading to weakened estrogenic activities in mammals. Synthetic musks (like tonalide, 1,3,4,6,7,8-hexahydro-4,6,6,7,8,8-hexamethylcyclopenta[g]-2-benzopyran (HHCB), and galaxolide), fragrances, and PCBs affect the nervous system in animals and are often proven to be carcinogenic compounds (Patel et al., 2020). Some other ECs, along with their effects on aquatic organisms, are described in Table 7.3.

4.2 Health impact on humans

Health impacts of ECs on humans are a growing field of study. However, concerns have been raised regarding their adverse effects on the reproductive and endocrine systems. Low-semen count, miscarriage, growth retardation, birth defects, skin allergies, ovarian complications, respiratory disorders, congenital abnormalities, and cancer are all impacts of ECs on the reproductive systems of both male and female human bodies (Anjanapriya et al., 2021; Thomaidis et al., 2012). The excessive utilization of pharmaceuticals, steroids, and EDCs has reduced their effectiveness. Simultaneously, it has led to the spread and development of antibiotic-resistant bacteria, increasing risks and healthcare costs (Gogoi et al., 2018). According to a World Bank prediction, antibiotic-resistant bacteria are projected to kill 10 million people each year by 2050 and could even push 28 million individuals into extreme poverty (Patel et al., 2020). Human populations that are completely dependent on river water for domestic and drinking purposes could be more vulnerable to chronic exposure at nanogram levels. This often results in cumulative toxicological impacts on populations reliant on riverine systems. Chemicals like triclosan, used in PCPs, and parabens, used as preservatives in food and cosmetics, could lead to microbial resistance and reduced estrogenic activity (Patel et al., 2020). Long-term exposure to fragrances and synthetic musks could lead to diseases related to cancer, immunological disorders, and neurodevelopmental issues. Chopra and Kumar (2018) reported that synthetic musks and PCP compounds have been detected in urine, breast milk, and blood, which, over time, could result in chronic physiological effects. Certain ECs can cross the placenta during pregnancy, reaching the fetus and compromising its development, thereby causing transgenerational health impacts (Thomaidis et al., 2012; Wilkinson et al., 2016).

TABLE 7.3 Emerging contaminants, their impact on hormones, and affected organisms.

Compound	Usage/Source	Target hormone	Organisms affected
Antibiotics (penicillin, sulfonamides, tetracyclines, roxithromycin, clarithromycin, tylosin)	Pharmaceuticals, agriculture	Resistance induction, microbial community alteration; growth inhibition in algae	Bacteria, algae, higher food chain
Antihistamines (ranitidine)	Pharmaceuticals	Histamine pathways	Fish, mammals
Benzophenone, UV filters	PCPs (sunscreens)	Estrogen activity	Fish, mammals
Bisphenol A (BPA)	Plastics	Thyroxine, cortisol	Mammals, birds, fish
Butylated hydroxyanisole (BHA)	Food antioxidant	Estrogens	Mammals, Fish
Caffeine	PCP, pharmaceuticals	Endocrine disruption	Fish (goldfish)
Carbamazepine	Antiepileptic drug	Neurotransmission	Fish, mammals
Ciprofloxacin, sulfonamides	Antibiotics	Bacterial cell wall, resistance	Aquatic bacteria, higher trophic levels
DDT	Insecticide	Estrogen, androgen	Mammals, birds, fish, reptiles, amphibians
Diclofenac	NSAIDs (antiinflammatory drugs)	Prostaglandin metabolism	Fish, mammals
Estrone, 17β-estradiol, 17α-ethinylestradiol	Synthetic contraceptives, hormones	Estrogens (feminization, endocrine disruption)	Fish, mammals
Fragrances (musk, tonalide, galaxolide, HHCB)	PCPs, perfumes, plastics	Estrogens, neurotoxicity	Fish, mammals, rodents
Gemfibrozil	Pharmaceutical (lipid regulator)	Growth inhibition	Algae

Continued

TABLE 7.3 Emerging contaminants, their impact on hormones, and affected organisms—cont'd

Compound	Usage/Source	Target hormone	Organisms affected
Nonylphenol	Surfactants, plastics	Estrogens	Mammals, birds, fish, reptiles, amphibians
Parabens	Preservatives in cosmetics and foods	Weak estrogenic activity	Mammals, birds
PCBs	Industrial (electrical equipment, coolants)	Estrogen, thyroid	Mammals, birds, fish
Phthalates	Plastics, solvents	Estrogens	Mammals, birds, fish, reptiles, amphibians
Propranolol	Pharmaceutical (β-blocker)	Reduces viable eggs	Fish (Japanese medaka)
Triclosan, triclocarban	Antimicrobial agents, PCPs (toothpaste, soaps)	Various, disrupts lipid synthesis, acts toxic/biocidal	Algae, fish, microbial communities

4.3 Mechanism for health impacts

ECs, especially EDCs, could interfere with, mimic, or block the natural hormones present in the body and further bind to specific receptors, thereby altering developmental and metabolic processes. The ECs present in the environment could either persist or, through the process of degradation and fragmentation, transform into more active toxic metabolites. Degradation occurs either during natural processes such as photolysis or during the treatment of wastewater in WWTPs. ECs can also accumulate in edible crops via irrigation, posing cancer and hormonal risks to exposed populations. The ECs can enter the bodies of aquatic organisms or humans via ingestion, which can include the consumption of contaminated water and food prepared from crops irrigated with contaminated water bodies. Another route of entry is skin/dermal absorption, which could be facilitated by water contact during occupational activities or other traditional and recreational activities such as swimming and bathing (Thomaidis et al., 2012; Wilkinson et al., 2016).

5 Advancements in analytical methods for ECs' detection

Until now, we have come to know that EC comprises a wide range of compounds, which include PFAS, EDCs, microplastics, nanomaterials, PCPs, pharmaceuticals, and pesticides. These contaminants are evidenced in various environmental matrices, such as air, soil, water, and food, at trace levels ranging from ng/L to mg/L. The occurrence at such trace levels makes them challenging to detect. Most of the ECs were detected only after the first decade of the 21st century (Tables 7.1 and 7.2). Recent technological and instrumental advancements have made it possible to detect such compounds. This section deals with these advancements one by one:

5.1 Mass spectrometry

The evolution of mass spectrometry (MS) is significant. It facilitates precise detection and high throughput for known and unknown contaminants. Most studies have incorporated chromatography techniques for studying the presence of ECs in the environment. Gas chromatography (GC) can identify, classify, and analyze many chemical compounds. GC equipped with MS yields precise and accurate outputs. Extraction of EC compounds from environmental matrices is performed using techniques such as silylation, alkylation, and acylation (Bukhari et al., 2025). Advanced spectrometry techniques such as high-resolution mass spectrometry (HRMS) and ultra-high-performance liquid chromatography equipped with MS/MS facilitate faster separation, analysis of nontarget compounds, detection of polar contaminants, and enhanced ionization for better performance in trace-level quantification of PFAS and other ECs (Geletu et al., 2025) Another spectrometry technique is Fourier transform near-infrared spectroscopy, which detects pharmaceutical compounds in contaminated river systems. FTIR spectroscopy can also detect microplastics of sizes greater than 20 μm, and Raman spectroscopy can detect and identify microplastics of sizes <20 μm (Murugan et al., 2023).

5.2 Biosensors and nanotechnology

Some of the novel technologies investigate PCPs from wastewater, such as oxidation techniques and molecularly imprinted sensors. Researchers have also studied the utilization of algae and white-rot fungi-based technologies to extract pharmaceuticals and PCPs (Bertolini and Conti, 2022). Nanomaterials, such as carbon or graphene nanotubes, are used in the detection of antibiotics, achieving a low range of detection down to picograms per milliliter (pg/ml). The detection of ECs using metal-organic frameworks and quantum dots enables fluorescence and optical-based detection (Karn et al., 2025)

5.3 Enzyme-linked immunosorbent assay

ECs, such as proteins, hormones, antibodies, peptides, and particularly EDCs present in water bodies, could be detected using enzyme-linked immunosorbent assay (ELISA). A few properties of ELISA, such as cost-effectiveness, specificity, and sensitivity, enable the detection of EDCs even when present at low concentrations in the nanogram per liter (ng/L) range. ELISA is also cost-effective when compared with spectrometry techniques such as GC–MS and LC–MS. Several studies have used ELISA for the detection of steroids, estrogens, bisphenols, and other EDCs with >80% recovery (Kim et al., 2007; Wang et al., 2023).

5.4 Green analytical techniques

Since the whole world is trying to do its bit to attain sustainability goals, so is the technological field. Though the above-mentioned techniques are energy-driven, there is a way that has proven to be more efficient and greener. For instance, tools such as Analytical Greenness (AGREE) assess the eco-friendly potential of methods. Microextraction techniques have been proven useful for EC quantification, requiring less sample and solvent volume to carry out the extraction procedure. The technique takes less time than conventional technologies, paving its way toward a greener solution for the extraction of contaminants from target food and water matrices (Acevedo, 2024). A portable solution has also been developed using a device named lab-on-a-chip. These kinds of devices integrate extraction, separation, and quantification steps into one chip, which lowers the environmental impact and could be the future for the quantification of ECs at the field level (Sihlahla and Mngadi, 2024).

6 Mitigation strategies for EC pollution in riverine ecosystem

Rivers act as conveyors for ECs from agricultural, industrial, and urban sectors, as well as discharges from wastewater treatment plants. Efficient mitigation strategies require a multisystem approach that integrates source control, modern treatment methods, and bioremediation techniques at contaminated sites (Bunke et al., 2019). This section briefly discusses each of these steps.

6.1 Source control

Source control means reducing the emissions of ECs at their origin. The reduction in emissions would help prevent their entry into rivers. This is one of the most proactive tools to combat the pollution caused by ECs. Traditional WWTPs and ETPs remove only 10%–15% of ECs. Advanced strategies such as coagulation, ultrafiltration (UF), and adsorption through powdered activated carbon (PAC) can remove 80%–95% of microplastics and pharmaceutical compounds (Li et al., 2024). Integrating PAC and UF eliminates pharmaceuticals like diclofenac and acetaminophen from effluents and also prevents membrane degradation (Acero et al., 2012). Attaching filters to washing machines can reduce the release of textile fibers by 75%–90% (Stapleton and Hai, 2023). The use of ecofriendly packaged PCPs, best management practices, and the safe disposal or recycling of plastic-packaged pharmaceuticals could curb domestic sources (Das et al., 2024).

6.2 Advanced treatment methods

The treatment methods are applied to ECs already present in river bodies or to the effluents loaded with ECs before discharge into rivers. Adsorption using granular activated carbon after UF can remove 95% of lipophilic ECs such as diclofenac and atrazine (Zhu and Xu, 2025). Degradation of persistent ECs using advanced oxidation technologies involving oxidizing agents like UV, H_2O_2, and ozone results in 70%–90% efficiency for antibiotics (Noorani et al., 2024). The oxidation techniques are suitable for low pollutant loads. Nanomaterials like graphene are used to remove ECs such as microplastics and PFAS (Wang et al., 2024).

6.3 Nature-based bioremediation techniques

Nature-based removal techniques for ECs involve microbial or algal degradation, constructed wetlands, riparian buffers, and a green approach to remove ECs from environmental matrices. For the removal of ECs from water, a photocatalyst approach was applied using a biopolymer of gellan gum integrated with TiO_2. This technique was proven to be a greener alternative for remediation, as the contaminants are degraded by visible light (Zhu and Xu, 2025). The consortia of microbes and aquatic plants (*Typha latifolia* and *Canna indica*) in floating treatment boards remove 60%–85% of microplastics and antibiotics through adsorption and degradation (Patel et al., 2025). PCPs and pharmaceuticals like sulfamethoxazole, trimethoprim, carbamazepine, triclosan, and ciprofloxacin can be removed by algae-based removal systems. Some antibiotics like ibuprofen, diclofenac, and triclosan are removed using biochar made from olive oil waste products (Zhou et al., 2024). Constructed wetlands are a sustainable technique deployed in hospital wastewater, landfill leachate treatment, and municipal wastewater for the removal of ECs. Continuous aeration enhances the removal of ECs by 99%. Hospital drugs such as pesor (50%) and atenolol (75%) are removed using constructed wetlands (Zhou et al., 2024).

7 Summary

In riverine ecosystems, contaminants beyond the known ones are referred to as ECs. These comprise a broad spectrum of synthetic chemicals that have been rapidly detected at trace to nanograms per liter (ng/L) levels. The aquatic ecosystem acts as a sink for ECs. Major sources in rivers include discharges from municipal wastewater treatment plants, pharmaceutical industries, hospitals, agricultural runoff, landfill leachate, urban runoff, and floods that transport complex debris from land to water. ECs have diverse physicochemical properties such as polarity, hydrophobicity, acidity, and fluorination, which lead to their dynamic interactions between soil–water, air–water, and nonpolar liquid–water interfaces. The occurrence of microplastics in river systems is toward of anthropogenic loads, especially around industrial sites where the input of such contaminants dominates. The health and ecological impacts of ECs include chronic toxicities, the development of antimicrobial resistance, and endocrine disruption. Locations with inefficient treatment structures often exceed the threshold values for these contaminants. Advancements in analytical techniques such as HRMS, ELISA, Raman spectroscopy, and the AGREE now open the now enable comprehensive surveillance. Mitigation strategies involve advancements in treatment plants to bridge the gap between inadequate treatment processes and the discharge of pollutants into river systems. Nonetheless, policies need to shift their priorities toward risk-based and harmonized frameworks.

8 Research outlook

ECs hold vast potential for future research due to their multidimensional, innovative, and interdisciplinary perspectives. Future research could expand the sampling networks downstream of WWTPs and upstream of river systems to study time-dependent exposure. Furthermore, since ECs include many contaminants, it would be imperative to develop compound-specific models that integrate hydrology with sorption mechanisms and precursor with product dynamics. This could open the field for prediction across riverine gradients. Second, research could explore more robust protocols for ecotoxicological and microbiome assays. In addition, to quantify biofilm-based effects, sublethal concentrations could be incorporated into regulatory risk assessment studies. Contaminants do not remain in the river forever; they follow a land-river-sea continuum. Therefore future studies could focus on extreme events and hydrodynamics that govern contaminant distribution across these three matrices. The integration of efforts from policy makers, civic attitudes toward proper waste disposal, and producer responsibility for the end stage of their product packaging would help reduce the menace of ECs in the ecosystem.

References

Acero, J.L., Benitez, F.J., Real, F.J., Teva, F., 2012. Coupling of adsorption, coagulation, and ultrafiltration processes for the removal of emerging contaminants in a secondary effluent. Chem. Eng. J. 1385-8947. 210, 1–8. doi:10.1016/j.cej.2012.08.043.

Acevedo, A., Green Metrics for Analytical Methodologies Measured for Detecting Emerging Contaminants in Food and Water. (2024),

Anjanapriya, S., SulaimanMumtaz, M., Mohideen, M.H.A.K., Radha, A., Sasirekha, N., Sawicka, B., Tamizhazhagan, V., 2021. Pharmaceutical pollution crisis in the world: a menace to ecosystem. Entomol. Appl. Sci. Lett. 2349-2864. 8 (1), 77–89. doi:10.51847/iuggphofkk.

Anjanapriya, S., SulaimanMumtaz, M., Mohideen, M.H.A.K., Radha, A., Sasirekha, N., Sawicka, B., Tamizhazhagan, V., 2021. Pharmaceutical pollution crisis in the world: a menace to ecosystem. Entomol. Appl. Sci. Lett. 8 (1-2021), 77–89. https://easletters.com/article/pharmaceutical-pollution-crisis-in-the-world-a-menace-to-ecosystem-9gffkqqwjn99shz.

Ashesh, A., Singh, S., Devi, N.L., Yadav, I.C., 2022. Organochlorine pesticides in multi-environmental matrices of India: a comprehensive review on characteristics, occurrence, and analytical methods. Microchem. J. 177, 107306. doi:10.1016/j.microc.2022.107306.

Baudisch, S.K., Goulart, F.D.A.B., Antonelli, J., Azevedo, J.C., Brauko, K.M., 2022. Emergent contaminants in spring rivers and their relation to the benthic macroinvertebrates. Rev. Bras. Recur. Hidricos. 2318-0331. 27. Available from: https://www.scielo.br/j/rbrh/a/hBv7b4V7jjGNpwSdVpdxPZN/?lang=en&format=pdf. doi:10.1590/2318-0331.272220210132.

Bertolini, M., Conti, F., 2022. Algae culture conditions and process parameters for phycoremediation and biomaterials production. Environ. Clim. Technol. 2255-8837. 26 (1), 1092–1105. doi:10.2478/rtuect-2022-0082.

Birch, G.F., Drage, D.S., Thompson, K., Eaglesham, G., Mueller, J.F., 2015. Emerging contaminants (pharmaceuticals, personal care products, a food additive and pesticides) in waters of Sydney estuary, Australia. Mar. Pollut. Bull. 0025-326X. 97 (1-2), 56–66. doi:10.1016/j.marpolbul.2015.06.038.

Biswas, P., Vellanki, B.P., 2021. Occurrence of emerging contaminants in highly anthropogenically influenced river Yamuna in India. Sci. Total Environ. 0048-9697. 782, 146741. doi:10.1016/j.scitotenv.2021.146741.

Biswas, P., Vellanki, B.P., Kazmi, A.A., 2024. Investigating biotic and abiotic degradation of emerging contaminants in surface water-sediment batch systems. J. Clean. Prod. 0959-6526. 467. Available from: https://www.sciencedirect.com/science/journal/09596526. doi:10.1016/j.jclepro.2024.143019.

Bujagić, I.M., Grujić, S., Laušević, M., Hofmann, T., Mićić, V., 2019. Emerging contaminants in sediment core from the Iron Gate I Reservoir on the Danube River. Sci. Total Environ. 662, 77–87. doi:10.1016/j.scitotenv.2019.01.205.

Bukhari, I., Haq, F., Kiran, M., Kundi, R., Kiran, A., 2025. Emerging contaminants in water systems: detection, environmental impacts, and sustainable removal solutions. Water, Air, & Soil Pollution. 0049-6979. 236 (8). doi:10.1007/s11270-025-08206-5.

Bunke, D., Moritz, S., Brack, W., López Herráez, D., Posthuma, L., Nuss, M., 2019. Developments in society and implications for emerging pollutants in the aquatic environment. Environ. Sci. Eur. 2190-4707. 31 (1). doi:10.1186/s12302-019-0213-1.

Chopra, S., Kumar, D., 2018. Pharmaceuticals and personal care products (PPCPs) as emerging environmental pollutants: toxicity and risk assessment. In: Advances in Animal Biotechnology and Its Applications. Springer Singapore, India, pp. 337–353. Available from: http://www.springer.com/in/book/9789811047015. doi:10.1007/978-981-10-4702-2_19.

Das, B., Sarkar, S., Paul, S., Zafar, A., Biswas, D., Panda, J.R., Chakraborty, T., 2024. Mitigation strategies of emerging pollutants from cosmetic industry effluent: a review. Eur. J. Theor. Appl. Sci. 2786-7447. 2 (3), 886–899. doi:10.59324/ejtas.2024.2(3).69.

Das, S.S., Nag, S.K., Kumari, K., Saha, K., Bandyopadhyay, S., Aftabuddin, M., Das, B.K., 2020. Occurrence and safety evaluation of antimicrobial compounds triclosan and triclocarban in water and fishes of the multitrophic niche of River Torsa, India. Arch. Environ. Contam. Toxicol. 79 (4), 488–499. doi:10.1007/s00244-020-00785-0.

De Castro-Català, N., Muñoz, I., Armendáriz, L., Campos, B., Barceló, D., López-Doval, J., Riera, J.L., 2015. Invertebrate community responses to emerging water pollutants in Iberian river basins. Sci. Total Environ. 503, 142–150. doi:10.1016/j.scitotenv.2014.06.110.

de Sousa, D.N.R., Mozeto, A.A., Carneiro, R.L., Fadini, P.S., 2014. Electrical conductivity and emerging contaminant as markers of surface freshwater contamination by wastewater. Sci. Total Environ. 1879-1026. 484 (1), 19–26. Available from: https://www.elsevier.com/locate/scitotenv. doi:10.1016/j.scitotenv.2014.02.135.

Evidence rising: the emerging pollutants poisoning our environment. (2019), Available from: https://www.unep.org/news-and-stories/story/evidence-rising-emerging-pollutants-poisoning-ourenvironment#:~:text=%E2%80%9CIt%20includes%20a%20variety%20of,gasoline%20additives%2C%20among%20others.%E2%80%9D.

Eze, C.G., Okeke, E.S., Nwankwo, C.E., Nyaruaba, R., Anand, U., Okoro, O.J., Bontempi, E., 2024. Emerging contaminants in food matrices: An overview of the occurrence, pathways, impacts and detection techniques of per-and polyfluoroalkyl substances. Toxicol. Rep. 12, 436–447. doi:10.1016/j.toxrep.2024.03.012.

Garrido, E., Camacho-Muñoz, D., Martín, J., Santos, A., Santos, J.L., Aparicio, I., Alonso, E., 2016. Monitoring of emerging pollutants in Guadiamar River basin (South of Spain): analytical method, spatial distribution and environmental risk assessment. Environ. Sci. Pollut. Res. 1614-7499. 23 (24), 25127–25144. Available from: http://www.springerlink.com/content/0944-1344. doi:10.1007/s11356-016-7759-x.

Geletu, A.K., Wondimu, K.T., Keru, I.U., Woldegebreil, T.A., Debelo, M.A., 2025. Trends in monitoring of emerging contaminants: advances and challenges in detection and assessment. Reg. Stud. Mar. Sci. 2352-4855. 89. Available from: https://www.sciencedirect.com/science/journal/23524855. doi:10.1016/j.rsma.2025.104358.

Gogoi, A., Mazumder, P., Tyagi, V.K., Chaminda, G.G.T., An, A.K., Kumar, M., 2018. Occurrence and fate of emerging contaminants in water environment: a review. Groundw. Sustain. Dev. 2352-801X. 6, 169–180. Available from: http://www.journals.elsevier.com/groundwater-for-sustainable-development/. doi:10.1016/j.gsd.2017.12.009.

Gomes, M.P., Brito, J.C.M., Vieira, F., Kitamura, R.S.A., Juneau, P., 2022. Emerging contaminants in streams of Doce River Watershed, Minas Gerais, Brazil. Front. Environ. Sci. 2296-665X. 9.

Available from: https://journal.frontiersin.org/journal/environmental-science. doi:10.3389/fenvs.2021.801599.

Hartl, M.G., Gubbins, E., Gutierrez, T., Fernandes, T.F., Review of existing knowledge–emerging contaminant: focus on nanomaterials and microplastics in the aquatic environment. (2015), Available from: https://pure.hw.ac.uk/ws/portalfiles/portal/9278426/CREW_Emerging_Contaminants.pdf.

He, S., Dong, D., Zhang, X., Sun, C., Wang, C., Hua, X., Zhang, L., Guo, Z., 2018. Occurrence and ecological risk assessment of 22 emerging contaminants in the Jilin Songhua River (Northeast China). Environ. Sci. Pollut. Res. 0944-1344. 25 (24), 24003–24012. doi:10.1007/s11356-018-2459-3.

Houtman, C.J., 2010. Emerging contaminants in surface waters and their relevance for the production of drinking water in Europe. J. Integr. Environ. Sci. 1943-8168. 7 (4), 271–295. doi:10.1080/1943815X.2010.511648.

Jiang, J.J., Lee, C.L., Fang, M.D., Tu, B.W., Liang, Y.J., 2015. Impacts of emerging contaminants on surrounding aquatic environment from a youth festival. Environ. Sci. Technol. 49 (2), 792–799. doi:10.1021/es503944e.

Juliano, C., Magrini, G.A., 2017. Cosmetic ingredients as emerging pollutants of environmental and health concern. A mini-review. Cosmetics. 2079-9284. 4 (2). Available from: https://www.mdpi.com/2079-9284/4/2/11/pdf. doi:10.3390/cosmetics4020011.

Karn, N., Kumar, V., Bhandari, S., Kumar, A., Sachan, S., 2025. Detection techniques for comprehensive identification of emerging contaminants. In: Emerging Contaminants and Associated Treatment Technologies. Springer Science and Business Media LLC, pp. 91–114. doi:10.1007/978-3-031-89591-3_3.

Kasonga, T.K., Coetzee, M.A.A., Kamika, I., Ngole-Jeme, V.M., Momba, M.N.B., 2021. Endocrine-disruptive chemicals as contaminants of emerging concern in wastewater and surface water: a review. J. Environ. Manag. 1095-8630. 277. Available from: https://www.sciencedirect.com/journal/journal-of-environmental-management. doi:10.1016/j.jenvman.2020.111485.

Keely, A., Antibiotics found in some of the world's rivers exceed 'safe' levels, global study finds. (2019), Available from: https://www.york.ac.uk/news-and-events/news/2019/research/antibiotics-found-in-some-of-worlds-rivers/.

Khalid, N.K., Devadasan, D., Aravind, U.K., Aravindakumar, C.T., 2018. Screening and quantification of emerging contaminants in Periyar River, Kerala (India) by using high-resolution mass spectrometry (LC-Q-ToF-MS). Environ. Monit. Assess. 1573-2959. 190 (6). Available from: https://www.wkap.nl/journalhome.htm/0167-6369. doi:10.1007/s10661-018-6745-9.

Khan, S., Naushad, M., Govarthanan, M., Iqbal, J., Alfadul, S.M., 2022. Emerging contaminants of high concern for the environment: current trends and future research. Environ. Res. 0013-9351. 207, 112609. doi:10.1016/j.envres.2021.112609.

Kim, A., Li, C.R., Jin, C.F., Lee, K.W., Lee, S.H., Shon, K.J., Park, N.G., Kim, D.K., Kang, S.W., Shim, Y.B., Park, J.S., 2007. A sensitive and reliable quantification method for Bisphenol A based on modified competitive ELISA method. Chemosphere. 0045-6535. 68 (7), 1204–1209. doi:10.1016/j.chemosphere.2007.01.079.

Koigoora, S., Kalva, S., Ambati, R., Gollapalli, R., Gokare, P., A., R., 2019. Emerging contaminants effect on aquatic ecosystem: human health risks. Agri. Res. Tech. Open Access J. 19 (4), 556104. doi:10.19080/ARTOAJ.2019.19.556104.

Kolpin, D.W., Furlong, E.T., Meyer, M.T., Thurman, E.M., Zaugg, S.D., Barber, L.B., Buxton, H.T., 2002. Pharmaceuticals, hormones, and other organic wastewater contaminants in U.S. streams, 1999-2000: a national reconnaissance. Environ. Sci. Technol. 0013-936X. 36 (6), 1202–1211. doi:10.1021/es011055j.

Konda, S., Nagendla, N.K., Williams, M., Mudiam, M.K.R., 2024. Investigating pharmaceuticals and personal care products in Musi River: a exploratory study in Hyderabad, India. J. Hazard. Mater. Adv. 2772-4166. 14. Available from: https://www.sciencedirect.com/science/journal/27724166. doi:10.1016/j.hazadv.2024.100420.

Kumar, G., Kumar, S., Paul, T., Pal, P., Shukla, S.P., Kumar, K., Jha, A.K., Pradeep, S., 2024. Ecotoxicological risk assessment of triclosan, an emerging pollutant in a riverine and estuarine ecosystems: a comparative study. Mar. Pollut. Bull. 0025-326X. (205), 116667. doi:10.1016/j.marpolbul.2024.116667.

Lambert, S., Wagner, M., 2018. Microplastics are contaminants of emerging concern in freshwater environments: an overview. In: Handbook of Environmental Chemistry, Vol. 58, Springer Verlag, Germany. pp. 1–23. Available from: http://www.springer.com/series/698. doi:10.1007/978-3-319-61615-5_1.

Li, X., Shen, X., Jiang, W., Xi, Y., Li, S., 2024. Comprehensive review of emerging contaminants: detection technologies, environmental impact, and management strategies. Ecotoxicol. Environ. Saf. 0147-6513. 278, 116420. doi:10.1016/j.ecoenv.2024.116420.

Matamoros, V., Arias, C.A., Nguyen, L.X., Salvadó, V., Brix, H., 2012. Occurrence and behavior of emerging contaminants in surface water and a restored wetland. Chemosphere. 1879-1298. 88 (9), 1083–1089. Available from: https://www.elsevier.com/locate/chemosphere. doi:10.1016/j.chemosphere.2012.04.048.

Mathew, G., Unnikrishnan, M.K., 2012. The emerging environmental burden from pharmaceuticals. Econ. Polit. Wkly. 47 (18), 31–34. http://www.jstor.org/stable/23214827.

Morin-Crini, N., Lichtfouse, E., Liu, G., Balaram, V., Ribeiro, A.R.L., Lu, Z., Stock, F., Carmona, E., Teixeira, M.R., Picos-Corrales, L.A., Moreno-Piraján, J.C., Giraldo, L., Li, C., Pandey, A., Hocquet, D., Torri, G., Crini, G., 2022. Worldwide cases of water pollution by emerging contaminants: a review. Environ. Chem. Lett. 1610-3661. 20 (4), 2311–2338. Available from: https://www.springer.com/journal/10311. doi:10.1007/s10311-022-01447-4.

Murugan, P., Sivaperumal, P., Balu, S., Arya, S., Atchudan, R., Sundramoorthy, A.K., 2023. Recent advances on the methods developed for the identification and detection of emerging contaminant microplastics: a review. RSC Adv. 2046-2069. 13 (51), 36223–36241. Available from: http://pubs.rsc.org/en/journals/journal/ra. doi:10.1039/d3ra05420a.

Ng, B., Quinete, N., Maldonado, S., Lugo, K., Purrinos, J., Briceño, H., Gardinali, P., 2021. Understanding the occurrence and distribution of emerging pollutants and endocrine disruptors in sensitive coastal South Florida ecosystems. Sci. Total Environ. 0048-9697. 757, 143720. doi:10.1016/j.scitotenv.2020.143720.

Noguera-Oviedo, K., Aga, D.S., 2016. Lessons learned from more than two decades of research on emerging contaminants in the environment. J. Hazard. Mater. 0304-3894. 316, 242–251. Available from: http://www.sciencedirect.com/science/journal/03043894. doi:10.1016/j.jhazmat.2016.04.058.

Noorani, K.R.P.M., Flora, G., Surendarnath, S., Stephy, G.M., Amesho, K.T., Chinglenthoiba, C., Thajuddin, N., 2024. Recent advances in remediation strategies for mitigating the impacts of emerging pollutants in water and ensuring environmental sustainability. J. Environ. Manag. 0301-4797. 351, 119674. doi:10.1016/j.jenvman.2023.119674.

Patel, N., Khan, M., Shahane, D., Rai, S., Chauhan, D., Kant, C., Chaudhary, V.K., 2020. Emerging pollutants in aquatic environment: source, effect, and challenges in biomonitoring and bioremediation-a review. Pollution 6 (1), 99–113. doi:10.22059/poll.2019.285116.646.

Patel, V., Sharma, S., Desai, C., Kikani, B., Madamwar, D., 2025. Synergistic removal of emerging contaminants using bacterial augmented floating treatment bed system (FTBs) of *Typha latifolia* and *Canna indica* for rejuvenation of polluted river water. Front. Microbiol. 1664-302X.

16. Available from: https://www.frontiersin.org/journals/microbiology#. doi:10.3389/fmicb. 2025.1512992.

Petrovic, M., Eljarrat, E., de Alda, M.J.L., Barcel, D., 2004. Endocrine disrupting compounds and other emerging contaminants in the environment: a survey on new monitoring strategies and occurrence data. Anal. Bioanal. Chem. 1618-2642. 378 (3), 549–562. doi:10.1007/s00216-003-2184-7.

Ramaswamy, B.R., Shanmugam, G., Velu, G., Rengarajan, B., Larsson, D.J., 2011. GC–MS analysis and ecotoxicological risk assessment of triclosan, carbamazepine and parabens in Indian rivers. J. Hazard. Mater. 186 (2-3), 1586–1593. doi:10.1016/j.jhazmat.2010.12.037.

Riva, F., Zuccato, E., Davoli, E., Fattore, E., Castiglioni, S., 2019. Risk assessment of a mixture of emerging contaminants in surface water in a highly urbanized area in Italy. J. Hazard. Mater. 0304-3894. 361, 103–110. doi:10.1016/j.jhazmat.2018.07.099.

Rodriguez-Narvaez, O.M., Peralta-Hernandez, J.M., Goonetilleke, A., Bandala, E.R., 2017. Treatment technologies for emerging contaminants in water: a review. Chem. Eng. J. 1385-8947. 323, 361–380. Available from: https://www.elsevier.com/inca/publications/store/6/0/1/2/7/3/index.htt. doi:10.1016/j.cej.2017.04.106.

Satyanarayana, G.N.V., Kumar, A., Pandey, A.K., Sharma, M.T., Natesan, M., Mudiam, M.K.R., 2023. Evaluating chemicals of emerging concern in the Ganga River at the two major cities Prayagraj and Varanasi through validated analytical approaches. Environ. Sci. Pollut. Res. 1614-7499. 30 (1), 1520–1539. Available from: https://www.springer.com/journal/11356. doi:10.1007/s11356-022-22226-2.

Shalini, K., Anwer, Z., Sharma, P.K., Garg, V.K., Kumar, N., 2010. A review on pharma pollution. Int. J. Pharmtech. Res. 2 (4), 2265–2270. https://sphinxsai.com/Oct_dec_2010_vol2_no.4/PharmTech_vol2_no.4_1_pdf/PT=24%20(2265-2270).pdf.

Sharma, A., Kumar, V., Shahzad, B., Tanveer, M., Sidhu, G.P.S., Handa, N., Kohli, S.K., Yadav, P., Bali, A.S., Parihar, R.D., Dar, O.I., Singh, K., Jasrotia, S., Bakshi, P., Ramakrishnan, M., Kumar, S., Bhardwaj, R., Thukral, A.K., 2019. Worldwide pesticide usage and its impacts on ecosystem. SN Appl. Sci. 2523-3971. 1 (11). Available from: https://springer.com/snas. doi:10.1007/s42452-019-1485-1.

Sihlahla, M., Mngadi, S., 2024. A review of occurrence of emerging contaminants and the advanced analytical techniques used for detection and removal of these pollutants in wastewater. In: Detection and Treatment of Emerging Contaminants in Wastewater. IWA Publishing, South Africa, pp. 203–226. Available from: https://iwaponline.com/ebooks/book/915/Detection-and-Treatment-of-Emerging-Contaminants. doi:10.2166/9781789063752_0203.

Singare, P.U., Dhabarde, S.S., 2014. Studies on pollution due to discharge of effluent from pharmaceutical industries of Dombivali industrial belt of Mumbai, India. Int. Lett. Chem. Phys. Astron. 3, 16–23.

Singh, S., Ashesh, A., Devi, N.L., Yadav, I.C., 2022. A comprehensive review on occurrence, source, effect, and measurement techniques of polycyclic aromatic hydrocarbons in India. Microchem. J. 183, 108005. doi:10.1016/j.microc.2022.108005.

Smital, T., Luckenbach, T., Sauerborn, R., Hamdoun, A.M., Vega, R.L., Epel, D., 2004. Emerging contaminants—Pesticides, PPCPs, microbial degradation products and natural substances as inhibitors of multixenobiotic defense in aquatic organisms. Mutat. Res. 552 (1-2), 101–117. doi:10.1016/j.mrfmmm.2004.06.006.

Stapleton, M.J., Hai, F.I., 2023. Microplastics as an emerging contaminant of concern to our environment: a brief overview of the sources and implications. Bioengineered 14 (1), 2244754. doi:10.1080/21655979.2023.2244754.

Stuart, M.E., Manamsa, K., Talbot, J.C., Crane, E.J., Emerging contaminants in groundwater. (2011), Available from: https://nora.nerc.ac.uk/id/eprint/14557/1/OR11013.pdf.

Thomaidis, N.S., Asimakopoulos, A.G., Bletsou, A.A., 2012. Emerging contaminants: a tutorial mini-review. Glob. Nest J. 1790-7632. 14 (1), 72–79. Available from: http://www.gnest.org/journal/Vol_14_no_1/72-79_823_Thomaidis_14-1.pdf. doi:10.30955/gnj.000823.

Tijani, J.O., Fatoba, O.O., Petrik, L.F., 2013. A review of pharmaceuticals and endocrine-disrupting compounds: sources, effects, removal, and detections. Water, Air, Soil Pollut. 1573-2932. 224 (11). Available from: http://www.kluweronline.com/issn/0049-6979/. doi:10.1007/s11270-013-1770-3.

Trasande, L., Shaffer, R.M., Sathyanarayana, S., Lowry, J.A., Ahdoot, S., Baum, C.R., Bernstein, A.S., Bole, A., Campbell, C.C., Landrigan, P.J., Pacheco, S.E., Spanier, A.J., Woolf, A.D., 2018. Food additives and child health. Pediatrics. 0031-4005. 142 (2). doi:10.1542/peds.2018-1410.

Unnikrishan, A., Khalid, N.K., Rayaroth, M.P., Thomas, S., Nazim, A., Aravindakumar, C.T., Aravind, U.K., 2024. Occurrence and distribution of steroid hormones (estrogen) and other contaminants of emerging concern in a south indian water body. Chemosphere. 1879-1298. 351. Available from: https://www.sciencedirect.com/science/journal/00456535. doi:10.1016/j.chemosphere.2024.141124.

Vimalkumar, K., Nikhil, N.P., Arun, E., Mayilsamy, M., Babu-Rajendran, R., 2021. Synthetic musks in surface water and fish from the rivers in India: seasonal distribution and toxicological risk assessment. J. Hazard. Mater. 0304-3894. 414, 125558. doi:10.1016/j.jhazmat.2021.125558.

Wagner, M., Scherer, C., Alvarez-Muñoz, D., Brennholt, N., Bourrain, X., Buchinger, S., Fries, E., Grosbois, C., Klasmeier, J., Marti, T., Rodriguez-Mozaz, S., Ralph Urbatzka, A.D.V., Winther-Nielsen, M., Reifferscheid, G., 2014. Microplastics in freshwater ecosystems: what we know and what we need to know. Environ. Sci. Eur. 2190-4707. 26 (1). doi:10.1186/s12302-014-0012-7.

Wang, R., Tang, H., Yang, R., Zhang, J., 2024. Emerging contaminants in water environments: progress, evolution, and prospects. Water Sci. Technol. 89 (10), 2763–2782. doi:10.2166/wst.2024.151.

Wang, Y., Wang, M., Zhou, L., Geng, X., Xu, Z., Zhang, H., 2023. Development of a competitive ELISA based on estrogen receptor and weak competitive molecule for the screening of potential estrogens in foods. Food Chem. 401, 134084.

Wilkinson, J.L., Hooda, P.S., Barker, J., Barton, S., Swinden, J., 2016. Ecotoxic pharmaceuticals, personal care products, and other emerging contaminants: a review of environmental, receptor-mediated, developmental, and epigenetic toxicity with discussion of proposed toxicity to humans. Crit. Rev. Environ. Sci. Technol. 1547-6537. 46 (4), 336–381. Available from: https://www.tandf.co.uk/journals/titles/10643389.asp. doi:10.1080/10643389.2015.1096876.

Williams, M., Kookana, R.S., Mehta, A., Yadav, S.K., Tailor, B.L., Maheshwari, B., 2019. Emerging contaminants in a river receiving untreated wastewater from an Indian urban centre. Sci. Total Environ. 647, 1256–1265. doi:10.1016/j.scitotenv.2018.08.084.

Yang, J., Qadeer, A., Liu, M., Zhu, J.M., Huang, Y.P., Du, W.N., Wei, X.Y., 2019. Occurrence, source, and partition of PAHs, PCBs, and OCPs in the multiphase system of an urban lake, Shanghai. Appl. Geochem. 1872-9134. 106, 17–25. Available from: http://www.journals.elsevier.com/applied-geochemistry. doi:10.1016/j.apgeochem.2019.04.023.

Zarębska, M., Bajkacz, S., Hordyjewicz-Baran, Z., 2024. Assessment of legacy and emerging PFAS in the Oder River: occurrence, distribution, and sources. Environ. Res. 0013-9351. 251 (Pt 1), 118608. doi:10.1016/j.envres.2024.118608.

Zhou, Q., Chen, H., Liu, G., Wang, X., 2024. Occurrence, sustainable treatment technologies, potential sources, and future prospects of emerging pollutants in aquatic environments: a review.

Front. Environ. Sci. 2296-665X. 12. Available from: https://journal.frontiersin.org/journal/environmental-science. doi:10.3389/fenvs.2024.1455377.

Zhu, X., Liu, S., Gao, X., Gu, Y., Yu, Y., Li, M., Yu, Y., 2024. Typical emerging contaminants in sewage treatment plant effluent, and related watersheds in the Pearl River Basin: ecological risks and source identification. J. Hazard. Mater. 476, 135046. doi:10.1016/j.jhazmat.2024.135046.

Zhu, Y., Xu, S., 2025. Advancing strategies for the control and treatment of emerging contaminants in water ecosystems. Water. 2073-4441. 17 (4), 465. doi:10.3390/w17040465.

Chapter 8

Industries as source of river pollution

Sonali Kumari[a,*], Nisha Verma[b,*], Ningombam Linthoingambi Devi[b], and Amrita Srivastava[a]
[a]*Department of Life Science, Central University of South Bihar, Gaya, Bihar, India*
[b]*Department of Environmental Science, Central University of South Bihar, Gaya, Bihar, India*

1 Introduction

In any country, the crucial sector for growth and development is industries, which lead to both direct and indirect adverse effects on our environment. Rapid industrialization has become a major source of pollution. These industries include iron and steel, plastics, chemicals and fertilizers, textiles, food processing, pharmaceuticals, soap, petroleum and petrochemicals, battery manufacturing, automobiles, tanneries, paper making, cosmetics, tobacco, and paint industries (Woldeamanuale, 2017).

The effluents released from these industries are highly toxic, including organic waste, inorganic salts, heavy metals, and mineral acids. Water used as a cleaning agent in industries is directly released around industrial areas and populated cities. When these effluents enter certain water bodies like lakes, rivers, and oceans, the result is the pollution of water, rendering it harmful and causing many environmental problems. These effluents reduce the quantity of dissolved oxygen, which affects aquatic life and ecosystems. This polluted water damages soil, crops, plants, and animals and causes diseases in humans such as cancer and gastroenteritis, which affect the kidneys, liver, and brain. Industrial effluents also enhance algal growth, which deteriorates the esthetic quality of lakes and rivers (Smith et al., 1999) . Pollution indices like the Heavy Metal Pollution Index, Heavy Metal Evaluation Index, and Degree of Contamination (Cd) classified the water as severely polluted, with Cd, Pb, Ni, and Cr exceeding WHO standards.

In developing countries, rapid urbanization, industrial growth, population increase, and various natural phenomena, such as weathering, rainfall,

* Sonali Kumari and Nisha Verma share equal authorship.

River Pollution. https://doi.org/10.1016/B978-0-443-45024-2.00011-7

and erosion, are major causes of the deterioration of water quality. Despite being a vital component of economic expansion, industrial development frequently poses serious environmental problems (Awan, 2013). According to the World Bank report, water quality is significant across a range of sectors and its impacts cut across nearly all the Sustaibable Development Goals (SDGs). It also demonstrated that pollution released upstream reduces economic growth in downstream areas, which, in turn, decreases the GDP growth in these regions, creating a downward impact on the overall GDP of the country. Almost half of the GDP is lost. Due to the polluted stretches in India, agricultural revenues decrease by 9%, and there is a 16% fall in agricultural yields in downstream areas (Damania et al., 2019).

Rivers are vital components of the Earth's hydrological and biogeochemical cycles, providing water for drinking, agriculture, industry, and energy production while sustaining aquatic ecosystems and biodiversity (Paul, 2017; Zhang et al., 2025). The acceleration of industrialization since the mid-20th century has profoundly altered the quality and ecological integrity of river systems worldwide. The expansion of chemical manufacturing, textile processing, metallurgy, and energy production has introduced complex pollutant mixtures such as heavy metals, persistent organic pollutants (POPs), dyes, surfactants, poly- and perfluoroalkyl substances (PFAS), microplastics, and thermal effluents into riverine environments (Piwowarska, 2024; Alfee, 2025; Awad et al., 2024). These contaminants disrupt aquatic trophic webs, reduce dissolved oxygen, alter microbial community structure, and, in many cases, lead to bioaccumulation and biomagnification through food chains, threatening human and ecosystem health (Maurya et al., 2019; Zeb et al., 2024). According to the United Nations Environment Programme (UNEP, 2023), nearly 50% of the world's wastewater, including industrial effluents, enters surface waters untreated, while <15% is reused. Such discharges impose cumulative pressures on rivers that are already constrained by urban sewage and agricultural runoff. The European Environment Agency (EEA, 2025) reports that although industrial releases of cadmium, chromium, nickel, and lead to European surface waters have declined markedly since 2010, nutrient discharges (nitrogen, phosphorus) and emerging pollutants such as PFAS show only marginal decreases. In the developing world, where industrial growth often outpaces environmental regulation, the situation remains more acute. The Citarum River in Indonesia, for example, receives wastewater from over 1000 textile and electroplating industries, yielding dissolved-metal and dye concentrations several orders of magnitude above permissible limits (Oladimeji et al., 2024). Similar conditions are observed in the Upper Atoyac basin in Mexico (Ibarraran et al., 2021), the Pearl River delta in China (Liu et al., 2024), and numerous African and South American basins where artisanal mining and chemical manufacturing discharge untreated wastewaters (Oladimeji et al., 2024).

India reflects a microcosm of these global trends. The country's rapid industrialization, coupled with expanding urban populations and limited effluent treatment infrastructure, has intensified pressure on its major rivers, particularly the Ganga, Yamuna, Godavari, Hindon, and Mithi (Paul, 2017; Singh et al., 2023). The Central Pollution Control Board (CPCB), through its National Water Quality Monitoring Programme, regularly lists over 350 polluted river stretches, many influenced by industrial estates, tanneries, dyeing units, and electroplating clusters (CPCB, 2024). In the Ganga River, heavy metals such as Cr, Ni, Pb, and Cd routinely exceed threshold values downstream of industrial centers like Kanpur and Varanasi (Gupta and Kumar, 2023). Similarly, studies on the Hindon River in Uttar Pradesh identify industrial effluent, especially from sugar, paper, and distillery plants, as a dominant contributor to high biochemical oxygen demand (BOD) and low dissolved-oxygen levels (Jadav et al., 2025).

The complexity of industrial river pollution arises from the multiplicity of contaminant sources, pathways, and transformations. Industrial effluents typically contain a mix of inorganic ions, organic matter, and trace metals that interact through adsorption, precipitation, and redox reactions within the water–sediment continuum (Debnath et al., 2021). Sediments serve as both sinks and secondary sources, gradually releasing adsorbed contaminants under changing pH, redox, or flow conditions (Piwowarska, 2024). Moreover, microbial and photochemical transformations can convert pollutants into more toxic intermediates, for instance, the reduction of Cr(VI) to Cr(III) or the transformation of chlorinated solvents to vinyl chloride (Zeb et al., 2024). The interplay of hydrology and chemistry thus determines pollutant fate, while seasonal monsoon dynamics in South Asia periodically resuspend contaminated sediments, creating pulses of reexposure for aquatic life and downstream users (Jadav et al., 2025). Beyond chemical toxicity, industrial pollution transforms river ecosystems at multiple scales. Elevated conductivity, color, and turbidity reduce light penetration and photosynthetic activity; nutrient enrichment triggers eutrophication and hypoxia; toxicants alter microbial communities and enzymatic pathways essential for organic matter turnover (Yaseen and Scholz, 2019). Chronic exposure to metals and PFAS impairs fish physiology, affects reproductive success, and poses dietary risks for human populations reliant on riverine fisheries (Maurya et al., 2019; Alfee, 2025). Emerging research links industrial pollutants to antibiotic-resistance gene propagation within aquatic microbiomes (Yu et al., 2025), revealing the cross-domain consequences of chemical stressors. The Indian experience underscores the need for integrative river-basin governance that bridges science, engineering, and policy. The National Mission for Clean Ganga, the Namami Gange initiative, and river-specific rejuvenation plans aim to combine end-of-pipe treatment, in situ bioremediation, and industrial zoning. Therefore, sustained improvement requires addressing legacy sediments, diffuse industrial runoff, and the mismatch between environmental compliance and industrial economics (Central Pollution Control Board (CPCB), 2024; World Bank, 2021).

2 Sources of industrial pollution in rivers

Industrial pollution in rivers originates through multiple pathways that reflect differences in industrial processes, waste handling, and regulatory control. The major categories include point-source discharges, diffuse or indirect inputs, mixed industrial clusters, and legacy pollutants stored in sediments (Fig. 8.1). Each represents a distinct mechanism by which contaminants enter and persist in aquatic ecosystems.

2.1 Point-source discharges

Point sources are direct, identifiable outlets, typically pipelines, drains, or channels, through which industries release wastewater into rivers or sewers. These discharges are continuous or intermittent, depending on production cycles, and typically contain high chemical oxygen demand (COD) and BOD, total dissolved solids (TDSs), and toxic metals such as chromium (Cr), nickel (Ni), zinc (Zn), copper (Cu), and lead (Pb) (Piwowarska, 2024; Paul, 2017). Because metals are nonbiodegradable, they accumulate in sediments and aquatic organisms, causing chronic toxicity and biomagnification. In Kanpur (India), tanneries discharge chromium-laden effluents into the Ganga, where Cr concentrations frequently exceed 2 $mg\,L^{-1}$, tenfold higher than CPCB standards (Patel et al., 2019). Comparable contamination has been observed in Ludhiana and Panipat, where plating and dyeing units pollute the Sutlej and Yamuna rivers (Central Pollution Control Board (CPCB), 2024). Globally, the Citarum River (Indonesia) exhibits similar conditions, receiving heavy-metal-rich effluents from over 1000 industrial facilities (Water Environment Partnership in Asia (WEPA), 2021). Such

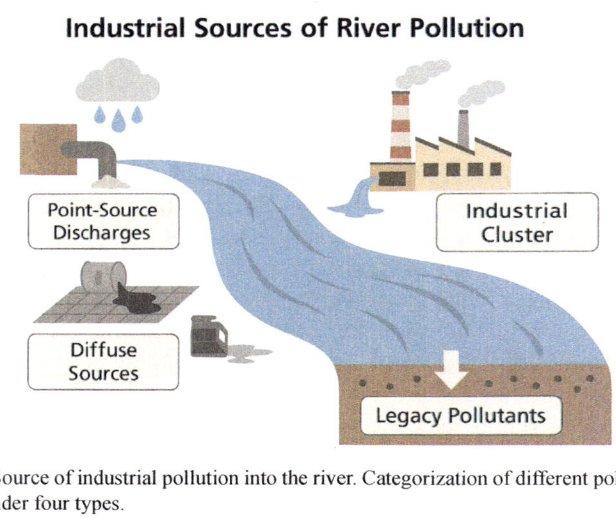

FIG. 8.1 Source of industrial pollution into the river. Categorization of different pollutants affecting rivers under four types.

findings highlight that inadequate or intermittent treatment remains the leading cause of river deterioration.

2.2 Diffuse or indirect sources

Diffuse industrial pollution arises from scattered, nonpoint pathways such as stormwater runoff, leachate migration, accidental spills, and atmospheric deposition (Oladimeji et al., 2024). Rainfall mobilizes oil, grease, and particulates from factory yards into storm drains, while unlined waste lagoons and ash ponds leach sulfates and metals into groundwater. For example, in the Damodar and Mahanadi basins, coal-ash leachates have raised Fe and Mn concentrations, imparting reddish discoloration to streams (Singh et al., 2023). Airborne particulates from smelters and power plants deposit trace metals (Pb, Cd, Zn) and acidifying compounds, which are later washed into rivers (Zeb et al., 2024). As point-source control improves, these indirect pathways constitute an increasingly important share of river contamination.

2.3 Industrial clusters and mixed pollution

Many small and medium enterprises operate in industrial estates or clusters, discharging combined effluent into Common Effluent Treatment Plants (CETPs) (Central Pollution Control Board (CPCB), 2024). Design limitations and poor operation frequently yield only partial treatment. In Tiruppur (Tamil Nadu), dyeing-unit effluents raised TDS in the Noyyal River above 6000 mg/L^{-1} even after CETP treatment (Islam and Rahman, 2025). In Vapi (Gujarat), mixed chemical and metal-finishing discharges elevated heavy-metal loads in the Daman Ganga estuary (Castillo-Suárez et al., 2023). Such complex wastewaters, containing both inorganic and synthetic organics, exhibit low biodegradability and form persistent by-products, requiring sector-specific pretreatment and continuous monitoring.

2.4 Sediment contamination and legacy pollutants

Rivers serve not only as conduits but also as repositories of historical contamination. Over decades, sediments accumulate heavy metals, hydrocarbons, dyes, and POPs, creating "legacy" or historical contaminants that remain active long after discharges cease (Piwowarska, 2024). Changes in pH, flow, or redox potential can remobilize these substances. Sediment-core studies from the Ganga and Yamuna reveal steadily rising Cr and Ni levels since the 1980s (Patel et al., 2019; Paul, 2017). Similar long-term residues occur in Europe's Rhine and Elbe, and the Cuyahoga River (United States), despite extensive remediation (Alfee, 2025; European Environment Agency (EEA), 2025). Hence, river restoration must include sediment monitoring and targeted remediation, such as dredging or capping.

3 Industrial contributions to river pollution: Global and Indian perspectives

Industrial sectors differ in pollutant profiles, wastewater chemistry, and ecological impact. The following summaries highlight sector-specific case studies, linking industry type, pollutant load (Table 8.1), and environmental response.

3.1 Metal processing, smelting, and electroplating industries

Smelting and metallurgical plants produce wastewaters and tailings enriched with Cr, Ni, Cu, Zn, Pb, Cd, and As, often accompanied by acidic pH due to sulfuric processes (Fig. 8.2). In the Bhilai industrial corridor (Chhattisgarh, India), drainage from steel and rolling-mill units has elevated Fe (5–8 $mg\,L^{-1}$) and Mn (2–3 $mg\,L^{-1}$) levels in the Kharun River, degrading irrigation water quality (Chakraborty et al., 2023). The Sukinda chromite belt (Odisha) exhibits hexavalent chromium concentrations of up to 12 $mg\,L^{-1}$ in surface runoff and 4 $mg\,L^{-1}$ in aquifers (Behera et al., 2024). The Zarafshan River (Uzbekistan), near copper-smelting complexes, shows $Cu > 1.5\,mg\,L^{-1}$ and sediment $Zn > 400\,mg\,kg^{-1}$ (European Environment Agency (EEA), 2025). In Sudbury (Canada), decades of Ni–Cu smelting have left residual Ni concentrations exceeding 3000 $\mu g\,g^{-1}$ in riverine sediments (Alfee, 2025).

TABLE 8.1 Major industrial sectors and their key pollutants.

Industry	Dominant pollutants	Impacts	References
Metallurgy/ electroplating	Cr, Ni, Cu, Zn, Pb, Cd, cyanide	Sediment toxicity, benthic mortality, biomagnification	Paul (2017); Piwowarska (2024)
Chemical/plastics	PFAS, solvents, catalysts, phthalates	Persistent organics, endocrine disruption	Alfee (2025); Awad et al. (2024)
Textile/dyeing	BOD, COD, salts, dyes, metals	Hypoxia, coloration, salinity alteration	Castillo-Suárez et al. (2023); Islam and Rahman (2025)
Pulp and paper/ food processing	BOD, nutrients (N, P), TSS	Eutrophication, microbial proliferation	Yaseen and Scholz (2019); Paul (2017)
Mining/power generation	Acid mine drainage, Fe, Mn, SO_4^{2-}, heat	Acidification, legacy metal stocks, thermal stress	Debnath et al. (2021); Singh et al. (2023)
Petrochemical/ refining	PAHs, BTEX, phenols, sulfides	Acute aquatic toxicity, carcinogenic residues	Maurya et al. (2019); Singh et al. (2023)

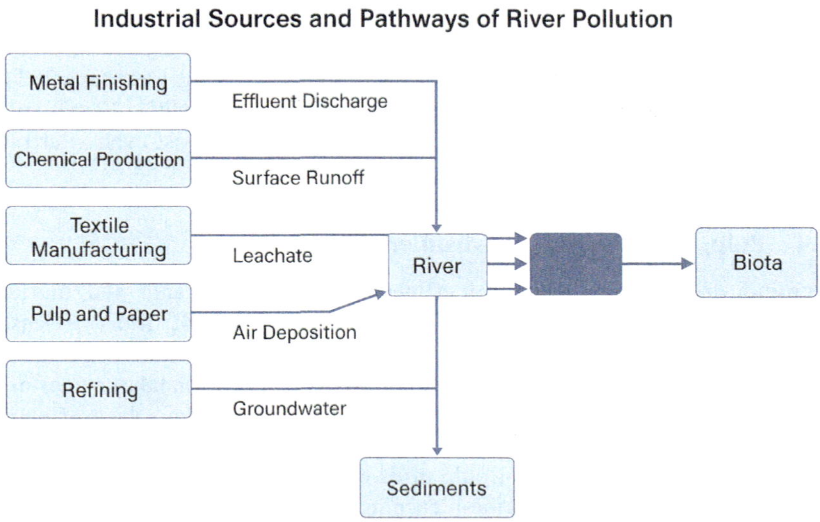

FIG. 8.2 Industrial sources and pathways of river pollution. Pathways of source-specific pollutants causing river pollution.

3.2 Chemical, pharmaceutical, and petrochemical industries

Complex organic effluents from petrochemical and pharmaceutical clusters contain PFAS, phenols, solvents, surfactants, and aromatic hydrocarbons. The Hyderabad pharmaceutical hub (Telangana, India) discharges antibiotic residues, including ciprofloxacin at concentrations of up to 14 mg L^{-1} in the Musi River, thereby promoting antimicrobial resistance (Kummerer et al., 2022). In Panipat (Haryana), refinery outfalls increase COD to 950 mg L^{-1} and phenols to 8 mg L^{-1} in the downstream Yamuna tributaries (Central Pollution Control Board (CPCB), 2024). The Yangtze River Delta (China) exhibits PFOS and PFOA concentrations of 500–1000 ng L^{-1} near fluoropolymer facilities (Zhang et al., 2025). In Houston (United States) and Rotterdam (Netherlands), petrochemical effluents release alkyl-benzene sulfonates and PAHs that persist in sediments (European Environment Agency (EEA), 2025). These pollutants resist biological oxidation and exhibit endocrine-disruptive effects even at sub-μg L^{-1} levels.

3.3 Textile, leather, and dye manufacturing

Textile and leather industries release azo dyes, salts, surfactants, and chromium-based mordants. In the Balotra–Pali belt (Rajasthan), effluents with TDS >8000 mg L^{-1} and color >2000 Pt–Co units are discharged into the Bandi River, reducing agricultural yields downstream (Central Pollution Control Board (CPCB), 2024). The Kanpur leather cluster, distinct from the Ganga mainstream examples, pollutes local drains such as the Sisamau Nala, where hexavalent

Cr > 4 mg L^{-1} and BOD > 900 mg L^{-1} have been reported (Srivastava et al., 2022). The Dhaleshwari River (Bangladesh) near Savar tanneries contains dissolved Cr > 3 mg L^{-1} and dark-brown COD > 2500 mg L^{-1} (The Guardian, 2024). In Morocco's Fez leather zone, effluents carry high salinity (EC > 12 mS cm^{-1}) and dye residues, demonstrating that salinity and color persist even after biological treatment (Yaseen and Scholz, 2019).

3.4 Pulp, paper, sugar, and distillery industries

Agroindustrial sectors discharge effluents with high organic and nutrient loads. The Saharanpur–Muzaffarnagar paper cluster (UP, India) releases BOD > 1000 mg L^{-1} and lignin-derived colorants into the Kali River, lowering DO to <2 mg L^{-1} (Jadav et al., 2025). In Belagavi (Karnataka), sugar-mill spent wash (BOD ≈ 6000 mg L^{-1}) contaminates the Markandeya River (Central Pollution Control Board (CPCB), 2024). Internationally, the Saint-Maurice River (Canada) downstream of pulp mills exhibits adsorbable organic halides up to 3 mg L^{-1}, reflecting chlorine-bleaching residues (Debnath et al., 2021). In Brazil's Paraná Basin, vinasse from ethanol distilleries adds K > 1000 mg L^{-1} and organic acids that alter aquatic microbial assemblages (Yaseen and Scholz, 2019).

3.5 Mining, mineral beneficiation, and power generation

Open-cast mining, ore washing, and ash-pond drainage release Fe, Mn, Al, SO_4^{2-}, and acidic effluent. In Goa's iron-ore belt, Mandovi River sediments contain Fe > 20,000 mg kg^{-1} and As > 60 mg kg^{-1} (Piwowarska, 2024). In Korba (Chhattisgarh), fly-ash seepage raises sulfate and boron levels in the Hasdeo River (Singh et al., 2023). International analogues include the Zambezi Basin (Zambia), where Cu–Co mining produces Cu > 10 mg L^{-1} and Co > 2 mg L^{-1} in tributaries (Alfee, 2025), and the Yellow River (China), affected by coal-mine drainage with Fe > 30 mg L^{-1} and pH ≈ 4 (Zeb et al., 2024). Thermal power plants also elevate temperature by 4°C–6°C, reducing dissolved oxygen and altering plankton in nearby reaches (Central Pollution Control Board (CPCB), 2024).

3.6 Petroleum refining and hydrocarbon processing

Oil refinery discharges typically contain PAHs, BTEX, phenols, sulfides, and oils. In Haldia (West Bengal), refinery drainage elevates total petroleum hydrocarbons to 2.8 mg L^{-1} in the Hooghly estuary (Central Pollution Control Board (CPCB), 2024). In Numaligarh (Assam), groundwater near effluent ponds contains benzene > 0.02 mg L^{-1} and phenols > 5 mg L^{-1}. Globally, the Venezuela Orinoco Delta and Nigeria's Niger Delta exhibit sediment TPH > 4 mg L^{-1} and widespread mangrove mortality (Oladimeji et al., 2024). The Kuwait Bay, post-Gulf War oil fires, retains PAH residues $>3\ mg\,kg^{-1}$ in benthic deposits decades later (European Environment Agency (EEA), 2025). Such hydrocarbon enrichment promotes carcinogenic and mutagenic effects in aquatic fauna (Maurya et al., 2019).

4 Pollutants from various industries

4.1 Textile industry

One of the largest and most well-established industries in the world is the textile sector. Textiles are necessary in the modern world for a variety of everyday applications (Verma et al., 2012). These days, a significant amount of the textile and dyeing industry is located in underdeveloped countries, sometimes with poor wastewater treatment infrastructure. It can be said that India might be the largest source of textile wastewater in South Asia. Dye is the primary chemical component utilized in the textile industry. A significant portion of the dyestuffs are lost to waste water during the dyeing process, and this waste eventually ends up in the environment. Furthermore, the production of textiles usually uses antimicrobial chemicals that are resistant to biological degradation, especially for natural fibers like cotton. The growth of methanogenic bacterial cultures is significantly hampered by azo dyes, which account for about 60% of dyes used in textiles (Hu and Wu, 2001). As a result, biodegradation is also difficult for the effluent. The toxic effects of dye can harm DNA once it enters the body through consumption. Both dyes and dye fixatives frequently contain heavy metals (Mirghani et al., 2008). Cobalt, nickel, copper, and chrome are examples of transition metals that are often used in these industries (Walters et al., 2005). Contaminants, including soaps, sulfur, naphthol, chromium compounds, nitrates, acetic acid, and heavy metals like cobalt, mercury, copper, arsenic, lead, cadmium, and nickel, as well as several auxiliary chemicals, make the effluent potentially hazardous. The temperature and pH of the mill discharge are often high, both of which are unsafe (Kant, 2012). Furthermore, color buildup disturbs the aquatic ecology by blocking sunlight (Georgiou et al., 2003; Merzouk et al., 2010). Hydrocarbons are released by drying ovens and mineral oils used for high-temperature drying and curing. During these processes, formaldehyde, softeners, acids, and other volatile materials may be emitted (Carmen and Daniela, 2012).

4.2 Dye industry

The US "Color Index" indicates that the number of commercial dyes has increased to tens of thousands, highlighting the rapid growth of the dye business in recent years (Cheng et al., 2003). These printing and dyeing wastewaters contain a variety of organic chemicals that are carcinogenic, teratogenic, and mutagenic. The dye composition is characterized by high concentration, intense color, and resistance to biodegradation (Yufeng et al., 2003).

The wastewater released from dye industries differs in several characteristics, including pH, COD, biological oxygen demand, total solids (TS), water utilization, and color in the wastewater generated by the wet process of the dye businesses. Approximately 60,000 tons of dyes, 80% of which are azo dyes, are released into the environment annually as nonbiodegradable waste worldwide (Yaseen and Scholz, 2019)

The water quality of waste water released from various processes is significantly impacted by fiber type, dye, and slurry. The waste water's chemical makeup includes benzene, naphthalene, anthraquinone, and other compounds, and it primarily comprises impurities such as fibers, inorganic salts, pastes, oil agents, auxiliaries, acids and alkalis, and colors due to the complicated types of dyes.

4.3 Rubber industry

In addition to producing rubber products as its main product, the rubber manufacturing units actually produce a significant quantity of noise, water, and air pollution as byproducts (Dlamini and Joubert, 1996). Different chemicals are employed throughout manufacturing operations and are released into the environment as effluent. The industry uses a variety of methods to begin the journey of latex, a raw form of rubber, and to move it through many steps and stages. Numerous rubber mixing compounds utilized in the industry are fire explosives, such as azodi carbonamide, which is employed as a blowing agent in some open-celled rubber, and sulfur and organic peroxides, which are typically used as curing agents. Within the industry, calendaring and extrusion are notoriously noisy processes. Their operational noise is so high that a worker's eardrums could be harmed by constant exposure to it. Chemical particulates like CO_2, CO, sulfates, lead, and other carcinogenic compounds make up the majority of the gases released by the rubber industry. These substances not only harm the environment but also cause a number of health issues in people, including skin cancer, eye irritation and damage, bronchitis, and lung cancer (Akrill et al., 2002), while the latex sheets are being formed and heated. Skin cancer is becoming a big concern due to the hazardous poisonous compounds, particularly carbon black, that are produced by these latex serums and heat fumes.

According to research, a significant amount of harmful substances, such as PAH, volatile organic compounds (VOCs), dust, and other particulate matter, are released into the atmosphere by the rubber business. These companies release effluents that are composed of 60%–70% serum and 30%–40% rubber. These serum components include lactic acid, which is produced in the latex, as well as carbohydrates, amino acids, and compounds that promote plant growth. Serum contains the same amount of the potent chemicals used in rubber production.

These consist of xylyl mercaptan, sodium sulfites, ammonia, axiloacid, bisulfate, and metabisulfite.

No other household purpose may be served with water tainted by such particles. It can potentially contaminate groundwater as well (Edirisinghe, 2013). Oil contaminants present in latex are also found in wastewater effluents from the rubber sector. When added to river bodies, these oils create an algal layer that covers the surface and prevents oxygen from dissolving in the water, suffocating marine life. Fish are deprived of oxygen and eventually die as a result. It also stops sunlight from penetrating and reaching river bottoms and river plants, interfering with photosynthesis (Mensch, 2003).

4.4 Paper industry

The annual production of paper in India is about 13 million tons. India produces about 2.6% of the paper used worldwide. The world's consumption of paper is rising, particularly in Africa and India. One of the biggest global concerns is the amount of raw resources used in the production of paper (Singh and Chandra, 2019; Yadav and Chandra, 2018).

Wastewater from these enterprises contains a variety of environmental contaminants. The pulping process produces 55%–60% of the lignocellulosic material to generate 40%–45% of pulp from the raw material. Following the industry's bleaching and digesting of wood, both organic and inorganic contaminants are discharged into the aquatic environment as waste. Because the pulp and paper business releases refractory organic compounds such as phenols, fatty acids, biocides, dioxins, furans, and resin acids, among others, the effluent from paper mills is harmful to animals and aquatic plants, making it one of the most polluting industries.

The chemical makeup of wastewater from paper mills includes high levels of total solids, BOD, COD, and pH. Chloride, phosphate, sulfate, nitrate, color, phenol, lignin, and so forth are examples of additional pollutants. Silanol, triester with boric acid, trimethyl- (H_3BO_3), D-lactic acid-DITMS, silane, tetracosane, heptacosane, heneicosane, eicosane (CAS), and other organic contaminants were all detected by GC–MS analysis. Most organic contaminants are classified as endocrine-disrupting substances and carcinogens (Kumar et al., 2020). Volatile sulfur compounds are emitted into the atmosphere during the pulping process; these include dimethyl disulfide, hydrogen sulfide, and methyl mercaptan (Smook, 1989).

4.5 Smelting industry

The nonferrous metal smelting plant is a crucial source of metals, such as Zn, Cu, Cd, Pb, and As, among industrial operations (Luo et al., 2023). As high temperatures are used in the smelting of nonferrous metals, certain metals may become suspended in the air after being gasified (Zhang et al., 2022), ultimately landing on the soil's surface. Similar to gasified metals, metal-containing particles are released into the surrounding atmosphere during the smelting of raw materials and waste materials (Fry et al., 2020). Furthermore, even far from the source, heavy metals will contaminate the soil elsewhere after being washed off by surface water, such as rainwater (Kumkrong et al., 2022). The terrestrial and aquatic environments are negatively impacted by mining and smelting operations, which are typically irreversible.

4.6 Fertilizer industry

Salts of potassium, ammonium, phosphate, and nitrate are the most common contaminants in nonorganic fertilizers. Natural radionuclides and heavy metals are also released from the fertilizer industry. Heavy metals like Cu, Pb, Hg, As, Ni, and Cd, as well as natural radionuclides like 238U, 210Po, and 232Th, are mostly found in it (Sonmez et al., 2007). The usage of nitrogen fertilizers at high levels can raise the nitrate amount in rivers and drinking water. Due to the movement of phosphorus fertilizer with surface flow, the level of phosphate in drinking water and rivers may rise. Soils are utilized to grow plants with a high degree of nitrogen fertilizer. It contains nitrosamines and other carcinogens, particularly when plants like spinach and lettuce leaves are consumed. The accumulation of NO3 and NO2 is detrimental (Çevre and Bakanlığı, 2022). The primary air pollutants released by fertilizer factories are nitrogen oxides, sulfur oxides, and particulate matter. Fluorine-containing gases are released into the environment from phosphate processing facilities. The most harmful contaminants that are introduced to water are heavy metals (particularly cadmium) and chemicals that include nitrogen and are soluble in water. Phosphate and sulfate ions make up the majority of the wastewater discharged from phosphate processing facilities (Popovici, 1994).

4.7 Pharmaceutical industry

The pharmaceutical industries can produce waste and byproducts that are harmful to the environment. Antibiotics, nonsteroidal antiinflammatory drugs, β-blockers, analgesics, antiretroviral medications, endocrine disruptors, and cancer treatment medications are the pharmaceuticals that are found in high concentrations in wastewater (Roberts and Thomas, 2006; Gros et al., 2010; Lian et al., 2017). The most alarming pharmaceutical compounds found in wastewater are antibiotics, due to their persistence, partial metabolism, and ease of movement between environments (Mukhtar et al., 2020).

5 Effects of remediation

Reducing water pollution and fulfilling water needs, particularly in low-income regions, requires the difficult but vital work of implementing remedial measures. In high-income regions like North America, Western Europe, and

Japan, 74% of wastewater is treated, compared to low-income places where just 4% of wastewater is treated (Jones et al., 2021). In contrast to low-income nations, high-income nations have the infrastructure and technology required to set up wastewater treatment facilities and enforce wastewater quality rules. Techniques for treatment can be divided into three categories: physical, chemical, and biological. Membrane filtration, adsorption, ion exchange, dissolved air floatation (DAF), screening, grit removal, comminution, sedimentation, and membrane filtration are examples of physical techniques. Biological treatment can be categorized as either anaerobic or aerobic based on oxygen availability. Membrane distillation, electrochemical techniques, precipitation, distillation, flocculation, solvent extraction, chemical oxidation, coagulation, and advanced oxidation processes are examples of chemical treatment.

There are quantifiable benefits to the environment when remediation technologies are used. Aquatic ecosystems and soil fertility are restored when treated industrial effluents show reduced levels of biochemical oxygen demand (BOD), heavy metals, and COD. By using biofilters and catalytic oxidation devices to capture particulate matter and VOCs, remediation also reduces air pollution. Furthermore, by boosting surface reactivity and quickening redox changes, contemporary techniques like nanoremediation and electrobioremediation improve the efficiency of pollutant breakdown. Integrated remediation, which combines physical, chemical, and biological approaches, has demonstrated the best success in lowering industrial pollution loads, even though the long-term viability of some expensive technologies is still being assessed.

6 Government policies

Countries all over the world have implemented various strategies to ensure sustainability as part of the transition to renewable energy. Under Germany's Energiewende policy, Germany has spearheaded the widespread adoption of solar and wind power, while Denmark generates 60% of its energy from wind. Countries that use carbon taxes and trading systems to control carbon emissions include Switzerland, Liechtenstein, Canada, and EU members. They also work to raise carbon prices and taxes. The world's biggest afforestation initiative is China's Great Green Wall, and Singapore's Green Building Masterplan strives for net-zero emissions.

6.1 Indian government policies/practices for emission reduction and sustainable development

Online Continuous Emission Monitoring System (OCEMS): To improve monitoring and ensure effective compliance through self-regulatory mechanisms, the CPCB ordered the installation of OCEMS in all 17 types of highly polluting industries. A total of 4315 units are spread over 17 industry groups, out of which 3734 have installed OCEMS, while 581 remain under closure orders.

The **Ministry of Environment, Forest and Climate Change (MoEF&CC)**, Government of India, is responsible for establishing and notifying environmental discharge standards for industries. The Environment (Protection) Act of 1986's Schedule-I contains the standards for the emission of environmental pollutants from different industries. When there are no specific criteria for an industrial sector, the general standards listed in Schedule-VI of the Environment (Protection) Rules, 1986, will apply.

- **The Hazardous and Other Wastes Rules, 2016:** In India, these regulations are essential for advancing environmentally friendly garbage management techniques. By handling, treating, and disposing of hazardous waste properly, they seek to reduce its negative effects on the environment and public health.
- **CETP Scheme**: A government-supported program has been implemented by the Ministry to allow MSEs to establish CETPs across the nation. In Gujarat and Tamil Nadu, textile and tannery clusters cleanse dye-laden effluents using CETPs.
- **Zero Liquid Discharge (ZLD) Policy**: Water recovery and reuse are encouraged by the ZLD wastewater treatment method, which guarantees that no liquid waste is released into the environment. Because it conserves water and promotes environmental sustainability, it is essential for industries, particularly those in water-scarce areas. Laws such as the Environmental Protection Act and the Water (Prevention and Control of Pollution) Act in India mandate ZLD.
- **NCAP, 2019**: The Ministry of Environment, Forests, and Climate Change launched the NCAP in January 2019 with the aim of involving all stakeholders to improve air quality in 131 cities across 24 States and Union Territories, including both nonattainment cities and Million Plus Cities. The project seeks to satisfy National Ambient Air Quality Standards or lower PM10 (particulate matter 10) concentrations by up to 40% by 2025–2026.
- **Namami Gange Programme (2014–present)**: This is an Integrated Conservation Mission; the goal of this mission is to preserve and revitalize the national river Ganga. This initiative promotes energy efficiency, waste reduction, and the use of clean technologies across industries.

References

Akrill, P., Cocker, J., Dixon, S., 2002. Occupational exposure in the rubber manufacturing industry. IARC Sci. Publ. 134, 265–269.

Alfee, S.L., 2025. A global review of river sediment contamination: heavy metals, emerging pollutants, and riverine pathways. J. Environ. Contam. Toxicol. 12, 45–63.

Awad, J., et al., 2024. Long-term management of PFAS-contaminated water using nature-based systems. Crit. Rev. Environ. Sci. Technol. 54, 233–260.

Awan, A.G., 2013. Relationship between environment and sustainable economic development: A theoretical approach to environmental problems. Int. J. Asian Soc. Sci. 3 (3), 741–761.

Behera, S., et al., 2024. Hexavalent chromium contamination in the Sukinda chromite mining area. Environ. Monit. Assess. 196, 321–338.

Carmen, Z. and Daniela, S., 2012. Textile organic dyes-characteristics, polluting effects and separation/elimination procedures from industrial effluents-a critical overview. In: Puzyn, T., Mostrag-Szlichtyng, A., (Eds.) (Vol. 3 , pp. 55–86). Rijeka: IntechOpen.

Castillo-Suárez, L., Rodríguez-Castro, J., Hernández-Mendoza, A., (2023). Heavy metal contamination and ecological risk assessment in the Daman Ganga estuarine environment influenced by industrial discharges. Environ. Monit. Assess. 195, p. 1124.

Central Pollution Control Board (CPCB), 2024. Annual Report 2022–23. New Delhi: central Pollution Control Board, India.

Çevre, Ş., Bakanlığı, İ.D., 2022. Çevresel göstergeler. Ankara: Çevresel Etki Değerlendirmesi, İzin ve Denetim Genel Müdürlüğü.

Chakraborty, A., et al., 2023. Assessment of heavy-metal contamination near Bhilai steel industrial zone. Environ. Adv. 14, 100451.

Cheng, Y., Zhou, Q.X., Ma, Q.Y., 2003. Advances in dye wastewater treatment technology. Environ. Pollut. Control Technol. Equip. 46, 56–60.

Damania, R., Desbureaux, S., Rodella, A.S., Russ, J. and Zaveri, E., 2019. Quality unknown: the invisible water crisis. World Bank Publications.

Debnath, A., et al., 2021. Metallic contamination of global river sediments and latest developments. J. Environ. Manag. 294, 112925.

Dlamini, K.D., Joubert, P.N., 1996. Industrial development, pollution and disease: the case of Swaziland. Pula: Botswana J. Afr. Stud. 10 (1), 71–83.

Edirisinghe, J.C., 2013. Community pressure and environmental compliance: case of rubber processing in Sri Lanka. J. Environ. Prof. Sri Lanka. 2279-1809. 1 (1), 14. doi:10.4038/jepsl.v1i1.5139.

European Environment Agency (EEA), 2025. Industrial pollutant releases to water in Europe (Indicator). Copenhagen: european Environment Agency.

Fry, K.L., Wheeler, C.A., Gillings, M.M., Flegal, A.R., Taylor, M.P., 2020. Anthropogenic contamination of residential environments from smelter As, Cu and Pb emissions: implications for human health. Environ. Pollut. 1873-6424. 262. Available from: https://www.journals.elsevier.com/environmental-pollution. doi:10.1016/j.envpol.2020.114235.

Georgiou, D., Aivazidis, A., Hatiras, J., Gimouhopoulos, K., 2003. Treatment of cotton textile wastewater using lime and ferrous sulfate. Water Res. 0043-1354. 37 (9), 2248–2250. doi:10.1016/s0043-1354(02)00481-5.

Gros, M., Petrović, M., Ginebreda, A., Barceló, D., 2010. Removal of pharmaceuticals during wastewater treatment and environmental risk assessment using hazard indexes. Environ. Int. 1873-6750. 36 (1), 15–26. Available from: https://www.elsevier.com/locate/envint. doi:10.1016/j.envint.2009.09.002.

Gupta, V., Kumar, A., et al., 2023. Heavy metal contamination in river water, sediment, groundwater and human blood in the Kanpur region of the Ganga River basin. Environ. Monit. Assess., 195, p. 706.

Hu, T.L., Wu, S.C., 2001. Assessment of the effect of azo dye RP2B on the growth of a nitrogen fixing cyanobacterium—*Anabaena* sp. Bioresour. Technol. 0960-8524. 77 (1), 93–95. doi:10.1016/S0960-8524(00)00124-3.

Ibarraran, M.E., Saldana-Vazquez, R.A., Perez-Garcia, T., 2021. The Cost of Pollution in the Upper Atoyac River Basin: A Systematic Review. arXiv preprint arXiv:2103.00095.

Islam, M.S., Rahman, M.M., 2025. Industrial effluent contamination and salinity stress in the Noyyal River basin, Tamil Nadu, India. J. Environ. Manag. Stud. 18 (2), 145–158.

Jadav, K., et al., 2025. Anthropogenic and climatic drivers of Hindon River degradation. Environ. Adv. 16, 100490.

Jones, E.R., Van Vliet, M.T.H., Qadir, M., Bierkens, M.F.P., 2021. Country-level and gridded estimates of wastewater production, collection, treatment and reuse. Earth Syst. Sci. Data. 1866-3516. 13 (2), 237–254. Available from: https://www.earth-system-science-data.net. doi:10.5194/essd-13-237-2021.

Kant, R., 2012. Textile dyeing industry an environmental hazard. Nat. Sci. 2150-4091. 04 (01), 22–26. doi:10.4236/ns.2012.41004.

Kumar, V., Singh, J., Kumar, P., 2020. Regression models for removal of heavy metals by water hyacinth (*Eichhornia crassipes*) from wastewater of pulp and paper processing industry. Environ. Sustain. 3 (1), 35–44.

Kumkrong, P., Dy, E., Tyo, D.D., Jiang, C., Gedara Pihilligawa, I., Kingston, D., Mercier, P.H.J., 2022. Investigation of metal mobility in gold and silver mine tailings by single-step and sequential extractions. Environ. Monit. Assess. 1573-2959. 194 (6). Available from: https://link.springer.com/journal/10661. doi:10.1007/s10661-022-10054-3.

Kümmerer, K., et al., 2022. Pharmaceutical residues in the environment: global perspectives. Chemosphere 301, 134759.

Lian, L., Yao, B., Hou, S., Fang, J., Yan, S., Song, W., 2017. Kinetic study of hydroxyl and sulfate radical-mediated oxidation of pharmaceuticals in wastewater effluents. Environ. Sci. Technol. 1520-5851. 51 (5), 2954–2962. Available from: http://pubs.acs.org/journal/esthag. doi:10.1021/acs.est.6b05536.

Liu, Z., Wu, Q., He, C., Zhang, P., Wang, X., Luo, Z., Yang, L., Huang, C., 2024. Ecological risk and chemical speciation of heavy metals in surface sediments of the Pearl River Estuary for comprehensive assessment. Front. Environ. Sci. 12, p. 1509277. https://doi.org/10.3389/fenvs.2024.1509277.

Luo, Y., Wang, Z., Zhang, Z.L., Huang, F.Y., Jia, W.J., Zhang, J.Q., Feng, X.Y., 2023. Characteristics and source analysis of potentially toxic elements pollution in atmospheric fallout around non-ferrous metal smelting slag sites—taking southwest China as an example. Environ. Sci. Pollut. Res. 1614-7499. 30 (3), 7813–7824. Available from: https://www.springer.com/journal/11356. doi:10.1007/s11356-022-22744-z.

Maurya, P.K., et al., 2019. Bioaccumulation and potential sources of heavy-metal contamination in fish. Sci. Rep. 9, 15834.

Mensch, U., 2003. OECD emission scenario document: additives in the rubber industry. OECD Report, 16, pp.241–249.

Merzouk, B., Madani, K., Sekki, A., 2010. Using electrocoagulation–electroflotation technology to treat synthetic solution and textile wastewater: two case studies. Desalination 250 (2), 573–577.

Mirghani, M.E.S., Qudsieh, I.Y., Elfaki, F.A., 2008. A new method for the determination of toxic dye using FTIR spectroscopy. IIUM Eng. J. 9 (2), 27–38.

Mukhtar, A., Manzoor, M., Gul, I., Zafar, R., Jamil, H.I., Niazi, A.K., Ali, M.A., Park, T.J., Arshad, M., 2020. Phytotoxicity of different antibiotics to rice and stress alleviation upon application of organic amendments. Chemosphere. 1879-1298. 258. Available from: https://www.elsevier.com/locate/chemosphere. doi:10.1016/j.chemosphere.2020.127353.

Oladimeji, T.E., et al., 2024. Impact of heavy metals from industrial wastewater: sources, fate, and toxicity. Pollut. Res. 43, 375–390.

Paul, D., 2017. Research on heavy metal pollution of river Ganga: A review. Ann. Agrar. Sci. 15 (2), pp.278-286. doi:https://doi.org/10.1016/j.aasci.2017.04.001.

Piwowarska, D., 2024. A global perspective on the nature and fate of heavy metals in river sediments. J. Environ. Manag. 340, 117663.

Popovici, N.N., 1994. Pollutants, wastes, and by-products of the fertilizer industry. In: Hodge, C.A., (Eds.), Pollution Control in Fertilizer Production, p. 9.

Roberts, P.H., Thomas, K.V., 2006. The occurrence of selected pharmaceuticals in wastewater effluent and surface waters of the lower Tyne catchment. Sci. Total Environ. 356 (1-3), 143–153. doi:10.1016/j.scitotenv.2005.04.031.

Singh, A.K., Chandra, R., 2019. Pollutants released from the pulp paper industry: aquatic toxicity and their health hazards. Aquat. Toxicol. 1879-1514. 211, 202–216. Available from: http://www.elsevier.com/wps/find/journaldescription.cws_home/505509/description#description. doi:10.1016/j.aquatox.2019.04.007.

Singh, B.J., Kumar, V., Sharma, A., 2023. Industrial wastewater management: challenges and opportunities. J. Environ. Manag. 343, 118271.

Smith, V.H., Tilman, G.D., Nekola, J.C., 1999. Eutrophication: impacts of excess nutrient inputs on freshwater, marine, and terrestrial ecosystems. Environ. Pollut. 100 (1-3), pp. 179–196.

Smook G.A., 1989. Overview of the pulp and paper industry from a chemical industry perspective. J. Chem. Technol. Biotechnol. 45 (1), 15–27. doi:10.1002/jctb.280450104.

Sonmez, I., Kaplan, M., Sonmez, S., 2007. Investigation of seasonal changes in nitrate contents of soils and irrigation waters in greenhouses located in Antalya-Demn region. Asian J. Chem. 19 (7), 5639.

Srivastava, S., et al., 2022. Assessment of chromium and organic pollution in Kanpur tannery drains. Environmental Nanotechnology, Monitoring & Management 18, 100676.

The Guardian, 2024. Alarming levels of 'forever chemicals' near Bangladesh garment factories. The Guardian, 29 May.

United Nations Environment Programme (UNEP), 2023. Wastewater: turning Problem to Solution. Nairobi: united Nations Environment Programme.

Verma, A.K., Dash, R.R., Bhunia, P., 2012. A review on chemical coagulation/flocculation technologies for removal of colour from textile wastewaters. J. Environ. Manag. 93 (1), 154–168.

Walters, A., Santillo, D., Johnston, P., 2005. An overview of textiles processing and related environmental concerns. Greenpeace Research Laboratories, Department of Biological Sciences, University of Exeter, UK.

Water Environment Partnership in Asia (WEPA). (2021). Outlook on water environmental management in Asia 2021. Institute for Global Environmental Strategies (IGES). https://wepa-db.net/wp-content/uploads/2023/02/WEPA_outlook2021_e.pdf

Woldeamanuale, T.B., 2017. Study on Assessment of Physico chemical properties of Industrial wastes. Int. J. Environ. Agric. Biotechnol. 2 (4), p. 238816.

World Bank, 2021. Industrial Wastewater Management in River Basins: lessons from Vietnam. Washington, DC: world Bank Open Knowledge Repository.

Yadav, S., Chandra, R., 2018. Detection and assessment of the phytotoxicity of residual organic pollutants in sediment contaminated with pulp and paper mill effluent. Environ. Monit. Assess. 1573-2959. 190 (10). Available from: https://link.springer.com/journal/10661. doi:10.1007/s10661-018-6947-1.

Yaseen, D.A., Scholz, M., 2019. Textile dye wastewater characteristics and treatment. Int. J. Environ. Sci. Technol. 16, 1193–1226.

Yu, K., Mohapatra, S., Chen, Y., Jiang, P., Tong, X., 2025. Interactive Effects of Climate Change and Contaminants in Aquatic Ecosystems on Environmental-Human Health. Curr. Pollut. Rep., 11 (1), p. 46.

Yufeng, Z., Jie, T., Xueying, Z., 2003. Progress of the researches on dyeing wastewater treatment techniques. Industrial Water Treatment (Tianjin) 23 (4), 23–27.

Zeb, M., et al., 2024. Heavy metals in riverine sediments: a systematic review. Environ. Pollut. 331, 121987.

Zhang, H., et al., 2025. Sources, distribution, and treatment of PFAS: a review. Environ. Res. 240, 117654.

Zhang, J., Sun, X., Deng, J., Li, G., Li, Z., Jiang, J., Wu, Q., Duan, L., 2022. Emission characteristics of heavy metals from a typical copper smelting plant. J. Hazard. Mater. 0304-3894. 424 (Pt A), 127311. doi:10.1016/j.jhazmat.2021.127311.

Chapter 9

River pollutants and their sources: Heavy metal pollution

Nisha Verma and Ningombam Linthoingambi Devi
Department of Environmental Science, Central University of South Bihar, Gaya, Bihar, India

1 Introduction

Increased industrial activity, industrial runoff, and environmental pollution from domestic and municipal waste have led to the deterioration of water equilibrium. The concentration of pollutants in water bodies is usually measured in water and sediments. Sediments are considered one of the important carriers and sources of environmental pollutants because, as a sink of pollutants, they play a significant role in the remobilization of contaminants in aquatic systems under favorable conditions and interactions between water and sediment (Hejabi et al., 2011). Among the inorganic compounds, heavy metals, like iron, manganese, lead, mercury, chromium, cadmium, nickel, cobalt, beryllium, and copper, are gaining more attention due to their nondegradable nature and accumulating behavior through the trophic level, which causes deleterious biological effects (Kar et al., 2008). Heavy metals are commonly defined as elements (metals or metalloids) with a specific density of $>5\ \text{gcm}^{-3}$ (Fernandez et al., 2013). They are toxic and poisonous at low concentrations, and they exist in the environment in both elemental and compound (organic and inorganic) forms (Duruibe et al., 2007). Some heavy metals, like Fe, Cu, Co, and Mn, are necessary for the metabolic system of humans and required in small quantities but may cause health issues in high concentrations (Ahirvar et al., 2023). Other metals, like Cr, Cd, Pb, and As, cause severe health implications even at lower concentrations (Ahirvar et al., 2023). The distribution of heavy metals is in four fractions: exchangeable bound, iron-manganese bound, organic matter, and residual species (Hejabi et al., 2011). In terms of chemical interactions, mobility, biological availability, and possible toxicity, they display general physical and chemical behavior (Hejabi et al., 2011). Bioavailability and ecotoxicity are the critical characteristics of heavy metals in the environment (Ahirvar et al., 2023). They can originate in the environment from both natural and anthropogenic sources. Natural causes include rock seepage into water, volcanic activity,

River Pollution. https://doi.org/10.1016/B978-0-443-45024-2.00001-4

and forest fires. Partitioning of toxic substances, concentrated in clay minerals with high absorption capabilities, between sedimentary rocks, precedent sediments, and water also causes pollution. They exist as ores in various chemical forms in rocks, which are then extracted as minerals. Heavy metal ores include sulfides (e.g., iron, arsenic, lead, lead-zinc, cobalt, gold, silver, nickel) and oxides (e.g., aluminum, manganese, gold, selenium, antimony) (Duruibe et al., 2007). The major cause of anthropogenic activities like mining, disposal of treated and untreated waste containing toxic metals, as well as metal chelates from different industries like tanneries, steel plants, battery industries, and thermal power plants, and the indiscriminate use of metal-containing pesticides and fertilizers have worsened the water quality and its equilibrium (Kar et al., 2008). Environmental pollution from heavy metals is common near mining sites and decreases with distance, as these metals are leached and transported by acidic water downstream or to the sea in sloped areas (Duruibe et al., 2007). Mining activities expose metal-bearing ores and often involve improper disposal methods, leading to contamination of water bodies and potentially violating drinking water standards in wells near mining sites (Duruibe et al., 2007). Untreated effluents, which are voluminous and highly contaminated, harm receiving water bodies and permanently pollute groundwater for drinking, irrigation, and domestic use (Srinivasa and Govil, 2008). Numerous health issues may arise if the concentrations of heavy metals in drinking water surpass the thresholds established by global organizations (Eid et al., 2024). Compared to other aquatic systems, rivers, being an open environmental space, are more vulnerable to heavy metal pollution (Zhang et al., 2021a, 2021b). During the last few decades, rivers have experienced tremendous water quality degradation along their course (Siddiqui and Pandey, 2019). The untreated sewage being flushed into rivers has increased by over ninefold in recent years compared to 1985 in India (Siddiqui and Pandey, 2019). Hundreds of industrial facilities, including textile, electroplating, power plants, sugar mills, distilleries, paper and pulp, and several other businesses, directly release waste into rivers in addition to municipal sewage (Siddiqui and Pandey, 2019). About 2500 MLD of industrial waste is generated in the Ganges Basin (Siddiqui and Pandey, 2019). Different studies have quantified metal pollution loads in rivers, including Gomti, India (Singh et al., 2005), Krotova, Bangladesh (Islam et al., 2015), Tigris River, Turkey (Varol, 2011), Brisbane River, Australia (Duodu et al., 2016), Liaohe River, China (Ke et al., 2017), and Wen-Rui River, China (Xia et al., 2018). Each heavy metal can be hazardous to most living things at high levels. Heavy metals can enter the human body through different pathways, including oral consumption, dermal contact, and inhalation, and these contaminants can be found in drinking water sources, such as surface water and groundwater. Heavy metal exposure can damage human blood cells, kidneys, liver, lungs, and central neurological and brain function. Parkinson's, Alzheimer's, muscular dystrophy, and multiple sclerosis may result from long-term exposure (Sadeghi et al., 2022). Therefore heavy metal contamination of surface water, groundwater, and sediments is a growing global environmental issue. Protecting the environment and human health is essential, necessitating the evaluation of water quality. The initial step in this process is assessing water quality and identifying sources of pollutants to reduce contamination levels. A variety of advanced instruments, including inductively coupled plasma mass spectrometry (ICP-MS), inductively coupled plasma optical emission spectroscopy (ICP-OES), atomic absorption spectroscopy (AAS), ultraviolet–visible (UV–Vis) spectrometer, and cyclic voltammetry, are employed for the quantification of heavy metals in water. The most efficient methods for detecting trace levels of heavy metal contamination in water are ICP-MS and (graphite furnace atomic absorption spectroscopy) GFAAS. Utilizing ICP-MS, we may ascertain metal concentrations in water as low as 0.1 μg/L (Reddy et al., 2012). Effective methodologies for assessing the environmental and human health concerns linked to heavy metals encompass the heavy metal pollution index, metal index, hazard quotient (HQ), hazard index (HI), and carcinogenic risk (CR) (Eid et al., 2024). Moreover, cluster analysis and principal component analysis are essential methodologies for categorizing heavy metal sources and elucidating hydrochemical processes in surface and groundwater (Eid et al., 2024).

2 Sources of heavy metals in river water

Heavy metals in water originate from both natural processes and anthropogenic activities, leading to extensive contamination of aquatic ecosystems. These metals can seep into aquatic environments through rock weathering and volcanic eruptions. Nonetheless, manmade sources are the primary contributors, encompassing industrial discharges, mining activities, agricultural runoff, inadequate waste management, and urban development. Industrial effluents, particularly from metal processing, battery production, and electroplating, can exhibit high levels of metals such as lead, cadmium, mercury, and arsenic. Moreover, the overapplication of fertilizers and pesticides in agriculture contaminates surface and groundwater systems with metals like copper and zinc. Upon release, these metals typically persist in the environment, accumulate in sediments, and infiltrate the food chain, presenting enduring risks to aquatic organisms and human health.

2.1 Natural sources of heavy metals in river water

A variety of natural geographical phenomena, such as rock weathering, mineral degradation, forest fires, volcanic eruptions, biogenic sources, sea-salt sprays, evaporation from soil and water surfaces, and wind-borne soil particles, exhibit heavy metals naturally (Jagaba et al., 2024). The emission of natural heavy metals is a local and global environmental hazard that must be addressed. Degassing from the Earth's crust, releases from volcanoes, and vaporization from natural water bodies are the foremost natural sources of mercury. Parent rock and weathering conditions might be the source of this factor, contributing to the emissions of heavy metals like Ni, Al, Sc, Fe, Co, Cu, Pb, Zn, Mg, and Ca (Wang et al., 2011). Iron (Fe), manganese (Mn), and titanium (Ti) likely represent the natural makeup of the Earth's crust, and strontium (Sr), an alkaline earth element, typically exists as $SrSO_4$ and $SrSO_3$ in geological formations (Luo et al., 2021). Volcanic activity and geothermal vents release Hg, As, Cd, and Pb into the atmosphere, while eolian dust transport mobilizes Fe, Mn, and Al over long distances (Jagaba et al., 2024). Forest fires recycle metals like Cr, Cu, and Zn from biomass back to the air, and biogenic processes volatilize elements such as Se and Hg (Wang et al., 2011). Sea spray aerosols contribute Mg, Na, Ca, and trace metals, while pedogenic processes and groundwater–rock interactions mobilize Cd, Mo, and V (Mokarram et al., 2020). In addition, glacial melting and hydrothermal systems act as natural reservoirs and pathways for heavy metal release (Jagaba et al., 2024).

2.2 Anthropogenic sources of heavy metals in river water

The primary contributors to rising heavy metals in the environment are anthropogenic sources. Human activities produce a diversity of heavy metals. The daily consumption of products that suit the population's needs is also one of the contributing factors to heavy metal emissions. Heavy metal contamination can be caused by various industrial activities and their effluents (Jagaba et al., 2024). Lead is released by smelting or treating metal ores, extraction, electroplating production, mining practices, coal-fired power and heat production, leather production, burning of fossil fuels, metallurgic activities, automobile exhaust, insecticides, waste incineration, and landfills. Heavy metals can enter the environment through anthropogenic causes such as the emission of agricultural, residential, and industrial waste (Jagaba et al., 2024). Industrial regions often have high Pb levels. Groundwater can contain lead (Pb) from fossil fuel combustion, lead-based material corrosion, municipal trash, mining, contaminated soil, air, crops, and dust. Lead-containing fertilizers, vehicle exhaust, and spent batteries are also sources (Vig et al., 2023). Pb, Zn, Cd, and As were considered as representing "anthropogenic toxic" pollution from metal activities/industrial effluents, and Pb and Zn indicated that mineral and rock erosion and anthropogenic pollution were all their sources (Wang et al., 2011). Nickel (Ni) and lead (Pb) mostly originate from the electronics sector and urban transportation, whereas zinc (Zn) is sourced from pigments, plastics, and commercial enterprises (Luo et al., 2021). Nickel (Ni) and arsenic (As) are associated with electronics production, while light industries release copper (Cu) and chromium (Cr). Cadmium (Cd) is linked to fertilizers, insecticides, and disinfectants (Luo et al., 2021). Flour, home appliances, paper mills, and sugar factories also contribute to the contamination of Cd and Ni (Mokarram et al., 2020). Mo primarily originates from wastewater coming from leather, flour, and tanning factories (Mokarram et al., 2020). Leaded petrol adds to lead contamination through air deposition. These metals primarily originate from anthropogenic activity in industries such as automotive, printing, metallurgy, and petroleum machinery (Luo et al., 2021). Mercury (Hg) emitted from anthropogenic activities is discharged into the atmosphere by fossil fuel burning, mining operations, smelting processes, and the incineration of solid waste (Emmanuel et al., 2022). Groundwater arsenic (As) contamination is attributed to geothermal activity, fossil fuel combustion, mining, mineral dissolution (notably pyrite oxidation), the application of arsenic-based chemicals in wood and agricultural preservatives, as well as reductive dissolution and desorption (Ahmed et al., 2022). Coal thermal power plants are critical sources of heavy metal pollution in groundwater, posing a significant health hazard. Concentrations of Al, As, Fe, Co, Mn, Ni, Cu, Cr, Cd, Pb, and Zn have been routinely assessed in groundwater (Vig et al., 2023). The wastewater produced from the overuse of pesticides and herbicides in agriculture is becoming a global concern, particularly in developing nations (Rad et al., 2022). Some fertilizers contain heavy metals; for example, superphosphate has elevated levels of cadmium, cobalt, copper, and zinc. Copper and iron sulfates exhibit high lead and detectable nickel concentrations (Gracia et al., 1996). Nitrate fertilizers contain moderate levels of cadmium and cobalt, whereas urea exhibits minimal metal concentration. Herbicides such as Saturn-G and Ordram have elevated levels of Fe, Mn, Zn, Pb, and Ni (Gracia et al., 1996).

3 Mobility of heavy metals in the aquatic environment

Heavy metals are naturally introduced into aquatic systems through the weathering, erosion, and hydrothermal alteration of igneous, sedimentary, and metamorphic rocks, which release elements such as nickel, cobalt, manganese, zinc, copper, and uranium (Seigel, 2002). Once released, these metals are redistributed by surface runoff and groundwater infiltration, eventually reaching rivers, lakes, and wetlands. In aquatic environments, their transport and fate depend on water chemistry, particularly pH, redox potential, temperature, and biological activity (Drever, 1988). Within rivers and lakes, metals occur in dissolved, colloidal, and particulate fractions. The dissolved form (<1 kDa) is the most mobile and bioavailable, posing the greatest ecological and health risks, while colloidal and particulate forms may accumulate in sediments or adsorb to aquatic vegetation (Fu and Wang, 2011). These sediments act as both sinks and secondary sources, releasing metals back into the water column under changing redox or pH conditions. Once mobilized, metals may also be ingested by aquatic organisms, bioaccumulating and biomagnifying through the food chain. While trace metals like copper and zinc serve essential biological roles, excess concentrations are toxic to aquatic flora, fauna, and humans consuming contaminated fish or water. Human activities intensify this natural input, making rivers and lakes hotspots for heavy metal contamination. Fertilizers and pesticides introduce cadmium, zinc, arsenic, and mercury into waterways via runoff (Kobielska et al., 2018). Industrial discharges, sewage effluents, and mining tailings release concentrated loads of lead, cadmium, chromium, and arsenic into aquatic systems, often generating acid mine drainage that greatly increases solubility and transport potential. Coal burning, smelting, and unregulated waste disposal also contribute to river and groundwater contamination (Adriano, 2001). Atmospheric deposition represents another pathway linking land-based emissions to aquatic pollution. Metals such as arsenic, mercury, cadmium, and lead, released during fossil fuel combustion and waste incineration, are deposited directly into rivers and lakes through rainfall or dry particulate settling (Fig. 9.1) (Seigel, 2002). This mechanism enables long-range transport, extending contamination even to remote aquatic systems. Overall, the movement of heavy metals in aquatic environments is controlled by the interplay of natural and anthropogenic drivers, with rivers, lakes, and wetlands functioning as dynamic reservoirs and transport pathways. Metals cycle between dissolved phases, sediments, and biota, creating conditions for persistent contamination and bioaccumulation. This continuous cycling ensures prolonged exposure, making aquatic ecosystems critical focal points for monitoring and managing heavy metal pollution (Kobielska et al., 2018).

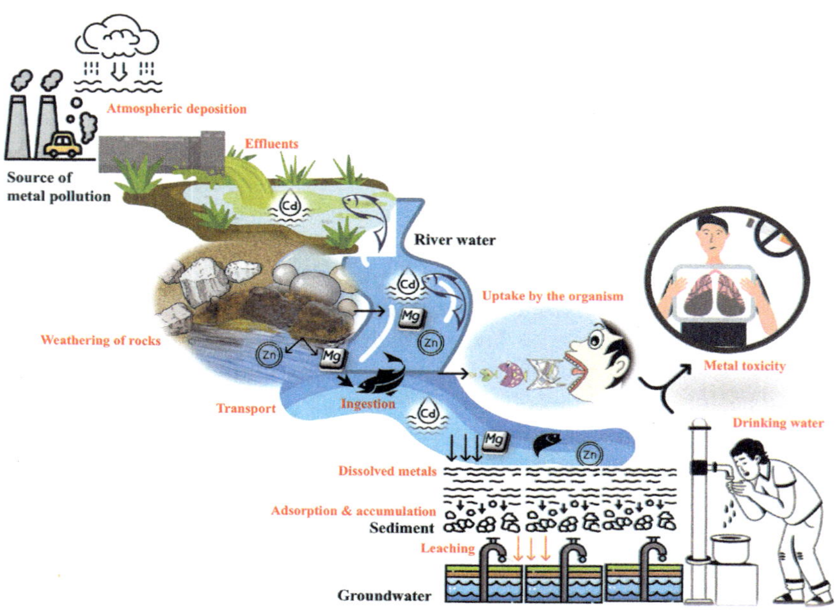

FIG. 9.1 Metals' mobility in the environment.

4 Metal bioavailability and toxicity in the aquatic life of river water

The toxicity of metals in river water is not governed solely by exposure dose but is strongly influenced by their physical and chemical speciation, which dictates potential toxicity, bioavailability, and mobility (Magalhaes et al., 2015). Speciation refers to the occurrence of metals in different forms, controlled by riverine conditions such as ionic strength, hardness, organic matter, pH, redox potential, and valence state (Templeton et al., 2000). For example, inorganic mercury and methylmercury are both present in rivers but exhibit distinct toxicity patterns, with methylmercury showing greater mobility and bioaccumulation in aquatic food webs (Templeton et al., 2000). In river systems, metals occur across fractions: dissolved (<1 kDa), colloidal (1 kDa–0.45 μm), and particulate (>0.45 μm), with the dissolved phase being the most hazardous due to its high mobility and direct bioavailability (Magalhaes et al., 2015). River water pH and redox potential play central roles in determining solubility and toxicity: acidic conditions promote the release of stable cations (e.g., Cu^{2+}, Zn^{2+}), increasing toxicity, while alkaline conditions favor the precipitation of oxides and hydroxides, reducing mobility (Fig. 9.2). Similarly, under reducing conditions common in eutrophic or sediment-rich rivers, Fe^{3+} is converted to the more soluble Fe^{2+}, enhancing its transport (Weiner, 2008). River hardness also modifies toxicity, as Ca^{2+} and Mg^{2+} compete for binding sites, thereby reducing the bioavailability of toxic metals. In contrast, soft, acidic rivers increase metal solubility and toxicity, particularly in polluted catchments (Magalhaes et al., 2015). Salinity gradients in estuarine reaches further affect speciation, with Cl^- and SO_4^{2-} ions promoting the formation of less bioavailable metal

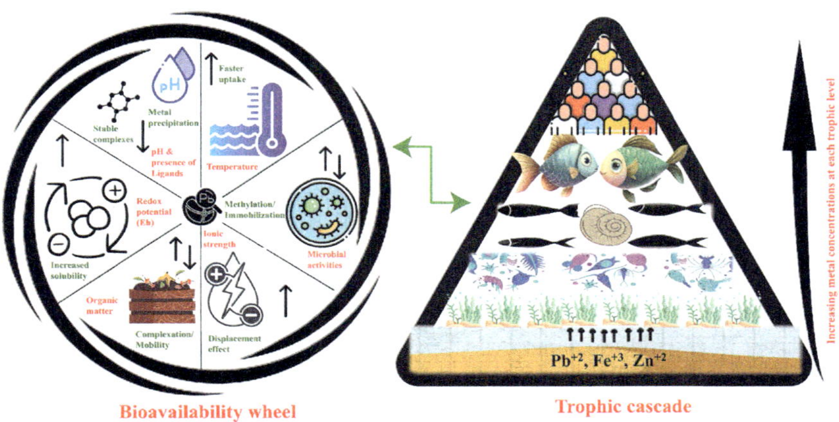

FIG. 9.2 Linking bioavailability and trophic transfer of heavy metals. Environmental factors (left) influence metal solubility and mobility, while bioavailable forms accumulate in aquatic organisms and are passed on to humans (right).

complexes (Cornelis and Nordberg, 2007). In addition, riverine organic matter, such as humic substances, often binds metals, reducing their free ionic forms and limiting toxicity, though some organic–metal complexes may remain harmful. Specific examples include chromium, where Cr(VI) in river water forms highly soluble and toxic chromates, while Cr(III) is typically insoluble and less harmful (Singh et al., 2013). Overall, river water chemistry acts as a dynamic regulator of metal behavior, determining whether pollutants remain sequestered in sediments, are transported downstream, or become bioavailable to aquatic organisms.

5 Common heavy metal pollution in river water across the world

5.1 Lead (Pb)

Lead (Pb) is recognized by the World Health Organization (WHO) as one of the ten most poisonous environmental chemicals, with no safe level of exposure in drinking water. Its main sources include leaded paints, plumbing materials, industrial effluents, toys, and contaminated waste disposal. Chronic exposure impairs neurological development, reduces IQ in children, and elevates blood pressure in adults, with infants and pregnant women being the most vulnerable (USEPA, 2023). Case studies across aquatic systems confirm Pb as a widespread contaminant (Table 9.1). In rivers, Pb has been reported in Poland (Sojka and Jaskuła, 2022), the Yellow River in China (Xie et al., 2022), the Gomti and Ganges in India (Khan et al., 2021; Samal et al., 2023), and the Ulhas River near Mumbai (Botle et al., 2025). Groundwater contamination is equally concerning, with high Pb levels in Nigeria (Adeyemi et al., 2021), Tamil Nadu (Raja et al., 2021), Delhi (Alsubih et al., 2021), Kanpur (Gupta et al., 2023), and arsenic–Pb cocontamination in Bihar (Pal et al., 2024). Collectively, these findings underscore Pb as both a neurotoxin and carcinogen, often exceeding permissible limits and posing major public health risks.

5.2 Arsenic (As)

Arsenic (As) is a ubiquitous toxic metalloid that exists mainly in drinking water as inorganic arsenite [As(III)] and arsenate [As(V)], both of which are carcinogenic Hopenhayn, 2006. Sources include natural weathering, volcanic activity, coal burning, pesticide application, and smelting industries (Villaescusa and Bollinger, 2008; Alinejad et al., 2016). Chronic exposure is linked to neurological impairment, cardiovascular dysfunction, black foot disease, skin lesions, and cancers. The WHO guideline for arsenic in drinking water is 10 μg/L. Environmental monitoring identifies arsenic as one of the most pervasive pollutants. In rivers, elevated concentrations occur in China's Xikuangshan and Han basins (Xie and Ren, 2022; Li et al., 2022), the Yellow River (Xie et al., 2022), the Yamuna River in India (Rajan and Nandimandalam, 2024), and Bangladeshi rivers (Pandey

TABLE 9.1 Concentration of heavy metals in the world.

Rivers	Fe	Mn	Co	Ni	Cu	Zn	Cr	Pb	Cd	As	References
Halda River, Bangladesh	424	610	5		25.54	940.8	130	2200	10	12	(Dey et al., 2021)
Korota River, Bangladesh	–	–	–	39	73	–	83	35	11	46	(Islam et al., 2015)
Nile River, Egypt	570	113		110	7	15	–	18	450	–	(Goher et al., 2014)
Njaba River, Nigeria	–	–	–	53	33	97	74	85	18	15	(Eze et al., 2021)
Winongo River, Indonesia	590	–	–	–	40	–	10	140	0.01	–	(Fadillah et al., 2023)
Yangtze River, East China	–	–	–	1.17	2.53	38.02	–	0.84	2.1	4.34	(Chen et al., 2024)
Yellow River, China	–	–	0.06	1.89	1.78	1.79	0.08	0.01	<0.01	2.37	(Xie et al., 2022)

and Madhuri, 2024). Groundwater studies reveal widespread contamination in Bangladesh (Shaibur et al., 2024), the Jiangsu Coastal Zone (Wang et al., 2021), the Campania Plain in Italy (Triassi et al., 2023), Wadi Ranyah in Saudi Arabia (Mohamed et al., 2025), and particularly Bihar, India, where >90% of samples exceeded WHO limits (Pal et al., 2024). Thus arsenic remains the most severe chronic groundwater contaminant, threatening millions with long-term toxicity.

5.3 Mercury (Hg)

Mercury (Hg) is a naturally hazardous volatile element, existing as elemental Hg^0, inorganic salts, and organic derivatives like methylmercury, which bioaccumulates to toxic levels in aquatic food chains (Verma et al., 2018). It has no nutritional value and is ranked third in global toxicity after As and Pb. Sources include volcanic eruptions, gold mining, fossil fuel combustion, chlor-alkali industries, and waste incineration (Kontos and Foteinis, 2020). Methylmercury, in particular, is a potent neurotoxin, bioaccumulating in fish and seafood. The WHO limit for Hg in drinking water is 0.001 mg/L. Environmental studies highlight both natural and anthropogenic sources. Elevated Hg levels have been reported in the Han, Yellow, and Xikuangshan rivers of China (Li et al., 2022; Xie et al., 2022; Xie and Ren, 2022) and in Yihe River tributaries (Duong Thanh et al., 2025). In groundwater, contamination is most evident in Wadi Ranyah, Saudi Arabia (Mohamed et al., 2025). Mercury overload causes thyroid, gastrointestinal, reproductive, and neurological toxicity, underscoring its global significance as a pollutant.

5.4 Zinc (Zn)

Zinc (Zn) is an essential micronutrient required for immune function, wound healing, and cell division, but excessive intake from drinking water (>3 mg/L, WHO) leads to nausea, vomiting, cramps, and interference with iron and copper metabolism (Kumar et al., 2021; Jamshaid et al., 2018). Deficiency, on the contrary, impairs immunity and fertility (Sankhla et al., 2019). Zn contamination largely originates from galvanization, steel processing, paints, rubber, plastics, and wastewater discharge (Sharrett et al., 1982). Globally, Zn enrichment has been reported in Polish rivers (Sojka and Jaskuła, 2022), the Ganges sediments (Samal et al., 2023), and the Chao Phraya River in Thailand, with seasonal variation (Niampradit et al., 2024). Groundwater studies document Zn pollution in Delhi (Alsubih et al., 2021) and the Jiangsu Coastal Zone in China (Wang et al., 2021). Though less acutely toxic than Cd or Pb, uncontrolled Zn discharges harm aquatic life and disrupt nutrient homeostasis, necessitating careful monitoring.

5.5 Manganese (Mn)

Manganese (Mn) is an essential trace mineral involved in enzyme activation, bone formation, and immunity (Zhang et al., 2021a, 2021b). However, excessive Mn intake, especially via contaminated drinking water, leads to neurological disorders resembling Parkinson's disease, including tremors, depression, and memory loss. The WHO allowable level is 1.06 mg/L. Mn contamination is primarily geogenic but is amplified by mining, fertilizer application, and industrial discharges (Liu et al., 2022). In rivers, high Mn levels have been found in the Yamuna (Rajan and Nandimandalam, 2024), Chao Phraya (Niampradit et al., 2024), Ulhas (Botle et al., 2025), and Bangladeshi rivers (Pandit et al., 2025). In groundwater, Mn frequently exceeds safe limits in Bangladesh (Shaibur et al., 2024), Bihar (Pal et al., 2024), Delhi (Alsubih et al., 2021), and areas near coal-based power plants in India (Vig et al., 2023a). Thus, Mn exemplifies an "essential yet toxic" element requiring strict regulation in drinking water.

5.6 Iron (Fe)

Iron (Fe) is crucial for hemoglobin synthesis and oxygen transport, but both deficiency and overload cause health problems. Deficiency leads to anemia, fatigue, and cognitive decline, while overload contributes to liver cirrhosis, cancer, diabetes, and cardiovascular disorders (Marousek et al., 2022; Zhang et al., 2021a, 2021b). The WHO drinking water guideline is 0.3 mg/L. Industrial steel production and natural leaching frequently elevate Fe levels in water bodies. River studies reveal Fe contamination in the Yamuna (Rajan and Nandimandalam, 2024), Ulhas (Botle et al., 2025), Chao Phraya (Niampradit et al., 2024), and Bangladeshi rivers (Pandey et al., 2024). Groundwater Fe levels are also high in Bangladesh (Shaibur et al., 2024), Bihar (Pal et al., 2024), Delhi (Alsubih et al., 2021), Kanpur (Gupta et al., 2023), and regions near coal power plants (Vig et al., 2023). Elevated Fe impairs taste, stains plumbing, and, when combined with Mn, creates complex health risks.

5.7 Cadmium (Cd)

Cadmium (Cd), a naturally occurring but highly toxic heavy metal, is classified as a Group 1 carcinogen. Acute exposure causes vomiting, diarrhea, and lung damage, while chronic exposure leads to kidney dysfunction, osteoporosis, reproductive toxicity, and cancer (Marousek et al., 2022). The WHO limit in drinking water is 0.003 mg/L. Cd primarily originates from mining, smelting, electroplating, and battery disposal (Liu et al., 2022). In rivers, Cd has been identified in the Gomti (Khan et al., 2021), Ganges (Samal et al., 2023), Han and Yellow Rivers (Li et al., 2022; Xie et al., 2022), and Yihe River tributaries (Dang et al., 2025). In groundwater, Cd pollution has been documented in Tamil Nadu (Raja et al., 2021), Delhi (Alsubih et al., 2021), Kanpur (Gupta et al., 2023), and Saudi Arabia (Mohamed et al., 2025). Given its persistence and bioaccumulation, cadmium remains one of the most hazardous pollutants, requiring stringent industrial effluent controls.

5.8 Chromium (Cr)

Chromium occurs in two major forms: trivalent Cr(III), an essential nutrient involved in glucose metabolism, and hexavalent Cr(VI), a potent carcinogen. While Cr(III) deficiency can cause impaired glucose tolerance and hyperlipidemia, Cr(VI) exposure results in cancers, skin ulcerations, and gastrointestinal disorders (Marousek et al., 2022; Kazemi et al., 2022). The WHO permissible limit for total Cr in drinking water is 0.05 mg/L. Industrial processes, including electroplating, tanning, and pigment production, are the primary sources of Cr(VI). Studies report Cr enrichment in the Gomti (Khan et al., 2021), Ganges (Samal et al., 2023), Yellow River (Xie et al., 2022), and Chao Phraya River (Niampradit et al., 2024). Groundwater contamination is widespread in Tamil Nadu (Raja et al., 2021), Delhi (Alsubih et al., 2021), Kanpur (Gupta et al., 2023), and Saudi Arabia (Mohamed et al., 2025). Given its toxicity, monitoring Cr, particularly Cr(VI), is critical in industrial regions.

5.9 Nickel (Ni)

Nickel (Ni) is widely distributed in the Earth's crust and is used in alloys, electroplating, and fossil fuel combustion. While trace amounts of Ni may play a role in vitamin and enzyme metabolism, its biological role remains uncertain (Adhikari et al., 2022). Overexposure to Ni causes skin dermatitis, respiratory issues, and an increased risk of cancers (Jamshaid et al., 2018). The WHO drinking water guideline for Ni is 0.02 mg/L. Ni contamination is consistently reported in rivers such as the Gomti (Khan et al., 2021), Ganges (Table 9.2) (Samal et al., 2023), Yellow River (Xie et al., 2022), and Chao Phraya (Niampradit et al., 2024). In groundwater, Ni has been detected in Nigeria (Adeyemi et al., 2021), Tamil Nadu (Raja et al., 2021), and Saudi Arabia (Mohamed et al., 2025). The recurrence of Ni across aquatic environments links it to industrial sources, highlighting its toxicological and carcinogenic significance.

6 Heavy metal pollution and its synergistic effects

Emerging contaminants (ECs) are chemicals occurring naturally or through anthropogenic activities that have the potential to enter the environment and pose detrimental biological and human health repercussions (Kumar et al., 2022). ECs such as microplastics, persistent organic pollutants, pathogens, and engineered metal nanoparticles are increasingly recognized for their synergistic interactions with heavy metals in aquatic environments (Kumar et al., 2022). Microplastics, due to their large surface area and hydrophobicity, can adsorb heavy metals like Pb, Cd, and As, facilitating their transport across long distances and bioaccumulation in aquatic organisms. These interactions can alter the mobility, toxicity, and bioavailability of metals (Yao et al., 2020). Similarly, organic pollutants such as PAHs (polyaromatic hydrocarbons), PCBs (polychlorinated biphenyls), and pesticides may form complexes with metals, enhancing

TABLE 9.2 Concentration of heavy metals in India.

Rivers	Fe	Mn	Co	Ni	Cu	Zn	Cr	Pb	Cd	As	References
Brahmani River, Odisha	5-95	1.5–102	4.0–5.6	9-52	1-4.7	0.4–80.1	–	10-27	0.4–4.0	–	(Reza et al., 2010)
Cauvery River		1250	2230	5250	1120	1070	320	9950	Na	–	(Begum et al., 2009)
Ganga River, Varanasi	5158.58	4.83	4.29	2.58	4.23	32.69	–	5.09	–	–	(Mattta et al., 2023)
Gomti River, Lucknow	152.3	52.4	–	–	–	–	7.6	83.9	19.5	6.1	(Khan et al., 2022)
Netravati River, Karnataka	–	436	200	–	400	280	880	948	–	–	(Gayathri et al., 2021)
Satlej River, Punjab	1408	248	12.9	16.2	22.2	480.1	13.3	28.3	0.9	–	(Setia et al., 2021)
Yamuna River, Delhi	126.67	–	26.12	164	75	1516.67	26.67	15.82	–	–	(Asim and Nageshwara, 2022)

their combined toxic effects on organisms through oxidative stress, endocrine disruption, and DNA damage (Singh et al., 2017). Coexposure scenarios often lead to nonadditive effects, either synergistic or antagonistic, which single-compound risk models cannot predict. Furthermore, metallic nanoparticles (e.g., silver, zinc oxide, titanium dioxide) used in cosmetics, paints, and electronics are being detected in water systems (Singh et al., 2017). These nanoparticles may undergo aggregation, dissolution, or redox reactions, affecting metal speciation and persistence. Their nanoscale size allows for cellular uptake, yet long-term effects on human health, reproductive systems, and microbial ecology remain poorly understood (Yao et al., 2020).

7 Toxicological effects of heavy metals

Over the past 50 years, rapid industrial development and the increasing use of metal-containing items have significantly heightened human exposure to heavy metals (Fu and Xi, 2020). Although the human body requires several metals, such as iron, zinc, copper, and manganese, excessive amounts of any one metal can be harmful (Fu and Xi, 2020). More importantly, nonessential metals such as lead, mercury, arsenic, cadmium, chromium, aluminum, and nickel have no biological utility and are toxic even at modest levels (Charkiewicz et al., 2023). Direct skin contact, inhalation of contaminated air, and consumption of contaminated food and drink are among the various ways these metals enter the human body (Martin et al., 2021). Once inside, they often accumulate in vital organs, including the brain, liver, kidneys, and bones, thereby disrupting normal biological processes (Mohod and Dhote, 2013). Lead, for example, builds up in bones and the brain, causing renal damage, cognitive impairments, and developmental abnormalities. Mercury targets the kidneys and brain, leading to memory loss and neurological issues (Fu and Xi, 2020). Cadmium induces renal dysfunction and bone demineralization, while arsenic is a recognized carcinogen that damages the skin, lungs, and liver (Martin et al., 2021). Specifically, in its hexavalent state (Cr^{+}), chromium is associated with lung cancer and DNA damage (Georgaki and Charalambous, 2023). Aluminum is linked to neurological disorders such as Alzheimer's, while nickel can trigger allergic reactions and respiratory cancers (Martin et al., 2021). Essential metals can be harmful when present in excess; iron overload damages the liver and heart, copper accumulation in the brain causes neurological damage (as seen in Wilson's disease), and high manganese exposure is associated with Parkinson-like symptoms. Primarily through the generation of reactive oxygen species, these metals induce oxidative stress, DNA damage, interference with enzyme and hormone activities, and disruption of neurotransmitter synthesis (Fu and Xi, 2020). Ultimately, heavy metal exposure is connected to a range of severe medical conditions, including cancers, renal and liver diseases, neurological disorders, and endocrine dysfunctions (Mohod and Dhote, 2013).

Marine ecosystems are diverse and ever-changing, and contaminants entering coastal waters and estuaries can have a profound impact on aquatic life, sometimes resulting in mass mortality (Baby et al., 2011). Among pollutants, heavy metals are of major concern. Heavy metals from residential, industrial, and other human activities have significantly polluted oceanic and coastal waters, disrupting ecological balance and threatening aquatic life (Baby et al., 2011). Metal uptake in aquatic organisms occurs in two stages: rapid surface binding followed by gradual transit into cells. Internal transport is the limiting step in epithelial tissues, and it can take place by diffusion or active transport via carrier proteins (Baby et al., 2011). Metal pollutants in water frequently remain dissolved or suspended before settling or being absorbed by marine organisms (Briffa et al., 2020). Fish are extremely sensitive to these contaminants due to their continuous exposure. Metals such as Fe, Zn, Pb, Cd, Cu, and Mn can be hazardous to fish, either directly or through synergy (Briffa et al., 2020). Toxic chemicals have a longer retention duration in the food chain compared to other components (Baby et al., 2016). Toxic metals, such as Hg, can accumulate in aquatic systems and pose a risk to humans and animals who consume fish. Heavy metals bioaccumulate, making them hazardous (Pandey and Madhuri, 2014). This accumulation is dangerous, especially in human-influenced settings. Traffic and industrial zones surround many zoos on the edges of cities, causing heavy metal pollution that might endanger wild animals in these protected settings (Briffa et al., 2020). Heavy metals in fish tissues degrade fish quality and pose major health hazards to fish-consuming populations in aquaculture (Pandey and Madhuri, 2014). Especially harmful is methylmercury (methyl Hg), which is up to a thousand times more soluble in lipids than water and concentrates in muscle, brain, and central nervous system tissues. Over time, these poisonous metals accumulate in their systems, causing sickness and endangering aquatic life. Mercury levels in fish can be 10,000–100,000 times greater than in the surrounding water, threatening aquatic ecosystems and human health (Pandey and Madhuri, 2014).

8 Advanced techniques for the detection of heavy metals

Currently, water quality assessment primarily relies on chemical, biological, and physical methods (Qi et al., 2024). Chemical techniques, like titration and electrochemical analysis, are accurate but require extensive reagents and are not suitable for real-time monitoring. Biological methods, including biosensors and enrichment analysis, often lack sensitivity and precision (Guo et al., 2020). Physical methods, such as hyperspectral and molecular spectroscopy, offer real-time, noninvasive analysis by detecting the absorption or emission spectra of substances (Kaplan et al., 2024). The integration of the Internet of Things, cloud computing, and machine learning with water quality sensors has enabled smart water monitoring systems capable of real-time, automated decision-making (Zharikova et al., 2022). These systems can track multiple contaminants simultaneously, sending alerts when concentrations exceed safe limits (Table 9.3).

TABLE 9.3 Analytical techniques for heavy metal detection in water.

Technique	Target heavy metals	Detection limit	Principle	Advantages	Limitations	References
AAS (flame/GFAAS)	Pb, Cd, Zn, Cu, Fe, Mn, Ni	1–50 μg/L (flame), <1 μg/L (GFAAS)	Measures the absorption of light by free metal atoms in a flame or furnace	Economical, accurate for single metals	Time-consuming for multielement analysis	(Jin et al., 2020)
ICP-MS	Pb, As, Cd, Cr, Ni, Hg, Zn, Cu	ng/L	Ionizes the sample with plasma and separates ions by mass in a mass spectrometer	Ultra-sensitive, detects multiple metals simultaneously	Expensive, requires trained personnel	(Skoog et al., 2019)
ICP-OES/ICP-AES	Fe, Mn, Zn, Cu, Cr, Pb, Ni	0.1–10 μg/L	Emits light at characteristic wavelengths from excited atoms in the plasma	Rapid, suitable for multimetal analysis	Less sensitive than ICP-MS	(Llaver et al., 2021)
UV–Vis spectrophotometry	Cr, As, Ni, Fe, Cu (after complexation)	0.01–1 mg/L	Measures the absorbance of metal–complex species at specific UV/visible wavelengths	Simple, low-cost	Low sensitivity; needs derivatization	(Yuan et al., 2011)
XRF (X-ray fluorescence)	Pb, Cd, Cr, Zn, Ni, Cu	μg/g (solids)	Emits characteristic secondary X-rays from excited atoms	Nondestructive, fast for solids	Not ideal for aqueous samples without preconcentration	(Sibal et al., 2018)
Electrochemical sensors (ASV, DPASV)	Pb, Cd, Hg, Zn, Cu	ng/L–μg/L	Measures current generated during redox reactions of metals at electrodes	Portable, selective, field-suitable	Susceptible to interference from organic matter	(Puniya et al., 2022)
HPLC (with detectors)	As (III/V), Cr (III/VI), Hg, Ni, Cu	~0.1–10 μg/L	Separates metal species by liquid chromatography; detected via UV or ICP-MS	Good for speciation, precise	Requires complexation or coupling, not for total metal load	(Hu et al., 2003)

8.1 Physical methods

Physical approaches examine water pollutants by measuring and observing their physical qualities, such as how much light they absorb, how they fluoresce, how cloudy they are, how well they conduct electricity, and their spectral signatures (Qi et al., 2024). These techniques are rapid, noninvasive, and increasingly used for real-time and remote sensing applications. Spectroscopic methods such as UV–Vis, infrared, Raman spectroscopy, and laser-induced breakdown spectroscopy are commonly used to detect organic pollutants, dyes, hydrocarbons, and sometimes heavy metals by analyzing their light absorption or emission at specific wavelengths (Akter et al., 2005; Guo et al., 2020; Han et al., 2022). We currently use remote sensing technology, such as hyperspectral and multispectral photography from satellites or drones, for indicators of pollution such as chlorophyll-a (indicative of eutrophication), turbidity, or thermal anomalies due to industrial discharge, to look for signs of pollution in surface water bodies (Kaplan et al., 2024). Turbidity meters, total dissolved solids meters, and conductivity probes help to assess general water quality parameters and are used in both field and laboratory settings (Dewangan et al., 2023). Physical methods are especially advantageous for continuous, large-scale monitoring and early detection of pollution trends in rivers, lakes, and reservoirs. Recent advances include portable spectrophotometers, fiber-optic sensors, and AI-enhanced remote sensing models for real-time pollutant tracking across vast regions (Zharikova et al., 2022).

8.2 Chemical methods

Chemical methods rely on specific chemical reactions between pollutants and reagents that yield measurable changes such as color, pH, redox state, or ionic strength, which are used for quantitative or qualitative analysis. These methods are highly reliable and widely applied in the detection of heavy metals, nitrates, phosphates, ammonia, sulfates, cyanide, and organic compounds (e.g., pesticides, phenols, and PAHs) (Yaroshenko et al., 2020). The Nessler method (for ammonia) and the molybdenum blue method (for phosphate) are two common colorimetric methods used to detect nutrients in water (Thompson and Morrison, 1951). chromatographic methods like gas chromatography and liquid chromatography, often coupled with mass spectrometry, are used to detect volatile and nonvolatile organic pollutants, including pesticides, pharmaceuticals, and industrial solvents. Titrimetric analysis is still used to check for alkalinity, hardness, and chloride levels (Daliborca et al., 2007). AAS, ICP-MS, and ICP-OES are common methods for measuring heavy metals (Table 9.3) (Akter et al., 2005; Baralkiewicz et al., 2007). Chemical sensors and ion-selective electrodes have also been developed for field detection of ions such as nitrate, fluoride, and heavy metals (Tang et al., 2018). However, traditional chemical processes often require expert workers, dangerous chemicals, and specialized laboratory equipment. In response, recent advances have led to the development of microfluidic devices (μPADs) and paper-based analytical tools, allowing for on-site, eco-friendly, and rapid chemical analysis of multiple water pollutants with low detection limits (Burratti et al., 2024).

8.3 Biological methods

Biological methods detect water pollutants by observing their effects on living organisms or using biological components like enzymes, DNA, or antibodies as sensors (Guo et al., 2020). They are especially useful for identifying complex pollutants, including pathogens, toxins, heavy metals, and pesticides. Whole-cell biosensors (e.g., *Vibrio fischeri* luminescence tests) and enzyme inhibition assays detect toxicity or specific chemicals (Olenin et al., 2011). DNA- and aptamer-based biosensors offer high precision using color or electrical signals. They achieve this by employing colorimetric or electrochemical readouts (Nurliyana et al., 2018). Culture techniques, enzyme-linked immunosorbent assays, and, more recently, polymerase chain reaction (PCR) and qPCR technologies are used to identify pathogens like *E. coli*, *Salmonella*, or *Cryptosporidium*. Biological methods have advanced significantly in recent years (Silva et al., 2015). Nanobiosensors, bioengineered microorganisms, and lab-on-a-chip devices combine biological sensing with microfluidics to create portable environmental diagnostics that can be performed at the point of care (Cairns, 2017).

9 Remediation and removal techniques for the removal of heavy metals in river water

Heavy metals are nonbiodegradable and tend to accumulate in aquatic systems and biological organisms. Therefore their removal from water is crucial to safeguard ecological and human health. Remediation techniques can be broadly categorized into conventional, advanced, and emerging technologies, each with distinct advantages and limitations (Fig. 9.3).

9.1 Conventional methods

9.1.1 Chemical precipitation

Chemical precipitation is a widely used method for removing heavy metals from industrial wastewater, particularly from the paper and electroplating industries (Akpor and Muchie, 2010). It involves adding chemical agents like lime, alum, or iron salts that react with metal ions to form insoluble precipitates (e.g., metal hydroxides), which can be separated by sedimentation or filtration. The reaction is typically optimized by adjusting pH (usually 9–11), temperature, and ion concentration (Zamora-Ledezma et al., 2021):

$$M^{2+} + 2OH^{-} \leftrightarrow M(OH)_2 \quad (1)$$

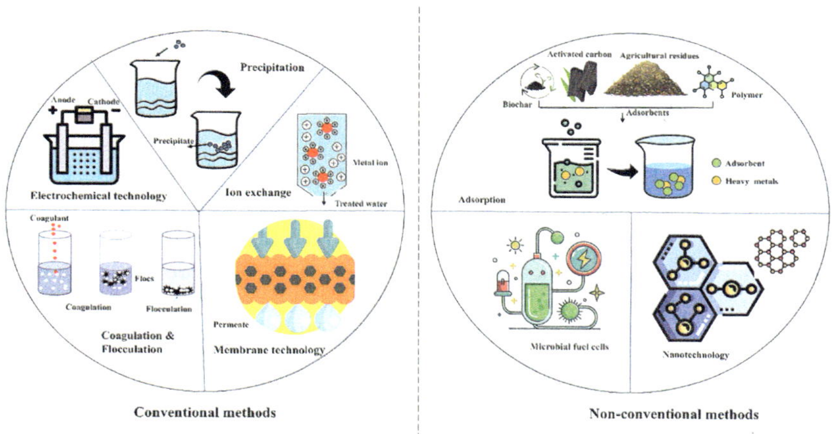

FIG. 9.3 Methods for the removal of heavy metals.

where M^{2+} and OH^- are the metal ions and the precipitant, respectively, and $M(OH)_2$ is the metal hydroxide. Chemical precipitation can be classified into hydroxide and sulfide precipitation (Singh et al., 2023). Coagulants improve removal in hydroxide precipitation, while sulfide precipitation forms dense sludge requiring careful handling due to the toxicity of H_2S (Zamora-Ledezma et al., 2021). Pre- and posttreatment are often necessary for effective results.

9.1.2 *Coagulation/flocculation*

Coagulation–flocculation is an effective physicochemical method for heavy metal removal. Coagulants like aluminum sulfate, ferric sulfate, and polyaluminum chloride promote the aggregation of fine particles into flocs, which then settle and are removed as sludge (Akpor and Muchie, 2010). This process reduces turbidity and pollutants and is used in various stages of wastewater treatment due to its versatility (Zamora-Ledezma et al., 2021).

9.1.3 *Ion exchange*

Ion exchange is a reversible process where an insoluble resin exchanges its ions with metal ions in solution (Ahmed et al., 2022). Common ion exchangers include synthetic resins, inorganic matrices, and hybrid materials. Resins with sulfonic acid groups bind metal ions, which can later be recovered by elution. The process is efficient and cost-effective for metal removal and recovery (Zamora-Ledezma et al., 2021):

$$nRSO_3^- - H^+ + M^{n+} \leftrightarrow nRSO_3^- - M^{n+} + nH^+ \quad (2)$$

where RSO_3^- and Mn^+ represent the anionic group attached to the ion exchange resin and the metal cation, respectively, and n is the coefficient of the reaction component and depends on the oxidation state of the metal ions.

9.1.4 Membrane technologies

Membrane filtration is a selective barrier-based technique that removes contaminants through size exclusion, charge repulsion (Donnan effect), and adsorption (Ahmed et al., 2022). Membranes are made from organic polymers or inorganic materials like ceramics and zeolites. Key types include microfiltration, ultrafiltration, nanofiltration, and reverse osmosis (RO), with RO removing up to 99% of metals like As^+, Cu^{2+}, and Cd^{2+} (Singh et al., 2023). Membrane performance depends on pore size, surface charge, hydrophilicity, and flow rate. Nanocomposite membranes also show high metal removal efficiency (80%–98%) for pollutants like As, Cr, Cd, Pb, and Zn (Zamora-Ledezma et al., 2021).

9.1.5 Electrochemical technologies

Electrochemical treatments are effective methods for removing heavy metals by converting them into elemental form through anodic and cathodic reactions. Techniques include electrocoagulation, electrodeposition, and electroflotation. Electrocoagulation generates coagulants in situ via electrolytic oxidation, reducing the need for added chemicals and producing minimal sludge (Zamora-Ledezma et al., 2021). Electrodeposition recovers metals like Pb, Cd, Cu, and precious metals (Ag, Au) through cathodic reduction. Electroflotation uses electrolysis-generated bubbles to float metal–pollutant complexes for removal. These methods are simple, eco-friendly, energy-efficient, and suitable for automation, making them practical alternatives for wastewater treatment (Zamora-Ledezma et al., 2021).

9.2 Nonconventional methods

9.2.1 Adsorption-based techniques

Adsorption has become one of the most widely used techniques for the removal (and sometimes detection) of heavy metals and organic pollutants Adsorbents such as activated carbon, zeolites, and clays are effective, but the use of engineered materials like magnetic nanoparticles, graphene oxide, and metal–organic frameworks has significantly enhanced adsorption efficiency and reusability (Zolkefli et al., 2020).

9.2.2 Biochar-based technologies

Biochar, a carbon-rich product derived from the pyrolysis of biomass, has shown excellent performance in adsorbing heavy metals like lead, cadmium, and arsenic due to its high surface area, porosity, and functional groups (He et al., 2022). Modified biochar, doped with iron, sulfur, or magnetic particles, is being developed to improve selectivity and facilitate recovery. Besides removal, biochar can also be incorporated into sensor platforms to improve signal conductivity and surface reactivity (Bao et al., 2022).

9.2.3 Nanotechnology

Nanomaterials, due to their high surface-to-volume ratios and unique chemical reactivity, have revolutionized water quality monitoring and treatment (Shi et al., 2023). Nanobiosensors offer ultra-sensitive detection for pathogens and metals, while nanocomposites like silver nanoparticles, titanium dioxide (TiO_2), and zinc oxide (ZnO) are being used for the photocatalytic degradation of organic pollutants (Burratti et al., 2024). Magnetic nanoparticles enable the easy separation of adsorbed contaminants, making the process efficient and reusable (Pinheiro et al., 2018). Carbon nanotubes, quantum dots, and nanosheets are also incorporated into analytical platforms for fluorescence and electrochemical detection with high specificity (Wang et al., 2013).

9.2.4 Plant-mediated remediation techniques

Phytoremediation is a green, cost-effective method that uses plants to remove or stabilize heavy metals from water and soil (Zamora-Ledezma et al., 2021). Key processes include phytoextraction, where plants absorb and store metals like Zn, Cu, and Pb in their tissues; rhizofiltration, where plant roots remove metals from water; and phytostabilization, which limits metal mobility by trapping them in the root zone (Ahmed et al., 2022). Other mechanisms include phytodegradation (breaking down pollutants), phytovolatilization (releasing volatile contaminants through leaves), and phytostimulation (enhancing microbial activity via root exudates). Though slower, it is eco-friendly, low-cost, and ideal for large or low-contamination sites (Akpor and Muchie, 2010).

10 Risk assessments

Risk assessment is a scientific and systematic process used to evaluate the likelihood and severity of adverse effects caused by the presence of toxic substances, such as heavy metals, in the environment. It plays a crucial role in identifying potential health risks, setting regulatory standards, and designing appropriate remediation strategies for polluted water bodies (Fig. 9.4). The risk assessment framework typically involves four major steps: hazard identification, dose–response assessment, exposure assessment, and risk characterization (Seelam et al., 2021).

In the context of human health, the process begins with estimating the average daily dose (ADD) of a metal based on its concentration in water, the rate of water ingestion, exposure duration, body weight, and other factors (Wongsasuluk et al., 2014). This dose is then compared to a reference dose, which is the estimated daily exposure level not expected to cause adverse effects during a lifetime. The ratio of these two values is called the HQ. If the HQ is <1,

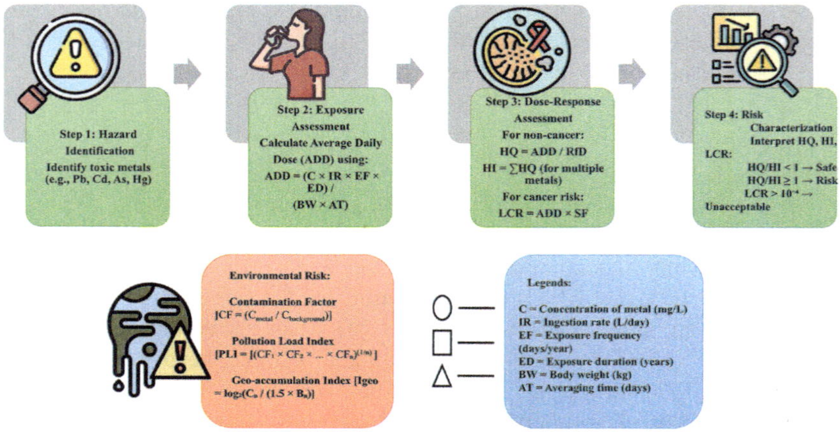

FIG. 9.4 Risk assessment framework for heavy metals.

the exposure is considered safe, while values equal to or greater than 1 indicate a potential risk (Khan et al., 2017). When multiple metals are involved, the HI is used, which is the sum of the individual HQs, reflecting the cumulative non-CR. Carcinogenic metals, such as arsenic or hexavalent chromium, are assessed for their CR using the lifetime cancer risk (LCR), which is calculated by multiplying the ADD by a chemical-specific slope factor. Acceptable cancer risk levels typically range between one in a million (10^{-6}) and one in ten thousand (10^{-4}) (Khan et al., 2017). Risks higher than this threshold are considered significant and require urgent attention.

Environmental risk assessment is equally important and evaluates the effects of heavy metals on soil and aquatic ecosystems. Indicators such as the contamination factor (CF) and pollution load index (PLI) help quantify how much a metal's concentration exceeds its natural background levels (Yi and Zhang, 2021). The geoaccumulation index (I_{geo}) is another widely used tool that classifies contamination levels based on logarithmic scaling, considering natural variability (Yi and Zhang, 2021).

11 Environmental regulations and standards

Environmental regulations and standards play a crucial role in controlling heavy metal contamination in water and safeguarding human and ecological health. Prominent organizations such as the WHO, Bureau of Indian Standards (BIS), Central Pollution Control Board (CPCB), and the US Environmental Protection Agency (US EPA) have defined maximum permissible limits for various heavy metals in drinking water, surface water, and wastewater based on toxicological data and risk assessments (Crawford-Brown, 2011). For example, WHO and US EPA recommend a maximum contaminant level of 0.01 mg/L for arsenic, 0.01 mg/L for lead, 0.001 mg/L for cadmium, and 0.002 mg/L for mercury in drinking water. The BIS drinking water standards (IS 10500:2012) also align closely, allowing up to 0.01 mg/L for arsenic, 0.01 mg/L for lead, 0.003 mg/L for cadmium, and 0.001 mg/L for mercury. The CPCB standards under the Environment Protection Rules (1986) define permissible discharge limits for heavy metals into inland surface waters, which are generally less stringent due to the dilution and dispersion capacity of natural water bodies (Table 9.4). It allows up to 0.1 mg/L of lead, 0.2 mg/L of arsenic, 1.0 mg/L of chromium (total), and 0.01 mg/L of cadmium in treated industrial effluent before discharge into surface waters (Table 9.4). These values vary for different receiving environments, like marine coastal areas or public sewers. Standards for wastewater reuse or agricultural irrigation (e.g., US EPA guidelines or India's CPCB norms) also define limits to prevent metal uptake by crops or

TABLE 9.4 Permissible limits of heavy metals in drinking water.

Metals	USEPA (mg/L)	WHO (mg/L)	CPCB (mg/L)	BIS (mg/L)
Lead (Pb)	0 (Goal), 0.015 (AL)	0.01	0.05	0.01 (acceptable), 0.05 (permissible)
Arsenic (As)	0.01	0.01	0.05	0.01 (acceptable), 0.05 (permissible)
Mercury (Hg)	0.002	0.001	0.001	0.001 (acceptable), no relaxation
Cadmium (Cd)	0.005	0.003	0.01	0.003
Chromium (Cr)	0.1 (total Cr)	0.05 (total Cr)	0.05	0.05
Nickel (Ni)	0.1	0.02	Not specified	0.02
Zinc (Zn)	5.0 (SMCL)	3.0	5.0	5.0
Iron (Fe)	0.3 (SMCL)	0.3	0.3	0.3
Manganese (Mn)	0.05 (SMCL)	0.4	0.1	0.1
Copper (Cu)	1.3 (AL)	2.0	0.05	0.05
Aluminum (Al)	0.05–0.2	0.2	0.2	0.03 (acceptable), 0.2 (permissible)

*__USEPA AL__: Action level (e.g., Pb and Cu).*__SMCL__: Secondary maximum contaminant level, based on esthetics (taste, color).*__CPCB__: Inland surface water used for drinking after conventional treatment.*__BIS__: Indian Standard IS 10500:2012 specifies both acceptable and permissible limits.

contamination of groundwater. The European Union and other regions may set even more conservative thresholds, especially for drinking water, with periodic reviews every 5 years based on new scientific evidence. Despite having strong regulatory frameworks, implementation in countries like India remains weak due to limited infrastructure, irregular monitoring, outdated laboratories, and poor enforcement, especially among small industries. State-wise variations and lack of real-time monitoring further hinder effectiveness. To overcome these issues, there is a need for better institutional capacity, investment in water testing, stricter enforcement, and public awareness. Collaboration between central and state agencies, supported by science and technology, is crucial for ensuring compliance and improving water quality.

12 Future research directions and policy recommendations

To address current challenges, future research must focus on real-time, low-cost, and portable detection systems for heavy metals in rural and remote areas. Technologies like biosensors, electrochemical detectors, and remote sensing should be developed for on-site monitoring. There is also a growing need to understand mixture toxicity, how combinations of heavy metals, organics, and nanoparticles affect humans and ecosystems over time. Moreover, long-term, low-dose exposure studies and climate change impacts (e.g., altered metal mobility due to temperature and rainfall changes) remain underexplored.

On the policy front, emphasis should be placed on decentralized water treatment systems utilizing affordable technologies, such as constructed wetlands, biofilters, and solar disinfection, particularly in underserved regions. Governments should tighten enforcement of industrial discharge norms, strengthen state pollution control boards, and incentivize pollution prevention measures. Promoting interdisciplinary collaboration, enhancing public participation, and aligning efforts with the UN Sustainable Development Goals will be crucial for sustainable water governance. Regular review of national water quality standards and investment in data-driven policymaking should also be prioritized.

13 Conclusion

The pervasive contamination of aquatic systems by heavy metals represents one of the most pressing and complex environmental challenges of the 21st century. Their nonbiodegradable nature, persistent geochemical cycling, and high toxicity at trace levels render them formidable pollutants in both surface and subsurface water systems. Originating from both lithogenic and anthropogenic pathways, heavy metals infiltrate aquatic environments via rock weathering, volcanic emissions, industrial effluents, mining activities, agrochemical applications, and urban runoff. Once introduced, they undergo dynamic partitioning across water, sediments, and biota, governed by multifaceted physicochemical factors, including pH, redox potential (Eh), ionic strength, organic matter content, and complexation behavior. The speciation and mobility of heavy metals, particularly in dissolved and colloidal forms, govern their bioavailability and toxicity. Metals like As, Pb, Cd, Cr, Hg, Ni, and Fe often exceed WHO limits, e.g., As up to 7.801 mg/L in Bihar and elevated Cr, Cd, and Pb in Kanpur, posing serious risks such as cancer, kidney damage, and neurological damage, especially for vulnerable populations. Widespread contamination across India, China, Nigeria, and Bangladesh stems from industrial, agricultural, and natural sources. While advanced detection methods (ICP-MS, GFAAS, electrochemical sensors) improve heavy metal analysis, risk assessment tools like HQ, HI, LCR, CF, PLI, and Igeo are essential for translating data into actionable health and ecological insights. Remediation strategies include conventional physicochemical methods and sustainable options like biochar, nanocomposites, and phytoremediation, offering improved selectivity and compatibility. Nanotechnology holds promise for real-time detection, targeted removal, and postremediation monitoring. Despite regulations (WHO, BIS, US EPA), enforcement remains weak in developing countries. EC interactions (e.g., with microplastics or nanoparticles) further complicate toxicity. Thus future efforts must integrate geochemistry, toxicology, AI modeling, and materials science. Establishing real-time monitoring networks and cost-effective, locally adapted solutions is vital. Long-term water sustainability demands a shift from reactive remediation to proactive, science-informed prevention driven by policy, innovation, and stakeholder engagement.

References

Adeyemi, A.A., Ojekunle, Z.O., 2021. Concentrations and health risk assessment of industrial heavy metals pollution in groundwater in Ogun state, Nigeria. Sci. Afr. 24682276. 11. Available from: https://www.journals.elsevier.com/scientific-african. doi:10.1016/j.sciaf.2020.e00666.

Adhikari, S., Yanuar, E., Ng, D.Q., 2022. Widespread nickel contamination in drinking water supplies of elementary schools in Taichung, Taiwan. Environ. Sci. Pollut. Res. 16147499. 29 (9), 12531–12539. Available from: https://link.springer.com/journal/11356. doi:10.1007/s11356-021-15137-1.

Adriano, D.C., 2001. Trace Elements in Terrestrial Environments: Biogeochemistry, Bioavailability, and Risks of Metals, Vol. 860. Springer, New York.

Ahirvar, B.P., Das, P., Srivastava, V., Kumar, M., 2023. Perspectives of heavy metal pollution indices for soil, sediment, and water pollution evaluation: an insight. Total Environ. Res. Themes. 27728099. 6, 100039. doi:10.1016/j.totert.2023.100039.

Ahmed, S.F., Kumar, P.S., Rozbu, M.R., Chowdhury, A.T., Nuzhat, S., Rafa, N., Mahlia, T.M.I., Ong, H.C., Mofijur, M., 2022. Heavy metal toxicity, sources, and remediation techniques for contaminated water and soil. Environ. Technol. Innov. 23521864. 25. Available from: http://www.journals.elsevier.com/environmental-technology-and-innovation/. doi:10.1016/j.eti.2021.102114.

Akpor, O.B., Muchie, M., 2010. Remediation of heavy metals in drinking water and wastewater treatment systems: processes and applications. Int. J. Phys. Sci. 19921950. 5 (12), 1807–1817. Available from http://www.academicjournals.org/ijps/PDF/pdf2010/4%20Oct/Akpor%20and%20Muchie.pdf.

Akter, K.F., Chen, Z., Smith, L., Davey, D., Naidu, R., 2005. Speciation of arsenic in ground water samples: a comparative study of CE-UV, HG-AAS and LC-ICP-MS. Talanta. 00399140. 68 (2), 406–415. doi:10.1016/j.talanta.2005.09.011.

Alinejad, A., Farsani, S.F., Bahmani, Z., Barafrashtehpour, M., Sarsangi, V., Khodadadi, R., Fakhri, Y., 2016. Evaluation of heavy metals level (arsenic, nickel, mercury and lead) effecting on health in drinking water resource of Kohgiluyeh county using geographic information system (GIS). Int. J. Med. Res. Health Sci. 5 (8), 233–241.

Alsubih, M., El Morabet, R., Khan, R.A., Khan, N.A., Ul Haq Khan, M., Ahmed, S., Qadir, A., Changani, F., 2021. Occurrence and health risk assessment of arsenic and heavy metals in groundwater of three industrial areas in Delhi, India. Environ. Sci. Pollut. Res. 16147499. 28 (44), 63017–63031. Available from: https://link.springer.com/journal/11356. doi:10.1007/s11356-021-15062-3.

Asim, M., Nageswara Rao, K. 2021. Assessment of heavy metal pollution in Yamuna River, Delhi-NCR, using heavy metal pollution index and GIS. Environ. Monit. Assess. 0167-6369. 193 (2). doi:10.1007/s10661-021-08886-6.

Baby, J., Raj, J.S., Biby, E.T., Sankarganesh, P., Jeevitha, M.V., Ajisha, S.U., Rajan, S.S., 2011. Toxic effect of heavy metals on aquatic environment. Int. J. Biol. Chem. Sci. 1991-8631. 4 (4). doi:10.4314/ijbcs.v4i4.62976.

Bao, Z., Shi, C., Wenying, T., Lijiao, L., Li, Q., 2022. Recent developments in modification of biochar and its application in soil pollution control and ecoregulation. Environ. Pollut. 02697491. 313, 120184. doi:10.1016/j.envpol.2022.120184.

Baralkiewicz, D., Gramowska, H., Hanc, A., Krzyzaniak, I., 2007. A comparison of ICP-OES and ICP-MS in the determination of elements in lake water. At. Spectrosc. 01955373. 28 (5), 164–170.

Begum, A., Ramaiah, M., Harikrishna, Khan, I., Veena, K., 2009. Heavy metal pollution and chemical profile of cauvery river water. J. Chem. 2090-9063. 6 (1), 47–52. doi:10.1155/2009/154610.

Botle, A., Salgaonkar, S., Tiwari, R., Barabde, G., 2025. Heavy metal pollution in River Ulhas, Maharashtra, India: unraveling contamination dynamics through pollution indices, multivariate analysis, and health risk assessment. Int. J. Environ. Res. 1735-6865. 19 (3). doi:10.1007/s41742-025-00754-2.

Briffa, J., Sinagra, E., Blundell, R., 2020. Heavy metal pollution in the environment and their toxicological effects on humans. Heliyon. 24058440. 6 (9), e04691. doi:10.1016/j.heliyon.2020.e04691.

Burratti, L., Sgreccia, E., Bertelà, F., Galiano, F., 2024. Metal nanostructures in polymeric matrices for optical detection and removal of heavy metal ions, pesticides and dyes from water. Chemosphere. 00456535. 362, 142636. doi:10.1016/j.chemosphere.2024.142636.

Cairns, J.E., 2017. Biological Monitoring in Water Pollution. Elsevier.

Charkiewicz, A.E., Omeljaniuk, W.J., Nowak, K., Garley, M., Nikliński, J., 2023. Cadmium toxicity and health effects—a brief summary. Molecules. 14203049. 28 (18). Available from: http://www.mdpi.com/journal/molecules. doi:10.3390/molecules28186620.

Chen, X., Xinyi, F., Guolian, L., Zhang, J., Haibin, L., Xie, F., 2024. Source-specific probabilistic health risk assessment of heavy metals in surface water of the Yangtze River Basin. Sci. Total Environ. 00489697. 926, 171923. doi:10.1016/j.scitotenv.2024.171923.

Cornelis, R., Nordberg, M., 2007. General Chemistry, Sampling, Analytical Methods, and Speciation Handbook on the Toxicology of Metals. Third ed. Elsevier, Belgium, pp. 11–38. Available from: https://www.sciencedirect.com/book/9780123694133. doi:10.1016/B978-012369413-3/50057-4.

Crawford-Brown, D., Crawford-Brown, S., 2011. The precautionary principle in environmental regulations for drinking water. Environ. Sci. Policy. 14629011. 14 (4), 379–387. doi:10.1016/j.envsci.2011.02.002.

Daliborca, C.V., Fantin, D., Benea, M., Putz, M.V., 2007. Titrimetric methods for water quality assessment and analytical applications. Chem. Bull. "POLITEHNICA" Univ. Timiş. 52 (66), 89–94.

de Paiva Magalhães, D., da Costa Marques, M.R., Baptista, D.F., Buss, D.F., 2015. Metal bioavailability and toxicity in freshwaters. Environ. Chem. Lett. 16103661. 13 (1), 69–87. Available from: https://www.springer.com/journal/10311. doi:10.1007/s10311-015-0491-9.

Dewangan, S.K., Shrivastava, S., Kadri, M., Saruta, S., Yadav, S., Minj, N., 2023. Temperature effect on electrical conductivity (EC) & total dissolved solids (TDS) of water: a review. Int. J. Res. Anal. Rev. 10 (2), 514–520.

Dey, M., Akter, A., Islam, S., Chandra Dey, S., Choudhury, T.R., Fatema, K.J., Begum, B.A., 2021. Assessment of contamination level, pollution risk and source apportionment of heavy metals in the Halda River water, Bangladesh. Heliyon. 24058440. 7 (12). Available from: http://www.journals.elsevier.com/heliyon/. doi:10.1016/j.heliyon.2021.e08625.

Drever, J.I., 1988. The Geochemistry of Natural Waters, Vol. 437. Prentice hall, Englewood Cliffs.

Duodu, G.O., Goonetilleke, A., Ayoko, G.A., 2016. Comparison of pollution indices for the assessment of heavy metal in Brisbane River sediment. Environ. Pollut. 18736424. 219, 1077–1091. Available from: https://www.elsevier.com/inca/publications/store/4/0/5/8/5/6. doi:10.1016/j.envpol.2016.09.008.

Duong Thanh, N., Dang, N.H., Nguyen, L.T.M., Hoang, C.T., Vu, T.T., Nguyen, V.D., Tran, T.D., Bui, V.V., Le Van, N., Cao Thi Thu, T., Bui, L.T.T., Nguyen Thi, T., Tran, L.H., Tran, T.D., 2025. Seasonal hydro-geochemical controls on heavy metal partitioning and ecological risk in the Red River estuaries, northern Vietnam.

Duruibe, J.O., Ogwuegbu, M.O.C., Egwurugwu, J.N., 2007. Heavy metal pollution and human biotoxic effects. Int. J. Phys. Sci. 2 (5), 112–118.

Eid, M.H., Eissa, M., Mohamed, E.A., Ramadan, H.S., Tamás, M., Kovács, A., Szűcs, P., 2024. New approach into human health risk assessment associated with heavy metals in surface water and groundwater using Monte Carlo Method. Sci. Rep. 20452322. 14 (1). Available from: https://www.nature.com/srep/. doi:10.1038/s41598-023-50000-y.

Emmanuel, U.C., Chukwudi, M.I., Monday, S.S., Anthony, A.I., 2022. Human health risk assessment of heavy metals in drinking water sources in three senatorial districts of Anambra State, Nigeria. Toxicol. Rep. 22147500. 9, 869–875. Available from: http://www.journals.elsevier.com/toxicology-reports/. doi:10.1016/j.toxrep.2022.04.011.

Eze, V., Ndife, C., Muogbo, M., 2021. Carcinogenic and non-carcinogenic health risk assessment of heavy metals in Njaba River, Imo State, Nigeria. Braz. J. Anal. Chem. 21793425. doi:10.30744/brjac.2179-3425.ar-05-2021.

Fadlillah, L.N., Utami, S., Rachmawati, A.A., Jayanto, G.D., Widyastuti, M., 2023. Ecological risk and source identifications of heavy metals contamination in the water and surface sediments from anthropogenic impacts of urban river, Indonesia. Heliyon. 24058440. 9 (4). Available from: http://www.journals.elsevier.com/heliyon/. doi:10.1016/j.heliyon.2023.e15485.

Fenglian, F., Wang, Q., 2011. Removal of heavy metal ions from wastewaters: a review. J. Environ. Manag. 03014797. 92 (3), 407–418. doi:10.1016/j.jenvman.2010.11.011.

Fernandez-Luqueno, F., Lopez-Valdez, F., Gamero-Melo, P., Luna-Suarez, S., Aguilera-Gonzalez, E.N., Martínez, A.I., Pérez-Velázquez, I.R., 2013. Heavy metal pollution in drinking water-a global risk for human health: a review. Afr. J. Environ. Sci. Technol. 7 (7), 567–584.

Gayathri, S., Krishnan, K.A., Krishnakumar, A., Maya, T.V., Dev, V.V., Antony, S., Arun, V., 2021. Monitoring of heavy metal contamination in Netravati river basin: overview of pollution indices and risk assessment. Sustain. Water Resour. Manag. 7 (2), 20. . doi:10.1007/s40899-021-00502-2.

Georgaki, M.N., Charalambous, M., 2023. Toxic chromium in water and the effects on the human body: a systematic review. J. Water Health. 19967829. 21 (2), 205–223. Available from: https://watermark.silverchair.com/jwh0210205.pdf. doi:10.2166/wh.2022.214.

Gimeno-García, E., Andreu, V., Boluda, R., 1996. Heavy metals incidence in the application of inorganic fertilizers and pesticides to rice farming soils. Environ. Pollut. 02697491. 92 (1), 19–25. doi:10.1016/0269-7491(95)00090-9.

Goher, M.E., Hassan, A.M., Abdel-Moniem, I.A., Fahmy, A.H., El-Sayed, S.M., 2014. Evaluation of surface water quality and heavy metal indices of Ismailia Canal, Nile River, Egypt. Egypt. J. Aquat. Res. 20903278. 40 (3), 225–233. Available from: http://www.elsevier.com/journals/egyptian-journal-of-aquatic-research/1687-4285. doi:10.1016/j.ejar.2014.09.001.

Guo, Y., Liu, C., Rongke, Y., Duan, Q., 2020. Advances on water quality detection by UV-Vis spectroscopy. Appl. Sci. 2076-3417. 10 (19), 6874. doi:10.3390/app10196874.

Gupta, V., Kumar, D., Dwivedi, A., Vishwakarma, U., Malik, D.S., Paroha, S., Gupta, N., 2023. Heavy metal contamination in river water, sediment, groundwater and human blood, from Kanpur, Uttar Pradesh, India. Environ. Geochem. Health 45 (5), 1807–1818.

Han, L., Sun, Y., Wang, S., Su, T., Cai, W., Shao, X., 2022. Understanding the water structures by near-infrared and Raman spectroscopy. J. Raman Spectrosc. 0377-0486. 53 (10), 1686–1693. doi:10.1002/jrs.6399.

He, M., Xu, Z., Hou, D., Gao, B., Cao, X., Ok, Y.S., Rinklebe, J., Bolan, N.S., Tsang, D.C.W., 2022. Waste-derived biochar for water pollution control and sustainable development. Nat. Rev. Earth Environ. 2662138X. 3 (7), 444–460. Available from: https://www.nature.com/natrevearthenviron/. doi:10.1038/s43017-022-00306-8.

Hejabi, A.T., Basavarajappa, H.T., Karbassi, A.R., Monavari, S.M., 2011. Heavy metal pollution in water and sediments in the Kabini River, Karnataka, India. Environ. Monit. Assess. 0167-6369. 182 (1-4), 1–13. doi:10.1007/s10661-010-1854-0.

Hopenhayn, C., 2006. Arsenic in drinking water: impact on human health. Elements 2 (2), 103–107.

Islam, M.S., Ahmed, M.K., Raknuzzaman, M., Habibullah -Al- Mamun, M., Islam, M.K., 2015. Heavy metal pollution in surface water and sediment: a preliminary assessment of an urban river in a developing country. Ecol. Indic. 1470160X. 48, 282–291. Available from: http://www.elsevier.com/locate/ecolind. doi:10.1016/j.ecolind.2014.08.016.

Jagaba, A.H., Lawal, I.M., Birniwa, A.H., Affam, A.C., Usman, A.K., Soja, U.B., Saleh, D., Hussaini, A., Noor, A., Aliyu Yaro, N.S., 2024. Sources of Water Contamination by Heavy Metals Membrane Technologies for Heavy Metal Removal from Water. CRC Press, Nigeria, pp. 3–27. Available from: https://www.taylorfrancis.com/chapters/edit/10.1201/9781003326281/review-effect-ionic-zwitterionic-materials-membrane-performance-heavy-metal-removal-application-syed-ibrahim-gnani-peer-mohamed-arun-isloor?context=ubx. doi:10.1201/9781003326281-2.

Jamshaid, M., Khan, A.A., Ahmed, K., Saleem, M., 2018. Heavy metal in drinking water its effect on human health and its treatment techniques-a review. Int. J. Biosci. 12 (4), 223–240.

Jin, M., Yuan, H., Liu, B., Peng, J., Xu, L., Yang, D., 2020. Review of the distribution and detection methods of heavy metals in the environment. Anal. Methods. 1759-9660. 12 (48), 5747–5766. doi:10.1039/d0ay01577f.

Kaplan, G., Yalcinkaya, F., Altıok, E., Pietrelli, A., Nastro, R.A., Lovecchio, N., Ieropoulos, I.A., Tsipa, A., 2024. The role of remote sensing in the evolution of water pollution detection and monitoring: a comprehensive review. Physics and Chemistry of the Earth, Parts A/B/C. 14747065. 136, 103712. doi:10.1016/j.pce.2024.103712.

Kar, D., Sur, P., Mandai, S.K., Saha, T., Kole, R.K., 2008. Assessment of heavy metal pollution in surface water. Int. J. Environ. Sci. Technol. 1735-1472. 5 (1), 119–124. doi:10.1007/bf03326004.

Kazemi, A., Esmaeilbeigi, M., Sahebi, Z., Ansari, A., 2022. Health risk assessment of total chromium in the qanat as historical drinking water supplying system. Sci. Total Environ. 00489697. 807 (Pt 2), 150795. doi:10.1016/j.scitotenv.2021.150795.

Ke, X., Gui, S., Huang, H., Zhang, H., Wang, C., Guo, W., 2017. Ecological risk assessment and source identification for heavy metals in surface sediment from the Liaohe River protected area, China. Chemosphere. 00456535. 175, 473–481. doi:10.1016/j.chemosphere.2017.02.029.

Khan, R., Saxena, A., Shukla, S., 2022. Assessment of the impact of COVID-19 lockdown on the heavy metal pollution in the River Gomti, Lucknow city, Uttar Pradesh, India. Environ. Qual. Manag. 1088-1913. 31 (4), 41–49. doi:10.1002/tqem.21746.

Khan, R., Saxena, A., Shukla, S., Sekar, S., Senapathi, V., Jianhua, W., 2021. Environmental contamination by heavy metals and associated human health risk assessment: a case study of surface water in Gomti River Basin, India. Environ. Sci. Pollut. Res. 0944-1344. 28 (40), 56105–56116. doi:10.1007/s11356-021-14592-0.

Khan, Z.I., Ahmad, K., Rehman, S., Siddique, S., Bashir, H., Zafar, A., Sohail, M., Ali, S.A., Cazzato, E., De Mastro, G., 2017. Health risk assessment of heavy metals in wheat using different water qualities: implication for human health. Environ. Sci. Pollut. Res. 16147499. 24 (1), 947–955. Available from: http://www.springerlink.com/content/0944-1344. doi:10.1007/s11356-016-7865-9.

Kobielska, P.A., Howarth, A.J., Farha, O.K., Nayak, S., 2018. Metal–organic frameworks for heavy metal removal from water. Coord. Chem. Rev. 00108545. 358, 92–107. Available from: http://www.journals.elsevier.com/coordination-chemistry-reviews/. doi:10.1016/j.ccr.2017.12.010.

Kontos, T. D., Foteinis, S., 2020. Mercury emissions from anthropogenic activities and their environmental impacts: a review. Environ. Sci. Pollut. Res. 27, 29827–29840. https://doi.org/10.1007/s11356-020-09257-7

Kumar, M., Isloor, A.M., Todeti, S.R., Nagaraja, H.S., Ismail, A.F., Susanti, R., 2021. Effect of binary zinc-magnesium oxides on polyphenylsulfone/cellulose acetate derivatives hollow fiber membranes for the decontamination of arsenic from drinking water. Chem. Eng. J. 13858947. 405. Available from: https://www.elsevier.com/inca/publications/store/6/0/1/2/7/3/index.htt. doi:10.1016/j.cej.2020.126809.

Kumar, V., Agrawal, S., Bhat, S.A., Heloisa Pinê Américo-Pinheiro, J., Shahi, S.K., Kumar, S., 2022. Environmental impact, health hazards, and plant-microbes synergism in remediation of emerging contaminants. Clean. Chem. Eng. 27727823. 2, 100030. doi:10.1016/j.clce.2022.100030.

Li, X., Bing, J., Zhang, J., Guo, L., Deng, Z., Wang, D., Liu, L., 2022. Ecological risk assessment and sources identification of heavy metals in surface sediments of a river–reservoir system. Sci. Total Environ. 00489697. 842, 156683. doi:10.1016/j.scitotenv.2022.156683.

Liu, L., Wu, J., He, S., Wang, L., 2022. Occurrence and distribution of groundwater fluoride and manganese in the weining plain (China) and their probabilistic health risk quantification. Expo. Health. 2451-9766. 14 (2), 263–279. doi:10.1007/s12403-021-00434-4.

Llaver, M., Fiorentini, E.F., Oviedo, M.N., Quintas, P.Y., Wuilloud R.G., 2021. Elemental speciation analysis in environmental studies: latest trends and ecological impact. Int. J. Environ. Res. Public Health 18(22), 12135.

Luo, P., Xu, C., Kang, S., Huo, A., Lyu, J., Zhou, M., Nover, D., 2021. Heavy metals in water and surface sediments of the Fenghe River Basin, China: assessment and source analysis. Water Sci. Technol. 0273-1223. 84 (10-11), 3072–3090. doi:10.2166/wst.2021.335.

Maroušek, J., Maroušková, A., Zoubek, T., Bartoš, P., 2022. Economic impacts of soil fertility degradation by traces of iron from drinking water treatment. Environ. Dev. Sustain. 1387-585X. 24 (4), 4835–4844. doi:10.1007/s10668-021-01636-1.

Martin, S., Griswold, W., Gopalakrishnan, K., 2021. Human health effects of heavy metals: exposure pathways and toxicological impacts. Environ. Toxicol. Pharmacol., 88, 103756. https://doi.org/10.1016/j.etap.2021.103756

Matta, G., Kumar, A., Nayak, A., Kumar, P., Kumar, A., Naik, P.K., Singh, S.K., 2023. Assessing heavy metal index referencing health risk in Ganga River system. Int. J. River Basin Manag. 18142060. 21 (4), 759–769. Available from: http://www.tandfonline.com/loi/trbm20. doi:10.1080/15715124.2022.2098756.

Mohamed, A., Othman, A., Asmaoy, A., Galal, W.F., Mohammed, M.A.A., 2025. Assessment of heavy metal pollution of groundwater at the upper stream of Wadi Ranyah, Saudi Arabia, using multivariate statistical approach. Appl. Water Sci. 21905495. 15 (4). Available from: https://www.springer.com/journal/13201. doi:10.1007/s13201-025-02397-2.

Mohod, C.V., Dhote, J., 2013. Review of heavy metals in drinking water and their effect on human health. International Journal of Innovative Research in Science, Engineering and Technology 2 (7), 2992–2996.

Mokarram, M., Saber, A., Sheykhi, V., 2020. Effects of heavy metal contamination on river water quality due to release of industrial effluents. J. Clean. Prod. 09596526. 277, 123380. doi:10.1016/j.jclepro.2020.123380.

Niampradit, S., Kiangkoo, N., Mingkhwan, R., Kliengchuay, W., Worakhunpiset, S., Limpananont, Y., Hongsibsong, S., Inthorn, D., Tantrakarnapa, K., 2024. Occurrence, distribution, and ecological risk assessment of heavy metals in Chao Phraya River, Thailand. Sci. Rep. 2045-2322. 14 (1). doi:10.1038/s41598-024-59133-0.

Nurliyana, M.R., Sahdan, M.Z., Wibowo, K.M., Muslihati, A., Saim, H., Ahmad, S.A., Sari, Y., Mansor, Z., 2018. The detection method of escherichia coli in water resources: a review. J. Phys. Conf. Ser. 1742-6588. 995 (1), 012065. doi:10.1088/1742-6596/995/1/012065.

Olenin, S., Elliott, M., Bysveen, I., Culverhouse, P.F., Daunys, D., Dubelaar, G.B.J., Gollasch, S., Goulletquer, P., Jelmert, A., Kantor, Y., Mézeth, K.B., Minchin, D., Occhipinti-Ambrogi, A., Olenina, I., Vandekerkhove, J., 2011. Recommendations on methods for the detection and control of biological pollution in marine coastal waters. Mar. Pollut. Bull. 0025326X. 62 (12), 2598–2604. doi:10.1016/j.marpolbul.2011.08.011.

Pal, S., Singh, S.K., Singh, P., Pal, S., Kashiwar, S.R., 2024. Spatial pattern of groundwater arsenic contamination in Patna, Saran, and Vaishali districts of Gangetic plains of Bihar, India. Environ. Sci. Pollut. Res. 16147499. 31 (41), 54163–54177. Available from: https://www.springer.com/journal/11356. doi:10.1007/s11356-022-25105-y.

Pandey, G., Madhuri, S., 2014. Heavy metals causing toxicity in animals and fishes. Res. J. Animal, Vet. and Fishery Sci. 2 (2), 17–23.

Pandit, D., Haque, M.M., Bhuyan, M.S., Harun-Al-Rashid, A., Barman, P.P., Roy, R., Sarker, B., Saifullah, M.K., Kunda, M., 2025. A comprehensive scenario of heavy metals pollution in the rivers of Bangladesh during the last two decades. Environ. Sci. Pollut. Res., 32 (16), 10240–10257.

Pinheiro, P.C., Daniel-da-Silva, A.L., Nogueira, H.I.S., Trindade, T., 2018. Functionalized inorganic nanoparticles for magnetic separation and SERS detection of water pollutants. Eur. J. Inorg. Chem. 10990682. 2018 (30), 3443–3461. Available from: http://onlinelibrary.wiley.com/journal/10.1002/(ISSN)1099-0682c. doi:10.1002/ejic.201800132.

Punia, P., Bharti, M.K., Dhar, R., Thakur, P., Thakur, A., 2022. Recent advances in detection and removal of heavy metals from contaminated water. ChemBioEng Rev. 21969744. 9 (4), 351–369. Available from: https://www.onlinelibrary.wiley.com/journal/10.1002/(ISSN)2196-9744. doi:10.1002/cben.202100053.

Qi, X., Lian, Y., Xie, L., Wang, Y., Lu, Z., 2024. Water quality detection based on UV-Vis and NIR spectroscopy: a review. Appl. Spectrosc. Rev. 0570-4928. 59 (8), 1036–1060. doi:10.1080/05704928.2023.2294458.

Qiufen, H., Yang, G., Zhao, Y., Yin, J., 2003. Determination of copper, nickel, cobalt, silver, lead, cadmium, and mercury ions in water by solid-phase extraction and the RP-HPLC with UV-Vis detection. Anal. Bioanal. Chem. 1618-2642. 375 (6), 831–835. doi:10.1007/s00216-003-1828-y.

Rad, S.M., Ray, A.K., Barghi, S., 2022. Water pollution and agriculture pesticide. Clean Technol. 25718797. 4 (4), 1088–1102. Available from: https://www.mdpi.com/journal/cleantechnol. doi:10.3390/cleantechnol4040066.

Raja, V., Lakshmi, R.V., Sekar, C.P., Chidambaram, S., Neelakantan, M.A., 2021. Health risk assessment of heavy metals in groundwater of industrial township Virudhunagar, Tamil Nadu, India. Arch. Environ. Contam. Toxicol. 14320703. 80 (1), 144–163. Available from: https://www.link.springer.de/link/service/journals/00244/index.htm. doi:10.1007/s00244-020-00795-y.

Rajan, S., Nandimandalam, J.R., 2024. Environmental health risk assessment and source apportion of heavy metals using chemometrics and pollution indices in the upper Yamuna river basin, India. Chemosphere. 00456535. 346, 140570. doi:10.1016/j.chemosphere.2023.140570.

Reddy, V.H., Prasad, P.M.N., Reddy, A.V.R., Reddy, Y.V.R., 2012. Determination of heavy metals in surface and groundwater in and around Tirupati, Chittoor (Di), Andhra Pradesh, India. Der Pharma Chemica 0975413X. 4 (6), 2442–2448. Available from http://derpharmachemica.com/vol4-iss6/DPC-2012-4-6-2442-2448.pdf.

Reza, R., Singh, G., 2010. Heavy metal contamination and its indexing approach for river water. Int. J. Environ. Sci. Technol. 1735-1472. 7 (4), 785–792. doi:10.1007/bf03326187.

Sadeghi, H., Fazlzadeh, M., Zarei, A., Mahvi, A.H., Nazmara, S., 2022. Spatial distribution and contamination of heavy metals in surface water, groundwater and topsoil surrounding Moghan's tannery site in Ardabil, Iran. Int. J. Environ. Anal. Chem. 10290397. 102 (5), 1049–1059. Available from: https://www.tandf.co.uk/journals/titles/03067319.asp. doi:10.1080/03067319.2020.1730342.

Samal, P., Singarasubramanian, S.R., Manoj, M.C., Srivastava, J., Dsouza, N., Balakrishna, K., Chauhan, M.M., Ali, S., 2023. Heavy metal contamination assessment and its associated human health risk evaluation in the Mahanadi River sediments, India. Int. J. Environ. Sci. Technol. 1735-1472. 20 (10), 10673–10694. doi:10.1007/s13762-022-04630-w.

Sankhla, M.S., Kumar, R., Prasad, L., 2019. Zinc Impurity in Drinking Water and Its Toxic Effect on Human Health, Indian Congress of Forensic Medicine & Toxicology.

Seelam, P.K., Golla, B., Vemula, S.R., 2021. Human health risk assessment of heavy metals in environmental matrices: concepts, methodologies, and applications. Environ. Chem. Lett. 19, 3605–3625. https://doi.org/10.1007/s10311-021-01243-8

Setia, R., Dhaliwal, S.S., Singh, R., Kumar, V., Taneja, S., Kukal, S.S., Pateriya, B., 2021. Phytoavailability and human risk assessment of heavy metals in soils and food crops around Sutlej river, India. Chemosphere. 18791298. 263. Available from: https://www.elsevier.com/locate/chemosphere. doi:10.1016/j.chemosphere.2020.128321.

Shaibur, M.R., Howlader, M., Ahmmed, I., Sarwar, S., Hussam, A., 2024. Water quality index and health risk assessment for heavy metals in groundwater of Kashiani and Kotalipara upazila, Gopalganj, Bangladesh. Appl. Water Sci. 21905495. 14 (5). Available from: https://www.springer.com/journal/13201. doi:10.1007/s13201-024-02169-4.

Sharrett, A.R., Carter, A.P., Orheimt, R.M., Feinleib, M., 1982. Daily intake of lead, cadmium, copper, and zinc from drinking water: the Seattle study of trace metal exposure. Environ. Res. 10960953. 28 (2), 456–475. doi:10.1016/0013-9351(82)90142-6.

Shi, J., Zhang, J., Wang, C., Liu, Y., Li, J., 2023. Research progress on the magnetite nanoparticles in the fields of water pollution control and detection. Chemosphere. 00456535. 336, 139220. doi:10.1016/j.chemosphere.2023.139220.

Sibal, L.N., Espino, M.P.B., 2018. Heavy metals in lake water: a review on occurrence and analytical determination. Int. J. Environ. Anal. Chem. 10290397. 98 (6), 536–554. Available from: https://www.tandf.co.uk/journals/titles/03067319.asp doi:10.1080/03067319.2018.1481212.

Siddiqui, E., Pandey, J., 2019. Assessment of heavy metal pollution in water and surface sediment and evaluation of ecological risks associated with sediment contamination in the Ganga River: a basin-scale study. Environ. Sci. Pollut. Res. 0944-1344. 26 (11), 10926–10940. doi:10.1007/s11356-019-04495-6.

Siegel, F.R., 2002. Environmental Geochemistry of Potentially Toxic Metals, Vol. 32. Springer, Berlin.

Silva, D.M., Domingues, L., 2015. On the track for an efficient detection of Escherichia coli in water: a review on PCR-based methods. Ecotoxicol. Environ. Saf. 01476513. 113, 400–411. doi:10.1016/j.ecoenv.2014.12.015.

Singh, H.P., Mahajan, P., Kaur, S., Batish, D.R., Kohli, R.K., 2013. Chromium toxicity and tolerance in plants. Environ. Chem. Lett. 16103661. 11 (3), 229–254. Available from: https://www.springer.com/journal/10311. doi:10.1007/s10311-013-0407-5.

Singh, K.P., Malik, A., Sinha, S., 2005. Water quality assessment and apportionment of pollution sources of Gomti river (India) using multivariate statistical techniques – a case study. Anal. Chim. Acta. 00032670. 538 (1-2), 355–374. Available from: http://www.journals.elsevier.com/analytica-chimica-acta/. doi:10.1016/j.aca.2005.02.006.

Singh, N., Gupta, V.K., Kumar, A., Sharma, B., 2017. Synergistic effects of heavy metals and pesticides in living systems. Front. Chem. 22962646. 5. Available from: http://journal.frontiersin.org/journal/chemistry. doi:10.3389/fchem.2017.00070.

Singh, R., Kumar, A., Sharma, P., Gupta, N., 2023. Chemical precipitation methods for heavy metal removal from wastewater: mechanisms, advantages, and limitations. J. Water Process Eng. 52, 103537. https://doi.org/10.1016/j.jwpe.2023.103537

Skoog, D.A., Holler, F.J., Crouch, S.R., 2019. Textbook "Principles of Instrumental Analysis". Cengage Learning, pp. 6.

Sojka, M., Jaskuła, J., 2022. Heavy metals in river sediments: contamination, toxicity, and source identification—a case study from Poland. Int. J. Environ. Res. Public Health. 1660-4601. 19 (17), 10502. doi:10.3390/ijerph191710502.

Srinivasa, G.S., Govil, P.K., 2008. Distribution of heavy metals in surface water of Ranipet industrial area in Tamil Nadu, India. Environ. Monit. Assess. 15732959. 136 (1-3), 197–207. doi:10.1007/s10661-007-9675-5.

Tang, X., Wang, P.-Y., Buchter, G., 2018. Ion-selective electrodes for detection of lead (II) in drinking water: a mini-review. Environments. 2076-3298. 5 (9), 95. doi:10.3390/environments5090095.

Templeton, D.M., Ariese, F., Cornelis, R., Danielsson, L.G., Muntau, H., Van Leeuwen, H.P., Łobiński, R., 2000. Guidelines for terms related to chemical speciation and fractionation of elements. Definitions, structural aspects, and methodological approaches (IUPAC recommendations 2000). Pure Appl. Chem. 00334545. 72 (8), 1453–1470. Available from: http://www.degruyter.com/view/j/pac. doi:10.1351/pac200072081453.

Thompson, J., Morrison, G., 1951. Determination of organic nitrogen. control of variables in the use of nessler's reagent. Anal. Chem. 0003-2700. 23 (8), 1153–1157. doi:10.1021/ac60056a029.

Triassi, M., Cerino, P., Montuori, P., Pizzolante, A., Trama, U., Nicodemo, F., D'Auria, J.L., De Vita, S., De Rosa, E., Limone, A., 2023. Heavy metals in groundwater of southern Italy: occurrence and potential adverse effects on the environment and human health. Int. J. Environ. Res. Public Health. 1660-4601. 20 (3), 1693. doi:10.3390/ijerph20031693.

U.S. Environmental Protection Agency (USEPA), 2023. Lead in drinking water. Retrieved from: https://www.epa.gov/ground-water-and-drinking-water/basic-information-about-lead-drinking-water.

Varol, M., 2011. Assessment of heavy metal contamination in sediments of the Tigris River (Turkey) using pollution indices and multivariate statistical techniques. J. Hazard. Mater. 03043894. 195, 355–364. doi:10.1016/j.jhazmat.2011.08.051.

Verma, R.K., Sankhla, M.S., Kumar, R., 2018. Mercury contamination in water & its impact on public health. Int. J. Forensic Sci. 1 (2), 72–78.

Vig, N., Ravindra, K., Mor, S., 2023. Heavy metal pollution assessment of groundwater and associated health risks around coal thermal power plant, Punjab, India. Int. J. Environ. Sci. Technol. 17352630. 1735-1472. 20 (6), 6259–6274. Available from: https://www.springer.com/journal/13762. doi:10.1007/s13762-022-04284-8.

Villaescusa, I., Bollinger, J.C., 2008. Arsenic in drinking water: sources, occurrence and health effects (a review). Rev. Environ. Sci. Biotechnol. 15691705. 7 (4), 307–323. doi:10.1007/s11157-008-9138-7.

Wang, C., Yu, C., 2013. Detection of chemical pollutants in water using gold nanoparticles as sensors: a review. Rev. Anal. Chem. 2191-0189. 32 (1), 1–14. doi:10.1515/revac-2012-0023.

Wang, L., Wang, Y., Xu, C., An, Z., Wang, S., 2011. Analysis and evaluation of the source of heavy metals in water of the River Changjiang. Environ. Monit. Assess. 0167-6369. 173 (1-4), 301–313. doi:10.1007/s10661-010-1388-5.

Wang, Z., Su, Q., Wang, S., Gao, Z., Liu, J., 2021. Spatial distribution and health risk assessment of dissolved heavy metals in groundwater of eastern China coastal zone. Environ. Pollut. 02697491. 290, 118016. doi:10.1016/j.envpol.2021.118016.

Weiner, E.R., 2008. Applications of Environmental Aquatic Chemistry: A Practical Guide, CRC Press.

Wongsasuluk, P., Chotpantarat, S., Siriwong, W., Robson, M., 2014. Heavy metal contamination and human health risk assessment in drinking water from shallow groundwater wells in an agricultural area in Ubon Ratchathani province, Thailand. Environ. Geochem. Health. 0269-4042. 36 (1), 169–182. doi:10.1007/s10653-013-9537-8.

Xia, F., Qu, L., Wang, T., Luo, L., Chen, H., Dahlgren, R.A., Zhang, M., Mei, K., Huang, H., 2018. Distribution and source analysis of heavy metal pollutants in sediments of a rapid developing urban river system. Chemosphere. 18791298. 207, 218–228. Available from: https://www.elsevier.com/locate/chemosphere. doi:10.1016/j.chemosphere.2018.05.090.

Xie, F., Yu, M., Yuan, Q., Meng, Y., Qie, Y., Shang, Z., Luan, F., Zhang, D., 2022. Spatial distribution, pollution assessment, and source identification of heavy metals in the Yellow River. J. Hazard. Mater. 03043894. 436, 129309. doi:10.1016/j.jhazmat.2022.129309.

Xie, Q., Ren, B., 2022. Pollution and risk assessment of heavy metals in rivers in the antimony capital of Xikuangshan. Sci. Rep. 2045-2322. 12 (1). doi:10.1038/s41598-022-18584-z.

Yao, N., Li, C., Yu, J., Xu, Q., Wei, S., Tian, Z., Yang, Z., Yang, W., Shen, J., 2020. Insight into adsorption of combined antibiotic-heavy metal contaminants on graphene oxide in water. Sep. Purif. Technol. 13835866. 236, 116278. doi:10.1016/j.seppur.2019.116278.

Yaroshenko, I., Kirsanov, D., Marjanovic, M., Lieberzeit, P.A., Korostynska, O., Mason, A., Frau, I., Legin, A., 2020. Real-time water quality monitoring with chemical sensors. Sensors. 1424-8220. 20 (12), 3432. doi:10.3390/s20123432.

Yi, Y., Zhang, S., 2021. Assessment of heavy metal contamination and ecological risk in sediments using contamination factor and pollution load index approaches. Ecotoxicol. Environ. Saf. 208, 111657. https://doi.org/10.1016/j.ecoenv.2020.111657

Yuan, X., Chapman, R.L., Wu, Z., 2011. Analytical methods for heavy metals in herbal medicines. Phytochem. Anal. 10991565. 22 (3), 189–198. Available from: http://onlinelibrary.wiley.com/journal/10.1002/(ISSN)1099-1565. doi:10.1002/pca.1287.

Zamora-Ledezma, C., Negrete-Bolagay, D., Figueroa, F., Zamora-Ledezma, E., Ni, M., Alexis, F., Guerrero, V.H., 2021. Heavy metal water pollution: a fresh look about hazards, novel and conventional remediation methods. Environ. Technol. Innov. 23521864. 22, 101504. doi:10.1016/j.eti.2021.101504.

Zhang, J., Li, X., Guo, L., Deng, Z., Wang, D., Liu, L., 2021a. Assessment of heavy metal pollution and water quality characteristics of the reservoir control reaches in the middle Han River, China. Sci. Total Environ. 00489697. 799, 149472. doi:10.1016/j.scitotenv.2021.149472.

Zhang, S., Tian, Y., Guo, Y., Shan, J., Liu, R., 2021b. Manganese release from corrosion products of cast iron pipes in drinking water distribution systems: effect of water temperature, pH, alkalinity, SO_42- concentration sand disinfectants. Chemosphere. 00456535. 262, 127904. doi:10.1016/j.chemosphere.2020.127904.

Zharikova, E.P., Yu Grigoriev, J., Grigorieva, A.L., 2022. Artificial intelligence methods for detecting water pollution. IOP Conf. Ser. Earth Environ. Sci. 1755-1307. 988 (2), 022082. doi:10.1088/1755-1315/988/2/022082.

Zhushan, F., Shuhua, X., 2020. The effects of heavy metals on human metabolism. Toxicol. Mech. Methods. 1537-6516. 30 (3), 167–176. doi:10.1080/15376516.2019.1701594.

Zolkefli, N., Sharuddin, S.S., Yusoff, M.Z.M., Hassan, M.A., Maeda, T., Ramli, N., 2020. A review of current and emerging approaches for water pollution monitoring. Water (Switzerland). 20734441. 12 (12), 1–30. Available from: http://www.mdpi.com/journal/water. doi:10.3390/w12123417.

Section II

Impact of pollution

Chapter 10

Effect on aquatic flora and fauna

Kanhaiya Kumar, Madhusmita Kalita, and Kasi Marimuthu
Department of Environmental Science, Tezpur University, Tezpur, India

1 Introduction

Water is an essential resource for the survival of human beings on Earth. Rivers, ponds, and lakes are the major components of a freshwater ecosystem (Mishra, 2023). They support habitats of diverse aquatic flora and fauna, including human livelihoods, and regulate biogeochemical cycles (Canning and Death, 2019). Rivers are channels that connect terrestrial and aquatic ecosystems. Aquatic biodiversity provides resources for livelihoods, such as irrigation, fisheries, and drinking water supply (Khosrovyan, 2024). However, rapid population growth, urbanization, industrialization, and the expansion of agricultural activities have severely deteriorated the water quality of the riverine ecosystem (Li et al., 2023). Particularly in developing countries, people frequently discharge untreated domestic wastewater, sewage, agricultural waste, and industrial effluents into nearby water bodies (Kesari et al., 2021; Singh et al., 2023). When this wastewater, industrial effluents, and other chemical contaminants from human activity interact with river water, it causes riverine pollution (Nishmitha et al., 2025).

The contamination of river water due to the release of anthropogenic substances, such as toxic chemicals and harmful microorganisms, which leads to changes in the physicochemical properties of river water and negatively affects the ecosystem, is known as riverine pollution (Uddin and Jeong, 2021). The sources of these pollutants are varied, including industrial wastewater, agricultural activities, mining, atmospheric deposition, and natural geogenic phenomena (Xu et al., 2022). Surface runoff facilitates the transport of pollutants from their sources to the river and alters the pH, dissolved oxygen, and nutrient levels of river water, thereby disturbing the equilibrium of the riverine ecosystem (Morin-Crini et al., 2022; Hairom et al., 2021). Riverine pollution not only disrupts the river's physicochemical properties but also adversely impacts human health and livelihoods (Nyagongo et al., 2024).

The impacts of pollution on aquatic flora and fauna can be both direct and indirect. It can cause physiological, neurological, and behavioral changes

River Pollution. https://doi.org/10.1016/B978-0-443-45024-2.00002-6

directly, and indirectly, it causes disturbances in their habitats, alters food availability, and disrupts the food web (Anetor et al., 2022; Bhendekar et al., 2024). The most common impacts of riverine pollution are nutrient enrichment, eutrophication, sudden declines in dissolved oxygen, increases in fish mortality rates, rises in temperature, and biodiversity loss (Tiwari and Pal, 2022). The key contaminants responsible for the ecological degradation of riverine ecosystems are excess nutrients, pesticides, organic pollutants, heavy metals, and microplastics (Ghirardelli et al., 2021).

It is essential to understand how pollutants affect aquatic flora and fauna for the assessment of freshwater ecosystem health and the development of mitigation plans (Xu et al., 2022; Bhendekar et al., 2024). In this chapter, we focus on the biological and ecological effects of riverine pollution on aquatic flora and fauna, with special emphasis on aquatic plants, planktonic organisms, invertebrates, and fish communities. It also includes the impacts of pollution from major river systems, including Indian rivers like the Ganga and Brahmaputra, to highlight region-specific issues and research findings.

2 Overview of riverine pollution and ecological stress in aquatic organisms

Most of the industries and industrial areas were established near river channels and water bodies. These water bodies, including rivers and marine waters, act as the ultimate sink for environmental pollution (Khan et al., 2021). The nature and type of pollutants depend on the source of origin, land use patterns, and regional hydrogeological conditions (Akhtar et al., 2021).

2.1 Types of pollutants responsible for riverine pollution

There are various types of pollutants present in river water or pond water, released from multiple sources, as shown in Fig. 10.1, which can negatively affect the aquatic flora and fauna. Some of the major pollutants are discussed in the following.

2.1.1 Organic pollutants

In river water, organic matter mainly comes from domestic sewage, cattle farming, agricultural waste, industrial waste, and car washing. These pollutants include pharmaceuticals, detergents, persistent organic pollutants (POPs), and petroleum products like oil and grease (Ghirardelli et al., 2021). When this organic matter decomposes, it consumes more oxygen, leading to a decline in the dissolved oxygen in river water, and sometimes it creates hypoxic conditions (Eriksen et al., 2022). Due to the presence of organic matter, microbial activity also increases, which can change the biogeochemical cycle of river water. Some organic matters like POPs and pharmaceuticals are toxic in nature and can remain in the water for a long time and move to longer distances (Yun et al., 2024).

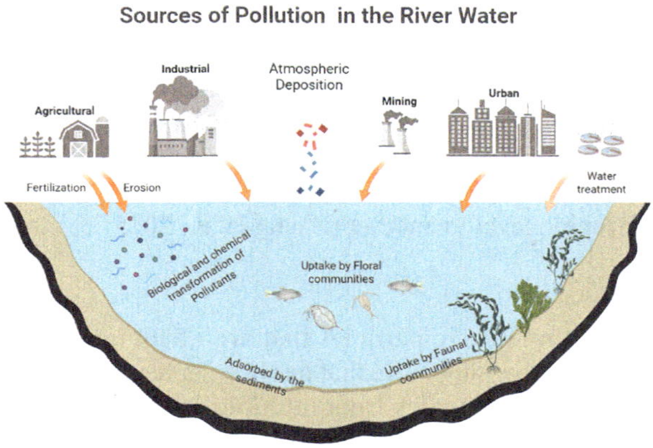

FIG. 10.1 Sources of pollution in river water. Various sources contributing to river pollution.

2.1.2 Nutrients (NPK)

The presence of excess nutrients in river water can also affect the life of aquatic flora and fauna (Tiwari and Pal, 2022). When excess nutrients, mainly nitrogen, phosphorus, and potassium from fertilizers, cattle farming, sewage, and industries, are transported to the river ecosystem through agricultural runoff and direct discharge, they cause rapid growth of algae and macrophytes, which leads to oxygen depletion and the death of fish species (Akhtar et al., 2021; Tiwari and Pal, 2022).

2.1.3 Heavy metals

Heavy metal pollution in riverine water is a serious environmental issue. These metals, particularly arsenic, lead, mercury, cadmium, and nickel, can cause significant effects on aquatic organisms (Das et al., 2024). The sources of heavy metals in river water can be natural or anthropogenic. Naturally, they are present in rocks and minerals and can contaminate river water through geogenic processes, while anthropogenically, they originate from industrial effluents, sewage treatment plants, mining activities, agricultural runoff, electronic waste, domestic waste, etc. (Kapoor and Singh, 2021). These metals do not degrade naturally like organic pollutants and can bioaccumulate in aquatic organisms and biomagnify in the aquatic food web (Danovaro et al., 2023). Trace quantities of metals like iron, zinc, and copper are essential for the growth and development of phytoplankton, but when the concentration of these metals exceeds physiological thresholds, toxicity develops (Wei et al., 2025).

2.1.4 Pesticides

Pesticides are synthetic organic compounds that are used to kill pests like insects, weeds, fungi, and other unwanted organisms in agriculture and aquaculture.

They are mainly used in agricultural fields to eliminate unwanted organisms that threaten crops and vegetables. They are also used for domestic purposes to kill insects and mosquitoes (Kumar et al., 2023a). Through agricultural runoff and rainwater, these pesticides are ultimately transported to river water. They are highly toxic in nature and can kill aquatic flora and fauna (Riaz et al., 2021). They can also bioaccumulate in the tissues of aquatic organisms and biomagnify from one trophic level to another (Zhang et al., 2025).

2.1.5 Microplastics

Microplastics are tiny plastic particles that are <5 mm in size. Microplastic contamination is an emerging issue that has been detected in every component of the environment. It originates from multiple sources, including industrial effluents, domestic wastewater, domestic solid waste, fishing nets, discarded plastic items, cosmetic products, and atmospheric deposition (Du et al., 2021). Microplastics can be categorized as primary or secondary. Primary microplastics are intentionally manufactured as small particles for various applications, such as in cosmetic products, while secondary microplastics are formed due to the breakdown of larger plastic particles through mechanical force, UV radiation, and abrasion (Nugnes et al., 2022). Microplastics provide an enhanced surface area for polycyclic aromatic hydrocarbons and polychlorinated hydrocarbons. In addition, other contaminants, such as heavy metals, organic pollutants, antibiotics, and pathogenic bacteria, are often present on the surface of microplastics. Due to their light weight, small size, buoyancy, and persistence in nature, microplastics can easily be transported to rivers and lakes.

2.2 Mechanism and pathway of pollutants entry into the riverine system

The major pathways through which environmental pollutants enter river water are surface runoff, point-source discharge, nonpoint-source discharge, atmospheric deposition, leaching, subsurface flow, and accidental spills (Akhtar et al., 2021). During rainfall, fertilizers and pesticides present in agricultural fields, solid waste on urban surfaces, landfills, and roads are transported to rivers and nearby water bodies through rainwater (Singh et al., 2022). Runoff mainly carries dissolved and suspended pollutants from the land surface. Another pathway for the transport of pollutants to river water is point-source discharge (Xu et al., 2022). Point source refers to a single fixed point or location where pollutants are discharged into the aquatic environment, and after discharge, these pollutants form a channel and ultimately enter the river system (Kumar et al., 2023a). In nonpoint-source discharge, pollutants are released from multiple points or locations and form a channel to enter river water (Kapoor and Singh, 2021). Sometimes, septic tanks and pipelines that carry effluents and sewage from households, industries, and landfills leak, discharging pollutants onto land that infiltrates groundwater through vertical downward movement. Through subsurface base flow connections, these pollutants can also enter river water (Rotaru and Răileanu, 2008). Pollutants from the leakage of septic tanks and pipelines can be directly transported to river water through surface flow (Ratnayake et al., 2025). Atmospheric deposition is also responsible for riverine pollution. Pollutants fall from the atmosphere in the form of particulate matter, which may contain toxic heavy metals, microplastics, pesticides, and other organic compounds (Shi et al., 2021). River sediment also acts as a pathway to transport pollutants from one point to another. It can transfer pollutants through surface adsorption and travel long distances (Debnath et al., 2021). Accidental spills of toxic chemicals, oil, and grease can also discharge pollutants into river water, severely impacting the aquatic ecosystem. An accidental oil spill can block oxygen exchange by forming a surface layer of oil (Ogolo et al., 2022).

3 Effects of riverine pollution on aquatic flora

Phytoplankton, macrophytes, and riparian vegetation are the key organisms that respond to riverine pollution in the aquatic ecosystem. Pollutants such as heavy metals, organic matter, hydrocarbons, and pesticides affect the physiological and chemical composition of aquatic flora and fauna communities, thereby impacting their ecosystem structure and function (Fig. 10.2) (Tan et al., 2023).

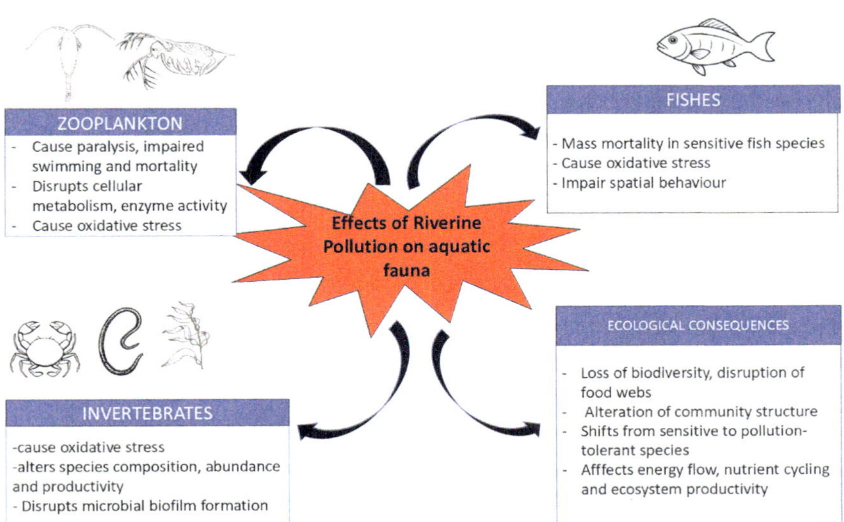

FIG. 10.2 Effects of riverine pollution on the aquatic fauna community. Consequences of river pollution.

3.1 Effects on phytoplankton

Phytoplankton are very small, floating, and photosynthetic organisms that can be observed under a microscope. Phytoplankton respond quickly to any changes in water quality, such as variations in water temperature, light, nutrients, and toxic contaminants (Ogamba et al., 2023). Phytoplankton can be significantly affected by various metals and heavy metals, including bismuth, zinc, chromium, cobalt, and barium. These metals also function as micronutrients and can positively influence the growth and metabolic activity of phytoplankton at low concentrations. However, at higher concentrations, these heavy metals can reduce the growth of phytoplankton and negatively impact their photosynthetic processes (Walkiewicz et al., 2022). A study was conducted in which phytoplankton, such as Chlorophyceae, Bacillariophyceae, and Cyanophyceae, were exposed to heavy metals like cobalt, zinc, and copper. The study observed that at low doses, the growth and population of phytoplankton increased, while at higher doses, the heavy metals inhibited their growth by reducing their pigment content. Interestingly, the metal chromium can indirectly increase the biomass of phytoplankton by reducing the population of zooplankton that feed on phytoplankton (Nawaz et al., 2024). Pesticides can directly affect phytoplankton by inhibiting chlorophyll synthesis, reducing cellular activity, decreasing protein and flavonoid content, and increasing stress-related enzyme activity (Wongsa et al., 2023). Some of the major effects on phytoplankton are outlined in the following.

3.1.1 Excess nutrient discharge

Discharge of excess amounts of nitrogen and phosphorus in river water can be responsible for the bloom of certain groups of algae, such as Cyanobacteria and Chlorophyceae. Their rapid growth leads to a reduction in species diversity, as these species become the dominating taxa (Oduor et al., 2023). A similar condition was observed in the River Irpin in Ukraine, where the growth of phytoplankton due to nutrient enrichment and a high load of chlorophyll A was negatively correlated with the growth of macrophytes (Nezbrytska et al., 2024).

3.1.2 Light limitation

The suspended particles from urban runoff, agricultural waste, industrial waste, and floods lead to blocking sunlight from reaching aquatic plants, which affects the photosynthesis phenomenon (Nan et al., 2023).

3.1.3 Altered species composition

Some species, like cyanobacteria and euglenophytes, are tolerant species and are able to grow in polluted water compared to other phytoplankton, and their overgrowth can alter the community composition (Sultana et al., 2024). At higher concentrations of microplastic contamination in the aquatic ecosystem, the population of Aphanocapsa and Pseudanabaena increases while reducing the population of Crucigenia and Chlamydomonas. This leads to a clear change in the phytoplankton community structure (Hitchcock, 2022). In a phosphorus-enriched eutrophic lake, nitrogen was added intentionally to observe the changes in phytoplankton community composition, and the results showed that the population of toxic cyanobacteria, like Microcystis and Planktothrix, dominates at low-to-moderate nitrogen concentrations, while high nitrogen concentrations favor the growth of green algae. The combination of nutrient enrichment leads to changes in phytoplankton species diversity (Dashkova et al., 2022).

3.1.4 Physiological stress

Heavy metals, pesticides, and polyaromatic hydrocarbons can affect photosynthesis by damaging chloroplast structure, inhibiting enzyme activity for carbon fixation, and suppressing chlorophyll synthesis (Gupta et al., 2019). Polycyclic aromatic hydrocarbons can accumulate in the cells of phytoplankton and cause oxidative stress and DNA damage, which ultimately inhibit the reproduction and growth of phytoplankton. Pesticides such as organophosphates and atrazine can inhibit cell division (Ali et al., 2020). An increase in suspended particulate matter in river water has been found to elevate malondialdehyde, superoxide dismutase, and catalase activities in *Chlorella pyrenoidosa*, while a decrease in lipid content has also been reported (Lin et al., 2024). Exposure to emerging contaminants such as microplastics and pharmaceutical waste can lead to the generation of reactive oxygen species (ROS), compromised cell membranes, and thylakoids, which ultimately affect primary productivity (Wongsa et al., 2023).

3.2 Effects on macrophytes

Macrophytes play an essential role in the aquatic ecosystem. They provide shelter for various aquatic organisms, stabilize the river bed by holding the sediment with their roots, and absorb excess nutrients (Kumar et al., 2023b). However, riverine pollutants and other contaminants can disturb these functions. Due to the excess discharge of nutrients, excessive growth of macrophytes occurs, resulting in reduced sunlight penetration for submerged plants, which affects the diversity of submerged plant species (Jin et al., 2020). Macrophytes are known as good absorbers of nutrients and pollutants. They can uptake and remove nitrogen and phosphorus from water, which helps to reduce algal blooms. They can also absorb heavy metals, hydrocarbons, and other toxic contaminants (Kumar et al., 2023b). However, under high contamination and prolonged exposure to these pollutants, their enzymatic activity is inhibited, the photosynthesis process is reduced, ROS are generated, necrosis occurs, growth is stunted, and even plant death can result. Pesticides can block photosystem II, resulting in reduced carbon fixation and oxygen production (Tan et al., 2023).

4 Effects of riverine pollution on aquatic fauna

Riverine pollution and aquatic fauna are tightly interlinked as a direct chain of cause and effect, where the influx of pollutants into the ecosystem alters the physicochemical parameters of water, which in turn affects aquatic fauna. Aquatic fauna, which refers to all animal organisms inhabiting a water body, require optimal water quality parameters for their health and survival, as their entire or some of their life processes, such as respiration, feeding, reproduction, and movement, depend on aquatic conditions (Sayer et al., 2025). Over the years, pollutants from various sources, mostly due to anthropogenic activities, have degraded water quality by decreasing the amount of dissolved oxygen, changing the pH of the water, increasing turbidity, and introducing toxic chemicals. Such alterations not only disrupt the physicochemical balance of water bodies but also affect the life processes in aquatic fauna, leading to physiological stress, behavioral alterations, and population decline (Amoatey and Baawain, 2019). Global estimates, according to IUCN (2025), indicate that approximately 24% of aquatic fauna are threatened, with pollution being a major critical factor (Sayer et., 2025). Although all types of pollutants have detrimental effects on all forms and types of aquatic fauna, such impacts are not uniform and vary significantly across different trophic levels (Ban et al., 2025).

4.1 Effects on zooplankton

Zooplankton are microscopic to small heterotrophic organisms, comprising various groups such as protozoans, rotifers, cladocerans, and copepods. They are primary consumers in an aquatic ecosystem, as they feed on phytoplankton and bacteria, thereby serving as a link between primary producers and higher trophic levels such as fish larvae and other aquatic invertebrates (Sharma et al., 2025). Zooplankton communities are highly sensitive to changes in water quality due to their short life cycles and rapid reproductive rates, making them reliable indicators of riverine pollution. Optimal physicochemical properties of water, including temperature, pH, dissolved oxygen, and nutrient concentration, are important for the composition and abundance of zooplankton (Xiong et al., 2019). Alterations in water quality due to pollution create unfavorable conditions for the survival of zooplankton (Sharma et al., 2025). The effects of various pollutants on zooplankton can be categorized into direct toxicity, indirect toxicity, and ecological consequences.

4.1.1 Direct toxicity effects

Chemical pollutants, such as pesticides, industrial chemicals, and heavy metals, have immediate lethal and sublethal effects on zooplankton. Insecticides, such as organophosphates and pyrethroids, which are designed to kill arthropods, exert toxic effects on zooplankton (Hanazato, 2001). These substances cause physiological stress on zooplankton even at low concentrations due to the permeable nature of their body surface, which can readily absorb dissolved pesticides from water. The chemicals present in insecticides interfere with neuromuscular transmission by inhibiting acetylcholinesterase activity, a key enzyme of the nervous system (Weng et al., 2025). This leads to excessive accumulation of acetylcholine at nerve synapses and, as a result, causes paralysis, impaired swimming, and mortality. Prolonged exposure to pollutants can cause rapid population crashes due to mortality, paralysis, and reproductive failure (Hanazato, 2001). Metal contamination, such as copper, cadmium, and zinc, can disrupt cellular metabolism, enzyme activity, and cause oxidative stress, as these metals interfere with different metalloenzymes that regulate essential physiological processes like respiration, digestion, and ion regulation (Weng et al., 2025). Emerging contaminants, such as microplastics, pharmaceuticals, and personal care products, cause ingestion, physical obstruction, and physiological stress. Microplastics, being small in size, are easily mistaken for food particles by zooplankton species. Once ingested, these particles accumulate in the gut, blocking the digestive tract and reducing feeding efficiency, energy intake, growth, and survival (Rathour et al., 2025).

4.1.2 Indirect toxicity effects

Agricultural runoff, sewage discharge, and industrial effluents lead to the enrichment of water bodies with excessive nutrients, primarily nitrogen and phosphorus, a phenomenon known as eutrophication. This has an indirect but negative effect on zooplankton. Initially, the influx of nutrients stimulates excessive growth of edible phytoplankton, thereby leading to an increase in zooplankton reproduction and abundance (Tiwari and Pal, 2022; Trabelsi et al., 2022). However, as eutrophication progresses, the phytoplankton community shifts toward inedible filamentous cyanobacteria (blue–green algae), which form large colonies that are difficult for zooplankton to consume and digest. Moreover, it creates a favorable condition for many species that produce potent hepatotoxins and neurotoxins, which can directly kill zooplankton or suppress their grazing activity. The decay of algal blooms by bacteria consumes dissolved oxygen, creating hypoxic or anoxic conditions throughout the water column that indirectly limit zooplankton survival (Oduor et al., 2023; Trabelsi et al., 2022).

4.1.3 Ecological consequences

Zooplankton play an important ecological role in energy transfer within aquatic ecosystems, acting as a crucial link between phytoplankton and fish or other large animals. They also improve water quality by consuming algae and facilitate nutrient cycling (Xiong et al., 2019). However, riverine pollution alters the ecological roles of zooplankton. Pollutants affect the community structure of zooplankton, thereby negatively impacting trophic interactions and energy flow within aquatic ecosystems (Sharma et al., 2025). Prolonged exposure to various pollutants results in changes in the composition and abundance of zooplankton (Hanazato, 2001). This change occurs due to selective pressure that favors tolerant species, such as rotifers, and eliminates sensitive taxa, such as cladocerans and copepods. This shift from a diverse, balanced community to a less diverse community disrupts food web dynamics (Xiong et al., 2019). As zooplankton are primary consumers that transfer energy from phytoplankton to higher trophic levels like fish, this, in turn, affects secondary consumers, particularly small fish and invertebrates that rely on them for nutrition. Reduced energy transfer at this level leads to decreased growth, reproduction, and survival of higher trophic organisms (Atkinson et al., 2021). Moreover, the accumulation of pollutants in zooplankton can lead to bioaccumulation and biomagnification, affecting predators and altering ecosystem nutrient cycling (Mchale et al., 2024). Overall, these ecological effects have a detrimental impact on the functional integrity and stability of riverine ecosystems.

4.2 Effects on invertebrates

Aquatic invertebrates represent the most diverse and ecologically important component of the riverine ecosystem. These include various groups, such as insects, mollusks, crustaceans, and annelids, and occupy different ecological niches, such as grazers, collectors, and predators. Riverine pollution exerts varying degrees of impact on different groups of aquatic invertebrates, depending on their habitat, physiology, and trophic role (Ban et al., 2025). These groups, mainly benthic, planktonic, and periphytic invertebrates, respond differently to various pollutants.

4.2.1 Benthic invertebrates

Benthic invertebrates refer to bottom-dwelling invertebrate groups such as insect larvae, mollusks, crustaceans, and nematodes. Because of their close association with the sediments, where pollutants accumulate, they are extremely sensitive and act as reliable indicators of pollution (Tampo et al., 2021). Heavy metals, such as cadmium, lead, and mercury, bind strongly to sediments and are absorbed by benthic fauna, causing oxidative stress, reduced enzyme activity, and bioaccumulation (Mchale et al., 2024).

4.2.2 Planktonic invertebrates

Planktonic invertebrates include zooplankton communities, which are sensitive to changes in water chemistry, nutrient load, and suspended solids. Riverine pollution alters species composition, abundance, and productivity (Xiong et al., 2019).

4.2.3 Periphytic invertebrates

Periphytic invertebrates comprise microorganisms such as algae, fungi, and detritus that live within the periphyton and are attached to submerged surfaces in aquatic environments. These invertebrates, along with other periphytic organisms, form a biofilm over submerged objects, creating the "aufwuchs" community. River pollution alters periphyton communities by negatively affecting their biomass, structure, and ecological functions (Agboola et al., 2019). Various suspended organic and inorganic pollutants lead to increases in turbidity, sedimentation, and nutrient enrichment. This also reduces light penetration and, as a result, impacts the survival of periphytic invertebrates (Nan et al., 2023). Moreover, chemical pollutants, such as pesticides and detergents, disrupt the microbial biofilm and suppress colonization by sensitive invertebrates such as mayfly and caddisfly larvae (Previsic et al., 2019).

4.3 Effects on fish

Fishes, occupying various trophic levels, are among the ecologically and economically important components of riverine ecosystems (Mustafa et al., 2024). The influx of pollutants into a riverine ecosystem acts as a stressor for fish, thereby affecting their health and survival. The impact of various pollutants on fish can be categorized into physiological, behavioral, and ecological responses.

4.3.1 Physiological response

Riverine pollution acts as a physiological stressor to fish, thereby making them vulnerable to survival. One of the most important water quality parameters for the survival of fish is dissolved oxygen (Satkar et al., 2024). Over the years, an influx of various organic pollutants from sewage has led to the depletion of available dissolved oxygen. This occurs due to the consumption of oxygen by bacteria as they decompose organic matter, creating an anoxic condition for fish and making them vulnerable to survival. Prolonged exposure to very low levels of dissolved oxygen often leads to mass mortality in some sensitive fish species (Mustafa et al., 2024). Direct industrial and municipal wastewater discharge into rivers leads to heavy metal contamination that is readily absorbed through the gills, skin, and digestive tract of fish. Heavy metals such as mercury (Hg), cadmium (Cd), lead (Pb), zinc (Zn), and copper (Cu) can bind to various metalloenzymes in fish, thereby disrupting their physiological and biochemical processes. For example, mercury exposure leads to neurological impairment as it interferes with heme synthesis, resulting in anemia and reduced oxygen transport in the body of fish (Garai et al., 2021). Exposure to heavy metals also causes oxidative stress by generating ROS, which lead to DNA fragmentation and cellular apoptosis (Shahjahan et al., 2022). Chemical pollutants, such as pesticides, pharmaceuticals, and plasticizers, act as endocrine-disrupting compounds. These can interfere with the hypothalamic–pituitary–gonadal axis, leading to altered sex steroid levels, reduced gonadosomatic index, and intersex development (Ray and Shaju, 2023). Moreover, the presence of microplastics in the riverine ecosystem negatively affects fish by causing digestive blockages, leading to reduced growth, behavioral changes, and impaired reproduction (Rathour et al., 2025).

4.3.2 Behavioral response

Riverine pollution affects several types of spatial behavior, such as navigational strategies, distributional patterns, exploration, and predator avoidance. Various fish species rely on their sight and vision to catch their prey. However, this behavior is affected due to low visibility in the water column as a result of an increase in suspended solids caused by various pollutants (Jacquin et al., 2020). Various heavy metals present in the riverine ecosystem due to pollution impair the neurological signals of behavior, leading to erratic swimming and reduced muscle coordination (Garai et al., 2021). Polluted riverine ecosystems also disturb reproductive behavior in fishes, such as nest building, courtship, and spawning. This leads to reduced reproductive success and early life stage mortality (Mustafa et al., 2024). Prolonged exposure to contaminants can lead to stress-related aggression in fish and result in physical injury and vulnerability to predation (Jacquin et al., 2020). Such behavioral impairments are often reversible and serve as early warning signs of environmental degradation (Satkar et al., 2024).

4.3.3 Ecological response

Riverine pollution has major impacts on the ecology of fish populations, influencing their survival, population structure, and community composition (Jacquin et al., 2020). The accumulation of various pollutants, such as heavy metals, organic matter, and nutrients, degrades spawning and nursery grounds, thereby reducing reproductive success (Dase et al., 2024). Organic pollution leads to oxygen depletion and favors the dominance of pollution-tolerant taxa, while sensitive taxa are eliminated. This results in reduced biodiversity and a less diverse community composition (Sultana et al., 2024; Das et al., 2024).

5 Conclusion

Riverine pollution poses multiple threats to the aquatic ecosystem by affecting its diversity, habitat, structure, and the function of aquatic flora and fauna. The continuous discharge of pollutants from industrial effluents, domestic sewage, and agricultural runoff deteriorates the quality of river water and disrupts the ecological balance of aquatic flora and fauna. Phytoplankton and zooplankton, which are the primary producers of the aquatic ecosystem, are highly sensitive to toxic pollutants and nutrient enrichment, leading to changes in their physiological and biochemical processes. Heavy metals and organic pollutants result in reduced photosynthetic processes, the generation of ROS, necrosis, stunted growth, and even plant death. Pesticides can block photosystem II, resulting in lower carbon fixation and oxygen production.

Invertebrates, microorganisms, aquatic plants, and algae play a vital role in nutrient cycling and the flow of energy, but under toxic and hypoxic conditions, they suffer from the bioaccumulation of toxic pollutants, which leads to changes in their behavior, reproductive failure, physiological stress, the generation of ROS, DNA damage, stunted growth, and other adverse effects. This ultimately impacts their biodiversity and food security. The continuous monitoring of pollutant discharge into river water and its restoration is essential to mitigate the impacts of riverine pollution on aquatic flora and fauna. Strict enforcement of environmental rules and regulations, community participation, and effective watershed management can help restore the health and integrity of the riverine ecosystem.

References

Agboola, O.A., Downs, C.T., O'Brien, G., 2019. Macroinvertebrates as indicators of ecological conditions in the rivers of KwaZulu-Natal, South Africa. Ecol. Indic. 106, 105465. doi:10.1016/j.ecolind.2019.105465.

Akhtar, N., Ishak, M.I.S., Bhawani, S.A., Umar, K., 2021. Various natural and anthropogenic factors responsible for water quality degradation: a review. Water. 2073-4441. 13 (19), 2660. doi:10.3390/w13192660.

Ali, H., Khan, E., Ilahi, I., 2020. Environmental chemistry and ecotoxicology of hazardous heavy metals: environmental persistence, toxicity, and bioaccumulation. J. Chem. 2020, 1–14. doi:10.1155/2019/6730305.

Amoatey, P., Baawain, M.S., 2019. Effects of pollution on freshwater aquatic organisms. Water Environ. Res. 15547531. 91 (10), 1272–1287. Available from: https://onlinelibrary.wiley.com/loi/15547531. doi:10.1002/wer.1221.

Anetor, G.O., Nwobi, N.L., Igharo, G.O., Sonuga, O.O., Anetor, J.I., 2022. Environmental pollutants and oxidative stress in terrestrial and aquatic organisms: examination of the total picture and implications for human health. Front. Physiol. 1664042X. 13. Available from: http://www.frontiersin.org/Physiology/archive/. doi:10.3389/fphys.2022.931386.

Atkinson, A., Lilley, M.K.S., Hirst, A.G., McEvoy, A.J., Tarran, G.A., Widdicombe, C., Fileman, E.S., Woodward, E.M.S., Schmidt, K., Smyth, T.J., Somerfield, P.J., 2021. Increasing nutrient stress reduces the efficiency of energy transfer through planktonic size spectra. Limnol. Oceanogr. 19395590. 66 (2), 422–437. Available from: http://aslopubs.onlinelibrary.wiley.com/hub/journal/10.1002/(ISSN)1939-5590/. doi:10.1002/lno.11613.

Ban, X., Guo, W., Fu, Y., 2025. Impact of environmental factors on aquatic ecosystem. Water 17 (10), 1453. doi:10.3390/w17101453.

Bhendekar, G., Bele, P., Wankhade, V., 2024. Effect of water pollution on Flora and Fauna. Int. J. Life Sci. 12 (2), 219–221.

Canning, A.D., Death, R.G., 2019. Ecosystem health indicators—freshwater environments. In: Encyclopedia of Ecology, Second ed., vols. 1–4. Elsevier, New Zealand, pp. 46–60. Available from: https://www.sciencedirect.com/book/9780444637680. doi:10.1016/B978-0-12-409548-9.10617-7.

Danovaro, R., Cocozza di Montanara, A., Corinaldesi, C., Dell'Anno, A., Illuminati, S., Willis, T.J., Gambi, C., 2023. Bioaccumulation and biomagnification of heavy metals in marine micropredators. Communications Biology. 23993642. 6 (1). Available from: https://www.springer.com/journal/42003. doi:10.1038/s42003-023-05539-x.

Das, M., Paul, M., Das, J., Mandal, B., 2024. Impact of heavy metals on fish reproduction: mechanisms, implications, and mitigation strategies. 9. 48–54.

Dashkova, V., Malashenkov, D.V., Baishulakova, A., Davidson, T.A., Vorobjev, I.A., Jeppesen, E., Barteneva, N.S., 2022. Changes in phytoplankton community composition and phytoplankton cell size in response to nitrogen availability depend on temperature. Microorganisms. 20762607. 10 (7). Available from: https://www.mdpi.com/2076-2607/10/7/1322/pdf?version=1656599476. doi:10.3390/microorganisms10071322.

Debnath, A., Singh, P.K., Sharma, Y.C., 2021. Metallic contamination of global river sediments and latest developments for their remediation. J. Environ. Manage. 298, 113378. doi:10.1016/j.jenvman.2021.113378.

Du, S., Zhu, R., Cai, Y., Xu, N., Yap, P.S., Zhang, Y., He, Y., Zhang, Y., 2021. Environmental fate and impacts of microplastics in aquatic ecosystems: a review. RSC Adv. 20462069. 11 (26), 15762–15784. Available from: http://pubs.rsc.org/en/journals/journal/ra. doi:10.1039/d1ra00880c.

Eriksen, T.E., Jacobsen, D., Demars, B.O.L., Brittain, J.E., Søli, G., Friberg, N., 2022. Effects of pollution-induced changes in oxygen conditions scaling up from individuals to ecosystems in a tropical river network. Sci. Total Environ. 18791026. 814. Available from: https://www.elsevier.com/locate/scitotenv. doi:10.1016/j.scitotenv.2021.151958.

Garai, P., Banerjee, P., Mondal, P., Ch, N., Saha, R., 2021. Effect of heavy metals on fishes: toxicity and bioaccumulation. J. Clin. Toxicol. 1–10. doi:10.35248/2161-0495.21.s18.001.

Ghirardelli, A., Tarolli, P., Kameswari Rajasekaran, M., Mudbhatkal, A., Macklin, M.G., Masin, R., 2021. Organic contaminants in Ganga basin: from the Green Revolution to the emerging concerns of modern India. iScience. 25890042. 24 (3). Available from: https://www.cell.com/iscience. doi:10.1016/j.isci.2021.102122.

Gupta, S., Rai, U.N., Tripathi, R.D., Inouhe, M., 2019. Impacts of heavy metals on the photosynthetic activity of aquatic plants and algae. Environ. Chem. Lett. 17 (1), 491–510. doi:10.1201/9780429298080.

Hairom, N.H.H., Soon, C.F., Mohamed, R.M.S.R., Morsin, M., Zainal, N., Nayan, N., Harun, N.H., 2021. A review of nanotechnological applications to detect and control surface water pollution. Environ. Technol. Innov. 24, 102032.

Hanazato, T., 2001. Pesticide effects on freshwater zooplankton: an ecological perspective. Environ. Pollut. 02697491. 112 (1), 1–10. doi:10.1016/s0269-7491(00)00110-x.

Hitchcock, J.N., 2022. Microplastics can alter phytoplankton community composition. Sci. Total Environ. 00489697. 819, 153074. doi:10.1016/j.scitotenv.2022.153074.

Jacquin, L., Petitjean, Q., Côte, J., Laffaille, P., Jean, S., 2020. Effects of pollution on fish behavior, personality, and cognition: some research perspectives. Front. Ecol. Evol. 2296-701X. 8. doi:10.3389/fevo.2020.00086.

Jin, S., Ibrahim, M., Muhammad, S., Khan, S., Gang, L., 2020. Light intensity effects on the growth and biomass production of submerged macrophytes in different water strata. Arab. J. Geosci. 1866-7511. 13 (18), doi:10.1007/s12517-020-05924-4.

Kapoor, D., Singh, M.P., 2021. Heavy metal contamination in water and its possible sources. In: Heavy Metals in the Environment. Elsevier, pp. 179–189. doi:10.1016/B978-0-12-821656-9.00010-9.

Kesari, K.K., Soni, R., Jamal, Q.M.S., Tripathi, P., Lal, J.A., Jha, N.K., Siddiqui, M.H., Kumar, P., Tripathi, V., Ruokolainen, J., 2021. Wastewater treatment and reuse: a review of its applications and health implications. Water, Air, Soil Pollut. 15732932. 232 (5). Available from: http://www.kluweronline.com/issn/0049-6979/. doi:10.1007/s11270-021-05154-8.

Khan, A.S., Anavkar, A., Ali, A., Patel, N., Alim, H., 2021. A review on current status of riverine pollution in India. Biosci. Biotechnol. Res. Asia. 24562602. 18 (1), 9–22. doi:10.13005/bbra/2893.

Khosrovyan, A., 2024. Biodiversity and ecosystem services in rivers. Water. 2073-4441. 16 (15), 2091. doi:10.3390/w16152091.

Kumar, P., Kumar, R., Thakur, K., Mahajan, D., Brar, B., Sharma, D., Sharma, A.K., 2023a. Impact of pesticides application on aquatic ecosystem and biodiversity: a review. Biol. Bull. 50 (6), 1362–1375. doi:10.1134/S1062359023601386.

Kumar, S., Singh, R., Kumar, D., Bauddh, K., Kumar, N., Kumar, R., 2023b. An introduction to the functions and ecosystem services associated with aquatic macrophytes. In: Aquatic Macrophytes: Ecology, Functions and Services. Springer Nature, India, pp. 1–20. Available from: 10.1007/978-981-99-3822-3_1. doi:10.1007/978-981-99-3822-3.

Lin, J., Liu, J., Xing, H., Chen, S., Nan, Y., He, J., Hu, B., Wei, Y., Guo, P., 2024. Effect of suspended particulate matter on physiological, biochemical and photosynthetic characteristics of Chlorella pyrenoidosa in the Jinjiang Estuary (Fujian, China). Ecotoxicology. 0963-9292. 33 (2), 151–163. doi:10.1007/s10646-024-02734-2.

McHale, M.E., Sheehan, K.L., 2024. Bioaccumulation, transfer, and impacts of microplastics in aquatic food chains. J. Environ. Expo. Assess. 3 (3). doi:10.20517/jeea.2023.49.

Mishra, R.K., 2023. Fresh water availability and its global challenge. Br. J. Multidiscip. Adv. Stud. 2517-2778. 4 (3), 1–78. doi:10.37745/bjmas.2022.0208.

Morin-Crini, N., Lichtfouse, E., Liu, G., Balaram, V., Ribeiro, A.R.L., Lu, Z., Stock, F., Carmona, E., Teixeira, M.R., Picos-Corrales, L.A., Moreno-Piraján, J.C., Giraldo, L., Li, C., Pandey, A., Hocquet, D., Torri, G., Crini, G., 2022. Worldwide cases of water pollution by emerging contaminants: a review. Environ. Chem. Lett. 16103661. 20 (4), 2311–2338. Available from: https://www.springer.com/journal/10311. doi:10.1007/s10311-022-01447-4.

Mustafa, S.A., Al-Rudainy, A.J., Salman, N.M., 2024. Effect of environmental pollutants on fish health: an overview. Egypt. J. Aquat. Res. 50 (2), 225–233. doi:10.1016/j.ejar.2024.02.006.

Nan, Y., Guo, P., Xing, H., Chen, S., Hu, B., Liu, J., 2023. Effects of suspended particles in the Jinjiang River Estuary on the physiological and biochemical characteristics of Microcystis flos-aquae. Environ. Sci. Pollut. Res. 1614-7499. 30 (19), 56687–56699. doi:10.1007/s11356-023-26367-w.

Nawaz, A., Šotek, P.E., Molnárová, M., 2024. Reciprocal Effects of Metal Mixtures on Phytoplankton. Phycology 4 (1), 117–138. doi:10.3390/phycology4010007.

Nezbrytska, I., Bilous, O., Sereda, T., Ivanova, N., Pohorielova, M., Shevchenko, T., Dubniak, S., Lietytska, O., Zhezherya, V., Polishchuk, O., Kazantsev, T., Prychepa, M., Kovalenko, Y., Afanasyev, S., 2024. Effects of war-related human activities on microalgae and macrophytes in freshwater ecosystems: a case study of the Irpin River Basin, Ukraine. Water. 2073-4441. 16 (24), 3604. doi:10.3390/w16243604.

Nishmitha, P.S., Akhilghosh, K.A., Aiswriya, V.P., Ramesh, A., Muthuchamy, M., Muthukumar, A., 2025. Understanding emerging contaminants in water and wastewater: a comprehensive review on detection, impacts, and solutions. J. Hazard. Mater. Adv. 27724166. 18. Available from: https://www.sciencedirect.com/science/journal/27724166. doi:10.1016/j.hazadv.2025.100755.

Nugnes, R., Lavorgna, M., Orlo, E., Russo, C., Isidori, M., 2022. Toxic impact of polystyrene microplastic particles in freshwater organisms. Chemosphere 00456535. 299, 134373. doi:10.1016/j.chemosphere.2022.134373.

Nyagongo, A.A., Ndibalema, V.G., Lalika, M.C., 2024. Assessing the impact of human activities on the health condition of the Lukosi River catchment using selected physico-chemical parameters as indicators of water quality. Front. Water. 6, 1378849. doi:10.3389/frwa.2024.1378849.

Oduor, N.A., Munga, C.N., Ong'anda, H.O., Botwe, P.K., Moosdorf, N., 2023. Nutrients and harmful algal blooms in Kenya's coastal and marine waters: a review. Ocean Coast. Manage. 09645691. 233, 106454. doi:10.1016/j.ocecoaman.2022.106454.

Ogamba, E.N., Iyiola, A.O., Yarkwan, B., Adetola, B.O., 2023. Potentials, threats, and sustainable conservation strategies of plankton and macrophytes. In: Sustainable Utilization and Conservation of Africa's Biological Resources and Environment. Springer Nature, Singapore, pp. 85–117. doi:10.1007/978-981-19-6974-4_4.

Ogolo, N.A., Anih, O.C., Onyekonwu, M.O., 2022. Sources and effects of environmental pollution from oil and gas industrial operations. Arab. J. Chem. Environ. Res. 9 (01), 98–121.

Previšić, A., Rožman, M., Mor, J., Acuña, V., Serra-Compte, A., Petrović, M., Sabater, S., 2019. Aquatic macroinvertebrates under stress: bioaccumulation of emerging contaminants and metabolomics implications. Sci. Total Environ. 704, 135333. doi:10.1016/j.scitotenv.2019.135333.

Rathour, A., Dhar, A., Pathania, S., 2025. Impact of microplastics pollution on human health and aquatic life: a review. Polym. Int. doi:10.1002/pi.6782.

Ratnayake, M.S., Senarathne, S.L., Diyabalanage, S., Bandara, C., Wickramarathne, S., Chandrajith, R., 2025. Processes of groundwater contamination in coastal aquifers in Sri Lanka: a geochemical and isotope-based approach. Water. 2073-4441. 17 (11), 1571. doi:10.3390/w17111571.

Ray, S., Shaju, S.T., 2023. Bioaccumulation of pesticides in fish resulting toxicities in humans through food chain and forensic aspects. Environ. Anal. Health Toxicol. 2671-9525. 38 (3), e2023017. doi:10.5620/eaht.2023017.

Riaz, U., Rafi, F., Naveed, M., Mehdi, S.M., Murtaza, G., Niazi, A.G., Mehmood, H., 2021. Pesticide pollution in an aquatic environment. In: Freshwater Pollution and Aquatic Ecosystems. Apple Academic Press, pp. 131–163. doi:10.1201/9781003130116.

Rotaru, A., Răileanu, P., 2008. Groundwater contamination from waste storage works. Environ. Eng. Manag. J. 7 (6), 731–735.

Satkar, S.G., Kumar, A., Anjana, A., Bhusare, S., Sahu, A., Gautam, R.K., 2024. Impact of pollution and toxic stress on fish health: mechanisms, consequences, and mitigation strategies. Uttar Pradesh J. Zool. 0256-971X. 45 (6), 29–45. doi:10.56557/upjoz/2024/v45i63949.

Sayer, C.A., Fernando, E., Jimenez, R.R., Macfarlane, N.B.W., Rapacciuolo, G., Böhm, M., Brooks, T.M., Contreras-MacBeath, T., Cox, N.A., Harrison, I., Hoffmann, M., Jenkins, R., Smith, K.G., Vié, J.C., Abbott, J.C., Allen, D.J., Allen, G.R., Barrios, V., Boudot, J.P., Carrizo, S.F., Charvet, P., Clausnitzer, V., Congiu, L., Crandall, K.A., Cumberlidge, N., Cuttelod, A., Dalton, J., Daniels, A.G., De Grave, S., De Knijf, G., Dijkstra, K.D.B., Dow, R.A., Freyhof, J., García, N., Gessner, J., Getahun, A., Gibson, C., Gollock, M.J., Grant, M.I., Groom, A.E.R., Hammer, M.P., Hammerson, G.A., Hilton-Taylor, C., Hodgkinson, L., Holland, R.A., Jabado, R.W., Bignoli, D.J., Kalkman, V.J., Karimov, B.K., Kipping, J., Kottelat, M., Lalèyè, P.A., Larson, H.K., Lintermans, M., Lozano, F., Ludwig, A., Lyons, T.J., Máiz-Tomé, L., Molur, S., Ng, H.H., Numa, C., Palmer-Newton, A.F., Pike, C., Pippard, H.E., Polaz, C.N.M., Pollock, C.M., Raghavan, R., Rand, P.S., Ravelomanana, T., Reis, R.E., Rigby, C.L., Scott, J.A., Skelton, P.H., Sloat, M.R., Snoeks, J., Stiassny, M.L.J., Tan, H.H., Taniguchi, Y., Thorstad, E.B., Tognelli, M.F., Torres, A.G., Torres, Y., Tweddle, D., Watanabe, K., Westrip, J.R.S., Wright, E.G.E., Zhang, E., Darwall, W.R.T., 2025. One-quarter of freshwater fauna threatened with extinction. Nature. 14764687. 638 (8049), 138–145. Available from: https://www.nature.com/nature/. doi:10.1038/s41586-024-08375-z.

Shahjahan, M., Taslima, K., Rahman, M.S., Al-Emran, M., Alam, S.I., Faggio, C., 2022. Effects of heavy metals on fish physiology – a review. Chemosphere. 18791298. 300. Available from: https://www.elsevier.com/locate/chemosphere. doi:10.1016/j.chemosphere.2022.134519.

Sharma, R., Singh, S., Reddy, P., 2025. Effects of water quality and seasonality on Zooplankton responses. World J. Biol. Pharm. Health Sci. 21, 619–627. doi:10.30574/wjbphs.2025.21.3.0339.

Shi, M., Geng, B., Zhao, T., Wang, F., 2021. Influence of atmospheric deposition on surface water quality and DBP formation potential as well as control technology of rainwater DBPs: a review. Environ. Sci. Water Res. Technol. 2053-1400. 7 (12), 2156–2165. doi:10.1039/d1ew00520k.

Singh, B.J., Chakraborty, A., Sehgal, R., 2023. A systematic review of industrial wastewater management: evaluating challenges and enablers. J. Environ. Manage. 348, 119230. doi:10.1016/j.jenvman.2023.119230.

Singh, R., Andaluri, G., Pandey, V.C., 2022. Chapter 1-Cities' Water pollution-Challenges and Controls. In: Algae and Aquatic Macrophytes in Cities. doi. 10.1016/B978-0-12-824270-4.00015-8.

Sultana, S., Khan, S., Shaika, N.A., Hena, S.M., Mahmud, Y., Haque, M.M., 2024. Ecology of freshwater harmful euglenophytes: a review. Heliyon. 24058440. 10 (8). Available from: https://www.sciencedirect.com/science/journal/24058440. doi:10.1016/j.heliyon.2024.e29625.

Szymańska-Walkiewicz, M., Glińska-Lewczuk, K., Burandt, P., Obolewski, K., 2022. Phytoplankton sensitivity to heavy metals in Baltic Coastal Lakes. Int. J. Environ. Res. Public Health 19 (7), 4131. doi:10.3390/ijerph19074131.

Tampo, L., Kaboré, I., Alhassan, E.H., Ouéda, A., Bawa, L.M., Djaneye-Boundjou, G., 2021. Benthic macroinvertebrates as ecological indicators: their sensitivity to the water quality and human disturbances in a Tropical River. Front. Water. 26249375. 3. Available from: https://www.frontiersin.org/journals/water. doi:10.3389/frwa.2021.662765.

Tan, H.W., Pang, Y.L., Lim, S., Chong, W.C., 2023. A state-of-the-art of phytoremediation approach for sustainable management of heavy metals recovery. Environ. Technol. Innovat. 23521864. 30. Available from: http://www.journals.elsevier.com/environmental-technology-and-innovation/. doi:10.1016/j.eti.2023.103043.

Tiwari, A.K., Pal, D.B., 2022. Nutrients contamination and eutrophication in the river ecosystem. In: Ecological Significance of River Ecosystems. Elsevier, pp. 203–216. doi:10.1016/B978-0-323-85045-2.00001-7.

Trabelsi, A., Guermazi, W., Leignel, V., Al-Enezi, Y., Karam, Q., Ali, M., Ayadi, H., Belmonte, G., 2022. Effects of eutrophication on plankton abundance and composition in the Gulf of Gabès (Mediterranean Sea, Tunisia). Water 14 (14), 2230. doi:10.3390/w14142230.

Uddin, M.J., Jeong, Y.K., 2021. Urban river pollution in Bangladesh during last 40 years: potential public health and ecological risk, present policy, and future prospects toward smart water management. Heliyon. 24058440. 7 (2). Available from: http://www.journals.elsevier.com/heliyon/. doi:10.1016/j.heliyon.2021.e06107.

Wei, Y., Song, L., Ma, Y., Mu, J., Yi, W., Sun, J., Qu, K., Cui, Z., 2025. Sedimentary heavy metal interactions with phytoplankton and zooplankton across the Bohai Sea. J. Environ. Manage. 03014797. 381, 125226. doi:10.1016/j.jenvman.2025.125226.

Weng, C., Ding, D., Cui, Z., Keming, Q., Wei, Y., Haiyan, H., 2025. Effects of heavy metal concentration on zooplankton community composition and abundance in the Yellow Sea coast. Mar. Environ. Res. 01411136. 204, 106870. doi:10.1016/j.marenvres.2024.106870.

Wongsa, J., Liamchang, R., Ngearnpat, N., Issakul, K., 2023. Cypermethrin insecticide residue, water quality and phytoplankton diversity in the lychee plantation catchment area. AIMS Environ. Sci. 2372-0352. 10 (5), 609–627. doi:10.3934/environsci.2023034.

Xiong, W., Ni, P., Chen, Y., Gao, Y., Li, S., Zhan, A., 2019. Biological consequences of environmental pollution in running water ecosystems: a case study in zooplankton. Environ. Pollut. 02697491. 252 (Pt B), 1483–1490. doi:10.1016/j.envpol.2019.06.055.

Xu, H., Gao, Q., Yuan, B., 2022. Analysis and identification of pollution sources of comprehensive river water quality: evidence from two river basins in China. Ecol. Indic. 135, 108561. doi:10.1016/j.ecolind.2022.108561.

Yuan, L., Wujuan, M., Ji, L., Qiusheng, H., Yang, P., Xie, S., Bi, Y., 2023. Urbanization and agriculture intensification jointly enlarge the spatial inequality of river water quality. Sci. Total Environ. 00489697. 878, 162559. doi:10.1016/j.scitotenv.2023.162559.

Yun, M., Zhang, C., Wang, B., Huang, J., Sun, J., 2024. The effects and mechanism of organic matter degradation in river sediment driven by humic-reducing bacteria. J. Water Process Eng. 22147144. 67, 106150. doi:10.1016/j.jwpe.2024.106150.

Zhang, X., Jiang, Y., Zeng, Y., Guo, J., Long, L., Pan, Z., Tian, Y., Zhang, Y., Luo, X., Peng, P., Mai, B., 2025. Seasonal bioaccumulation and biomagnification of typical organochlorine pesticides in the Pearl River Estuary, South China. Environ. Pollut. 02697491. 383, 126855. doi:10.1016/j.envpol.2025.126855.

Chapter 11

Histopathological changes in fishes

H.A.C.C. Perera
Department of Zoology and Environmental Management, Faculty of Science, University of Kelaniya, Kelaniya, Sri Lanka

1 Introduction

Water is indeed essential for all animal life, ranging from microscopic organisms to larger whales. It provides a vital medium for numerous biological processes; furthermore, it acts as a solvent, provides a transport medium, and plays a crucial role in maintaining cellular structure and function. Therefore water is an invaluable and indispensable fundamental component that is crucial for sustaining life in all its forms on Earth. The significance of water lies in its unique properties, including its ability to dissolve vital nutrients, regulate temperature, absorb heat efficiently, and maintain a density that supports various biological processes. Therefore water is essential for all known life forms. Furthermore, these characteristics collectively make water an ideal medium for the survival and flourishing of a variety of living organisms (Sargen, 2019; Mahagamage et al., 2014).

The physical, chemical, and biological characteristics of water determine its suitability for supporting life. The quality of water is an absolutely critical factor that determines the well-being of an aquatic ecosystem. These properties impact the survival of individuals, the health, and the stability of entire ecosystems, highlighting that good water quality is an essential factor for maintaining the health of aquatic ecosystems, including aquatic organisms. Aquatic ecosystems are often used for the disposal of contaminated wastes. The discharge of waste from agriculture, industries (including untreated industrial waste, petrochemicals, solvents, oil, heavy metals, and agricultural runoff), urban activities, and construction activities into aquatic ecosystems results in the production of different types of organic and inorganic pollutants. Domestic effluents may further add a variety of organic and inorganic pollutants to the aquatic body. This gradually becomes more harmful to both aquatic environments and human health.

River Pollution. https://doi.org/10.1016/B978-0-443-45024-2.00007-5

Therefore the addition of undesirable components into aquatic ecosystems changes the water quality (Briggs, 2003; Alrumman et al., 2016; Rajamanickam and Narayanan, 2009).

2 Sources and impact of pollutants on water quality

There are two main sources of water pollution, i.e., point sources and nonpoint sources, which include industrial waste, sewage, mining, dumping, oil spills, fertilizers, etc. Pollutants have a significant negative impact on water quality, reducing the healthiness of the aquatic ecosystem, with pollutants ranging from chemicals, heavy metals, and even pathogens. Pathogen-polluted water may contain bacteria, viruses, and other microorganisms. These may disrupt aquatic processes, contributing to various health issues. Several studies have shown the impact of pollutants on water quality (Table 11.1) (Farkas et al., 2000; Subhendu, 2000; Malik et al., 2020).

TABLE 11.1 Effects of pollution on physicochemical properties of water.

Factor	Impact
Light	High turbidity of water reduces the penetration quantity of light.
Temperature	The temperature of water increased during thermal pollution, e.g., the discharge of heated water via industrial processes.
pH	The pH of the water will change due to the addition of waste generated from batteries, tanneries, vinegar production, and mining.
Dissolved oxygen	Discharges of sewage and effluent with high organic matter reduced the dissolved oxygen.
Carbon dioxide	Eutrophication and organic pollutants may affect dissolved oxygen depletion in aquatic ecosystems, thereby causing an increase in carbon dioxide.
Salinity	Salinity reduces the dissolved oxygen level in water, leading to increased osmotic pressure.
Color and turbidity	Color of the water also changed due to the accumulation of dye and pigment. The turbidity of the water increased due to soil erosion and heavy algal bloom.
Heavy metals	Various trace elements are present in natural water in very small amounts. However, an increase in the quantity of heavy metals can change the water quality, causing different anomalies in organisms.
Eutrophication	Water pollution from domestic sewage increases the organic load, affecting the rate of productivity of the aquatic ecosystem, e.g., complex growth of phytoplankton.

2.1 Impacts of pollutants on fish species

Increased development activities and urbanization due to the rise in human population cause water pollution across the world. Therefore pollution impairs the physical, chemical, and biological factors in the aquatic ecosystem, thereby altering its natural quality (Subhendu, 2000; Nkwoji et al., 2010). Increased pollution in aquatic ecosystems leads to severe morphological and physiological changes, as well as behavioral changes (e.g., impairing feeding, swimming, and sensory perception) in aquatic organisms (Zhou et al., 2008). Different pollutants affect various fish species differently, causing sudden deaths, biomagnification, or accumulation in tissues, which can damage vital organs such as gills, liver, and kidneys, reduce metabolism, and sometimes result in long-term health issues. Research has shown that the most common contaminants in aquatic environments are heavy metals, which accumulate in tissues and circulate through the food chain, and pesticides, which affect the life cycle of aquatic organisms (Qiu et al., 2004; Ventura et al., 2008). Nanoparticles also cause damage to the aquatic ecosystem, which may have serious impacts on both humans and aquatic organisms in many ways, including changes in water quality and food web structure (Wang et al., 2018).

2.2 Effects of pollutants on fish behavior

Fishes are considered living aquatic organisms that serve as invaluable indicators of aquatic ecosystem health, responding to early signs of water quality degradation. Fishes exhibit different kinds of responses when exposed to toxicants (Adams, 2002; Chakraborthy et al., 2012). Fish have been widely used for studies on behavior, cognitive mechanisms, and processes (Brown et al., 2006). Various types of inorganic and organic pollutants differently affect the behavior of fish, including their migration, swimming, feeding, reproduction, antipredatory behavior, boldness, aggression, and sociality (Little et al., 1990; Sloman and McNeil, 2012; Saaristo et al., 2018; Klaminder et al., 2019).

2.3 Water pollution-induced histopathological changes in fish organs/tissues

Pollutants causing irreversible damage to biota in riverine ecosystems are inevitable due to the increased human population (Pan et al., 2016). Since fishes represent higher positions in the aquatic food chain, they act as good bioindicators responsible for heavy metal accumulation. Therefore they accumulate a considerable amount of heavy metals present in the aquatic ecosystem (Ambedkar and Muniyan, 2011; Hemmadi, 2017). The most vulnerable life stages of fish to heavy metals and pesticides are fry, larvae, and fingerlings, which can lead to alterations in the gills, liver, kidneys, reproduction, etc. (Khoshnood, 2017). Pollutant-induced harmful effects can be identified using histopathological biomarkers (Bernet et al., 1999; Van der Oost et al., 2003). Fish biomarkers refer to detectable biological traits in fish that provide the health status of an individual fish. Therefore the impacts of aquatic pollution on fish or exposed organisms can be assessed using biomarkers (Colin et al., 2016). Furthermore, some researchers have demonstrated the use of multibiomarkers in fish at molecular, biochemical, and cellular levels, which provide a more reliable picture of the impact of pollutants (Vieira et al., 2017). When fish are exposed to pollutants, factors such as the type of pollutants, concentration and duration of exposure, species type, and environmental conditions (e.g., pH, temperature, salinity) influence the structural changes in fish.

2.4 Histopathology

Fish gills, kidneys, and liver are important organs of fishes that play a major role in respiration, excretion, and the accumulation and biotransformation of various xenobiotics. Fishes are very sensitive to water pollution, and they respond differently based on their life cycle stages, sizes, and ages (Burger et al., 2002; Mishra and Shukla, 2003). When they are exposed to heavy metals or pollutants, these contaminants enter the food web, leading to direct exposure for consumers (Savassi et al., 2020). Some researchers have shown that a decline in dissolved oxygen in water affects the physiological changes of aquatic fauna (Shahid et al., 2022). Therefore their responses to these short-term or long-term stressors can be assessed using histopathological changes in their tissues. Such structural changes have been studied globally using different fish species to evaluate the effects of pollutants/toxicants on fish health and to identify the causative agents (Malik et al., 2012; Nascimento et al., 2012; Reddy and Rawatt, 2013). These biomarkers can be used at all stages of biological organization, starting from the molecular level to the overall attributes of the organism. However, each biomarker shows a characteristic or unique response when exposed to pollutants (Nascimento et al., 2012; Pathiratne et al., 2010).

2.5 Histopathological changes in gills

The fish gills are one of the main organs that target different pollutants. Some researchers have shown that damaged gills may cause imbalances in osmotic pressure and gas exchange due to necrosis, edema, telangiectasia, and epithelial separation (Balah et al., 1993). The fish gills are important organs with multiple functions related to gas exchange, acid-base balance, osmotic regulation, and the excretion of nitrogenous waste. Different fish species exhibit a variety of changes in their gills, such as epithelial separation, gill hyperplasia, lamellar fusion, aneurysm, and necrosis, when exposed to pollutants. Therefore these abnormalities may disrupt gas exchange (Fig. 11.1). When fish are exposed to oils, phenols, and mercury, they show epithelial lifting in the gills, which is one of the early responses. When both epithelial separation and lifting of gill lamellae occur together, it increases the intracellular spaces (Poleksic and Mitrovic-Tutundzic, 1994).

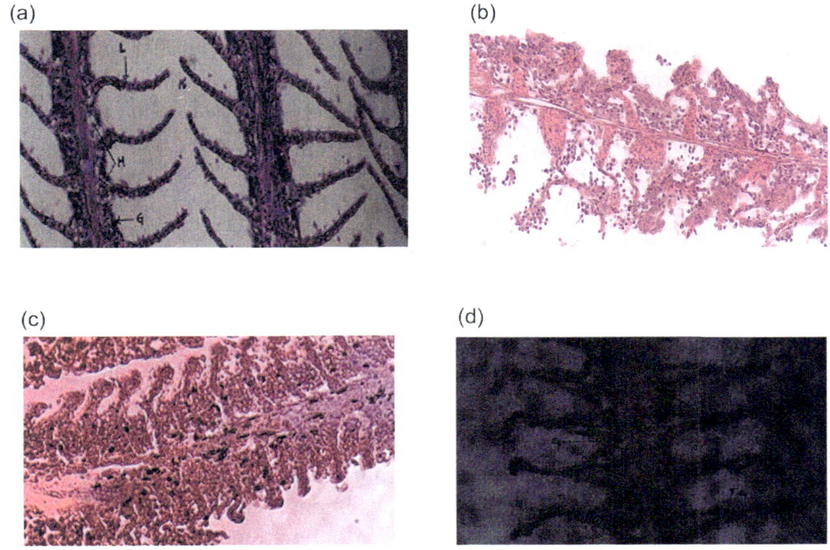

FIG. 11.1 Histological changes in fish gills. (A) Normal gill histology; L- Gill Lamellae, M-Mucous cells, G- Gill filament. (B) and (C) club shaped deformities and mucous cells. (D) Epithelial separation (x400).

Both interstitial edema and leukocyte infiltrations are more common abnormalities in gills, when fish are exposed to chemical pollutants or pesticides, formalin, and harmful algal blooms like red tides, which might affect their osmotic pressure regulation (Yang and Albright, 1992; Strzyżewska et al., 2016). Further, De Brito Carvalho et al. (2020) showed the interstitial edema and of leukocyte infiltration of fish gills. According to Karges and Woodward (1984), two types of gill abnormalities have observed when they exposed to microorganisms and ammonia. Further, increased number of mucous cells observed there by reducing respiratory area. Some species have shown gill lamellar fusion associated with hyperplasia, response to low water quality (Sollid and Nilsson, 2006). De Brito Carvalho et al. (2020) showed microscopic damages and lost pillar cells together with fused lamellae having uniform space with blood. Therefore this type of blood congestion can disrupt the gas exchange causing lamellar disorders. Naeemi et al. (2013) observed massive lamellar hyperplasia due to fusion of secondary lamellar which is one of the defense mechanisms shown by fish to reduce the surface area.

Moreover, some researchers have shown blood leakage in gill lamellae with ruptured pillar cells associated with dilated blood vessels, which is an irreversible change called aneurysm, when fish are exposed to cadmium and organic pollutants (Stehr et al., 1998) and (De Brito Carvalho et al., 2020). Other histopathological changes observed by researchers in fish gills include necrosis when fish are exposed to toxins or subjected to long-term exposure to different irritants. In more polluted areas, necrosis was observed in all fish species

(De Brito Carvalho et al., 2020). Another phenomenon associated with gill histopathology in fish is the presence of more mucous cells when fish are exposed to pollutants, formalin, and disturbed aquatic ecosystems. This is a nonspecific response exhibited by fish against acidic conditions (Mallatt, 1985; Perera and Pathiratne, 2005). Some researchers have noted ruptured pillar cells associated with aneurysm (the rupture of the pillar cells) accompanied by heavy blood flow, which resulted from the direct impacts of contaminants (Naeemi et al., 2013). Research conducted in Sepetiba Bay, Brazil, has shown a higher number of branchial lesions in fish gills due to exposure to heavy metals and solvents, suggesting that histological changes serve as effective biomarkers (De Brito Carvalho et al., 2020).

2.6 Histopathological changes in liver

The use of liver histopathology is one of the widely utilized biomarkers to assess the overall health (ecosystem or organism) of fish. Liver histology plays an integral role in interpreting the deleterious impacts caused by toxicants, providing insights into the degree of damage, inflammation, and other pathological conditions resulting from the presence of pollutants or contaminants. Using routine histopathological techniques, changes in the fish liver, such as tissue damage (localized cell death) and cellular degeneration (liver cells, vacuolization) due to environmental stressors (poor water quality, temperature changes, heavy metals, pesticides) or diseases (viral, bacterial, or parasitic infections), can be used to identify sources of pollution or disease-causing agents. The fish liver plays a major role in the uptake, accumulation, and excretion of toxic elements, thereby processing and eliminating various toxins (Pedlar et al., 2002). Studies conducted globally have shown increased vacuoles in the cytoplasm, cell necrosis, ischemia, shrinkage of cells, regression of hepatocytic microvilli, loss of glycogen, and fatty degeneration (Kalaiyarasi et al., 2017) (Fig. 11.2).

2.7 Histopathological changes in kidneys

Fishes live in an aquatic environment with fluctuating salt concentrations; therefore maintaining a constant internal environment is important. In this process, fish kidneys play a vital role in osmoregulation, there by degrading toxic substances using control mechanisms that could otherwise cause structural damage to kidney cells (Dhirender and Jain, 2019). The fish kidneys assist in regulating water loss or gain, as freshwater fish gain water from the environment while marine fish lose water from their bodies. This process can be triggered when they are exposed to different types of environmental contaminants, impairing homeostasis, function, and causing structural damage. Therefore histopathological changes in the kidneys (e.g., damage to nephrons, disrupted filtration) provide a good indicator of environmental pollution. Some studies have shown shrinkage of glomeruli, degeneration of proximal, distal, and collecting tubules,

(a)

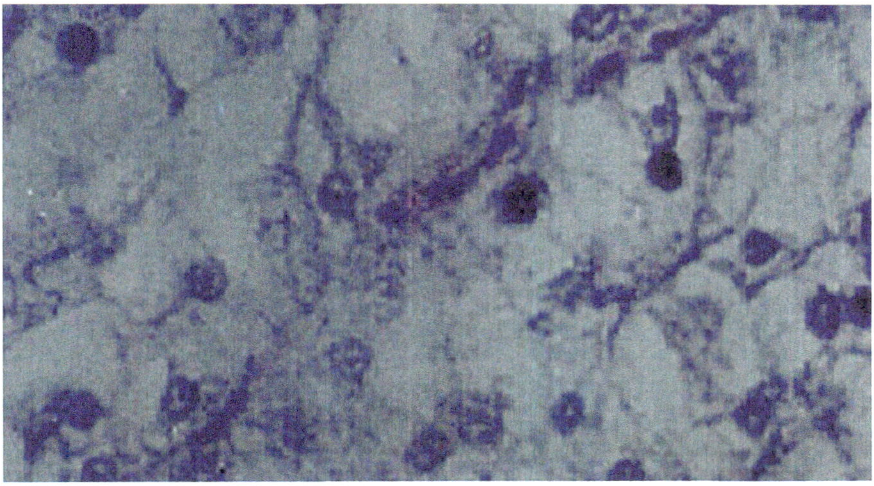

(b)

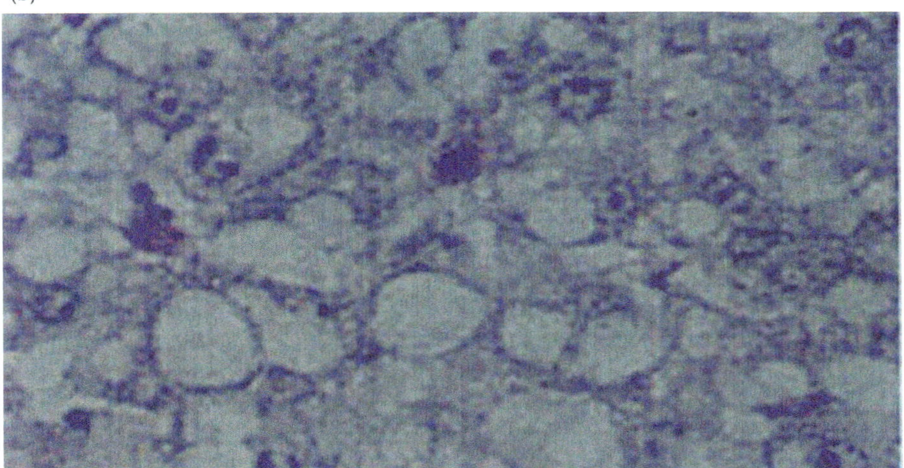

FIG. 11.2 Histological changes in fish liver. (A) H; Hepatocytes. (B): V; vacuolated area (x400).

the appearance of vacuoles, loss of nuclei, congestion of glomeruli, and atrophy of renal tubules (Dhanapakiam and Premlatha, 1994; Sastry and Gupta, 1996; Dhirender and Jain, 2019). Furthermore, some researchers have reported renal dysfunction in kidneys due to damaged nephrons, as well as disturbed pH and body fluid volume associated with erythropoiesis. Moreover, some studies have highlighted more alterations in kidney tissues, namely, severe clogging and regeneration of tubules, absence of Bowman's space, hypertrophied nuclei, necrosis of the hematopoietic interstitial tissue, and nephrotoxicity in response to water pollution and heavy metals (Hussain et al., 2019).

2.8 Other histopathological changes in fish due to exposure to pollutants

In addition to those changes, melanomacrophage centers (clusters of cells) can be seen in spleen, kidney, and liver tissues, which are important in assessing both environmental and fish health.

Except for the gills, kidneys, and liver histopathological changes, there can also be alterations in muscles, spleen, brain, and gonads, causing edema, muscle fiber splitting, lymphoid depletion, neural degeneration, and reproductive impairment. Some researchers have shown damaged muscle layers in fish exposed to carbaryl (Singh and Mehrotra, 1999; Ramesh and Nagarajan, 2013). Furthermore, some studies have demonstrated the formation of depleted glycogen in tissues in response to increased toxicants (Ramesh and Nagarajan, 2013). According to Ramesh and Nagarajan (2013), necrosis, lesions, and cellular degeneration of muscle tissues were observed in Nile tilapia, and some fishes (*Cyprinus carpio*) exhibited remarkable changes and separation of muscle bundles with edema when exposed to cadmium and lead. In addition, inflammatory cells and atrophy of muscle bundles were also observed by different researchers when fish were exposed to various pollutants. Moreover, some researchers have found myolysis in the muscle tissues of *O. spilurus*, and pathological changes such as destruction and vacuolation were observed in Tilapia muscle cells when exposed to chromium (Ramesh and Nagarajan, 2013). Degeneration in the muscle tissues of fish due to oxidative stress was also observed (İrkin et al., 2022).

Instead of the above-mentioned histopathological changes in fish due to exposure to pollutants, the brain also shows some alterations such as necrosis (cell death), edema (swelling of brain tissues), neural degeneration (damage to neuron cells), vacuolization (fluid-filled spaces), hemorrhage (bleeding within the brain tissue), and a decrease in the size of the brain tissue (atrophy). These changes depend on the duration of exposure, the concentration of the pollutant, and the type of pollutant. Therefore these histopathological changes disrupt the function of the brain due to damaged neurons. In addition to that, fishes show behavioral abnormalities such as altered swimming patterns associated with reduced activities and increased levels of mortality due to neurological dysfunctions. According to Lakshmaiah (2017), highly damaged neural bundles in the brain were observed, thereby causing impairment of behavioral functions such as loss of equilibrium and erratic swimming patterns. Other histopathological alterations observed in different fish species include vascular dilation, convulsions, scattered neuron cells, and loss of differentiation of brain cells (Lakshmaiah, 2017).

Not only the brain and muscle tissue of fish but also the fish spleen is a good indicator of the overall health of the ecosystem. The spleen is one of the major organs responsible for antigen trapping and plays a major role in the immune system. Since teleost fishes do not have a medullary cavity in their bones, the spleen acts as their fundamental hematopoietic organ. According to David and Kartheek (2015), when exposed to environmental stressors, fishes exhibit necrosis (death of spleen cells), degeneration of tissues (structural damage to cells), hemosiderin accumulation (damage to red blood cells), an increased number of macrophages, and bleeding, among other effects.

Some researchers have shown abnormal enlargement of the spleen, a decline in lymphocytes, and pulp proliferation. Moreover, the accumulation of hemosiderin, which is a breakdown product of hemoglobin, was observed by authors due to the high rate of fish erythrocyte destruction. Sporadic movement of hemoglobin from the spleen was also observed (Gogal et al., 1999; David and Kartheek, 2015). According to Valigurova et al. (2025), extracellular hemolysis with the deposition of hematogenous pigments in the parenchyma of spleen cells was also recorded.

Impacts of pollution on the reproductive organs of fish have been observed by different authors using various fish species, showing impaired germ cell development, spermatogenesis, and changes in oocyte development and maturation; these may ultimately affect fish development and reproductive success. More generally, pollutant-induced changes such as necrosis, degeneration, fibrosis, and even alterations in the gonadosomatic index (GSI) value of the fish have also been observed. The impact of pollutants on gonads, including the reduction in gonad size and a decline in fecundity, may compromise the reproductive success of fish. Since the GSI is an important factor in fish biology that provides a good understanding of the duration of a particular fish species' spawning season, it might affect the management of fisheries and aquaculture.

3 Conclusions

Histopathology is the study of microscopic structural changes in different tissues (gills, liver, kidneys, spleen, brain, etc.). There can be changes in fish tissues when they are exposed to toxicants, environmental stressors, and diseases. Therefore histopathology is a valuable tool in environmental monitoring and fish health management, as histopathological changes provide early warning signs of pollutant exposure in aquatic organisms. This is an important intervention to mitigate long-term damage to the biota and overall ecosystem health. Fishes are excellent biomarkers for assessing water quality and ecosystem health because they are highly sensitive to environmental changes; they represent different life stages and occupy various trophic levels of the food pyramid. This chapter elaborates on the histopathological changes associated with fish gills, kidneys, liver, gonads, brain, and spleen when they are exposed to different pollutants. Furthermore, it highlights a variety of changes in fish tissues, such as necrosis, tissue degeneration, edema, telangiectasia, epithelial separation, gill hyperplasia, lamellar fusion, aneurysm, vacuolization, ischemia, shrinkage of cells, and regression of hepatocytic microvilli. Therefore studying histopathological changes in fishes enables the implementation of proper actions to improve the ecological stability of water bodies.

References

Adams, S.M., 2002. Biological indicators of aquatic ecosystem stress. Am. Fish Soc. 3, 104–112.

Alrumman, S.A., El-kott, A.F., Kehsk, M.A., 2016. Water pollution: source and treatment. Am. J. Environ. Eng. 6 (3), 88–98.

Ambedkar, G., Muniyan, M., 2011. Accumulation of metals in the five commercially important freshwater fishes available in vellar river, Tamil Nadu, India. Arch. Appl. Sci. Res. 3, 261–264.

Balah, A.M., El-Bouhy, Z.M., Easa, M.E., 1993. Histologic and histopathologic studies on the gills of Tilapia nilotica, *Oreochromis niloticus* under the effect of some heavy metals. Zagazig Vet. J. 21, 351–364.

Bernet, D., Schmidt, H., Meier, W., Burkhardt-Holm, P., Wahli, T., 1999. Histopathology in fish: proposal for a protocol to assess aquatic pollution. J. Fish Dis. 0140-7775. 22 (1), 25–34. doi:10.1046/j.1365-2761.1999.00134.x.

Briggs, D., 2003. Environmental pollution and the global burden of disease. Br. Med. Bull. 1471-8391. 68 (1), 1–24. doi:10.1093/bmb/ldg019.

Brown, T., Laland, D.E., Krause, D. (eds), 2006. Fish Cognition and Behavior, 1st edn. Wiley-Blackwell, Hoboken, NJ.

Burger, J., Gaines, K.F., Boring, C.S., Stephens, W.L., Snodgrass, J., Dixon, C., McMahon, M., Shukla, S., Shukla, T., Gochfeld, M., 2002. Metal levels in fish from the Savannah river: potential hazards to fish and other receptors. Environ. Res. 0013-9351. 89 (1), 85–97. Available from: http://www.elsevier.com/inca/publications/store/6/2/2/8/2/1/index.htt. doi:10.1006/enrs.2002.4330.

Chakraborty, P., Bhuyan, S.K., Jen, J.K., Pillai, B.R., Kumar, P., 2012. Studies on the growth of *Otolithes ruber* and *Pennaha macrophthalmus*: pardeep Coast, Orissa, India. Indian J. Fish 59 (2), 89–93.

Colin, N., Porte, C., Fernandes, D., Barata, C., Padrós, F., Carrassón, M., Monroy, M., Cano-Rocabayera, O., de Sostoa, A., Piña, B., Maceda-Veiga, A., 2016. Ecological relevance of biomarkers in monitoring studies of macro-invertebrates and fish in Mediterranean rivers. Sci. Total Environ. 1879-1026. 540, 307–323. Available from: https://www.elsevier.com/locate/scitotenv. doi:10.1016/j.scitotenv.2015.06.099.

David, M., Kartheek, R.M., 2015. Histopathological alterations in spleen of freshwater fish *Cyprinus carpio* exposed to sublethal concentration of sodium cyanide. Open Vet. J. 5 (1), 1–5.

de Brito Carvalho, T.L.A., Nascimento, A.A.D., Fábio Dos Santos Gonçalves, C., Antônio José Dos Santos, M., Sales, A., 2020. Assessing the histological changes in fish gills as environmental bioindicators in Paraty and Sepetiba bays in Rio de Janeiro, Brazil. Lat. Am. J. Aquat. Res. 0718-560X. 48 (4), 590–601. doi:10.3856/vol48-issue4-fulltext-2351.

Dhanapakiam, P., Premlatha, J., 1994. Histopathological changes in the kidney of *Cyprinus carpio* exposed to malathion and sevin. J. Environ. Biol. 0254-8704. 15 (4), 283–287. Available from: http://www.jeb.co.in/index.php?page=journal_archives.

Dhirender, Jain, S., 2019. histopathological changes in the kidney of a freshwater teleost channa punctatus after 24 hours of exposure to lead nitrate. J. Exp. Zool. India 22, 15–17.

Farkas, A., Salánki, J., Varanka, I., 2000. Heavy metal concentrations in fish of Lake Balaton. Lakes Reserv.: Sci., Policy Manag. Sustain. Use. 1320-5331. 5 (4), 271–279. doi:10.1046/j.1440-1770.2000.00127.x.

Gogal, R.M., Smith, B.J., Robertson, J.L., Smith, S.A., Holladay, S.D., 1999. Tilapia (*Oreochromis niloticus*) dosed with azathioprine display immune effects similar to those seen in mammals, including apoptosis. Vet. Immunol. Immunopathol. 0165-2427. 68 (2-4), 209–227. doi:10.1016/s0165-2427(99)00024-0.

Hemmadi, V., 2017. A critical review on integrating multiple fish biomarkers as indicator of heavy metals contamination in aquatic ecosystem. Int. J. Bioassays. 2278-778X. 6 (9), 5494. doi:10.21746/ijbio.2017.9.5.

Hussain, B., Fatima, M., Al-Ghanim, K.A., Mahboob, S., 2019. Environmentally induced nephrotoxicity and histopathological alternations in *Wallago attu* and *Cirrhinus mrigla*. Saudi J. Biol. Sci. 1319-562X. 26 (4), 752–757. Available from: http://www.ksu.edu.sa/sites/Colleges/CollegeofScinces/BotanyDepartment/SJBS/Default.aspx. doi:10.1016/j.sjbs.2019.02.003.

İrkin, L.C., Öztürk, Ş., Aldık, R., 2022. Histopathological evaluation of muscle tissue of Horse Mackerel (*Trachurus trachurus*, Linnaeus, 1758) in Çanakkale Strait. Acta Aquat. Turc. 18 (3), 303–313. doi:10.22392/actaquatr.1037585.

Kalaiyarasi, T., N Jayakumar, P.J., Ahilan, B., Subbura, J., 2017. Histological changes in the gill and liver of marine spotted catfish, *Arius maculatus* from sewage disposal site, Therespuram off Thoothukudi, Southeast coast of India. J. Entomol. Zool. Stud. 5, 1710–1715.

Karges, R.G., Woodward, B., 1984. Development of lamellar epithelial hyperplasia in gills of pantothenic acid-deficient rainbow trout, Salmo gairdneri Richardson. J. Fish Biol. 0022-1112. 25 (1), 57–62. doi:10.1111/j.1095-8649.1984.tb04850.x.

Khoshnood, Z., 2017. Effects of environmental pollution on fish: a short review. Transylv. Rev. Syst. Ecol. Res. 2344-3219. 19 (1), 49–60. doi:10.1515/trser-2017-0005.

Klaminder, J., Jonsson, M., Leander, J., Fahlman, J., Brodin, T., Fick, J., Hellström, G., 2019. Less anxious salmon smolt become easy prey during downstream migration. Sci. Total Environ. 1879-1026. 687, 488–493. Available from: https://www.elsevier.com/locate/scitotenv. doi:10.1016/j.scitotenv.2019.05.488.

Lakshmaiah, G., 2017. Brain histopathology of the fish *Cyprinus carpio* exposed to lethal concentrations of an organophosphate Insecticide phorate. Int. J. Adv. Res. Dev. 2, 668–672.

Little, E.E., Archeski, R.D., Flerov, B.A., Kozlovskaya, V.I., 1990. Behavioural indicators of sublethal toxicity in rainbow trout. Arch. Environ. Contam. Toxicol. 19, 380–385.

Mahagamage, Y., Mahagamage, M.G.Y.L., Manage, P.M. (2014). *Water Quality Index (CCME-WQI) Based Assessment Study of Water Quality in Kelani River Basin, Sri Lanka*. https://www.researchgate.net/publication/281223991

Malik, D.S., Sharma, A.K., Sharma, A.K., Thakur, R., Sharma, M., 2020. A review on impact of water pollution on freshwater fish species and their aquatic environment. In: Advances in Environmental Pollution Management: Wastewater Impacts and Treatment Technologies. Agriculture and Environmental Science Academy, pp. 10–28. doi:10.26832/aesa-2020-aepm-02.

Malik, G.M., Raval, H.V., Ahmed Khali, H.K., 2012. Toxic effects of effluent on mortality and behavior changes on fresh water fish Poecilia reticulate. J. Environ. Res. Develop. 7 (2A), 1036–1039.

Mallatt, J., 1985. Fish gill structural changes induced by toxicants and other irritants: a statistical review. Can. J. Fish. Aquat. Sci. 0706-652X. 42 (4), 630–648. doi:10.1139/f85-083.

Mishra, R., Shukla, S.P., 2003. Endosulfan effects on muscle malate dehydrogenase of the freshwater catfish *Clarias batrachus*. Ecotoxicol. Environ. Saf. 0147-6513. 56 (3), 425–433. Available from: http://www.elsevier.com/inca/publications/store/6/2/2/8/1/9/index.htt. doi:10.1016/S0147-6513(03)00006-X.

Naeemi, A., Jamili, S., Shabanipour, N., Mashinchian, A., Shariati Feizabadi, S., 2013. Histopathological changes of gill, liver and kidney in Caspian kutum exposed to Linear

Alkylbenzene Sulfonate. Iran. J. Fish. Sci. 1562-2916. 12 (4), 887–897. Available from: http://www.jifro.ir/browse.php?a_id=1348&slc_lang=en&sid=1&ftxt=1.

Nascimento, A.A., Araújo, F.G., Gomes, I.D., Mendes, R.M.M., Sales, A., 2012. Fish gills alterations as potential biomarkers of environmental quality in a eutrophized tropical river in South-Eastern Brazil. Anat., Histol., Embryol.. 0340-2096. 41 (3), 209–216. doi:10.1111/j.1439-0264.2011.01125.x.

Nkwoji, J.A., Yakub, A., Ajani, G.E., Balogun, K.J., Renner, K.O., Igbo, J.K., Ariyo, A.A., Bello, B.O., 2010. Seasonal variations in the water chemistry and benthic macroinvertebrates of a South Western Lagoon, Lagos, Nigeria. J. Anim. Sci. 6, 85–92.

Pan, B., Yuan, J., Zhang, X., Wang, Z., Chen, J., Lu, J., Yang, W., Li, Z., Zhao, N., Xu, M., 2016. A review of ecological restoration techniques in fluvial rivers. Int. J. Sediment Res. 1001-6279. 31 (2), 110–119. Available from: http://www.elsevier.com. doi:10.1016/j.ijsrc.2016.03.001.

Pathiratne, A., Pathiratne, K.A.S., de Seram, P.K.C., 2010. Assessment of biological effects of pollutants in a hyper eutrophic tropical water body, Lake Beira, Sri Lanka using multiple biomarker responses of resident fish, Nile tilapia (*Oreochromis niloticus*). Ecotoxicology 19 (6), 1019–1026. doi:10.1007/s10646-010-0483-2.

Pedlar, R.M., Ptashynski, M.D., Evans, R., Klaverkamp, J.F., 2002. Toxicological effects of dietary arsenic exposure in lake whitefish (*Coregonus clupeaformis*). Aquat. Toxicol. 0166-445X. 57 (3), 167–189. doi:10.1016/S0166-445X(01)00198-9.

Perera, H.A.C.C., Pathiratne, A., 2005. Effects of short term exposure to therapeutic levels of formalin on health status of Nile tilapia, *Oreochromis niloticus*. J. Natl. Sci. Found. Sri Lanka. 1391-4588. 33 (4), 239–245. Available from: https://jnsfsl.sljol.info/. doi:10.4038/jnsfsr.v33i4.2113.

Poleksic, V., Mitrovic-Tutundzic, V., 1994. Fish gills as a monitor of sublethal and chronic effects of pollution. In: Sublethal and Chronic Effects of Pollutants on Freshwater Fish. Cambridge University, Cambridge, pp. 339–352.

Qiu, X., Zhu, T., Li, J., Pan, H., Li, Q., Miao, G., Gong, J., 2004. Organochlorine pesticides in the air around the Taihu Lake, China. Environ. Sci. Technol. 38 (5), 1368–1374. doi:10.1021/es035052d.

Rajamanickam, V., Narayanan, M., 2009. Heavy metal induced histopathological alterations in selected organs of the *Cyprinus carpio* L. (common carp). Int. J. Environ. Res. 3 (1), 95–100.

Ramesh, Nagarajan, K., 2013. Histopathological changes in the muscle tissue of the fish *Clarias Batrachus* exposed to untreated and treated sago effluent. Adv. Biosci. Bioeng. 1, 74–80.

Reddy, P.B., S.S, R., 2013. Assessment of aquatic pollution using histopathology in fish as a protocol. Int. Res. J. Environ. Sci. 2 (8), 79–82.

Saaristo, M., Brodin, T., Balshine, S., Bertram, M.G., Brooks, B.W., Ehlman, S.M., McCallum, E.S., Sih, A., Sundin, J., Wong, B.B.M., Arnold, K.E., 2018. Direct and indirect effects of chemical contaminants on the behaviour, ecology and evolution of wildlife. Proc. R. Soc. B Biol. Sci. 1471-2954. 285 (1885). Available from: http://rspb.royalsocietypublishing.org/. doi:10.1098/rspb.2018.1297.

Sargen, M., Biological roles of water: why is water necessary for life? Science in the News. (2019), Available from: https://sitn.hms.harvard.edu/uncategorized/2019/biological-roles-of-water-why-is-water-necessary-for-life/.

Sastry, K.V., Gupta,, Combined toxic effect of heavy metal cadmium and pesticide dimethoate on liver and kidney of freshwater teleost fish, Channa punctatus. The academy of environ. (1996),

Savassi, L.A., Paschoalini, A.L., Arantes, F.P., Rizzo, E., Bazzoli, N., 2020. Heavy metal contamination in a highly consumed Brazilian fish: immunohistochemical and histopathological assessments. Environ. Monit. Assess. 1573-2959. 192 (8). Available from: https://link.springer.com/journal/10661. doi:10.1007/s10661-020-08515-8.

Shahid, S., Sultana, T., Sultana, S., Hussain, B., Al-Ghanim, K.A., Al-Bashir, F., Riaz, M.N., Mahboob, S., 2022. Detecting aquatic pollution using histological investigations of the gills, liver, kidney, and muscles of *Oreochromis niloticus*. Toxics. 2305-6304. 10 (10). Available from: http://www.mdpi.com/journal/toxics. doi:10.3390/toxics10100564.

Singh, S., Mehrotra, A., 1999. Histopathological changes induced by carbaryl in the intestine of fresh water fish Nandus nandus. J. Ecotoxicol. Environ. Monit. 11 (2), 129–132.

Sloman, K.A., McNeil, P.L., 2012. Using physiology and behaviour to understand the responses of fish early life stages to toxicants. J. Fish Biol. 0022-1112. 81 (7), 2175–2198. doi:10.1111/j.1095-8649.2012.03435.x.

Sollid, J., Nilsson, G.E., 2006. Plasticity of respiratory structures—Adaptive remodeling of fish gills induced by ambient oxygen and temperature. Respir. Physiol. Neurobiol. 1569-9048. 154 (1-2), 241–251. doi:10.1016/j.resp.2006.02.006.

Stehr, C.M., Johnson, L.L., Myers, M.S., 1998. Hydropic vacuolation in the liver of three species of fish from the U.S. West Coast: lesion description and risk assessment associated with contaminant exposure. Dis. Aquat. Organ. 0177-5103. 32 (2), 119–135. Available from: http://www.int-res.com/journals/dao/dao-home/. doi:10.3354/dao032119.

Strzyżewska, E., Szarek, J., Babinska, I., 2016. Morphologic evaluation of the gills as a tool in the diagnostics of pathological conditions in fish and pollution in the aquatic environment: a review. Veterinarni Medicina 61 (3), 123–132.

Subhendu, D., 2000. Effects of aquatic pollution on fish and fisheries. Pollution—An International problem for fisheries. Can. J. Fish. Aquat. Sci. 66, 400–480.

Valigurová., A., Hodová, I., Vetešník, L., Šimková, A., 2025. Histoarchitecture of the liver and spleen in the interspecific hybrids of common carp (*Cyprinus carpio*) and gibel carp (*Carassius gibelio*). J. Vertebr. Biol. 74 (24098), 1–32. doi:10.25225/jvb.24098.

Van der Oost, R., Beyer, J., Vermeulen, N.P.E., 2003. Fish bioaccumulation and biomarkers in environmental risk assessment: a review. Environ. Toxicol. Pharmacol. 13 (2), 57–149. doi:10.1016/S1382-6689(02)00126-6.

Ventura, B.C., Angelis, D.F., Maria, A.M., Morales, M., 2008. Mutagenic and genotoxic effects of the Atrazine herbicide in *Oreochromis niloticus* (Perciformes, Cichlidae) detected by the micronuclei test and the comet assay. Pestic. Biochem. Physiol. 90, 42–51.

Vieira, C.E.D., Costa, P.G., Cabrera, L.C., Primel, E.G., Fillmann, G., Bianchini, A., Bueno Dos Reis Martinez, C., 2017. A comparative approach using biomarkers in feral and caged Neotropical fish: implications for biomonitoring freshwater ecosystems in agricultural areas. Sci. Total Environ. 1879-1026. 586, 598–609. Available from: https://www.elsevier.com/locate/scitotenv. doi:10.1016/j.scitotenv.2017.02.026.

Wang, R., Song, B., Wu, J., Zhang, Y., Chen, A., Shao, L., 2018. Potential adverse effects of nanoparticles on the reproductive system. Int. J. Nanomed. 1178-2013. 13, 8487–8506. Available from: https://www.dovepress.com/getfile.php?fileID=46882. doi:10.2147/IJN.S170723.

Yang, C.Z., Albright, L.J., 1992. Effects of the harmful diatom Chaetoceros concavicornis on respiration of rainbow trout *Oncorhynchus mykiss*. Dis. Aquat. Organ. 0177-5103. 14, 105–114. doi:10.3354/dao014105.

Zhou, Q., Zhang, J., Fu, J., Shi, J., Jiang, G., 2008. Biomonitoring: an appealing tool for assessment of metal pollution in the aquatic ecosystem. Anal. Chim. Acta. 0003-2670. 606 (2), 135–150. doi:10.1016/j.aca.2007.11.018.

Chapter 12

Land-use change and change in meandering pattern of rivers due to pollution

Vinod Kumar[a], Surbhi Kharwar[b], and Ekta Shukla[c]
[a]*Department of Botany, Pandit Prithi Nath College (Affiliated to CSJM University, Kanpur), Kanpur, Uttar Pradesh, India*
[b]*Department of Botany, University of Lucknow, Lucknow, Uttar Pradesh, India*
[c]*Department of Botany, Sunbeam College for Women Bhagwanpur (An Autonomous College), Varanasi, Uttar Pradesh, India*

1 Introduction

Rivers have been diminished, simplified, and degraded globally by the concentration of agriculture, transportation, and development in valley bottoms over decades and centuries, substantially limiting their ecological health and value (Skidmore and Wheaton, 2022). Significant hydrological alterations are the result of large investments in water technologies and infrastructures for irrigation and hydropower across the globe, as well as land-use changes (Ekka et al., 2020). The term "river morphology" is used to describe both the cross-sectional shape of a river channel and the process by which a river's section changes as a result of sedimentation and erosion from the banks and beds (Shahid et al., 2017). Hydrogeomorphologically, the study of river meandering provides information on the tendency of rivers to reach and form a state of equilibrium (Zahoor and Mushtaq, 2023).

Agricultural intensification, urbanization, and industrial discharge all contribute to the pollution of rivers, changing both the quality of the water and the riparian and aquatic ecosystems. A large number of anthropogenic contaminants are associated with the finest sediment sizes (<63 micrometers), which help transport and deposit these contaminant substances within river channels. As a result of advances in water treatment methods over the past several decades, there has been a significant reduction in the accumulation and discharge of heavy metals and organic pollutant compounds. However, legacy contaminants stored in older sediments remain a substantial ecological concern, as they can

River Pollution. https://doi.org/10.1016/B978-0-443-45024-2.00006-3

be remobilized during high-flow or flood events (Gerbersdorf et al., 2011). It is thus critical to investigate how alterations to land use, pollution, and meandering river channels influence the management of river basins. Understanding the changes in lateral channel migration, sediment transport pathways, and vegetative dynamics of floodplains from the perspective of the overall health of rivers (ecologically and botanically) is essential in determining how the system will respond, how to develop mitigation measures to reduce flooding and erosion, and how to apply scientific research to develop sustainable land and water management.

Rivers have long been regarded as the lifelines of civilizations, shaping landscapes and nurturing both ecological and human communities (Wang and He, 2022). Among the various types of rivers, meandering rivers are particularly distinctive for their sinuous channels, which migrate laterally across floodplains through a delicate balance of erosion and deposition (Leopold and Wolman, 1960). The development of these meanders reflects the constant interaction between hydrodynamics, sediment load, and riparian vegetation. However, the natural balance is constantly being disturbed by human activities, particularly pollution and land-use change (Allan, 2004; Poff et al., 1997).

Pollution influences river morphology in both physical and biological ways. Alterations in water chemistry affect sediment cohesion and flow dynamics, while changes in riparian vegetation caused by nutrient loads, heavy metals, or other pollutants influence bank resistance and sediment retention (Dosskey et al., 2010; Grabowski et al., 2022; Okeke et al., 2022). Runoff containing agricultural nutrients promotes the growth of invasive plant species that destabilize riverbanks, whereas industrial pollutants weaken root systems and diminish the protective role of vegetation (Edokpayi et al., 2020). Urbanization and deforestation modify hydrological behavior, intensifying flood peaks and altering sediment flow (Knighton, 1998). Collectively, these changes lead to increased lateral migration, channel instability, and altered meandering patterns. These processes transcend disciplinary boundaries, necessitating an integrated approach to understand the relationship between pollution and river meandering. Pollution emerges as both a physical and biogeochemical driver, influencing flow regimes, sediment dynamics, and vegetation responses (Naiman et al., 2005), which collectively shape the morphodynamic development of meandering rivers.

Despite their global significance, tropical river systems remain underresearched in this regard. The Barak River, located in northeastern India, exemplifies this gap, as it faces concurrent pressures from pollution, deforestation, and unregulated land use (Lambin et al., 2003; Annayat et al., 2022). With its floodplain subjected to high rainfall and intensive agriculture, the region offers an ideal setting to examine how pollution interacts with hydrological and vegetative controls to influence river morphology (Serrao et al., 2022; Kayitesi and Biswas, 2022; Kabir et al., 2024). Recent studies have documented pronounced lateral migration and the loss of agricultural land within the basin, indicating a strong link between human impacts and geomorphic instability (Nath and Ghosh, 2022; Annayat and Sil, 2020).

This chapter, therefore, seeks to explain how pollution influences river meandering by addressing three key questions: How does riparian vegetation, under pollution stress, affect bank stability and sediment dynamics? In what ways does pollution alter the hydrological regime? And how do these factors collectively influence meander development, particularly in the context of the Barak River in India? This chapter provides a comprehensive understanding of how pollution reshapes river morphology. Furthermore, by integrating global perspectives with local evidence, it advances an eco-geomorphological framework for the sustainable management of polluted meandering rivers.

1.1 Land-use change and fluvial systems

Land-use change refers to an alteration in the way a specific area of land is used or managed by humans, often driven by urban growth and resulting in various local and global effects, including biodiversity loss and habitat destruction (Sajikumar and Remya, 2015). More importantly, changes in land use directly influence riverbank vegetation, wetland plant communities, and the ecological stability of riparian corridors by altering hydrological pathways, sediment supply, and nutrient loading (Pelletier et al., 2015). In riverine areas, the processes of change (deforestation, agricultural development, and urban development) have altered many fluvial processes through changed discharge regimes, sediment loads, and channel stability (geomorphology), thus impacting the rates of their migration, the creation of floodplains, and the productivity of riparian vegetation (Latrubesse, 2015).

Vegetation within floodplain/riparian vector, vegetation, and wetlands has exhibited dramatic shifts due to changes in land use. Changes in the water cycle (hydrologic change), sediments moving through (sediment transport), and nutrients flowing into (water quality) a location have contributed to altered vegetation composition, disruption of the natural process for vegetation to regrow, and long-term unstable populations of riparian vegetation. Over time, cumulative shifts in riparian/floodplain vegetation can have a major impact on the riverine ecosystem's functioning, particularly in terms of stabilizing the banks of the stream or river, capturing sediments, and improving water quality. The interactive nature of vegetation, as well as the influence of altered flooding, sediment pollution entering the system, and continued geomorphic changes, shapes river systems' ecological health and morphodynamic makeup, defining the present and future state of rivers worldwide (Ciotti et al., 2021).

1.2 Hydrogeomorphology and landscape sensitivity

The word "hydromorphology" comes from "hydro," meaning water, and "morphology," meaning the physical features of the environment that hold the water in place. These environments include rivers, lakes, and transitional waters, such as estuaries, lagoons, and tidal sections of rivers. The hydrogeomorphic impacts of land use have important ramifications for the fluvial system at all scales, but particularly for small headwater catchments, which have stronger hillslope-channel coupling than points further downstream (Gajalakshmi and Anantharama, 2017).

The ability of river landscapes to be altered over time demonstrates how much and how quickly flooding, river banks, and flood plains react to disturbances, such as land-use changes, increased sediment deposition, altered hydrologic delivery systems, and pollution. Rapidly changing landscapes will lead to quick alterations in meander shape, the stability of the river bank, and the vegetation structure of the riparian zone (Stojković, 2017). Landscape sensitivity is an important way to assess ecological responses to disturbances and to predict geomorphic changes within disturbed watersheds.

1.3 Geographic Information Systems, geospatial data, and analytical approaches

Up to date, remote sensing and Geographic Information Systems (GISs) are the best methods for studying spatial distributions and the evolution of geographic phenomena in relation to land-use and land-cover changes, as well as for quantifying these changes. The application of remote sensing and GIS in the assessment of land-use changes is increasing over time (Myslyva et al., 2018; Navin and Agilandeeswari, 2020). The information derived from GIS will be used to identify the critical processes happening in rivers (i.e., meander migration, floodplain change, vegetation change, and pollution hotspots) and quantify the impacts land-use/hydrogeomorphic changes have on riparian habitat, sediment pathways, and water quality, to provide managers with the necessary data at both the ecological and basin scale to effectively manage rivers.

1.4 Flood regimes and flood generation

At the same time, regional gradients of flooding dynamics are also thought to influence patterns of forest composition among floodplains. Down-valley changes in geomorphology can produce shifts in the characteristics of local flood regimes within a single basin. Flood regimes can vary strongly across rivers within a physiographic system due to interactions among basin size, river-valley morphology, and flow magnitudes, which then interact locally with landforms to produce fine-scale variation in floodplain inundation characteristics (Van Appledorn and Baker, 2022). Flood events for each watershed are affected by the characteristics, intensity, and timing of flood regimes. Flood regime categories fall along a continuum, with one extreme being completely nival (rooftop) and the opposite extreme being completely pluvial (rainfall).

The characteristics, intensity, and frequency of floods influence sediment deposition, channel mobility, inundation frequency, and riparian vegetation zonation (Blöschl, 2022). Thus flood dynamics will determine how to properly assess the effects of land-use change and pollution on the ecological viability of rivers and floodplain vegetative communities.

1.5 Soil erosion, sediment yields, and alluvial stratigraphy

The sediment found on riverbeds and banks forms the channel of the respective river and plays a central role in determining channel structure, hydraulic capacity, and downstream sediment continuity along the river network (Dutta, 2016). In addition to channel alteration, additional sediment entering the stream negatively affects adjacent floodplain plants through changes in nutrient availability. As a result, changes in nutrient availability, soil texture, and soil moisture can influence the composition and productivity of floodplain plant communities.

The movable nature of banks controls how rapidly sediment will be deposited on a river channel. Therefore the characteristics of the adjacent overbank environment strongly influence channel morphodynamics, sediment deposition patterns, and the development of associated alluvial strata.

Observations represent an advance in the understanding of the rules that govern long-term sediment storage in floodplains and will be useful for formulating rules-based models for the formation of alluvial stratigraphy. The results differ from those found in studies of braided streams with noncohesive sediment, in which no information about hydrograph shape was transferred to the stratigraphy (Akmal et al., 2023).

The geomorphic processes are also responsible for the distribution of riparian vegetation. Sediment transport and sediment deposition are critical to understanding the interactions of pollutants, changing land uses, and erosion on long-term ecological changes within river basins.

1.6 Pollution and water quality responses

Contaminated waters experience much lower-quality degradation than noncontaminated waters; thus when pollutants from agriculture, industry, urban and suburban stormwater, and sediment-bound spills enter our rivers and streams, they impact the water chemistry, aquatic vegetation, riparian zones, and overall ecological function of the river or stream. Eutrophication may occur when nutrient concentrations increase, and heavy metals and pesticide residues build up in sediments, both resulting in substantial impacts on seed germination, root development, and species composition in floodplain environments (Akmal et al., 2023).

Pollution affects hydrogeomorphic processes, as eroded sediment may serve as a mechanism to transport (downstream) the adsorbed pollutants, and meander migration and flooding events have the potential to "redistribute" (remobilize) legacy contaminants that are buried (within) the alluvial deposits. All of these processes have significant ecological implications with respect to plant toxic responses, habitat suitability, and the long-term resilience of riverine vegetation.

2 River meandering: A geomorphic overview

River meandering is the result of self-organized feedback among flow velocity, sediment load, and bank resistance (Frascati and Lanzoni, 2009). Classical models describe this process as a function of differential erosion and deposition along the outer and inner banks of the channel (Leopold and Wolman, 1960). Acceleration of flow along the outer bank enhances erosion, while deceleration along the inner bank promotes deposition, gradually forming a sinuous, meandering planform. The hydrodynamic theories proposed by Ikeda et al. (1981) suggest that secondary currents, bed topography, and sediment sorting reinforce curvature-driven migration. Over time, this migration results in meander cutoffs, oxbow lake formation, and the development of floodplains (Knighton, 1998). However, the natural meandering equilibrium has been disrupted by anthropogenic influences, leading to alterations in sediment supply, discharge variability, and bank strength.

2.1 Eco-geomorphology and riparian feedbacks

Eco-geomorphology integrates biotic and abiotic interactions that shape river morphology (Thoms and Parsons, 2002). Vegetation stabilizes banks by reinforcing root structures, modifies hydrodynamics through flow resistance, and influences sedimentation patterns by trapping fine particles and organic matter (Naiman et al., 2005; Gurnell and Petts, 2002) and creates biogeomorphic feedbacks, a term that describes the mutual adjustments between geomorphological and ecological processes (Corenblit et al., 2007). Under natural conditions, riparian vegetation enhances channel stability and regulates the amplitude of meanders. However, pollution alters this response in several ways (Finotello et al., 2024b). Nutrient pollution stimulates excessive macrophyte growth, which can reduce flow velocity and sediment transport. In contrast, toxic contamination, such as heavy metals and hydrocarbons, impedes root development and vegetation density, thereby weakening bank cohesion (Dosskey et al., 2010; Njuguna et al., 2017; Wilkes et al., 2019). Sediment pollution resulting from erosion and runoff can smother seedbeds, inhibiting regeneration. These altered responses lead to dynamic changes in channel form, manifesting as either accelerated lateral migration or excessive channel stabilization (Camporeale et al., 2013).

3 Hydrological regime modification and pollution

Hydrological regimes determine how rivers transport water, sediment, and nutrients from their catchments. Poff et al., (1997) introduced the concept of the natural flow regime and identified key flow components—magnitude,

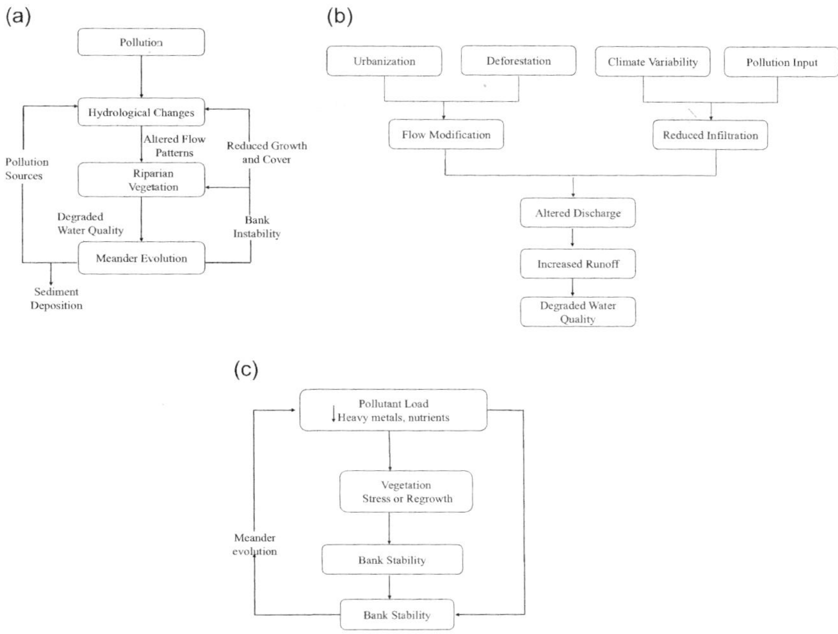

FIG. 12.1 (A) The model showing linkages between pollution stressors, vegetation degradation, hydrological alteration, and meander evolution; (B) Hydrological regime modification under anthropogenic influence, highlighting changes in discharge variability and sediment flux; (C) Feedback loops among pollution, sediment, and vegetation, demonstrating the self-reinforcing eco-geomorphological cycle.

frequency, duration, timing, and rate of change—as essential for maintaining ecological and geomorphic integrity. Pollution interacts with these components in multiple ways. Urban runoff increases surface flow and causes flash floods due to impervious surfaces, while agricultural pollutants modify infiltration and evapotranspiration through soil erosion and nutrient enrichment (Zhang et al., 2019). However, industrial effluents alter water chemistry, viscosity, and sediment cohesiveness. The combined effects of these changes modify the hydrological regime, shifting it from natural variability to artificial flow patterns that destabilize channel morphology (Knighton, 1998). Fig. 12.1A illustrates the conceptual relationships among pollution sources, hydrological alterations, vegetation responses, and meander development.

4 Mechanisms linking pollution and river meandering

Pollution affects river flow in three principal ways: through physicochemical alterations, biological responses, and hydrological changes. In terms of physicochemical alterations, pollutants modify the physical properties of riparian materials, leading to changes in soil cohesion, hydraulic conductivity, and erodibility (Nanson and Huang, 2008). For instance, increased salinity can cause soil dispersion, thereby enhancing susceptibility to erosion, while organic pollution may temporarily increase aggregation. In terms of biological responses, pollution affects plant physiology, reducing biomass, root tensile strength, and vegetative cover. As riparian vegetation deteriorates, bank protection diminishes, promoting lateral channel migration (Edokpayi et al., 2020). In the case of hydrological alterations, polluted catchments exhibit modified runoff patterns. Nutrient-enriched agricultural runoff increases sediment yield, while reduced infiltration due to soil compaction intensifies peak discharges. The resulting fluctuations enhance meander bend erosion and cut-off formation (Zhang et al., 2019). The interactions among these processes give rise to system-level feedbacks (Fig. 12.1B).

5 Feedback dynamics and system resilience

Feedback mechanisms play a central role in understanding the resilience of meandering rivers. Pollution-induced changes create a self-reinforcing cycle whereby increasing pollution weakens vegetation → weakened vegetation enhances erosion → erosion increases sediment load → sediment-bound pollutants intensify contamination → further degradation of vegetation and hydrology occurs, as illustrated in Fig. 12.1C. Such cyclical dynamics drive river systems toward alternative stable states, either highly migratory (characterized by excessive erosion and cut-off formation) or overly stable (with vegetation-choked channels and reduced mobility). The resilience of the system depends on the capacity of riparian vegetation to absorb disturbances and on the degree of hydrological variability.

Integrating the aforementioned mechanisms provides a foundation for eco-geomorphological modeling of polluted rivers. The proposed framework emphasizes the interdependence of hydrological, geomorphological, and ecological processes; the vulnerability of riparian zones as mediators between terrestrial and aquatic systems; and the central role of pollution in altering both physical and biological feedbacks.

6 Riparian vegetation controls

Riparian vegetation is a fundamental bio-geomorphic component that mediates interactions among hydrology, sediment, and channel morphology (Naiman et al., 2005). It stabilizes riverbanks, regulates sediment deposition, and provides hydraulic resistance, all of which collectively determine meander development and floodplain formation. Therefore the resilience of riparian zones is a key determinant of how rivers respond to anthropogenic disturbances such as pollution.

6.1 Vegetation and bank stabilization mechanisms

The stabilizing effect of vegetation arises primarily from the mechanical reinforcement provided by root systems and the hydraulic roughness generated by stems and foliage. Root networks bind soil particles, increase shear strength, and reduce erodibility. Studies have shown that root-reinforced soils can withstand two to five times greater shear stress than unvegetated soils (Abernethy and Rutherfurd, 2000). However, vegetation also increases hydraulic roughness, dissipating flow energy and promoting sediment deposition along channel margins and point bars. This process leads to the gradual accretion of floodplains and the stabilization of meander bends (Gurnell and Petts, 2002). In addition, plant transpiration reduces pore-water pressure, thereby increasing bank strength during high-flow events, a phenomenon referred to as the biostabilization capacity of riparian vegetation. Moreover, vegetation–hydrology interactions are nonlinear and sensitive to external disturbances. Under conditions of high runoff, vegetation may be uprooted or submerged, diminishing its stabilizing role. Conversely, moderate floods enhance vegetation abundance by depositing nutrient-rich sediments (Corenblit et al., 2007). This response complexity creates a self-regulating system that maintains meandering equilibrium under natural conditions.

6.2 Pollution-induced alterations in riparian dynamics

Pollution disrupts the stabilizing effect of vegetation through both physiological and structural pathways. Nutrient enrichment from agricultural runoff promotes the dominance of opportunistic species such as *Eichhornia crassipes* and *Ipomoea aquatica*, whose roots exhibit limited cohesion compared with native shrubs and trees. Heavy metal contamination impedes root elongation and microbial symbiosis, thereby weakening soil–root binding (Edokpayi et al., 2020). Furthermore, fine sediment pollution, a by-product of deforestation and surface erosion, smothers seeds and roots, thereby inhibiting regeneration. The reduction in vegetation density decreases hydraulic roughness, which, in turn, increases near-bank flow velocity and accelerates erosion (Nepf, 2012).

Studies have shown that pollution has led to vegetation degradation in the Mississippi and Ganges rivers, resulting in up to a 30% increase in the annual rate of lateral migration (Camporeale et al., 2013; Gurnell and Petts, 2002). In tropical basins such as the Barak River, where monsoonal hydrology amplifies erosive energy, even a slight decline in riparian vegetation can translate into severe geomorphic instability.

6.3 Vegetation–hydrology coupling under pollution stress

The relationship between vegetation and hydrology governs the spatial distribution and effectiveness of riparian zones. Hydrological alterations induced by pollution—such as increased runoff from impervious surfaces or nutrient-driven algal blooms—affect soil moisture, infiltration, and groundwater dynamics (Zhang et al., 2019). These changes disrupt vegetation zonation patterns, shifting community composition toward flood-tolerant but less stabilizing species. In agricultural catchments, excessive fertilizer inputs increase nitrogen and phosphorus concentrations in groundwater and surface runoff, promoting the proliferation of hydrophytes in floodplain depressions. These plants increase transpiration but often lack the root tensile strength required to stabilize riverbanks. In contrast, trees and shrubs with deep root systems suffer from toxicity-induced stress, leading to wilting and a reduction in mechanical reinforcement (Allan, 2004). Such hydrological–vegetation feedbacks have significant geomorphic consequences. A reduction in vegetative cover increases flow concentration along the banks, thereby enhancing shear stress and sediment deposition. This process leads to altered meandering stability, often resulting in highly dynamic, braided, or anabranching systems, depending on the balance among discharge, sediment supply, and vegetative response (Corenblit et al., 2009; Kleinhans, 2010).

6.4 Vegetation impacts on meandering

Recent advances in remote sensing and numerical modeling have made it possible to quantify vegetation–meandering relationships. By correlating Normalized Difference Vegetation Index (NDVI) data derived from satellite imagery with meander migration rates, the influence of riverbank health on channel dynamics can be assessed. Studies conducted along the Ganga and Brahmaputra rivers have revealed a strong negative correlation between NDVI and lateral migration ($r = -0.68$ to -0.84) (Nath and Ghosh, 2022; Annayat and Sil, 2020). In the Barak River Basin, calibrated NDVI data (1990–2020) exhibit a similar trend, with densely vegetated areas showing limited migration, while degraded riparian corridors display increased curvature and channel width. This indicates that riparian vegetation functions both as a controller and an indicator of meander stability. Fig. 12.2B summarizes these relationships, showing a progressive decline in vegetation accompanied by a corresponding increase in meander wavelength. Pollution-induced vegetation degradation weakens this control, resulting in increased morphological variability. Thus riparian vegetation serves as a sensitive indicator of ecosystem health, providing early warning signals of hydrogeomorphic change. Data from the Barak River reveal a clear correlation, highlighting that spatial patterns of vegetation degradation align with areas of meander amplification and floodplain instability (Annayat et al., 2022). These interactions form the basis for understanding the feedbacks associated with hydrological regime alterations, which are explored in the following section.

6.5 Hydrological regime modifications

Hydrological regimes govern the timing, quantity, and quality of water flow within river systems. The natural flow regime, characterized by periodic floods, seasonal low flows, and inter-annual variability, sustains channel morphology

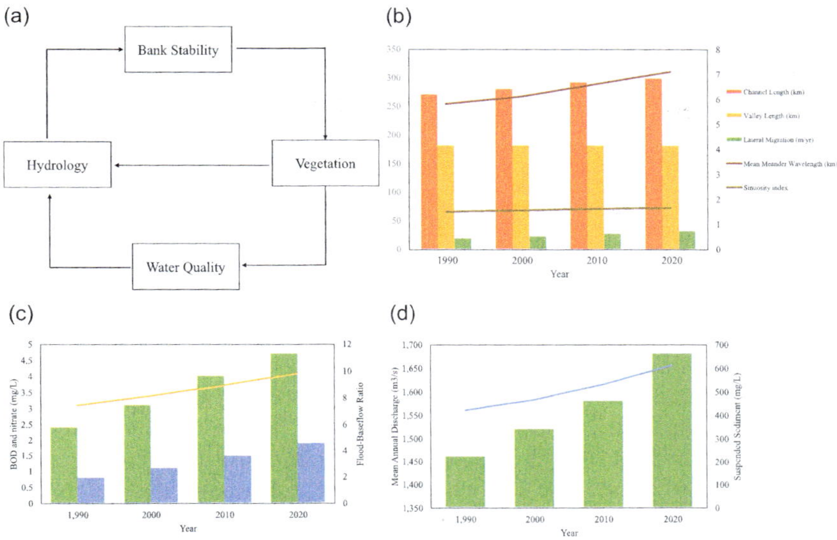

FIG. 12.2 (A) Integrated eco-geomorphological feedback system showing linkages among pollution, hydrology, vegetation, sediment dynamics, and meander morphology; (B) Meander migration and sinuosity changes; (C) and (D) Hydrological and pollution parameters for the Barak River (1990–2020).

and ecological balance (Poff et al., 1997). However, anthropogenic pollution—combined with urbanization, agricultural intensification, and climate change—has brought about substantial alterations to these regimes. Under pollution stress, hydrological changes manifest through variations in discharge magnitude, sediment transport, infiltration dynamics, and evapotranspiration. These alterations disrupt sediment–flow–vegetation feedbacks, resulting in modified meandering behavior (Allan, 2004; Corenblit et al., 2009).

6.6 Anthropogenic drivers of hydrological change

Pollution influences hydrological behavior through both direct and indirect pathways. Direct effects include alterations in water chemistry, increased suspended solids, and reduced infiltration capacity caused by soil clogging with fine sediments and organic residues (Knighton, 1998). Indirectly, land-use changes such as deforestation and urban expansion increase the extent of impervious surfaces, resulting in flashier hydrographs and higher peak discharges. Urban runoff carries heavy metals, oils, and suspended solids that increase flow turbidity and alter the roughness of the riverbed (Pochodyła-Ducka et al., 2023). Agricultural pollution introduces fertilizers and pesticides that affect infiltration and evapotranspiration by altering soil structure and moisture retention capacity (Zhang et al., 2019). Industrial effluents, however, modify the ionic composition of water, influencing sediment flocculation and the stability of the channel bed (Allan, 2004). Collectively, these processes transform hydrological regimes from gradual, seasonal variability to abrupt, high-intensity flow events, thereby destabilizing channel morphology.

6.6.1 Pollution-induced alterations in flow variability

In natural systems, the frequency and duration of floods maintain floodplain connectivity, facilitating sediment exchange and vegetation regeneration. Pollution alters these dynamics by increasing both the frequency and magnitude of runoff peaks. High nutrient loads promote algal blooms and macrophyte overgrowth, which obstruct flow paths and exacerbate local flood retention (Junk et al., 1989; Paul and Meyer, 2001). Conversely, excessive sediment and organic loads reduce channel depth, thereby increasing the frequency of overbank flooding. Fig. 12.1B schematically illustrates this process: pollutants originating from urban, agricultural, and industrial sources alter infiltration and runoff dynamics, resulting in irregular hydrographs. During flood peaks, increased sediment transport intensifies bank erosion, whereas sediment deficiency during low-flow phases promotes incision. The resulting inconsistency in discharge behavior disrupts the natural meander adjustment cycle (Hooke, 2003).

6.6.2 Sediment flux modification and channel instability

Pollution and hydrological alterations modify sediment dynamics by affecting both supply and transport capacity. Agricultural runoff increases the flux of fine sediments, whereas industrial pollution often enhances sediment cohesion through chemical flocculation. An imbalance between sediment supply and transport leads to channel instability, manifested as either aggradation or incision (Nanson and Huang, 2008).

In the Barak River Basin, simulated data (Fig. 12.2C and D) show a progressive increase in suspended sediment concentration (SSC), rising from an average of 420 mg/L in 1990 to 610 mg/L in 2020, accompanied by a 15% increase in mean annual discharge. During the same period, the ratio of flood to base-flow discharge increased from 7.4 to 9.8, indicating enhanced flow variability. These changes are strongly correlated ($r = 0.81$) with the observed rise in meander wavelength and lateral migration rates, suggesting a trend toward hydrological intensification and pollution amplification, characteristic of tropical river basins experiencing agricultural and urban expansion.

7 Hydrological vegetation feedbacks and meander morphology

Altered flow regimes modify the feedback between hydrology and riparian vegetation. Increased high-flow events lead to frequent inundation and vegetation uprooting, while reduced base flows limit seed germination and survival

(LeRoy et al., 1997). The resulting decline in vegetative cover, in turn, reduces bank resistance and enhances lateral erosion. This self-reinforcing feedback accelerates meander migration and floodplain reconfiguration (Finotello et al., 2024b). Studies have shown that increased runoff variability drives meandering systems toward greater curvature and higher cut-off frequency. For instance, in the Barak River, simulations indicate that every 10% increase in the flood-to-base flow ratio corresponds to a 5–7% rise in average meander amplitude. This nonlinear relationship highlights the sensitivity of meander dynamics to hydrological and vegetative feedbacks. Thus changes in hydrological regimes act as the primary mediator linking pollution to meandering behavior. Pollution modifies hydrology both quantitatively, through changes in discharge and sediment load, and qualitatively, through alterations in water chemistry and viscosity.

Under pollution stress, alterations in hydrological regimes constitute a multifaceted process involving changes in discharge variability, sediment dynamics, and vegetative response. These changes disrupt the delicate balance between erosional and depositional forces that sustain meandering morphology (Allan, 2004). In valleys such as the Barak River, intensified hydrological pulses combined with degraded riparian vegetation shift the system from dynamic equilibrium toward geomorphic instability.

8 Mechanistic linkages

Mechanistic relationships illustrate how pollution translates into morphological changes through feedback loops among hydrology, vegetation, and sediment. Pollution alters the natural equilibrium of river systems through a network of interdependent physical, chemical, and biological processes (Corenblit et al., 2024). The conceptual model presented here synthesizes these interactions into three primary feedback domains: hydrological, sedimentological, and biological, each contributing to the morphological trajectory of meandering rivers.

8.1 Physicochemical interactions: Pollution and sediment cohesion

The first mechanistic relationship operates through pollution-induced alterations in sediment properties. Fine-grained sediments, such as silt and clay, play a crucial role in maintaining bank stability. Their cohesion depends on electrochemical interactions among soil minerals, organic matter, and the chemistry of pore water (Nanson and Huang, 2008). Pollution modifies these interactions in several ways. Salinity and ionic alterations—arising from industrial effluents and agricultural runoff—increase the ionic concentration of river water. Depending on the type and concentration of ions, high salinity can either flocculate or disperse soil particles, thereby altering the erodibility of sediments (Allan, 2004). Organic contaminants, such as petroleum hydrocarbons and organic wastes, reduce surface tension and increase sediment lubrication, which, in turn, enhances bank slumping. However, the adsorption of heavy metals, such as lead, zinc, and cadmium, onto fine sediments increases their density and affects settling velocity. Collectively, these processes modify sediment cohesion, altering the threshold shear stress required for erosion (Zander et al., 2022).

8.2 Hydrological feedbacks: Altered flow dynamics

Pollution-induced hydrological alterations propagate through the system as changes in flow velocity, depth, and shear stress distribution. Polluted basins typically exhibit increased peak discharges during storm events due to impervious surfaces, reduced base flows during dry periods owing to lower infiltration, and elevated water density and viscosity caused by suspended sediments and contaminants (McGrane, 2016). These changes modify channel hydraulics, disrupting the balance between erosional and depositional forces. High peak flows intensify erosion along outer banks, while low flows promote deposition on inner bends, generating asymmetry that accelerates meander migration and increases channel curvature. Annayat and Sill (2020b) reported a 20% rise in peak discharge within a decade in the Barak River, accompanied by a 30% increase in sediment concentration, twice the rate of lateral migration.

8.2.1 Biological feedbacks: Vegetation response and bank stability

The third mechanistic relationship is linked to the response of riparian vegetation to pollution-induced stress. Vegetation interacts with hydrology and sediment dynamics through root reinforcement, transpiration, and organic matter cycling (Naiman et al., 2005). Under pollution stress, however, several feedbacks emerge: toxic contaminants reduce germination rates and root growth; nutrient enrichment favors fast-growing, shallow-rooted species over deep-rooted stabilizers; and sediment-bound pollutants inhibit microbial decomposition, thereby altering soil structure. These processes reduce bank shear strength, making banks more susceptible to collapse during high-flow events. Camporeale et al. (2013) demonstrated, based on field and simulation data, that riparian vegetation degradation accounts for up to 40% of the observed increase in meander migration rates within polluted basins.

8.2.2 Integrated feedback dynamics

Pollution-induced changes in sediment cohesion, hydrology, and vegetation create a self-reinforcing feedback loop. This loop is characterized by pollution-driven increases in runoff and sediment load, which suppress vegetation. The degradation of vegetation reduces bank stability, leading to greater erosion and the release of more sediment-bound pollutants, further deteriorating water quality and forming a closed-loop system (Allan, 2004). As illustrated in Fig. 12.2A, this represents the cyclical relationships that drive morphodynamic instability in polluted rivers.

Such feedbacks drive rivers toward alternative morphological states. **Highly migratory systems**, where erosion dominates due to vegetation degradation and sediment softening, are common in the Barak and Ganga systems. **Stable systems**, on the contrary, occur where excessive nutrient enrichment promotes macrophyte colonization, thereby reducing channel mobility (as observed in lower European rivers). Both outcomes represent deviations from the natural self-regulating meandering equilibrium (Annayat et al., 2022). Data from the Barak River (1990–2020) illustrate the strength of interactions among pollution, vegetation, and erosion parameters. Table 12.1 summarizes the Pearson correlation coefficients between key variables derived from remote sensing and modeled datasets. These correlations indicate that pollution acts as a central driver, indirectly influencing meander dynamics through vegetation-mediated erosion. The high correlation coefficients (>0.8) confirm the tightly coupled nature of these interactions within polluted river systems.

Pollution functions as a multiscalar stressor, influencing both physical and biological components. Vegetation acts as a response mediator, either mitigating or amplifying morphodynamic reactions. Hydrology, however, serves as the transmission mechanism through which the effects of pollutants are translated into morphological outcomes. Together, these elements form a coherent ecogeomorphological framework that links pollution to meander development. The subsequent case studies, presented in both global and regional contexts, elucidate these mechanisms and culminate in a detailed analysis of the Barak River Basin (Annayat et al., 2022).

TABLE 12.1 Correlation between pollution load, vegetation index, and erosion rate for the Barak River (1990–2020).

Parameter pair	Correlation (r)	Relationship description
Pollution load–vegetation index	−0.82	Increased pollution reduces vegetation cover
Vegetation index–erosion rate	−0.77	Vegetation decline enhances erosion
Pollution load–erosion rate	+0.84	Pollution increases sediment detachment
Sediment load–sinuosity index	+0.68	Higher sediment load increases channel curvature

9 Case studies and empirical evidence

Pollution-induced hydrological and vegetative changes have brought about significant alterations in river morphology. The following examples illustrate how these relationships manifest across climatic, geomorphic, and anthropogenic gradients. They offer insight into both the universality and variability of pollution–meandering interactions (Table 12.2).

9.1 The Ganges River, India

The Ganga River Basin exemplifies the interaction between pollution, vegetation, and hydrology within a vast tropical river system. Extending over 2500 km, the Ganga flows through a catchment characterized by intense agricultural activity, urban expansion, and high population density. The continual influx of nutrients and sediments from agriculture has altered the river's hydrological behavior, leading to increased variability in peak discharge and sediment load (Allan, 2004). Hydrological data from the 20th century indicate that during the monsoon months, the average SSC exceeds 1000 mg/L, with corresponding

TABLE 12.2 Comparative summary of restoration interventions and morphodynamic responses in major polluted meandering rivers.

River	Pollution type	Hydrological impact	Morphological response	Restoration measures	References
Ganges	Nutrient and sediment	Increased flood peaks, reduced infiltration	Enhanced meandering and cutoff formation	Riparian vegetation replanting, sediment control	Dutta et al. (2020)
Mississippi	Nutrient and flow regulation	Reduced flow variability, sediment starvation	Channel stabilization, incision	Controlled flooding, wetland rehabilitation	Kesel (2003)
Yangtze	Industrial and dam-induced	Sediment trapping, reduced variability	Reduced sinuosity, channel narrowing	Downstream reforestation, flow pulse restoration	Lai et al. (2017)
Po	Agricultural and nutrient	Decreased floodplain connectivity	Sinuosity decline, vegetation loss	Vegetation buffers, ecological corridors	Mandarino et al. (2019)
Rhine	Industrial and nutrient	Oxygen depletion, low flow stress	Sediment instability, algal dominance	Water quality improvement, riparian restoration	Moss (2008)

turbidity levels inhibiting aquatic vegetation growth (Gurnell and Petts, 2002). Remote sensing analyses between 1985 and 2020 reveal an increase in average meander wavelength from 6.4 to 7.2 km, indicating enhanced lateral migration and instability (Annayat and Sil, 2020a). Pollution-induced eutrophication has led to algal mat formation, reduced conveyance capacity, and increased bank erosion in several regions of Uttar Pradesh and Bihar. The Ganga River illustrates how sustained pollutant loads can trigger a feedback cycle linking water quality degradation, vegetation loss, and morphological evolution (Fig. 12.1A).

9.2 The Mississippi River, United States

Extensive dam construction and agricultural nutrient inputs have altered the hydrological and sedimentary regimes of the Mississippi River. While nutrient pollution has intensified eutrophication and hypoxia in the lower reaches, flow regulation by dams has substantially reduced flow variability and sediment supply in the upper reaches (Poff et al., 1997). These changes have promoted channel stabilization rather than migration. Since the mid-20th century, the frequency of meander cut-offs in the lower Mississippi has declined by >40% (Knox and Daniels, 2002). Low sediment loads and reduced overbank flooding pressures have led to channel incision and floodplain disconnection becoming common features. However, in agricultural tributaries such as the Yazoo and Atchafalaya, local nutrient enrichment has promoted excessive macrophyte growth, resulting in channel narrowing and aggradation. Thus the Mississippi exhibits a dual response: stabilization in regulated reaches and overgrowth-induced aggradation in unregulated ones.

9.3 The Yangtze River, China

The Yangtze River exemplifies pollution-induced hydrogeomorphic alterations within an industrial monsoonal basin. Yang et al., (2018) demonstrated that, over the past five decades, industrial effluent discharge and dam construction—particularly the Three Gorges Dam—have reduced downstream sediment flux by >70%. Simultaneously, nutrient enrichment has increased aquatic productivity in floodplain lakes, altering evapotranspiration and soil moisture regimes. This combined effect has led to a marked reduction in channel dynamism. Between 1970 and 2020, the average channel width of the Yangtze River decreased from 2.8 km to 2.2 km, while meander amplitude declined by 15%. Satellite imagery indicates progressive floodplain disconnection and bar stabilization. However, tributaries downstream of the dam exhibit enhanced local meandering due to compensatory sediment reworking, a process further intensified by pollution-induced vegetation degradation. This system exemplifies the complex interplay between hydrological regulation and pollution, wherein the suppression of sediment supply counteracts erosional responses, while pollution simultaneously alters the structure of riparian vegetation.

9.4 The Rhine and Po rivers, Europe

European rivers such as the Rhine and the Po have experienced similar pollution-related transformations, albeit on different temporal scales. In the Rhine, nutrient enrichment during the 20th century led to extensive algal blooms and oxygen depletion, resulting in the loss of benthic vegetation and destabilization of sedimentary beds (Naiman et al., 2005). Since the 1990s, restoration efforts—through riparian vegetation regeneration and controlled flood releases—have partially reversed these effects. In contrast, the Po River in Italy has exhibited a gradual decline in meandering. Agricultural intensification and dam construction have reduced flood frequency, sediment supply, and riparian vegetation diversity. Regional surveys indicate that since the 1950s, channel sinuosity has decreased by 25%, correlated with elevated nitrate concentrations and reduced discharge variability (Finotello et al., 2024b). This suggests that the combined effects of pollution and hydrological alteration produce predictable morphological outcomes, including channel narrowing, loss of sinuosity, and floodplain disconnection. Table 12.1 summarizes these trends, comparing morphological and ecological responses across major polluted river systems. Despite variations in climate, basin size, and management practices, pollution consistently interacts with hydrological regimes and vegetation structure, altering flow behavior. Global evidence also highlights the nonlinear nature of river responses, where even minor changes in pollutant loads or flow regulation can lead to disproportionate morphological outcomes.

In tropical basins such as the Barak River, these responses are further amplified by monsoonal hydrology and high sediment availability. The following section focuses on the Barak River in India, which serves as a regional case study illustrating the synergistic effects of pollution, vegetation degradation, and hydrological alterations on meander development and floodplain dynamics. This section presents an in-depth synthesis of hydrological, geomorphological, and ecological evidence from the Barak River Basin in northeastern India.

10 Case study: The Barak River, India

The Barak River, one of the major transboundary rivers of northeastern India, originates in the hills of Manipur and flows for approximately 900 km before entering Bangladesh, where it becomes part of the Meghna River system. The river drains an area of about 52,000 square km, encompassing diverse topographic and climatic zones, from the steep hill catchments of Manipur to the extensive floodplains of the Cachar and Karimganj districts in Assam. The Barak River is a monsoon-dominated system, receiving over 3000 mm of annual rainfall. This results in pronounced seasonal flow variability, with discharge increasing fivefold to sixfold during the monsoon months. The floodplain supports dense settlements, agricultural activities, and urban centers such as Silchar, making it both a hydrologically dynamic and anthropogenically stressed environment (Annayat et al., 2022).

The Barak River Basin has experienced increasing pollution pressure due to agricultural intensification (i.e., high fertilizer use and pesticide runoff in the Cachar plains), urban waste (untreated sewage from Silchar and adjoining municipalities), and sand mining and deforestation (which disrupt riparian zones and enhance sediment production). Furthermore, water quality monitoring (1990–2020) indicates a consistent rise in pollutant loads. Biochemical oxygen demand increased from approximately 2.5 mg/L in 1990 to about 4.8 mg/L in 2020, while nitrate concentrations doubled from 0.9 to 1.8 mg/L. Suspended sediment load also rose by nearly 45%, primarily during the monsoonal flood period (Nath and Ghosh, 2022).

10.1 Morphological evolution: Meander migration and sinuosity

Landsat-derived and GIS data reveal significant changes in the planform morphology of the Barak River between 1990 and 2020. The river's average sinuosity index increased from 1.49 to 1.65, while the mean channel width expanded by 12%. These results indicate a gradual yet consistent intensification of meandering activity. Annayat and Sil (2020b) further reported pronounced lateral migration in the southern floodplain near Silchar, resulting in an estimated loss of 10.2 square kilometers of agricultural land since 1990. The observed increase in curvature is strongly correlated with vegetation degradation and hydrological intensification ($r = 0.81$, Table 12.1). The underlying feedback mechanisms include pollutant-driven vegetation decline, increased sediment influx, and floodplain reworking.

10.2 GIS-based analysis and floodplain morphodynamics

GIS analysis (1990–2020) identified alternating zones of erosion and deposition along the Barak River's meander belts. Erosion was concentrated on the outer bends near Phulertal and Harinagar, whereas deposition dominated the inner banks near Dolu and Kushiyarkul. This reflects the integration of floodplain topography, erosion–deposition dynamics, and major settlement zones. The analysis revealed that areas with high pollutant accumulation closely correspond to erosion-dominated zones. These regions typically exhibit low vegetation index values and high SSCs, reinforcing the pollution–vegetation–hydrology feedback loop.

10.3 Hydrological regime transformation

Hydrological data indicate an increase in both peak discharge and flood frequency. Between 1990 and 2020, the mean annual flood discharge rose by approximately 14%, while base-flow discharge declined by 10%, suggesting a reduction in groundwater contribution. The flood-to-base flow ratio increased from 7.4 to 9.8 (Fig. 12.2C and D). When hydrological intensification coincides with a decline in riparian vegetation and increasing pollutant loads, it accelerates lateral channel migration and floodplain erosion. Thus the Barak River exhibits a mixed response, combining the naturally migratory tendencies of tropical rivers with the destabilizing effects of anthropogenic pollution (Annayat et al., 2022).

10.4 Eco-geomorphological feedbacks in the Barak River

In the Barak River, the interaction between pollution, vegetation, and hydrology forms an eco-geomorphological feedback system. The sequence operates as follows: nutrient enrichment and organic pollution degrade water quality and weaken vegetation; degraded vegetation reduces bank resistance, leading to increased erosion; intensified erosion mobilizes more pollutant-laden sediments; the resuspension of these sediments heightens turbidity and water toxicity, further inhibiting vegetation growth (Corenblit et al., 2009; Annayat et al., 2022). This self-reinforcing feedback loop (Fig. 12.2A) progressively amplifies meandering activity and floodplain instability.

11 Implications for floodplain management

The evolving morphology of the Barak River poses significant challenges for agriculture, infrastructure, and biodiversity. Frequent channel shifts have rendered large areas of agricultural land uncultivable. Pollution and sedimentation have degraded aquatic habitats, while the decline in riparian vegetation has increased the vulnerability of settlements to flooding. Therefore management and mitigation strategies must focus on both pollution control and geomorphic restoration (Corenblit et al., 2009; Annayat et al., 2022). Riparian vegetation rehabilitation, sediment management, and controlled floodplain reactivation can help restore hydrological balance and reduce the intensity of erosion. The Barak River exemplifies the interdependence of pollution, hydrology, and vegetation in shaping meandering dynamics. Quantitative analyses reveal that between 1990 and 2020, its sinuosity increased by 11%, lateral migration intensified by approximately 70%, vegetation density declined by around 22%, and sediment concentration rose by nearly 45% (Corenblit et al., 2009; Annayat et al., 2022). This demonstrates that anthropogenic pollution and hydrological alterations are primary drivers of morphological instability in tropical rivers (Tanaka et al., 2021). This section provides a comprehensive eco-geomorphological framework, and the preceding sections demonstrate that pollution, vegetation, and hydrology interact through a closely linked feedback system that governs the morphodynamic evolution of meandering rivers. Synthesizing global case studies with the Barak River analysis reveals several overarching themes relevant to both theory and practice. These may be broadly classified into (1) process relationships, (2) system responses, and (3) management implications (Kayitesi et al., 2022).

11.1 Process linkages: The eco-geomorphological triad

The integration of hydrological, geomorphological, and ecological processes forms the foundation of eco-geomorphology. In polluted meandering systems, these three domains do not operate independently but function as a tridimensional feedback system (Fig. 12.2A).

1. **Hydrology → Vegetation:** Altered flow regimes determine soil moisture, inundation frequency, and vegetation zonation. Pollution-induced hydrological changes (e.g., increased runoff and reduced infiltration) disrupt these zones, weakening the vegetative control over bank stability.
2. **Vegetation → Geomorphology:** Loss of vegetation reduces soil cohesion and hydraulic roughness, thereby promoting bank erosion and meandering. Conversely, in eutrophic systems, excessive vegetation growth can obstruct flow channels, leading to aggradation and flow diversion (Zhang et al., 2024).
3. **Geomorphology → Hydrology:** Morphological changes, such as cut-off formation and channel widening, influence hydrological behavior by altering flood storage capacity and base-flow recharge.

Within this triad, pollution functions as both a driver and an amplifier, altering the behavior of each component and reinforcing the negative feedback loops that destabilize the river system.

11.2 System response: From dynamic equilibrium to instability

Under natural conditions, meandering rivers maintain a state of dynamic equilibrium, that is, continuous morphological adjustment without long-term degradation. However, in polluted environments, this balance is often disrupted, giving rise to two distinct forms of instability (Ormerod et al., 1997). The first is **amplified meandering (hyperdynamism)**, observed in tropical rivers such as the Barak and the Ganga, where pollution-induced vegetation decline and increased sediment supply accelerate lateral migration. Enhanced flood intensity and sediment load promote cut-off frequency, while reduced root reinforcement and weakened bank resistance further exacerbate erosion, resulting in meander amplification, floodplain expansion, and land loss. The second form, **attenuated meandering (channel stabilization)**, has been observed in temperate rivers such as the Mississippi and the Yangtze under conditions of heavy flow regulation (Rogala et al., 2020). This occurs due to reduced sediment loads and diminished flow variability. The suppression of migration leads to nutrient enrichment and promotes excessive vegetation growth, resulting in channel narrowing and floodplain disconnection. These contrasting trajectories demonstrate that pollution can generate either hypermigratory or overstabilized morphologies, depending on climatic, hydrological, and management contexts (McCann, 2013; Kayitesi et al., 2022).

Pollution is emerging as a geomorphic agent alongside water, wind, ice, and gravity. Through its ability to modify sediment cohesion, hydrological regimes, and vegetation structure, pollution effectively reshapes the landscape. For instance, chemical alterations in sediments reduce the threshold shear stress required for erosion; biological toxicity destroys vegetation that stabilizes riverbanks; and hydrological disturbances from urban runoff increase discharge variability (Grabowski et al., 2011; Yin et al., 2018). Collectively, these changes generate morphodynamic responses comparable to those driven by tectonic or climatic forces. In this sense, pollution functions as an anthropogenic geomorphic driver capable of producing measurable changes in riverbed evolution within just a few decades (Zhang et al., 2019).

12 The Barak River in a global context

The Barak River exemplifies the interaction between tropical hydrology and anthropogenic pollution. Its meander amplification, increased curvature, and elevated sediment flux mirror patterns observed in the Ganga and Amazon basins, where high rainfall and population density intensify feedback mechanisms. However, two characteristics distinguish the Barak system (Annayat et al., 2022). The first is its high geomorphic sensitivity: its fine alluvial floodplain responds rapidly to even minor hydrological disturbances, amplifying pollutant effects. The second is its limited regulation: unlike dammed systems, the Barak retains natural flow variability, making it an ideal case for studying pollution-driven rather than regulation-driven morphodynamics. These features make the Barak River a valuable analog for tropical basins worldwide, particularly in South and Southeast Asia, where similar hydroclimatic and anthropogenic pressures prevail (Annayat et al., 2022).

12.1 Management and restoration framework

Integration of hydrological and geomorphological insights is essential for effective management and restoration strategies. This approach draws upon principles of ecohydrology and river basin management, emphasizing pollution mitigation, vegetation restoration, and hydrological rehabilitation. River management has traditionally been divided into separate domains of pollution control, flood management, and habitat restoration. Therefore effective restoration requires the adoption of an integrated framework that simultaneously addresses both physical and ecological processes (Perini and Sabbion, 2019). Pollution-induced alterations in meandering behavior arise from concurrent changes in hydrology, sediment dynamics, and vegetation. Consequently, interventions aimed solely at improving water quality—without addressing geomorphic and vegetative responses—are likely to yield only temporary benefits. Thus the objective of integrated management is to restore **river resilience**, defined as the capacity to absorb disturbances while maintaining morphological and ecological functions (Thoms and Parsons, 2002).

12.1.1 Components of the restoration framework

12.1.1.1 Pollution control and source reduction

Pollution management must begin at the source, with strategies encompassing industrial waste regulation (enforcing emission standards for heavy metals and hydrocarbons), agricultural runoff control (promoting best management practices such as contour farming, vegetative filter strips, and precision fertilization), and urban wastewater treatment (establishing decentralized treatment plants to reduce the discharge of untreated sewage into river channels) (Verdonschot and Arle, 2018; Lyu et al., 2020). Reducing pollutant loads weakens the self-reinforcing pollution–erosion feedback, thereby accelerating riparian vegetation recovery and enhancing sediment cohesion.

12.1.1.2 Riparian vegetation restoration

Riparian zone rehabilitation serves both as a pollution buffer and as a bank stabilizer. Restoration techniques include the replanting of deep-rooted native species such as *Ficus religiosa* and *Salix tetrasperma*, which strengthen root networks and enhance sediment retention; the establishment of vegetated buffer strips along floodplain margins to intercept nutrient and sediment runoff from agriculture; and the control of invasive macrophytes, particularly nutrient-loving species such as *E. crassipes*, which exacerbate channel obstruction. Riparian vegetation restoration not only stabilizes riverbanks but also enhances microhabitat diversity and improves biotic nutrient cycling (González et al., 2015).

12.1.1.3 Hydrological regime restoration

Restoring natural flow variability is crucial for sustaining meander dynamics. Environmental flows that mimic natural flood waves promote sediment redistribution and facilitate vegetation regeneration (Zeiringer et al., 2018). Reconnecting floodplains through controlled embankment breaches enhances overbank deposition and groundwater recharge. Upstream rainwater harvesting and increased infiltration reduce flash flooding and help stabilize base flows. In the Barak River, a 10% increase in base flow could reduce lateral migration rates by approximately 15% over a decade.

12.1.1.4 Sediment and bank management

Sediment management aims to restore the balance between erosion and deposition through various approaches, such as bank protection using bioengineering techniques (combining vegetation with structural measures like coir logs, fascines, and live staking); regulated sand mining (restricting extraction to non-critical zones and low-flow periods); and sediment reintroduction downstream of dams or barrages to maintain natural floodplain accretion rates (Thorolf Kondolf et al., 2020; Li et al., 2024). In this way, sediment and bank management help prevent the morphological extremes of erosion and overdeposition, thereby supporting the maintenance of meander equilibrium.

12.1.1.5 Monitoring and adaptive management

Adaptive management relies on continuous feedback between monitoring and policy action. Essential monitoring tools include remote sensing and GIS-based change detection for meander migration and vegetation cover, hydrological modeling for discharge and sediment flux, and bioindicators (e.g., benthic macroinvertebrates) for pollution assessment (Chatrabhuj et al., 2024). Together, these tools support a data-driven management cycle: monitoring → evaluation → adaptation → implementation.

13 Application to the Barak River Basin

A spatially regional management approach is required for the Barak River. Zone I—Upper Catchment (Manipur Hills): characterized by deforestation and high sediment load. Key management approaches include reforestation and erosion-control check dams. Zone II—Middle Basin (Silchar–Phulertal): affected by urban waste discharge and bank erosion. Management strategies include sewage treatment and the establishment of riparian buffer zones. Zone III—Lower Basin (Karimganj Border): characterized by agricultural runoff and silt deposition in floodplains (Das et al., 2024). Wetland restoration and the control of nutrient runoff are effective measures. Implementing these interventions concurrently can yield synergistic outcomes. Modeling studies indicate that a 20% reduction in pollutant load combined with a 25% increase in riparian vegetation could reduce the annual rate of bank erosion by approximately one-third.

13.1 Socioeconomic and policy considerations

Effective restoration requires institutional coordination among agencies responsible for water resources, agriculture, and forestry. Public awareness and community participation are equally crucial, as many sources of pollution originate from decentralized land-use practices (Carr et al., 2022). Developing incentive mechanisms, such as Payments for Ecosystem Services for riparian conservation, can encourage landowners to maintain vegetative buffer zones. Given the Barak River's downstream continuity as part of the Surma–Meghna system, transboundary cooperation with Bangladesh is also essential. River restoration in polluted meandering systems requires a shift from static engineering solutions to dynamic, process-based management approaches (Beechie et al., 2010). Pollution control, vegetation restoration, hydrological rehabilitation, and sediment balance must be pursued concurrently. The Barak River offers an opportunity to operationalize this integrated framework within a tropical context, providing a model for similar basins across South and Southeast Asia.

14 Conclusion

This chapter examines how pollution drives river meandering by affecting hydrology, sediment dynamics, and riparian vegetation. Pollution is not merely a water quality issue; it acts as a morphodynamic force, altering sediment properties, erosion thresholds, and channel stability while disrupting flow regimes. Riparian vegetation mediates these effects, with decline weakening banks and accelerating meander migration, and restoration enhancing roughness, sediment retention, and resilience.

Hydrological alterations, particularly increased flood-to-baseflow ratios, transmit pollutant impacts into geomorphic change. Data from the Barak River (1990–2020) show an 11% rise in sinuosity, nearly doubled lateral migration rates, and a 22% decline in vegetation, correlating with elevated pollutants and sediment loads. These trends illustrate universal eco-geomorphological feedbacks. The literature underscores the need for integrated management combining pollution control, vegetation restoration, and hydrological rehabilitation. Future research should employ quantitative modeling, remote sensing, and hydrodynamic simulations to assess pollution–meandering interactions under climate and socioeconomic variability. Sustainable river management must, therefore, integrate hydrology, ecology, and geomorphology within a unified framework.

Authors' contribution

ES, SK, and VK conceptualized the study and retrieved pertinent literature. SK and ES wrote the manuscript with contributions and thorough revisions from SK, VK, and ES. All authors have read and agreed to the published version of the manuscript.

Conflict of interest

The authors declare no conflicts of interest.

Acknowledgment

The authors would gratefully acknowledge the University of Lucknow, Lucknow; Pandit Prithi Nath (PG) College, Kanpur (affiliated to CSJM University, Kanpur); and Sunbeam College for Women, Bhagwanpur, Varanasi, for their encouragement and support.

References

Abernethy, B., Rutherfurd, I.D., 2000. The effect of riparian tree roots on the mass-stability of riverbanks. Earth Surface Processes and Landforms 25 (9), 921–937.

Akmal, N., Faez, M., Jalal, A., Chowdhury, K., Hafizah, N., 2023. Water pollution and the assessment of water quality parameters: a review. Desalination and Water Treatment 294, 79–88. doi:10.5004/dwt.2023.29433.

Allan, J.D., 2004. Landscapes and riverscapes: the influence of land use on stream ecosystems. Annual Review of Ecology, Evolution, and Systematics. 00664162. 35, 257–284. doi:10.1146/annurev.ecolsys.35.120202.110122.

Annayat, A., Sil, B., 2020a. Quantifying meander migration and bank erosion of the Ganges River using remote sensing. Geomorphology 351, 106951. doi:10.1016/j.geomorph.2019.106951.

Annayat, A., Sil, B., 2020b. Long-term morphodynamics of the Barak River using Landsat time series and GIS techniques. Hydrological Sciences Journal 65 (12), 2007–2021. doi:10.1080/02626667.2020.1801431.

Annayat, W., Ashwini, K., Sil, B.S., 2022. Channel migration detection and landuse/landcover changes for Barak River in Northeast India. Journal of the Geological Society of India 98, 981–990.

Annayat, W., Sil, B.S., 2020. Assessing channel morphology and prediction of centerline channel migration of the Barak River using geospatial techniques. Bull. Eng. Geol. Environ. 79 (10).

Beechie, T.J., Sear, D.A., Olden, J.D., Pess, G.R., Buffington, J.M., Moir, H., Roni, P., Pollock, M.M., 2010. Process-based principles for restoring river ecosystems. BioScience. 00063568. 60 (3), 209–222. doi:10.1525/bio.2010.60.3.7.

Blöschl, G., 2022. Flood generation: process patterns from the raindrop to the ocean. Hydrology and Earth System Sciences. 1607-7938. 26 (9), 2469–2480. doi:10.5194/hess-26-2469-2022.

Camporeale, C., Perucca, E., Ridolfi, L., Gurnell, A.M., 2013. Modeling the interactions between river morphodynamics and riparian vegetation. Reviews of Geophysics. 8755-1209. 51 (3), 379–414. doi:10.1002/rog.20014.

Carr, G., Barendrecht, M.H., Balana, B.B., Debevec, L., 2022. Exploring water quality management with a socio-hydrological model: a case study from Burkina Faso. Hydrol. Sci. J. 67 (6), 831–846.

Chatrabhuj, K.M., Mishra, U., Omar, P.J., 2024. Integration of remote sensing data and GIS technologies in river management system. Discover Geoscience. 2948-1589. 2 (1). doi:10.1007/s44288-024-00080-8.

Ciotti, D.C., Mckee, J., Pope, K.L., Kondolf, G.M., Pollock, M.M., 2021. Design Criteria for Process-Based Restoration of Fluvial Systems. BioScience. 15253244. 71 (8), 831–845. Available from http://bioscience.oxfordjournals.org/. doi:10.1093/biosci/biab065.

Corenblit, D., Piégay, H., Arrignon, F., González-Sargas, E., Bonis, A., Davies, N.S., Ebengo, D.M., et al., 2024. Interactions between vegetation and river morphodynamics. Part I: Research clarifications and challenges. Earth-Sci. Rev. 253, 104769. doi:10.1016/j.earscirev.2024.104769. (104586).

Corenblit, D., Steiger, J., Gurnell, A.M., Naiman, R.J., 2009. Plants intertwine fluvial landform dynamics with ecological succession and natural selection: a niche construction perspective for riparian systems. Global Ecology and Biogeography. 14668238. 18 (4), 507–520. doi:10.1111/j.1466-8238.2009.00461.x.

Corenblit, D., Tabacchi, E., Steiger, J., Gurnell, A.M., 2007. Reciprocal interactions and adjustments between fluvial landforms and vegetation dynamics in river corridors: a review. Earth-Science Reviews 84 (1–2), 56–86. doi:10.1016/j.earscirev.2007.05.004.

Das, T., Sil, B.S., Ashwini, K., 2024. Riverbank Erosion and Vulnerability Assessment for the Alluvial Section of Barak River in North-East India by In-situ Approach. KSCE Journal of Civil Engineering 28 (10), 4356–4367.

Dosskey, M.G., Vidon, P., Gurwick, N.P., Allan, C.J., Duval, T.P., Lowrance, R., 2010. The role of riparian vegetation in protecting and improving chemical water quality in streams. Journal of the American Water Resources Association. 17521688. 46 (2), 261–277. doi:10.1111/j.1752-1688.2010.00419.x.

Dutta, S., 2016. Soil erosion, sediment yield and sedimentation of reservoir: a review. Model. Earth Syst. Environ. 2363-6203. 2 (3). doi:10.1007/s40808-016-0182-y.

Dutta, V., Dubey, D., Kumar, S., 2020. Cleaning the River Ganga: Impact of lockdown on water quality and future implications on river rejuvenation strategies. Sci. Total Environ. 743, 140756. doi:10.1016/j.scitotenv.2020.140756.

Edokpayi, J.N., Odiyo, J.O., Durowoju, O.S., 2020. Impact of pollution on the aquatic environment. In: Dalu, S.E., Nhiwatiwa, N.A., Eds.. Emerging Pollutants in the Environment. IntechOpen, pp. 1–25. doi:10.5772/intechopen.90624.

Ekka, A., Pande, S., Jiang, Y., Zaag, P.V.D., 2020. Anthropogenic modifications and river ecosystem services: a landscape perspective. Water. 20734441. 12 (10). Available from http://www.mdpi.com/journal/water. doi:10.3390/w12102706.

Finotello, A., Ielpi, A., Lapôtre, M.G.A., Lazarus, E.D., Ghinassi, M., Carniello, L., Favaro, S., Tognin, D., D'Alpaos, A., 2024b. Vegetation enhances curvature-driven dynamics in meandering rivers. Nat. Commun. 20411723. 15 (1). Available from https://www.nature.com/ncomms/. doi:10.1038/s41467-024-46292-x.

Frascati, A., Lanzoni, S., 2009. Morphodynamic regime and long-term evolution of meandering rivers. Journal of Geophysical Research: Earth Surface. 21699011. 114 (2). Available from: http://onlinelibrary.wiley.com/journal/10.1002/(ISSN)2169-9011. doi:10.1029/2008JF001101.

Gajalakshmi, K., Anantharama, V., 2017. Hydrogeomorphology based land-use planning for sustainable development of hard rock terrain region of Karnataka. International Journal of Civil Engineering and Technology. 09766316. 8 (10), 698–705. Available from www.iaeme.com/ijciet/index.asp.

Gerbersdorf, S.U., Hollert, H., Brinkmann, M., Wieprecht, S., Schüttrumpf, H., Manz, W., 2011. Anthropogenic pollutants affect ecosystem services of freshwater sediments: the need for a "triad plus x" approach. J. Soils Sediments. 14390108. 11 (6), 1099–1114. doi:10.1007/s11368-011-0373-0.

González, E., Sher, A.A., Tabacchi, E., Masip, A., Poulin, M., 2015. Restoration of riparian vegetation: a global review of implementation and evaluation approaches in the international, peer-reviewed literature. J. Environ. Manag. 10958630. 158, 85–94. Available from https://www.sciencedirect.com/journal/journal-of-environmental-management. doi:10.1016/j.jenvman.2015.04.033.

Grabowski, R.C., Droppo, I.G., Wharton, G., 2011. Erodibility of cohesive sediment: the importance of sediment properties. Earth-Science Reviews 105 (3-4), 101–120.

Grabowski, R.C., Vercruysse, K., Holman, I., Azhoni, A., Bala, B., Shankar, V., Beale, J., Mukate, S., Poddar, A., Peng, J., Meersmans, J., 2022. The land–river interface: a conceptual framework of environmental process interactions to support sustainable development. Sustain. Sci. 18624057. 17 (4), 1677–1693. Available from http://www.springer.com/east/home?SGWID=5-102-70-144940151-0&changeHeader=true&SHORTCUT=www.springer.com/journal/11625. doi:10.1007/s11625-022-01150-x.

Gurnell, A.M., Petts, G.E., 2002. Island-dominated landscapes of large floodplain rivers, a European perspective. Freshwater Biology. 0046-5070. 47 (4), 581–600. doi:10.1046/j.1365-2427.2002.00923.x.

Hooke, J., 2003. River meander behaviour and instability: a framework for analysis. Transactions of the Institute of British Geographers 28 (2), 238–253. doi:10.1111/1475-5661.00089.

Ikeda, S., Parker, G., Sawai, K., 1981. Bend theory of river meanders. Water Resources Research 17 (5), 1222–1232. doi:10.1029/WR017i005p01222.

Junk, W.J., Bayley, P.B., Sparks, R.E., 1989. The flood-pulse concept in river-floodplain systems. In: Dodge, D.P. (Ed.). Proceedings of the International Large River Symposium (Can. Spec. Publ. Fish. Aquat. Sci. No. 106, pp. 110–127.

Kabir, A., et al., 2024. Analysis of land use change and its impact on channel morphology in tropical floodplain rivers. Int. J. River Basin Manag. 23, 1–18. doi:10.1080/15715124.2024.2342407.

Kayitesi, N., Kansiime, F., Wasswa, J., 2022. Impacts of land use land cover change and climate change on river hydro-morphology: a review of research studies in tropical regions. Journal of Hydrology 615, 128702.

Kayitesi, N.M., Biswas, S.R., 2022. Impacts of land-use and land-cover change and climate change on river hydro-morphology: a review of research studies in tropical regions. Journal of Hydrology XX, 101. (review article).

Kesel, R.H., 2003. Human modifications to the sediment regime of the Lower Mississippi River flood plain. Geomorphology 56 (3–4), 325–334. doi:10.1016/S0169-555X(03)00159-4.

Kleinhans, M.G., 2010. Sorting out river channel patterns. Progress in Physical Geography. 03091333. 34 (3), 287–326. Available from https://journals.sagepub.com/home/PPG. doi:10.1177/0309133310365300.

Knighton, D., 1998. Fluvial Forms and Processes: A New Perspective, Arnold Publishers.

Knox, J.C., Daniels, R.B., 2002. Human impacts on geomorphic processes. Geomorphology 49 (1–2), 1–3.

Lai, X., Yin, D., Finlayson, B.L., Wei, T., Li, M., Yuan, W., Chen, Z., 2017. 'Will river erosion below the Three Gorges Dam stop in the middle Yangtze?. J. Hydrol. 554, 24–31. doi:10.1016/j.jhydrol.2017.08.079.

Lambin, E.F., Geist, H.J., Lepers, E., 2003. Dynamics of land-use and land-cover change in tropical regions. Annu. Rev. Environ. Resour. 28, 205–241. doi:10.1146/annurev.energy.28.050302.105459.

Latrubesse, E.M., 2015. Large rivers, megafans and other Quaternary avulsive fluvial systems: a potential "who's who" in the geological record. Earth-Science Reviews. 00128252. 146, 1–30. Available from http://www.sciencedirect.com/science/journal/00128252. doi:10.1016/j.earscirev.2015.03.004.

Leopold, L.B., Wolman, M.G., 1960. River meanders. Bulletin of the Geological Society of America. 19432674. 71 (6), 769–793. Available from http://gsabulletin.gsapubs.org/content/by/year. doi:10.1130/0016-7606(1960)71[769:RM]2.0.CO;2.

LeRoy, P.O.F.F., Allan, J.D., Bain, M.B., Karr, J.R., Prestegaard, K.L., Richter, B.D., Stromberg, J.C., 1997. The natural flow regime: a paradigm for river conservation and restoration. BioScience 47 (11), 769–784.

Li, J., Wang, X., Wu, L., 2024. Research on sediment deposition characteristics and the vegetation restoration of ecological riverbanks in the deep waterway regulation scheme of Yangtze River. Water 20734441. 16 (16), 2350.

Lyu, T., Song, L., Chen, Q., 2020. Gang Pan, Lake and River Restoration: method, Evaluation and Management. Water. 2073-4441. 12 (4), 977. doi:10.3390/w12040977.

Mandarino, A., Maerker, M., Firpo, M., 2019. Channel planform changes along the Scrivia River floodplain reach in northwest Italy from 1878 to 2016. Quat. Res. 91 (2), 620–637. doi:10.1017/qua.2018.111.

McCann, C.J., 2013. Urbanization and its effects on channel morphology (Doctoral dissertation).

McGrane, S.J., 2016. Impacts of urbanisation on hydrological and water quality dynamics, and urban water management: a review. Hydrological Sciences Journal. 21503435. 61 (13), 2295–2311. Available from http://www.tandfonline.com/loi/thsj20. doi:10.1080/02626667.2015.1128084.

Moss, B., 2008. Water pollution by agriculture. Philos. Trans. R. Soc. B Biol. Sci. 363 (1491), 659–666. doi:10.1098/rstb.2007.2176.

Myslyva, T., Sheluto, B., Kutsaeva, O., Naskova, S., 2018. Use of Geospatial Analysis Methods in Land Management and Cadastre. Baltic Surveying. 2255-999X. 9, 56–62. doi:10.22616/j.balticsurveying.2018.019.

Naiman, R.J., Décamps, H., McClain, M.E., Likens, G.E., 2005. Riparia: Ecology, Conservation, and Management of Streamside Communities Riparia: Ecology, Conservation, and Management of Streamside Communities. Elsevier, United States, pp. 1–430. Available from https://www.sciencedirect.com/book/9780126633153. doi:10.1016/B978-0-12-663315-3.X5000-X.

Nanson, G.C., Huang, H.Q., 2008. Self-adjustment in rivers: evidence for least action as the governing principle. Earth Surface Processes and Landforms 33 (11), 1558–1578. doi:https://doi.org/10.1002/esp.1639.

Nath, D., Ghosh, P., 2022. Morphodynamic response of tropical rivers to climate and anthropogenic drivers: a case study of the Barak River. Environmental Earth Sciences 81 (2), 72–88. doi:https://doi.org/10.1007/s12665-021-09994-3.

Navin, M.S., Agilandeeswari, L., 2020. Comprehensive review on land use/land cover change classification in remote sensing. Journal of Spectral Imaging. 20404565. 9, 1–21. Available from https://www.impopen.com/subs/jsi/v9/I09_a8.pdf. doi:10.1255/jsi.2020.a8.

Nepf, H.M., 2012. Hydrodynamics of vegetated channels. J. Hydraul. Res. 50 (3), 262–279.

Njuguna, S.M., Onywere, S., Hora, R.R., Macharia, G., 2017. Assessment of macrophyte, heavy metal, and nutrient concentrations in the water of the Nairobi River, Kenya. Environmental Monitoring and Assessment 189, 250. doi:https://doi.org/10.1007/s10661-017-5912-4.

Okeke, C.A., Uno, J., Academe, S., Emenike, P.C., Abam, T. K., Omole, D.O., 2022. An integrated assessment of land use impact, riparian vegetation and lithologic variation on streambank stability in a peri-urban watershed (Nigeria). Sci. Rep. 12 (1), 10989.

Ormerod, S., Brookes, A., Shields, F.D., 1997. River Channel Restoration: guiding Principles for Sustainable Projects. The Journal of Applied Ecology. 00218901. 34 (4), 1106. doi:10.2307/2405297.

Paul, M.J., Meyer, J.L., 2001. Streams in the urban landscape. Annual Review of Ecology and Systematics 32, 333–365. doi:10.1146/annurev.ecolsys.32.081501.114040.

Pelletier, J.D., Murray, A.B., Pierce, J.L., Bierman, P.R., Breshears, D.D., Crosby, B.T., Ellis, M., Foufoula-Georgiou, E., Heimsath, A.M., Houser, C., Lancaster, N., Marani, M., Merritts, D.J., Moore, L.J., Pederson, J.L., Poulos, M.J., Rittenour, T.M., Rowland, J.C., Ruggiero, P., Ward, D.J., Wickert, A.D., Yager, E.M., 2015. Forecasting the response of Earth's surface to future climatic and land use changes: a review of methods and research needs. Earth's Future. 23284277. 3 (7), 220–251. Available from http://onlinelibrary.wiley.com/journal/10.1002/(ISSN)2328-4277. doi:10.1002/2014EF000290.

Perini, C.R., Sabbion, M., 2019. River restoration integrated with sustainable urban water management for resilient cities. Sustainability 12 (11), 4677.

Pochodyła-Ducka, E., Glińska-Lewczuk, K., Jaszczak, A., 2023. Changes in stormwater quality and heavy metals content along the rainfall–runoff process in an urban catchment. Water 15 (19), 3505.

Poff, N.L., Allan, J.D., Bain, M.B., Karr, J.R., Prestegaard, K.L., Richter, B.D., Sparks, R.E., Stromberg, J.C., 1997. The natural flow regime: a paradigm for river conservation and restoration. BioScience 47 (11), 769–784. doi:10.2307/1313099.

Rogala, J.T., Fitzpatrick, F.A., Hendrickson, J.S., 2020. Recent planform changes in the Upper Mississippi River. USGS Report, 6.

Sajikumar, N., Remya, R.S., 2015. Impact of land cover and land use change on runoff characteristics. J. Environ. Manag. 03014797. 161, 460–468. doi:10.1016/j.jenvman.2014.12.041.

Serrao, L., et al, 2022. Hydro-morphological disturbance and suitability for agricultural uses in a dynamic tropical wandering river. Water Resour. Res. 58 (4), e2021WR030674.

Shahid, M., Cong, Z., Zhang, D., 2017. Understanding the impacts of climate change and human activities on streamflow: a case study of the Soan River basin, Pakistan. Theor Appl Climatol 1–15. doi:10.1007/s00704-017-2269-4.

Skidmore, P., Wheaton, J., 2022. Riverscapes as natural infrastructure: meeting challenges of climate adaptation and ecosystem restoration. Anthropocene. 22133054. 38 (100334). doi:10.1016/j.ancene.2022.100334.

Stojković, S. 2017. GIS analysis of land use changes: case study: the Stara Pazova municipality, Serbia. Zbornik Radova - Geografski Fakultet Univerziteta U Beogradu. 1450-7552. (65-1a), 295–306. doi: 10.5937/zrgfub1765295g.

Tanaka, Y., Minggat, E., Roseli, W., 2021. The impact of tropical land-use change on downstream riverine and estuarine water properties and biogeochemical cycles: a review. Ecological Processes 10, 40.

Thoms, M.C., Parsons, M., 2002. Eco-geomorphology: an interdisciplinary approach to river science. In: Miller, A.J., Gupta, A. (Eds.), Fluvial Sedimentary Systems: Global and Regional Perspectives, pp. 115–141. IAHS Publication No. 276.

Thorolf Kondolf, M., Piegay, H., Landon, N., 2020. Sediment Augmentation for River Rehabilitation and Management — a Review. Land 10 (12), 1309.

Van Appledorn, M., Baker, M.E., 2022. Flood regimes alter the role of landform and topographic constraint on functional diversity of floodplain forests. Ecography 2023 (3), doi:10.1111/ecog.06519.

Verdonschot, P.F.M., Arle, J., 2018. Restoration in Integrated River Basin Management. In: Thorp, C.R. (Ed.) Riverine Ecosystem Management. Springer, pp. 273–299.

Wang, H., He, G., 2022. Rivers: linking nature, life, and civilization. River. 2750-4867. 1 (1), 25–36. doi:10.1002/rvr2.7.

Wilkes, M.A., Sear, D.A., Williams, R.J., 2019. Physical and biological controls on fine-sediment transport and storage in rivers. WIREs Water 6 (2), e1331. doi:10.1002/wat2.1331.

Yang, S.L., Milliman, J.D., Li, P., Xu, K.H., 2018. 50,000 dams later: erosion of the Yangtze River and its delta. Global and Planetary Change 162, 53–65. doi:10.1016/j.gloplacha.2018.01.004.

Yin, Z., Wang, P., Yu, Z., Liu, X., 2018. Water temperature, pH, and salinity effects on fluvial erosion of cohesive streambanks — implications for pollution and land-use change. Water 10 (3), 302.

Zahoor, I., Mushtaq, A., 2023. Water Pollution from Agricultural Activities: a Critical Global Review. International Journal of Chemical and Biochemical Sciences. 22269614. 23 (1), 164–176. Available from https://www.iscientific.org/wp-content/uploads/2023/05/19-IJCBS-23-23-24.pdf.

Zander, F., Shakeel, A., Kirichek, A., Chassagne, C., Gebert, J., 2022. Effects of organic matter degradation in cohesive sediment: linking sediment rheology to spatio-temporal patterns of organic matter degradability. J. Soils Sediments. 1439-0108. 22 (11), 2873–2882. doi:10.1007/s11368-022-03155-6.

Zeiringer, B., Seliger, C., Greimel, F., Schmutz, S., 2018. River hydrology, flow alteration, and environmental flow. In: Schmutz, S., Sendzimir, J. (Eds.), Riverine Ecosystem Management: Science for Governing Towards a Sustainable Future. Springer International Publishing, Cham, pp. 67–89. doi:10.1007/978-3-319-73250-3_4.

Zhang, Y., Tang, C., Wang, Y., 2019. Agricultural pollution and hydrological regime modification in Asian river basins: implications for morphodynamic change. Environmental Research Letters 14 (8), 085002. doi:10.1088/1748-9326/ab31f2.

Zhang, Z.-Q., Jiang, T.-B., Wu, T., 2024. Vegetation enhances curvature-driven dynamics in meandering rivers. Nat. Commun. 15, 46292.

Chapter 13

River salinization

Okram Zenita Devi[a] **and Ahanthem Rebika Devi**[b]
[a]*Department of Chemistry, Lady Irwin College, University of Delhi, New Delhi, India*
[b]*G.B. Pant National Institute of Himalayan Environment, Itanagar, India*

1 Introduction

Salinity refers to the total concentration of dissolved inorganic salts—such as sodium chloride, magnesium sulfate, potassium nitrate, and sodium bicarbonate (typically dominated by sodium chloride)—in water or soil, at levels that can affect human and natural assets, including plants, animals, aquatic ecosystems, water supplies, agriculture, and infrastructure (Cañedo-Argüelles et al., 2013; Williams and Sherwood, 1994). While salinity can, in principle, refer to any dissolved inorganic ions, in practice, it is primarily due to the presence of the following major ions: chloride (Cl^-), sodium (Na^+), magnesium (Mg^{2+}), calcium (Ca^{2+}), potassium (K^+), sulfate (SO_4^{2-}), carbonate (CO_3^{2-}), and bicarbonate (HCO_3^-) (Williams, 1987). Dissolved ions can also be expressed in terms of a solution's ionic activity, its ability to conduct electrical current, measured as electrical conductivity (EC), typically reported in Siemens per meter. EC is, therefore, commonly used as a proxy for salinity, although the relationship between the two depends on water temperature.

Primary salinity refers to the natural occurrence of salts in terrestrial and aquatic environments, including salt marshes, hypersaline lakes, tidal wetlands, and natural salt accumulations. This phenomenon results from natural processes, such as the weathering of rocks and the long-term deposition of salts by wind and rain over thousands of years. Secondary salinity, on the contrary, is caused by human activities that disturb the natural water balance, leading to the mobilization and accumulation of salts in the soil or waterways. It arises due to factors such as vegetation clearance, poor land management, urbanization, agriculture (both irrigated and dryland), and various industrial practices (Podmore, 2009).

1.1 Salinity classification

Salinity is commonly measured in ppt or as a percentage (%). The measurement of salinity, or dissolved salt content of water, is essential because aquatic

River Pollution. https://doi.org/10.1016/B978-0-443-45024-2.00014-2

TABLE 13.1 Water classification based on salinity concentrations.

Types of water	Salinity (ppt)	Level of salinity
Freshwater	≤0.5	Very low salinity, river
Oligohaline	0.5–5.0	Low salinity, upper estuary
Mesohaline	5.0–18.0	Moderate salinity, mid-estuary
Polyhaline	18.0–30.0	High salinity, lower estuary
Euhaline	>30.0	Typical seawater, near ocean interface

Data from Measuring Salinity of Water: https://www.horiba.com/ind/water-quality/applications/water-wastewater/measuring-salinity-of-water/.

organisms, livestock, and crops all have different tolerances to salinity levels. For example, saltwater organisms survive in salinity levels up to 40 ppt, yet many freshwater organisms cannot live in salinity levels above 1 ppt. Table 13.1 shows the level of salinity in different types of water, such as freshwater (Table 13.1).

1.2 Salinity measurement

Water salinity can be measured using these common methods:

Handheld refractometer: A portable, telescope-like device that provides quick and accurate salinity readings by measuring how light bends through a water sample. It is commonly used in aquaculture.

Hydrometer: Inexpensive tools for measuring water's specific gravity using Archimedes' principle. Ideal for aquarium use but unsuitable for soil salinity or high-precision applications. Always verify the calibration temperature (60°F/16°C or 77°F/25°C). Available as swing-arm or glass hydrometers.

Conductivity meter (EC sensor): Measures water's EC in milliSiemens per centimeter (mS/cm), correlating with salt concentration. It is fast, reliable, and suitable for high-salinity waters. Widely used in hydroponics, aquaculture, and environmental monitoring. Higher ion concentration produces higher conductivity readings.

Analytical methods: Direct ion measurement: Laboratory techniques, including titrimetric methods, ion chromatography, and spectrophotometry, can quantify specific dissolved ions (sodium, chloride, potassium, magnesium) for precise individual ion concentrations.

Total dissolved solids (TDSs): Represents total dissolved inorganic and organic substances. High TDS often indicates high salinity. It is measured using EC meters with conversion factors or through evaporation and weighing methods.

1.3 Freshwater salinization monitoring

Comprehensive freshwater salinization monitoring requires measuring EC alongside specific ionic concentrations (Na^{+}, K^{+}, Mg^{2+}, Ca^{2+}, Cl^{-}, SO_4^{2-}, SiO_3^{2-}) and accounting for complex ion interactions. Relying solely on EC measurements may miss critical ion-specific impacts affecting ecosystem health.

Freshwater salinization often coincides with nutrient and pesticide contamination from agriculture, food processing, and energy production. Agricultural activities contribute elevated sodium chloride (NaCl), nitrate (NO_3^{-1}), phosphate (PO_4^{-3}), potassium (K^{+}), calcium (Ca^{2+}), and magnesium (Mg^{2+}) concentrations, yet comprehensive documentation remains poor in many waterways.

Urban wastewater and industrial activities, particularly ethanol and agri-food production, compound the problem by releasing high levels of nutrients and salts. For example, sugarcane vinasse from ethanol production contains extremely high pollution levels (total nitrogen: 639 mg/L; total phosphorus: 150 mg/L; sodium: 60 mg/L; conductivity: 8420 μS/cm). Many river basins face overlapping agricultural and wastewater inputs, creating complex "chemical cocktails" that disrupt ecological balance, alter community composition, and favor salt- or nutrient-tolerant species while threatening sensitive species. Effective monitoring must encompass both salinity levels and comprehensive chemical composition to ensure conservation and sustainable management of aquatic ecosystems under increasing human and industrial pressures (Moyano-Salcedo et al., 2025).

1.4 River salinization

River salinization involves progressive salt concentration increases from natural processes (mineral weathering, salt leaching, climate change) and human activities (agricultural runoff, industrial/domestic wastewater, mining, urbanization, groundwater extraction, damming, seawater intrusion). Global examples include Poland's Kłodnica River (industrial/mining wastewater) (Zgórska et al., 2020), India's Kuttiyadi River (seawater intrusion) (Dipu and Josna, 2020), and the Amu Darya River (ADR) Basin (evaporation and intensive agriculture) (Hao et al., 2023). In arid regions, irrigated agriculture causes secondary salinization through salt accumulation from evaporation and plant transpiration.

1.5 Global relevance

River salinization affects diverse climates worldwide, such as Australia's Murray-Darling Basin (irrigation/land clearing) (Loch and Gregg, 2018), Western US rivers like the Colorado and Rio Grande (agriculture/urban use) (Garcia, 2017; Butler and Von Guerard, 1996), the India–Pakistan Indus River Basin (overirrigation/poor drainage) (Qureshi and Perry, 2021), and Europe's Ebro River (industrial/agricultural discharges) (Tedeschi et al., 2001).

2 Causes of river secondary salinization

2.1 Human-induced river salt contamination

Unlike primary salinization, which arises from natural weathering and arid climates, secondary salinization results from human activities that alter the hydrological balance or introduce additional salts into river catchments. This increasingly recognized water quality issue has profound ecological, agricultural, and socioeconomic consequences worldwide.

2.1.1 Irrigation-induced salinity

Irrigation represents the most significant contributor to secondary salinization, particularly in arid and semiarid regions. When water is applied to agricultural land, only a fraction is absorbed by crops; the remainder infiltrates the soil, carrying dissolved salts. Over time, this "leakage" raises groundwater tables, bringing salts into root zones and surface waters, reducing soil fertility, and degrading freshwater ecosystems. Recharge rates in irrigated zones exceed those in dryland areas due to combined rainfall and irrigation inputs. Rising water tables concentrate salts in upper soil profiles through evaporation, which are eventually mobilized during runoff events or through base flow into adjacent rivers (Cañedo-Argüelles et al., 2013; Podmore, 2009).

Notable examples include Central Asia's Amu Darya (Létolle and Chesterikoff, 1999) and Syr Darya Rivers, South Africa's Breede River (Scherman et al., 2003), Spain's Ebro River (Isidoro et al., 2006), and Turkey's Great Menderes (Koç, 2008). In Australia, removing native deep-rooted vegetation has exacerbated the issue by allowing saline groundwater to rise and discharge into streams, which is particularly problematic given the continent's naturally saline groundwater.

2.1.2 Mining activities

Mining, particularly potash and coal extraction, significantly contributes to secondary salinization. Potash mining generates salt-rich tailings containing NaCl, KCl, and $MgCl_2$. These waste products, stored in large impoundments, leach into surface and groundwater systems through runoff and infiltration. In Catalonia's Bages region, potash extraction exposes natural salts that dissolve with rain and humidity, contaminating the Llobregat River (Arroyo, 2025). Mountain top coal mining contributes through the weathering exposure of sulfide-bearing rock, leaching sulfate into surface waters, contrasting with chloride-dominated effluent from salt mining (Hartman et al., 2005; Pond et al., 2008).

2.1.3 Deicing salt application

In temperate and cold climates, widespread deicing salt application represents a well-documented secondary salinization source. Predominantly sodium chloride, these salts are applied in massive quantities and transported to water bodies via runoff and snowmelt.

North American road salt usage increased from 0.9 million tons in the 1960s to 14 million tons annually by the early 2000s. Runoff chloride concentrations have exceeded 18 g/L, with stream concentrations near major roads reaching levels 31 times higher than those at upstream sites. Some rural streams show chloride levels approaching estuarine salt fronts (Hanes et al., 1970; Environment Canada, 2001).

2.1.4 Industrial discharges and urbanization

Pollution-intensive industries, including chemicals, fertilizers, pulp and paper, aluminum, and steel, routinely discharge salt-laden effluents into nearby water bodies. Urban runoff from impervious surfaces compounds this problem, particularly in rapidly growing cities without adequate wastewater treatment infrastructure.

Unplanned urbanization and industrialization in South Asia have significantly degraded river quality. Industries discharge alkalis, acids, dyes, fluoride, ammonia, cyanide, and chromium compounds directly into the water. In Bangladesh, the Buriganga River suffers from untreated industrial effluents, human waste, and saltwater intrusion. In India, over 80% of water bodies are reportedly polluted by untreated sewage and industrial waste, with rivers like the Ganga and Yamuna reaching critical contamination levels. Sri Lanka's inadequate solid waste management infrastructure has led to toxic runoff and widespread surface water pollution (Sarker et al., 2021).

2.1.5 Seawater intrusion

Saline water intrusion into coastal freshwater bodies disrupts trophic structures and reduces biodiversity. It strengthens trophic interactions, leading to the mixing of saline and freshwater fish species, causing many freshwater species to decline or disappear (Dipu and Josna, 2020; Brucet et al., 2012).

2.1.6 Other contributing factors

Secondary salinization also results from sewage treatment plant effluents, salt dilution techniques, and reduced river discharge due to damming and flow regulation.

Secondary salinization is a global phenomenon resulting from human land-use changes, hydrological modifications, and increasing pollution loads. While irrigation, mining, industrial discharge, urbanization, and road deicing are frequently reported causes, numerous other anthropogenic pathways exist. Rising river salinity poses major threats to freshwater biodiversity, agriculture, and drinking water supplies. Addressing this issue requires integrated watershed management, stricter regulatory oversight, and sustainable land and water use practices. Understanding these main drivers through global case studies informs policy and guides future research on river ecosystem health.

Case study I
Saline water intrusion in Kuttiyadi River, Kerala, India (Dipu and Josna, 2020)

Objective

The Kuttiyadi River, originating from the Western Ghats at Thirumeni Hills in Kerala, India, serves as a critical water resource for northern Kerala, supporting hydroelectric power generation, agricultural irrigation, and ecological maintenance. However, growing saltwater intrusion poses serious threats to freshwater quality, particularly downstream. This study assessed the extent and impact of saline water intrusion in the Kuttiyadi River, focusing on surface and groundwater quality analysis. The river is crucial for Vadakara Municipality and surrounding areas, especially during the dry premonsoon season.

Research was conducted at seven surface water sampling sites (SW1–SW7) and several nearby groundwater wells: SW1 (estuary region, 200 m from the sea), SW2 (5.1 km upstream), SW3 (10.6 km upstream), SW4 (15.8 km upstream), SW5 (22.5 km upstream), SW6 (26.4 km upstream), and SW7 (31.5 km upstream). These locations were selected to study spatial salinity gradients from the estuary upstream and to evaluate correlations between salinity and water quality parameters in both surface and groundwater systems.

Scope

The study included systematic water sampling during the premonsoon season to capture baseline conditions. Comprehensive analysis of physicochemical parameters included salinity, EC, TDSs, chloride, sulfate, sodium, potassium, calcium, magnesium, and biochemical oxygen demand (BOD). The research evaluated groundwater contamination due to the lateral intrusion of saline water into freshwater aquifers and conducted correlation analysis between salinity and associated ions to understand contamination relationships and sources.

Audience

Environmental scientists, hydrologists, Kerala State Water Authority officials, and policymakers address regional water management challenges. Local municipal planners from Vadakara Municipality collaborate to develop sustainable water infrastructure. Environmental non-governmental organizations promote community engagement in water conservation, while students and educators in environmental sciences contribute research insights.

Rationale

The Kuttiyadi River, stretching 74 km from the Western Ghats to the Arabian Sea, is the lifeline of Vadakara and its surrounding villages. Increased saline intrusion during the dry premonsoon months has rendered parts of the river unsuitable for drinking, irrigation, and domestic use. Sand mining, low river flow, and sea-level rise further exacerbate the degradation of water quality.

Expected results and deliverables

The study developed a comprehensive water quality database and a detailed salinity profile along the river. A geospatial map delineated the saltwater intrusion zone, extending 22.5 km inland. Based on the findings, the study offered policy recommendations and mitigation strategies, including the construction of regulatory structures and the control of unsustainable sand mining.

Safety considerations

Sampling followed APHA (2017) protocols to ensure accuracy and safety throughout data collection. Field personnel received personal protective equipment to maintain safety standards during sample collection. Well water samples were collected with proper permission from property owners to ensure ethical and legal compliance.

Methodology

The research methodology involved comprehensive field collection of surface and groundwater samples across locations SW1–SW7 and various wells. On-site measurements of pH, EC, and salinity were conducted using portable meters for immediate water quality assessments. Laboratory analysis determined concentrations of key ions, including chloride, sulfate, sodium, potassium, calcium, and magnesium. Biochemical oxygen demand testing assessed organic contamination levels. Statistical correlation analysis utilized Pearson's correlation coefficients to identify relationships between water quality parameters. GIS mapping of saline intrusion patterns was developed using spatial interpolation tools to visualize the extent and distribution of saltwater contamination.

Key challenges and solutions

Key challenges included limited water flow during premonsoon sampling, human interference requiring public engagement, differentiating natural versus anthropogenic salt sources through ionic analysis, and accessing private wells through community involvement.

Results and discussion

Salinity concentrations ranged from 0.02 ppt at SW7 to 33.42 ppt at SW1, demonstrating a clear salinity decline with increasing distance from the estuary (Table 13.2). Location SW5, situated 22.5 km upstream, recorded 6.83 ppt salinity, indicating significant inland saltwater penetration extending up to 22.5 km inland. EC followed similar spatial patterns, ranging from 31.32 μS/cm (SW7) to

TABLE 13.2 Physicochemical characteristics of the Kuttiyadi River (premonsoon season).

Parameter	SW1 (Estuary)	SW2 (5.1 km)	SW3 (10.6 km)	SW4 (15.8 km)	SW5 (22.5 km)	SW6 (26.4 km)	SW7 (31.5 km)
Salinity (ppt)	33.42	24.81	15.72	10.34	6.83	0.07	0.02
EC (μS/cm)	32,840	25,430	19,780	13,540	9210	312	31.32
TDS (mg/L)	50,520	38,730	28,410	19,460	12,510	196	20.52
Chloride (mg/L)	16,913	12,780	9231	5850	2340	178	4.03
Sodium (mg/L)	11,162.5	8642	5902	3620	1740	109	2.14
Potassium (mg/L)	950.4	720.3	502.1	370.5	220.6	33.2	7.6
Calcium (mg/L)	316.8	245.3	182.4	129.7	70.8	13.4	0
Magnesium (mg/L)	1202	930.4	741.2	680.1	790.7	42.1	0.96
Total Hardness (mg/L)	5742	4350	3110	2208	1290	51.4	7.92
BOD (mg/L)	6.4	4.2	3.5	2.9	2.3	1.2	0.8
pH	7.0	7.1	7.2	7.3	7.4	7.6	7.7

Dipu, S., Josna, R.P., 2020. Saline water intrusion in urban coastal area: a case study of KuttiyadiRiver, Kerala, India. Am. J. Water Resour. 8 (3), 128–133.

32,840 μS/cm (SW1), exhibiting strong positive correlations with TDS and salinity ($r > 0.9$). Chloride levels varied widely, from 4.03 mg/L at SW7 to 16,913 mg/L at SW1, often exceeding irrigation and drinking water standards. Sodium concentrations ranged from 2.14 mg/L (SW7) to 11,162.5 mg/L (SW1). All samples, except SW6 and SW7 showed elevated sodium concentrations, making the water unsuitable for drinking and irrigation due to saltwater intrusion. Total hardness increased toward the estuary, ranging from 7.92 mg/L at SW7 to 5742 mg/L at SW1. BOD values were lower in upstream locations, suggesting less organic pollution but highlighting increased salinity stress in downstream areas.

Learning and knowledge outcomes

The study confirmed significant saltwater intrusion extending 22.5 km inland during the premonsoon season, with salinity strongly correlating with chloride, sulfate, sodium, calcium, and magnesium concentrations. Groundwater wells showed signs of contamination, indicating the vulnerability of shallow aquifers. Water from sites SW1–SW5 was unsuitable for irrigation and domestic use. Mitigation recommendations include constructing tidal barrages at strategic locations and banning uncontrolled sand dredging, which deepens riverbeds and exacerbates intrusion.

Conclusion

This study highlights the serious threat posed by saline water intrusion in the Kuttiyadi River, with impacts extending over 22 km inland. Surface and groundwater quality deterioration due to elevated salinity occurs particularly during summer periods when river discharge is minimal. Regulatory infrastructure, policy reforms, and community participation are essential to prevent further degradation and ensure sustainable water use in the region.

Case study II
Five decades of freshwater salinization in the ADR Basin (Hao et al., 2023)

Objective

The Amu Darya is one of Central Asia's largest rivers, flowing through Afghanistan, Tajikistan, Turkmenistan, and Uzbekistan into the Aral Sea basin. This study examined the spatial and temporal dynamics of river water salinization in the ADR basin over 50 years (1970–2019), focusing on arid environments. The research aimed to identify and quantify natural and anthropogenic factors driving salinization, particularly regarding increasing agricultural activities and climate change. Using long-term hydrological and water chemistry data from 12 monitoring sites, the study sought to understand correlations between river discharge and salinity trends and to develop a conceptual framework for future water management policies in Central Asia.

Scope

This research covered the entire ADR course, from upstream alpine regions in Tajikistan to downstream delta areas near the Aral Sea. The study incorporated river discharge, return flow salinity, fertilizer use, water withdrawals, and climatic factors, including temperature and precipitation. Advanced statistical and machine learning models were applied to quantify and spatially attribute salinity changes. The scope extended to developing predictive insights and management strategies replicable in other arid and semiarid river basins.

Audience

The study targets environmental scientists, hydrologists, and water resource managers concerned with river basin management and water quality in arid regions. Policymakers and governmental agencies responsible for water distribution and irrigation practices in Central Asia would benefit from the findings. In addition, researchers and development organizations working on UN Sustainable Development Goals, especially those related to clean water, agriculture, and climate resilience, will find the results crucial for informing interventions.

Rationale

The ADR is a vital transboundary river in Central Asia, supporting agriculture and the drinking water supply for millions. Recent decades have seen rapid agricultural intensification, increasing population density, and climate change, placing enormous pressure on this fragile water system. Despite being water-rich upstream, downstream areas have witnessed sharp flow declines and alarming salinity increases. These changes threaten regional food security and socioeconomic stability, necessitating a data-driven understanding of how human interventions contribute to secondary river salinization.

Expected results and deliverables

Expected outcomes included detailed assessments of discharge and salinity trends across different periods and locations, identification of dominant salinization drivers, and development of a conceptual model capturing river salinization processes in arid regions. Deliverables included predictive maps and data visualizations, such as violin plots and salinity-distance linear models, to guide regional water governance strategies. The study aimed to inform evidence-based policymaking to mitigate salinization impacts and support sustainable development planning.

Safety considerations

This research utilized secondary data from governmental and scientific databases, eliminating the need for fieldwork or exposure to hazardous conditions.

Safety considerations primarily involved the ethical use of data, proper dataset attribution, and adherence to open-source licensing regulations. Computational analyses were conducted using statistical software and machine learning platforms in a secure, remote environment.

Actions taken/tools used

The research involved sequential steps, starting with acquiring long-term river discharge and water chemistry data from hydro-meteorological archives and scientific repositories. Data were cleaned using the Pauta criterion (3σ) to remove outliers and ensure integrity. Linear regression and violin plots were used to assess temporal and distributional salinity changes. Salinity–discharge relationships were analyzed using general hyperbolic models and Spearman's rank correlation. A random forest machine learning model was employed to identify primary salinization drivers. Spatial trends were visualized, and a conceptual model was developed, integrating findings across natural and anthropogenic domains.

Challenges and solutions

Key challenges included the inconsistency and incompleteness of historical datasets, particularly in matching discharge data with water chemistry records, which were addressed through rigorous data filtering and probabilistic outlier detection methods. Disentangling the overlapping effects of natural processes and human activities was resolved using machine learning approaches, enabling variable importance ranking and better causality attribution. Spatial heterogeneity in salinity trends was addressed by analyzing upstream, midstream, and downstream sections separately.

Results and discussion

Analysis reveals pronounced spatial and temporal variability in river discharge and salinity, largely shaped by anthropogenic factors, particularly irrigation withdrawals and agrochemical use. River discharge trends show clear spatial gradients, with flow volumes decreasing markedly from upstream to downstream. Upstream regions showed a slight discharge decline from 220 to 204 m^3/s between the 1970s and 2000s, primarily attributed to natural variability. Midstream sections experienced sharp reductions from 962 to 307 m^3/s, signaling substantial agricultural water withdrawals. The most dramatic reductions occurred downstream, where discharge fell from 481 to 62 m^3/s at M6 and from 595 to 28 m^3/s at M7.

Salinity patterns exhibit distinct spatial and temporal variation. Between the 1970s and 1980s, salinity levels exceeded 1000 mg/L at downstream stations, surpassing safe thresholds for drinking and irrigation. However, from 2000 to 2015, certain areas showed improvement, with significant salinity decline at M6 (−75.84 mg/L/year) and the Kafirnigan tributary (−19.49 mg/L/year)

(Table 13.3). Nonetheless, salinity increased at midstream stations, raising concerns about water quality deterioration.

TABLE 13.3 Summary of key datasets used in the ADR salinity study.

Variable	Station(s)	Time period	Observed trend	Remarks
River discharge (m^3/s)	M1 (upstream)	1970s–2000s	Slight decrease (220 → 204)	Low human activity; mainly natural control
River discharge (m^3/s)	M4 (midstream)	1970s–2000s	Sharp decline (962 → 307)	Agricultural water withdrawal increases downstream
River discharge (m^3/s)	M6 (downstream)	1970s–2000s	Sharp decline (481 → 62)	Strong human influence
River discharge (m^3/s)	M7 (downstream)	1970s–2000s	Sharp decline (595 → 28)	High intensity of irrigation and return flow impacts
River salinity (mg/L)	M5, M6, M7	1970s–1980s	Increased to > 1000 mg/L	Exceeds safe limits for drinking/irrigation
River salinity (mg/L)	M6	2000–2010	Decreased (slope: – 75.84 mg/L/year)	Improved possibly due to interventions
River salinity (mg/L)	S2 (Kafirnigan River)	2000–2015	Decrease (slope: – 19.49 mg/L/year)	Tributary shows positive trend
River salinity (mg/L)	M2, S1	2000–2015	Increase (slope: + 16.22, + 22.26 mg/L/year)	Trending toward unsafe salinity levels
River salinity (mg/L)	M3, M4, S3, S4	2000–2015	Stable (<1000 mg/L)	Within acceptable thresholds
Return flow salinity (mg/L)	Not site-specific	General (2000–2015)	High in middle and downstream	Major contributor to secondary salinization
Agricultural water withdrawal	M4, M6	Entire study period	High levels correlated with increased salinity	Key anthropogenic factor
Nitrogen fertilizer use	M2	Entire study period	Identified as the primary driver via the random forest model	Indicates strong agrochemical influence

Hao, L., Wang, P., Gojenko, B., Jingjie, Y., Aifeng, L., Fadong, L., Kenjabaev, S., Kulmatov, R.,Khikmatov, F., 2023. Five decades of freshwater salinization in the Amu Darya River basin. J.Hydrol. Reg. Stud. 22145818. 47, 101375. doi:10.1016/j.ejrh.2023.101375.

Return flow and agricultural practices emerge as central to understanding salinization trends. High salinity in return flows highlights the impacts of inefficient irrigation and inadequate drainage systems. Random forest modeling identified nitrogen fertilizer application as a significant driver of salinity increases, emphasizing the need for integrated water resource management (IWRM) to mitigate both hydrological and water quality challenges in the ADR Basin.

Learning and knowledge outcomes

This study provided critical insights into river salinization drivers in arid environments, revealing that increased precipitation alone cannot mitigate salinization without sustainable water management practices. Machine learning techniques, like random forests, proved valuable for environmental attribution studies, while findings reinforced the need for fine-scale, subbasin analyses, including detailed irrigation return flow data. The research highlighted the importance of integrating climate, land use, and hydrological data for effective water resource planning.

Conclusion

The ADR Basin shows clear degradation in river discharge and water quality, particularly downstream. While some stations exhibit improving trends from management interventions, others experience rising salinity levels. These findings underscore urgent needs for IWRM, focusing on efficient irrigation, pollution control, and basin-wide monitoring. The work emphasizes intersectoral collaboration and systems-thinking approaches, such as the SDG nexus framework, to balance agricultural demands with environmental sustainability and sustain regional water availability.

3 Effects of river salinization

River salinization is an increasingly serious environmental issue affecting freshwater ecosystems, agriculture, public health, and infrastructure. It results from both natural processes and human activities that introduce excess salts into rivers, lakes, and groundwater systems (Cañedo-Argüelles et al., 2013; Schuler et al., 2019).

3.1 Impact on freshwater ecosystems

Salinity severely impacts freshwater biodiversity by altering species composition and threatening sensitive organisms. Even minor salt increases can shift aquatic communities toward salt-tolerant species, reducing overall biodiversity (Cañedo-Argüelles et al., 2013). Invertebrates, such as zooplankton, are especially vulnerable, and their decline disrupts food chains and ecosystem stability (Nielsen et al., 2003; Bailey et al., 2004).

3.2 Impact at different levels of ecosystem organization

At the organism level, secondary salinization disrupts the osmoregulatory balance of freshwater species, forcing them to expend more energy maintaining internal ion concentrations. Most freshwater organisms are osmoregulators that actively control internal salt levels. When external salinity exceeds tolerance, these mechanisms fail, leading to cell damage, reduced growth, or death (Hart et al., 1991; Pinder et al., 2005). These physiological costs affect survival, reproduction, and ultimately influence species composition and ecosystem resilience.

3.3 Ecosystem functioning and services

Salinization hampers ecosystem functioning by affecting oxygen solubility, organic matter decomposition, and microbial processes. Oxygen dissolves less effectively in saline water, stressing aquatic organisms and promoting hypoxic conditions. High salinity can mobilize nutrients like nitrates, leading to harmful algal blooms and further deoxygenation (Cañedo-Argüelles et al., 2013).

3.4 Threats to human health

Salinized water increases health risks. High-salinity drinking water has been linked to hypertension and cardiovascular conditions (Schuler et al., 2019). Salts can mobilize toxic heavy metals and radioactive elements, like radium, from soils and pipes into drinking water, elevating chronic exposure risks (Environmental Protection Agency, 2022).

3.5 Agricultural and soil impacts

Irrigation with saline water causes soil salt build-up, reducing crop yields and rendering land unproductive (Smedema and Shiati, 2002). Osmotic stress interferes with plant water uptake, while specific ion toxicity damages root systems, severely impacting food production in arid regions.

3.6 Economic and social consequences

Freshwater salinization incurs substantial economic costs. In the Colorado River Basin, annual damages reach \$300 million, including \$176 million in crop losses. California experiences \$1.7–7 billion in annual agricultural revenue losses from salinization (Schuler et al., 2019). River salinization has profound social implications, including reduced access to clean water and arable land, which leads to displacement and migration in vulnerable rural communities facing livelihood loss, food insecurity, and increased health risks.

3.7 Ion-specific toxicity

Different ions associated with salinization have distinct toxic effects on aquatic life. Potassium and magnesium can be more harmful than sodium or chloride under certain conditions, with toxicity depending on concentrations, ratios, and interactions with other contaminants (Environmental Protection Agency, 2022).

3.8 Conclusion

River salinization is a multifaceted issue threatening biodiversity, human health, food security, and the economy, demanding urgent attention through integrated watershed management, targeted regulations, and public engagement to prevent the continued degradation of freshwater systems worldwide.

Case study I
Impacts of salinization on riverine biota—A case from the Wipper River, Germany (Feld et al., 2023)

Objective

This case study aims to assess both the direct physiological and indirect ecosystem-level effects of river salinization on aquatic biota, particularly macroinvertebrates, in the Wipper River, a tributary of the Unstrut in central Germany. The study focuses on identifying how multiple environmental stressors, including salinity, oxygen depletion, nutrient enrichment, and hydromorphological degradation, interact to affect biodiversity and ecological health.

Scope

The research was conducted across nine sampling sites along the Wipper River, covering a range of salinity and habitat degradation conditions. It involved comprehensive biological, chemical, and physical monitoring, focusing primarily on the response of macroinvertebrate communities, especially those belonging to the sensitive Ephemeroptera, Plecoptera, and Trichoptera (EPT) taxa.

Audience

The findings are relevant to freshwater ecologists, river basin managers, environmental policymakers, and restoration practitioners who seek to understand and mitigate the impacts of salinization and other stressors in river ecosystems.

Rationale

Salinization in rivers is increasingly recognized as a critical stressor, but its indirect impacts, such as those mediated through vegetation growth and oxygen fluctuations, are often overlooked. The Wipper River, which is impacted by both saline inputs and nutrient loading, serves as a valuable model to explore these dynamics. Understanding how salinity affects biological communities not only through direct osmoregulatory stress but also via altered habitat and oxygen regimes is essential for holistic river management.

Expected results and deliverables

The study was expected to reveal strong correlations between water quality (including salinity) and biological health indicators. Specifically, it aimed to demonstrate that salinity, while directly harmful, also promotes macrophyte overgrowth, which, in turn, causes oxygen depletion at night, stressing aquatic life. The key deliverables include a stress-gradient analysis, correlation of macroinvertebrate indices with salinity, and management recommendations to address both direct and indirect salinity impacts.

Safety considerations

All fieldwork was conducted under standard ecological monitoring guidelines, with minimal disturbance to aquatic habitats. Equipment and methods were selected to ensure researcher safety and ecological integrity.

Actions taken/tools used/simulations and analyses

Macroinvertebrate samples and water quality parameters were collected at nine locations along the Wipper. Water chemistry, particularly salinity, nutrient levels, and oxygen content, was measured using multiparameter probes. Continuous oxygen loggers were installed at key sites rich in salt-tolerant macrophytes (e.g., *Stuckenia pectinata*) to track diel oxygen fluctuations. The Index of General Degradation (IGD) and EPT% were used as bioindicators. A principal component analysis (PCA) was performed to disentangle the effects of multiple stressors and identify the primary environmental gradients influencing biological conditions (Feld et al., 2023).

Challenges and solutions

One major challenge was the cooccurrence of multiple stressors, such as salinity, nutrients, and habitat degradation. To address this, PCA helped isolate salinity as a key gradient. Another difficulty was measuring fluctuating oxygen levels linked to plant respiration; this was overcome through high-frequency oxygen logging. Furthermore, the study distinguished between direct physiological effects and indirect ecological effects by linking biological metrics to continuous oxygen and habitat data.

Results and discussion

The PCA identified two major gradients: one related to water quality, including salinity and nutrient enrichment, and another associated with habitat degradation. Macroinvertebrate health (IGD and EPT%) was strongly correlated with water quality but showed only weak correlations with physical habitat features (Table 13.4). In macrophyte-dominated sites, especially those with high salinity

TABLE 13.4 Principal component analysis (PCA): main stressor gradients.

Principal component (PC)	Dominant variables	Ecological interpretation
PC1: Water quality gradient	High salinity, elevated chloride, and nutrient concentrations	Indicates pollution-driven deterioration of water quality; strongly correlated with macroinvertebrate metrics (IGD, EPT%)
PC2: Habitat degradation	Physical habitat structure, hydromorphological alterations	Reflects physical disturbance (e.g., channel modification), weakly linked to biological metrics

TABLE 13.5 Night-time oxygen levels at salinized, macrophyte-rich site.

Time	Dissolved O_2 (mg/L)	Ecological Impact
Daytime	8–10 mg/L	High due to photosynthesis
Night-time	<6 mg/L (frequent)	Respiratory oxygen depletion stress on macroinvertebrates

Feld, C.K., Lorenz, A.W., Peise, M., Fink, M., Schulz, C.J., 2023. Direct and indirect effects of salinisation on riverine biota: a case study from river Wipper, Germany. Hydrobiologia. 15735117. 850 (14), 3043–3059. Available from: https://www.springer.com/journal/10750.doi:10.1007/s10750-023-05229-z.

and dense *S. pectinata*, night-time oxygen levels frequently fell below 6 mg/L, indicating oxygen stress likely caused by plant respiration at night compared to day time oxygen levels of 8–10 mg/L (Table 13.5).

The findings from the Wipper River reveal a strong environmental gradient influenced by salinity. Downstream sites exhibited salinity as high as 1200 mg/L, along with reduced nighttime oxygen levels below 6 mg/L, a sign of biological oxygen demand linked to excessive macrophyte respiration.

Biotic indicators, such as the IGD and EPT taxa (%), improved progressively upstream, indicating recovery in ecosystem health with decreasing salinity. For instance, EPT percentage rose from 12% at S1 to 63% at S9, reflecting increased macroinvertebrate diversity. PCA showed that water chemistry, particularly salinity and oxygen, had a greater influence on biota than habitat degradation. These findings confirm both direct osmotic stress and indirect hypoxic effects as key pathways of ecological degradation. The combination of salinity and hypoxia poses a dual threat to riverine fauna, suggesting that river management should target both chemical and biological stressors, especially where aquatic vegetation amplifies salinization effects.

Learning and knowledge outcomes

The Wipper River study illustrates how salinization impacts freshwater biota through both direct and indirect pathways. While organisms like macroinvertebrates suffer from osmotic stress at high salinities, the promotion of salt-tolerant vegetation can exacerbate oxygen depletion, especially at night. These interactions significantly shape community structure and biodiversity. Importantly, the study shows that monitoring only chemical parameters may underestimate ecosystem stress and that integrated biological and physical monitoring is crucial for accurate ecological assessment and effective river management.

Conclusion

Salinization affects river ecosystems through two main pathways. Directly, it causes osmotic and ionic stress on macroinvertebrates. Indirectly, it promotes the growth of salt-tolerant plants like *S. pectinata*, which increase nighttime respiration, leading to hypoxia.

Biological indicators, such as **IGD and EPT%,** showed a **strong link to water chemistry**, especially salinity and oxygen levels, but responded **less to habitat structure**. This suggests that **chemical stressors** have a greater impact on aquatic life than physical degradation in the Wipper River.

For management, both **ionic pollution** and **habitat issues** must be addressed. Most importantly, **hypoxia should be mitigated** through **macrophyte control** or **regulated flow**, ensuring oxygen levels remain stable. A combined approach is essential to restore and protect freshwater biodiversity.

Case study II
The assessment of eco toxicological impacts of industrial and municipal discharges along the Kłodnica River (Upper Silesia, Poland), combining physicochemical monitoring with bioassays using a battery of indicator organisms (*Vibrio fischeri, Daphnia magna, Lemna minor*) (Zgórska et al., 2020)

Objective

This study evaluated the impact of industrial and municipal discharges on the Kłodnica River ecosystem in Upper Silesia's Coal Basin, assessing chemical and ecological status relative to Water Framework Directive (WFD) objectives. Sixteen sampling sites were selected to capture influences from wastewater treatment plants and active coalmines, with emphasis on salinity, ionic composition, and bioassay responses using *V. fischeri*, *D. magna*, and *L. minor* as indicator organisms.

Scope

Multiple sampling sites along the Kłodnica River, including upstream and downstream industrial discharge zones, were analyzed.

Analysis included both physicochemical parameters (e.g., salinity, major ions, nutrients, chemical oxygen demand [COD]).

Ecotoxicological assays across three trophic levels: Bacteria (***V. fischeri***), Crustacea (***D. magna***), and aquatic plant (***L. minor***).

Audience

The research addresses environmental scientists, ecotoxicologists, policymakers implementing WFD regulations, industrial operators, municipal authorities, and NGOs focused on aquatic ecosystem protection and environmental governance.

Rationale

The rationale stems from the limitations of traditional chemical-only monitoring approaches, which often fail to detect sublethal or synergistic toxic effects significantly impacting aquatic ecosystems. By integrating bioassays with physicochemical data, the study provides a more comprehensive and ecologically relevant assessment of water quality and ecosystem health. This combined approach is particularly crucial in Upper Silesia, a heavily industrialized area where the Kłodnica River consistently fails to meet ecological standards, requiring updated insights for effective management and remediation strategies.

Expected results and deliverables

The project is expected to produce several key deliverables and findings that contribute to understanding the ecological impact of industrial activities on the Kłodnica River. Deliverables will include comprehensive data on water chemistry, covering parameters such as salinity, ion concentrations, and nutrient levels. In addition, bioassay toxicity metrics will be generated, including EC_{50} values for *V. fischeri* and *D. magna*, and percentage growth inhibition for *L. minor*. A correlative analysis will be conducted to link chemical quality indicators with observed toxicological responses, providing a holistic view of ecological stress. The study will also assess compliance with WFD objectives, identifying areas that fall short of regulatory standards.

The expected findings include confirmation of poor ecological quality in river sections impacted by industrial discharges. It is anticipated that bioassay sensitivity will vary by trophic level, with plant-based assays, such as *L. minor*, expected to show the highest responsiveness to chemical stressors. Furthermore, the results are likely to highlight the significant role of salinity in shaping bioassay outcomes, both as a direct stressor and as a confounding factor in interpreting toxicity.

Safety considerations

The safety considerations for this research encompass comprehensive standard laboratory protocols for handling industrial effluents and contaminated samples to protect personnel and ensure accurate results. Special protocols are implemented for live organisms used in bioassays, addressing both ethical considerations and biosafety concerns throughout the experimental procedures. Personnel are trained to remain mindful and exercise heightened caution during sampling activities near industrial outflows to minimize exposure risks and maintain safety standards. The study ensures strict compliance with local chemical safety regulations and animal welfare guidelines to meet all regulatory requirements and uphold ethical research practices throughout the investigation.

Actions taken/workflow/tools used/simulations and analyses

1. **Sampling** at multiple sites: upstream and downstream of industrial discharges.
2. **On-site physicochemical measurements**: salinity, conductivity, pH, nutrients, and COD.
3. **Laboratory-based bioassays**:
   - ***V. fischeri*** bioluminescence inhibition test (Microtox)
   - Acute toxicity test using *D. magna*
   - ***L. minor*** phytotoxicity/growth inhibition assay
4. **Data analysis**:
   - EC_{50} and percent inhibition values
   - Correlation/regression between chemical metrics and bioassay responses
   - Comparison with WFD "Good Status" standards
5. **Interpretation** of whether biological results align with physicochemical readings and WFD classification.

Challenges and solutions

Challenge: Variable salinity complicates the interpretation of toxicity results.

Solution: Statistical analysis to isolate salinity effects, acknowledging positive stimulation (hormesis) in Vibrio's response.

Challenge: Disparate sensitivities among bioassays.

Solution: The battery approach, combined organism responses provided comprehensive insight.

Challenge: Some assays showed no acute toxicity.

Solution: Focused on sublethal effects, recognizing that only ***L. minor*** showed >25–50% inhibition.

Results and discussion

This study confirms that industrial and mining activities are the main sources of increased salinity in the Kłodnica River. Extensive coal mining and heavy industry discharge saline mine waters and process effluents, leading to high

EC and elevated levels of major ions (Cl^-, SO_4^{2-}, Na^+, Ca^{2+}, Mg^{2+}), especially downstream.

Physicochemical data from 16 sampling sites reveal sharp rises in EC and salinity immediately downstream of active coal mine outflows, including KWK Wujek (Site 4), KWK Halemba (Site 8), KWK Sośnica-Makoszowy (Site 10), and CZOK Gliwice (Site 11). Conductivity increased from 0.83 mS/cm at Site 3 to 6.27 mS/cm at Site 4, while salinity rose from 0.4 g/L to 3.4 g/L. Similar patterns appeared at other discharge points, with conductivity reaching 16.27 mS/cm (Site 10) and salinity hitting 9.6 g/L at Site 11, indicating substantial inputs of saline mine waters rich in chlorides, sulfates, sodium, and other major ions.

Biological responses in bioassays emphasize salinity's impact on aquatic life. The aquatic plant *L. minor* showed marked growth inhibition, especially at downstream sites with high salinity: Site 4 (25% inhibition), Site 11 (50%), and Sites 12 and 13 (89% and 56%, respectively) (Table 13.6). This suggests saline discharges interfere with nutrient uptake and osmotic balance, making *L. minor* a sensitive bioindicator of salinity-related phytotoxicity. The freshwater crustacean *D. magna* exhibited no acute toxicity across all samples, indicating greater tolerance to increased salinity levels. However, the absence of chronic toxicity testing means potential long-term sub-lethal effects cannot be ruled out.

Interestingly, the bioluminescent bacterium *V. fischeri* showed increased light emission in several samples with higher conductivity, such as Site 9, where stimulation reached 51%. These hormetic responses likely result from ionic stress caused by high concentrations of sodium and potassium ions in mine waters, potentially masking actual toxic effects and complicating toxicity interpretation.

D. magna showed no acute toxicity across all samples.

V. fischeri showed increased bioluminescence (hormetic effect) at some sites, especially Site 9.

Data gaps for some parameters are due to unreported values in the original text.

Learning and knowledge outcomes

This study demonstrated the added value of using multitiered bioassays alongside traditional chemical monitoring to gain a more comprehensive understanding of water quality and ecological risk. Among the tested species, *L. minor* emerged as a particularly sensitive indicator, effectively reflecting phytotoxicity in areas impacted by industrial discharge. The findings also provided clear evidence of salinity's confounding role in ecotoxicological assessments, underscoring the importance of interpreting bioassay results in the context of ionic stress. Collectively, these insights support the development of more balanced and robust monitoring protocols aligned with the WFD, integrating both chemical and biological indicators to improve ecological status assessments.

TABLE 13.6 The key results related to salinity and its ecological impact on the Kłodnica River.

Sampling site	Location description	Conductivity (mS/cm)	Salinity (g/L)	*Lemna minor* growth inhibition (%)	*Daphnia magna* toxicity	*Vibrio fischeri* bioluminescence stimulation (%)	Notes
3	Upstream KWK Wujek discharge	0.83	0.4	Not reported	No toxicity	Not reported	Baseline upstream site
4	Downstream KWK Wujek discharge (D1)	6.27	3.4	25%	No toxicity	Not reported	Increased salinity postdischarge
7	Upstream KWK Halemba discharge	4.36	2.3	Not reported	No toxicity	Not reported	Baseline upstream site
8	Downstream KWK Halemba discharge (D2)	13.57	7.8	Not reported	No toxicity	Not reported	High salinity and conductivity
9	Mid-river site downstream industrial area	Not reported	Not reported	Not reported	No toxicity	51%	Strong bacterial stimulation
10	Downstream KWK Sośnica-Makoszowy discharge (D3)	16.27	Not reported	Not reported	No toxicity	Not reported	Peak conductivity at mining site
11	Downstream CZOK Gliwice discharge (D4)	Not reported	9.6	50%	No toxicity	Not reported	High salinity and phytotoxicity
12	Near river mouth (final study section)	Not reported	Not reported	89%	No toxicity	Not reported	Highest Lemna growth inhibition
13	Near river mouth (final study section)	Not reported	Not reported	56%	No toxicity	Not reported	High Lemna growth inhibition

Zgórska, A., Bondaruk, J., Dudziak, M., Hamerla, A., 2020. Impact of Industrial Discharge on Aquatic Ecosystems of the Klodnica River with Reference to Water Framework Directive Objectives. Pol. J. Environ. Stud. 29 (4), 2945–2953.

Conclusion

The data highlight that salinity, driven largely by industrial and mining effluents, plays a critical ecological role in the Kłodnica River. Elevated salinity directly stresses sensitive species like *L. minor* while confounding bioassay results through nonlinear responses in bacteria. This complexity emphasizes the need to integrate ion-specific toxicity thresholds and consider salinity as a key factor in water quality monitoring to prevent underestimation of ecological risks and misclassification of water status under the WFD.

4 Mitigation strategies

4.1 Sustainable irrigation and land-use practices

One effective strategy to combat river salinization involves the strategic shift from deep-rooted perennial pastures to irrigated annual crops. This transition reduces evapotranspiration rates, particularly during fallow periods, allowing more rainfall and irrigation water to percolate through the soil and recharge groundwater. Consequently, this practice minimizes the risk of saline groundwater entering river systems (Podmore, 2009).

4.2 Integrated water management

IWRM strategies aim to balance water use efficiency with environmental needs. The implementation of precision irrigation systems, real-time soil moisture monitoring, and the restoration of natural flow regimes can maintain river health while reducing salt accumulation. Reinstating environmental flows also assists in diluting saline concentrations and improving overall riverine ecosystem resilience (Williams, 2001).

4.3 Wastewater treatment and alternative disposal

Treated wastewater must comply with established quality standards before discharge into rivers. Various techniques, such as ion-exchange using plate dolomite, have been explored for reducing salinity levels in industrial and mining effluents, though their implementation must consider cost-effectiveness, legal feasibility, and environmental impact (Richter, 2010). Where technically and economically feasible, evaporation basins and agricultural reuse represent preferred alternatives to direct river disposal.

4.4 Nonriver disposal and dilution strategies

During periods of high river discharge, saline waste can be diluted more effectively. The strategic timing of discharges to coincide with these high-flow events is a widely used mitigation technique. Alternatively, transferring water from less saline sources can improve dilution capacity, though such interventions may have downstream ecological and political consequences (O'Keeffe and De Moor, 1988).

4.5 Use of salt-tolerant crops and soil amendments

Deploying salt-tolerant crops helps sustain agricultural output under saline conditions. Soil amendments like gypsum, compost, or biochar can enhance soil structure and reduce salt mobility. These approaches are particularly crucial in semiarid and arid regions, where irrigation-induced salinization is widespread (Podmore, 2009).

4.6 Regulatory frameworks and monitoring protocols

Effective mitigation of river salinization requires enforceable, ion-specific water quality standards. Different anthropogenic activities introduce ions like Na^+, Cl^-, SO_4^{2-}, and Mg^{2+}, each with distinct environmental effects. Current regulations often focus only on sodium and chloride, neglecting the broader ionic interactions, which leads to inadequate freshwater protection (Cañedo-Argüelles et al., 2013; Schuler et al., 2019).

Freshwater salinization is poorly monitored and weakly regulated. There is an urgent need for ecologically informed thresholds, ion-specific monitoring, and legally binding limits (Cañedo-Argüelles et al., 2013). Limited public awareness and political commitment also hinder effective action in addressing river salinization. Improved education, expanded research programs, and enhanced outreach activities are essential components driving evidence-based policy and management (Cañedo-Argüelles et al., 2013).

4.7 Role of technology in monitoring and mitigation

Advances in technology are revolutionizing salinity management. Satellite remote sensing enables large-scale monitoring of river salinity by detecting surface reflectance and vegetation stress, offering near-real-time data on salinization trends and land degradation (Alatorre et al., 2016). Satellite imaging tools, like Landsat and Sentinel, help in mapping salinity-affected zones, facilitating early intervention.

Artificial intelligence (AI) and machine learning models are increasingly utilized to predict water quality based on real-time environmental and hydrological data. AI models can forecast salinity levels, detect anomalies, and optimize irrigation scheduling to prevent excess salt runoff into water bodies (Ma et al., 2022). These predictive tools enhance decision-making for water managers and farmers alike.

In addition, sensor networks and Internet-of-Things devices provide continuous monitoring of parameters like EC, temperature, and ion concentrations at multiple points along a river system. These technologies allow for targeted and timely mitigation responses, reducing both ecological and economic risks.

4.8 Community engagement and policy interventions

Successful mitigation requires collective action involving farmers, local stakeholders, scientists, and policymakers. Education campaigns can raise awareness of salinization's risks, while financial incentives and technical support can drive the adoption of sustainable land and water use practices. Cross-sector policy integration is essential to ensuring compliance and maximizing ecological benefits (Williams, 2001; Cañedo-Argüelles et al., 2013).

4.8.1 Conclusion

River salinization is a complex, multidimensional problem that demands an integrated approach, combining sustainable land use, advanced monitoring technologies, regulatory reform, and community participation. Embracing innovations like satellite remote sensing, AI-driven water quality prediction, and precision agriculture can significantly enhance mitigation efforts, ensuring the long-term health of freshwater ecosystems and the communities that depend on them.

Case study I
River salinity mapping through machine learning and statistical modeling using Landsat 8 OLI imagery (Ansari et al., 2025)

Globally, societies have relied on river water to serve multiple purposes, including as a primary source of drinking water, irrigation, climate regulation, industrial use, and recreation (Cañedo-Argüelles et al., 2013). In recent years, there has been an unprecedented threat to rivers due to rising dissolved salt concentrations (Zaeri et al., 2023). River salinization has become a global environmental issue, as it directly causes negative impacts on soil, drinking water, and agriculture, particularly in developing countries (Perri et al., 2020; Al-jabery et al., 2016). It also has the potential to alter the structure and function of aquatic biodiversity and freshwater ecosystems and to have direct and detrimental physiological effects on freshwater organisms (Cunillera-Montcusí et al., 2022; Feld et al., 2023).

Both human activities and natural processes are the main drivers of freshwater salinization. Human-induced activities, including agriculture, resource extraction, urbanization, and the direct discharge of untreated wastewater (Gašparović and Singh, 2020), contribute to the degradation of surface water quality (Said and Khan, 2021). The production and application of fertilizers, namely, potash [potassium chloride (KCl)], constitutes another important source of salinization (Feld et al., 2023). Geological processes such as erosion, chemical weathering, acidification–alkalinization, and ion exchange in soils also contribute to salinization in rivers (Kaushal et al., 2021). Climate change and human activities have further amplified these geological processes (Cunillera-Montcusí et al., 2022).

Ansari et al. (2025) conducted a study on water salinity in the Karun River, Iran's longest river with the highest discharge volume. Salinity in rivers originates from various sources, including saline discharges from arid lands, irrigation practices, fish hatcheries, and urban sources. The salinization issue in the river stems from deposits of gypsum and halite (rock salt), which are exposed due to the construction of the Gotvand Dam and the formation of a reservoir in Gachsaran. The study and mapping of salinity dynamics provide insights into the environmental challenges faced by the river, as well as implications for agricultural practices and water quality management. The findings of the research offer crucial information to stakeholders regarding the impacts of salinity in rivers and their relationship to the ecosystem and local communities.

Objective

The primary objective of this study is to investigate the driving causes of river salinization, including both natural and human factors. It aims to identify environmental and socioeconomic effects and propose effective mitigation strategies for implementation at various scales. The study will examine salinity mapping in the Karun River using Landsat 8 OLI imagery. In addition, it will assess features for salinity estimation through various analysis methods and evaluate multiple regression for salinity prediction.

Scope

The scope of the study focuses on several key aspects related to the assessment and mapping of salinity levels in the Karun River basin in south-western Iran. The Karun River basin spans a geographical area of 67,000 sq. km. It was found that agricultural and industrial activities contribute to salinity levels in the river. The salinity in river water was studied using multitemporal satellite data, machine learning techniques, incorporating in situ salinity data points, and creating salinity maps. The study provides insights into the environmental challenges and issues faced by the Karun River, particularly concerning agricultural practices and water quality management. The concentration of salinity, which is important to protect freshwater biodiversity, is lacking in most states or confederations of states, and these regulations are usually not legally enforced (Cañedo-Argüelles et al., 2013).

Methodology

This section outlines a comprehensive methodology for mapping river salinity using multitemporal satellite data, along with in situ salinity data. According to the study, effectively mapping and analyzing salinity dynamics in river systems requires substantial integration of remote sensing data with advanced machine learning techniques.

The first step in mapping river salinity is the acquisition of multitemporal satellite data and the collection of in situ salinity data points. The estimation of salinization is a crucial step; features for sensitive indicators were identified, such as salinity indices and spectral bands from the satellite data. In the study of salinity modeling, various regression models were tested, including statistical models, kernel-based models, decision tree-based models, and neural network models. The evaluation of the model performance is vital, and the effectiveness of the regression models was assessed using statistical error metrics and measures of residual spatial autocorrelation. The overall methodology for mapping river salinity using satellite data and in situ data is illustrated in the flowchart.

Audience

The audience of the present study includes ecologists, as surface water salinization leads to a decline in current aquatic biodiversity. This also results in the invasion of new aquatic species that are more tolerant of saline conditions. Remote sensing specialists, environmental scientists, and hydrologists primarily focus on assessing and monitoring water quality. Policymakers and environmentalists are concerned about water resource management and the impacts of agricultural and industrial activities on river systems (Cunillera-Montcusí et al., 2022). River management, due to salinization, promotes the mass development of many salt-tolerant macrophyte species, leading to oxygen depletion and the killing of many fish (Feld et al., 2023).

Rationale

The study of river salinization has become one of the most significant research areas and yet remains an underrecognized water quality challenge globally. Due to diverse causes and consequences, freshwater salinization has become a serious and widespread environmental issue in humid regions. The salinization of freshwater in inland waters has a long history of being acknowledged as a significant source of water quality degradation (Kaushal et al., 2021). There is a need for more research to understand the impact of increasing salinity and to identify specific ion concentrations that affect aquatic ecology and ecosystem services (Stets et al., 2020).

Expected results and deliverables

Multitemporal satellite data has proven to be an effective approach for mapping salinity dynamics. It helps to identify the concentration of salinity across the city located near the river. The main causes of river salinity were identified, such as agricultural activities leading to another source of river salinization.

Actions taken/workflow/tools used

The study of river salinization using in situ data yields accurate results. However, in situ data are unsuitable for large-scale and temporal monitoring of freshwater salinization (Guo et al., 2023). Remote sensing data combined with geoinformatics offer significant potential and have transformed research across various environmental disciplines (Abdelfattah et al., 2008). These technologies have proven to be excellent tools for mapping, detecting, and assessing water quality, including salinity and waterlogged soils (Trung et al., 2021). In recent years, both national and international scientific communities have conducted extensive research, monitoring, and mapping of river salinization using satellite data. Due to the dynamic nature of river systems, salinity levels fluctuate both spatially and temporally. The spatial, spectral, and temporal resolution of satellite imagery must be sufficient to differentiate salinity levels, as they are sensitive to variations in colored dissolved organic matter and total suspended sediment, facilitating effective monitoring across different areas of interest (Ansari et al., 2025). Currently, various sensors and water quality indices have been employed for mapping salinity, with techniques adapted to different environmental and hydrological contexts (Perri et al., 2020) (Table 13.7).

TABLE 13.7 List of software, tools, methods, and indices used for the study of river salinity.

Sl. no.	Indices	Formula	References
1	Normalized Difference Water Index (NDWI)	$\frac{Green - NIR}{Green + NIR}$	(Yagmur et al., 2021)
2	Weighted Arithmetic Water Quality Index (WAWQI)	$\frac{\sum_{i=1}^{n} W_i Q_i}{\sum W_i}$	(Khatri et al., 2020)
3	Canadian Council of Ministers of the Environment Water Quality Index (CCMEWQI)	$100 - \frac{\sqrt{F_1^2 + F_2^2 + F_3^3}}{1.732}$	(Khatri et al., 2020)
4	Modified Normalized Difference Water Index (MNDWI)	$\frac{Green - MIR}{Green + MIR}$	(Nurmemet et al., 2015)
5	Salinity Index (SI)	$\sqrt{B1XB3}$	(Nurmemet et al., 2015)
6	Normalized Difference Salinity Index (NDSI)	$\frac{Red - Vegetation\, red\, edge}{Red + Vegetation\, red\, edge}$	(Matysik et al., 2025)

Stimulations and analyses

There are various simulations and analyses to achieve the objectives of mapping river salinity through remote sensing data, machine learning, and statistical modeling. The key aspects of the simulations and analyses are as follows: (1) Evaluation of model: Cloud-free, multitemporal Landsat satellite data were used for mapping salinity. The salinity was calculated within a 1-km buffer around five major cities, selected based on their geographical locations and agricultural activities that influence salinity. The salinity maps generated from multitemporal Landsat satellite data were validated using 102 in situ salinity data points. (2) Comparison of model performance: The study compared the performance of various regression models, including statistical and machine learning approaches. The support vector regression model outperformed others, providing a comprehensive understanding of the river's dynamics and the effectiveness of different modeling approaches in addressing environmental challenges related to water quality.

Challenges and solutions

Over the last decades, river salinization has risen at an unprecedented rate. The study of river salinity has faced several challenges due to the complex and dynamic nature of the river ecosystem. Salinity is also a dynamic variable that changes due to the interaction of land and water, requiring a large in situ sampling size because of its high spatial variability (Mohamed et al., 2023). Freshwater salinization syndrome has developed due to the complex interrelationships between salt ions and chemical, biological, and geologic parameters, as well as its consequences on the natural, social, and built environment (Kaushal et al., 2021). Traditional models have struggled to generalize across large spatial and temporal scales. The low- and medium-spatial-resolution satellite data often have mixed pixels, highlighting the need for very high spatial resolution satellite data to map, assess, and detect salinity complications. There is also a limitation in long-term in situ salinity measurement, which is necessary for establishing baseline conditions (Kumar et al., 2024).

The drawbacks and shortcomings of traditional approaches to mitigating water salinity are overcome with modern technologies. The investigation of spatiotemporal monitoring of water quality parameters in remote areas can be achieved through the application of multi- and hyperspectral satellite data, and several techniques have been developed to map salinity and create indices (Gašparović and Singh, 2020; Mohanavelu et al., 2021).

Results

The study on mapping river salinity using multitemporal satellite data resulted in several effective modeling approaches and the identification of factors influencing salinity in rivers. The analysis revealed that salinity levels in the Karun River increased from Gotvand to Ahvaz City, primarily due to agricultural activities and the influence of the Gachsaran Formation within the reservoir. The study identified 24 features, including various indices related to salinity, and identified the bands of satellite data to assess their sensitivity in estimating salinity. This finding underscores the impact of human activities and geological factors on river salinity (Ansari et al., 2025). The major water quality challenges are mainly linked to population densities and areas of economic growth. It is anticipated that the greatest increases in exposure to pollutants will happen in lower- and middle-income countries (du Plessis, 2022).

Learning and knowledge outcomes

Several important insights and knowledge have been contributed by the present study:

1. River salinization is a serious environmental problem; identifying and understanding it in current and future scenarios result from complex interactions between geological, hydrological, and anthropogenic factors.
2. Salinity is causing severe social, economic, and environmental damage, including impacts on agricultural productivity, increased water treatment costs, and ecological degradation.
3. Climate change is likely to exacerbate river salinization through altered precipitation patterns, increased evaporation rates, and sea-level rise affecting coastal river sections.
4. The salinization of freshwater habitats is linked to lower biodiversity measures, such as species richness and functional diversity, which cause shifts in aquatic community makeup. Freshwater organisms show different levels of sensitivity to salinity stress, so increasing salinity is likely to change community compositions either directly (by species loss due to toxicity) or indirectly through altered species interactions, whether as a single stressor or combined with multiple stressors (Berger et al., 2019).
5. Geospatial and modeling approaches have shown significant potential for assessing and predicting the spatial and temporal changes in river salinity, enabling more effective monitoring and management strategies.
6. Satellite data combined with in situ data provide a significant and cost-effective solution for large-scale monitoring and early warning detection of salinity in rivers.

References

Abdelfattah, M.A., Shahid, S.A., Othman, Y.R., 2008. A model for salinity mapping using Remote Sensing and Geographic Information Systems - A Case Study from Abu Dhabi, United Arab Emirates, UAE. In: *2nd International Salinity Forum (Vol. 123)., March*, 1–6.

Al-jabery, A.A.R., Hussein, A., Obeidi, A., 2016. The Use of Remote Sensing Techniques for the Detection of Salinization of the Euphrates River in the Provinces of Qadisiyah, Muthannain Iraq. Environ. Sci. 5 (8), 1–5. doi:https://doi.org/10.21275/ART2016212.

Alatorre, L.C., Beguería, S., Nunes, J.P., 2016. Monitoring soil salinization processes with remote sensing and GIS. Sci. Total Environ. 572, 445–457.

Ansari, M., Knudby, A., Homayouni, S., 2025. River salinity mapping through machine learning and statistical modeling using Landsat 8 OLI imagery. Adv. Space Res. 02731177. 75 (10), 6981–7002. doi:10.1016/j.asr.2025.03.037.

APHA, 2017. Standard Methods for the Examination of Water and Waste Water. American Public Health Association, Washington, DC.

Arroyo, A.S., 2025. The silent crisis of freshwater salinization, communication and outreach. Available from: https://www.idaea.csic.es/the-silent-crisis-of-freshwater-salinization/.

Bailey, S.A., Duggan, I.C., Van Overdijk, C.D.A., Johengen, T.H., Reid, D.F., Macisaac, H.J., 2004. Salinity tolerance of diapausing eggs of freshwater zooplankton. Freshw. Biol. 49 (3), 286–295. doi:10.1111/j.1365-2427.2004.01185.x.

Berger, E., Frör, O., Schäfer, R.B., 2019. Salinity impacts on river ecosystem processes: a critical mini-review. Philos. Trans. R. Soc. B Biol. Sci. 374 (1764), 1–10. doi:https://doi.org/10.1098/rstb.2018.0010.

Brucet, S., Boix, D., Nathansen, L.W., Quintana, X.D., Jensen, E., Balayla, D., Meerhoff, M., Jeppesen, E., 2012. Effects of temperature, salinity and fish in structuring the macroinvertebrate community in shallow lakes: implications for effects of climate change. PLoS ONE. 19326203. 7 (2). Available from: http://www.plosone.org/article/fetchObjectAttachment.action?uri=info%3Adoi%2F10.1371%2Fjournal.pone.0030877&representation=PDF. doi:10.1371/journal.pone.0030877Denmark.

Butler, D.L., Von Guerard, P., 1996. Salinity in the Colorado River in the Grand Valley, Western Colorado, 1994-95, [S. Department of the Interior], US Geological Survey.

Cañedo-Argüelles, M., Kefford, B.J., Piscart, C., Prat, N., Schäfer, R.B., Schulz, C.J., 2013. Salinisation of rivers: an urgent ecological issue. Environ. Pollut. 18736424. 173, 157–167. doi:10.1016/j.envpol.2012.10.011.

Cunillera-Montcusí, D., Beklioğlu, M., Cañedo-Argüelles, M., Jeppesen, E., Ptacnik, R., Amorim, C.A., Arnott, S.E., Berger, S.A., Brucet, S., Dugan, H.A., Gerhard, M., Horváth, Z., Langenheder, S., Nejstgaard, J.C., Reinikainen, M., Striebel, M., Urrutia-Cordero, P., Vad, C.F., Zadereev, E., Matias, M., 2022. Freshwater salinisation: a research agenda for a saltier world. Trends Ecol. Evol. 01695347. 37 (5), 440–453. Available from: https://www.sciencedirect.com/journal/trends-in-ecology-and-evolution. doi:10.1016/j.tree.2021.12.005.

Dipu, S., Josna, R.P., 2020. Saline water intrusion in urban coastal area: a case study of Kuttiyadi River, Kerala, India. Am. J. Water Resour. 8 (3), 128–133.

du Plessis, A. (2022). Persistent degradation: global water quality challenges and required actions. One Earth 5 (2), 129–131.

Environment Canada, 2001. Canadian Environmental Protection Act, 1999. Priority Substances List Assessment Report Road Salt. Environment Canada, Hull, Quebec.

Environmental Protection Agency, 2022. U.S. Environmental Protection Agency EPA Researching the Impacts of Freshwater Salinization Syndrome. Published November 29. https://www.epa.gov/sciencematters/epa-researching-impacts-freshwater-salinization-syndrome

Feld, C.K., Lorenz, A.W., Peise, M., Fink, M., Schulz, C.J., 2023. Direct and indirect effects of salinisation on riverine biota: a case study from river Wipper, Germany. Hydrobiologia. 15735117. 850 (14), 3043–3059. Available from: https://www.springer.com/journal/10750. doi:10.1007/s10750-023-05229-z.

Garcia, S., Tracing anthropogenic salinity inputs to the semi-arid Rio Grande river: a multi-isotope tracer approach (2017).

Gašparović, M., Singh, S.K., 2020. Spatio-temporal salinity monitoring of the ghaghara river using landsat time-series imagery and multiple regression analysis. The International Archives of the Photogrammetry, Remote Sensing and Spatial Information Sciences. 2194-9034. XLIII-B3-2020 3, 401–405. doi:10.5194/isprs-archives-xliii-b3-2020-401-2020.

Guo, B., Miao, L., Fan, Y., Hongwei, W., Yang, Y., Wang, C., 2023. A novel remote sensing monitoring index of salinization based on three-dimensional feature space model and its application in the Yellow River Delta of China. Geomatics, Natural Hazards and Risk. 1947-5705. 14 (1), 95–116. doi:10.1080/19475705.2022.2156820.

Hanes, R.E., Zelazny, L.W., Blaser, R.E., 1970. Effects of Deicing Salts on Water Quality and Biota: Litterature Review and Recommended Research, Highway Research Board, National Academy of Sciences.

Hao, L., Wang, P., Gojenko, B., Jingjie, Y., Aifeng, L., Fadong, L., Kenjabaev, S., Kulmatov, R., Khikmatov, F., 2023. Five decades of freshwater salinization in the Amu Darya River basin. J. Hydrol. Reg. Stud. 22145818. 47, 101375. doi:10.1016/j.ejrh.2023.101375.

Hart, B.T., Bailey, P., Edwards, R., Hortle, K., James, K., McMahon, A., Meredith, C., Swadling, K., 1991. A review of the salt sensitivity of the Australian freshwater biota. Hydrobiologia. 15735117. 210 (1-2), 105–144. doi:10.1007/BF00014327.

Hartman, K.J., Kaller, M.D., Howell, J.W., Sweka, J.A., 2005. How much do valley fills influence headwater streams? Hydrobiologia 532 (1), 91–102. doi:10.1007/s10750-004-9019-1.

Isidoro, D., Quílez, D., Aragüés, R., 2006. Environmental Impact of Irrigation in La Violada District (Spain) II. Nitrogen Fertilization and Nitrate Export Patterns in Drainage Water. J. Environ. Qual. 35 (3), 776–785.

Kaushal, S.S., Likens, G.E., Pace, M.L., Reimer, J.E., Maas, C.M., Galella, J.G., Utz, R.M., Duan, S., Kryger, J.R., Yaculak, A.M., Boger, W.L., Bailey, N.W., Haq, S., Wood, K.L., Wessel, B.M., Park, C.E., Collison, D.C., Aisin, B.Y.I., Gedeon, T.M., Chaudhary, S.K., Widmer, J., Blackwood, C.R., Bolster, C.M., Devilbiss, M.L., Garrison, D.L., Halevi, S., Kese, G.Q., Quach, E.K., Rogelio, C.M.P., Tan, M.L., Wald, H.J.S., Woglo, S.A., 2021. Freshwater salinization syndrome: from emerging global problem to managing risks. Biogeochemistry. 1573515X. 154 (2), 255–292. Available from: www.wkap.nl/journalhome.htm/0168-2563. doi:10.1007/s10533-021-00784-w.

Khatri, N., Tyagi, S., Rawtani, D., Tharmavaram, M., 2020. Assessment of river water quality through application of indices: a case study River Sabarmati, Gujarat, India. Sustain. Water Resour. Manag. 2363-5037. 6 (6). doi:10.1007/s40899-020-00459-8.

Koç, C., 2008. The environmental effects of salinity load in Great Menderes Basin irrigation schemes. Environ. Monit. Assess. 0167-6369. 146 (1-3), 479–489. doi:10.1007/s10661-008-0478-0.

Kumar, S., Imen, S., Sridharan, V.K., Gupta, A., McDonald, W., Ramirez-Avila, J.J., Abdul-Aziz, O.I., Talchabhadel, R., Gao, H., Quinn, N.W.T., Weiss, W.J., Poulose, T., Palmate, S.S., Lee, C.M., Baskaran, L., 2024. Perceived barriers and advances in integrating earth observations with water resources modeling. Remote Sens. Appl. Soc. Environ. 23529385. 33. Available from: https://www.sciencedirect.com/science/journal/23529385. doi:10.1016/j.rsase.2023.101119.

Létolle, R., Chesterikoff, A., 1999. Salinity of surface waters in the Aral Sea region. Int. J. Salt Lake Res. 1037-0544. 8 (4), 293–306. doi:10.1007/bf02442116.

Loch, A., Gregg, D., 2018. Salinity management in the Murray-Darling Basin: a transaction cost study. Water Resour. Res. 54 (11), 8813–8827.

Ma, X., Wang, C., Li, Y., 2022. Predicting river salinity using machine learning. Water Resour. Manag. 36, 1329–1344.

Matysik, M., Absalon, D., Janczewska, N., Woźnica, A., 2025. Detection of changes in water salinity based on patrol monitoring and remote sensing methods. Desalin. Water Treat. 19443986. 321, 100930. doi:10.1016/j.dwt.2024.100930.

Mohamed, S.A., Metwaly, M.M., Metwalli, M.R., AbdelRahman, M.A.E., Badreldin, N., 2023. Integrating Active and Passive Remote Sensing Data for Mapping Soil Salinity Using Machine Learning and Feature Selection Approaches in Arid Regions. Remote Sens. 20724292. 15 (7). Available from: http://www.mdpi.com/journal/remotesensing. doi:10.3390/rs15071751.

Mohanavelu, A., Naganna, S.R., Al-Ansari, N., 2021. Irrigation induced salinity and sodicity hazards on soil and groundwater: an overview of its causes, impacts and mitigation strategies. Agriculture (Switzerland) 11 (10), 983.

Moyano-Salcedo, A.J., Piana, T., Crabot, J., Kefford, B.J., Berger, E., Arnott, S.E., Thorslund, J., Meybeck, M., Kaushal, S.S., Schäfer, R.B., Cañedo-Argüelles, M., 2025. GLOBal river SALiniTy and associated ions (GlobSalt). Sci. Rep. 20452322. 15 (1). Available from: https://www.nature.com/srep/. doi:10.1038/s41598-025-96222-0.

Nielsen, D.L., Brock, M.A., Crosslé, K., Harris, K., Healey, M., Jarosinski, I., 2003. The effects of salinity on aquatic plant germination and zooplankton hatching from two wetland sediments. Freshw. Biol. 00465070. 48 (12), 2214–2223. Available from: http://onlinelibrary.wiley.com/journal/10.1111/(ISSN)1365-2427. doi:10.1046/j.1365-2427.2003.01146.x.

Nurmemet, I., Ghulam, A., Tiyip, T., Elkadiri, R., Ding, J.-L., Maimaitiyiming, M., Abliz, A., Sawut, M., Zhang, F., Abliz, A., Sun, Q., 2015. Monitoring Soil Salinization in Keriya River Basin, Northwestern China Using Passive Reflective and Active Microwave Remote Sensing Data. Remote Sens. 2072-4292. 7 (7), 8803–8829. doi:10.3390/rs70708803.

O'Keeffe, J.H., De Moor, F.C., 1988. Changes in the physico-chemistry and benthic invertebrates of the great fish river, South Africa, following an interbasin transfer of water. Regul. Rivers Res. Manag. 0886-9375. 2 (1), 39–55. doi:10.1002/rrr.3450020105.

Perri, S., Suweis, S., Holmes, A., Marpu, P.R., Entekhabi, D., Molini, A., 2020. River basin salinization as a form of aridity. Proc. Natl. Acad. Sci. U. S. A. 117 (30), 17635–17642. Available from: https://www.pnas.org/content/pnas/117/30/17635.full.pdf. doi:10.1073/pnas.2005925117.

Pinder, A.M., Halse, S.A., McRae, J.M., Shiel, R.J., 2005. Occurrence of aquatic invertebrates of the wheatbelt region of Western Australia in relation to salinity. Hydrobiologia 543 (1), 1–24. doi:10.1007/s10750-004-5712-3.

Podmore, C., 2009. Irrigation salinity–causes and impacts. Primefact 937 (1), 1–4.

Pond, G.J., Passmore, M.E., Borsuk, F.A., Reynolds, L., Rose, C.J., 2008. Downstream effects of mountaintop coal mining: comparing biological conditions using family- and genus-level macroinvertebrate bioassessment tools. J. N. Am. Benthol. Soc. 08873593. 27 (3), 717–737. doi:10.1899/08-015.1.

Qureshi, A.S., Perry, C., 2021. Managing Water and Salt for Sustainable Agriculture in the Indus Basin of Pakistan. Sustainability. 2071-1050. 13 (9), 5303. doi:10.3390/su13095303.

Richter, B.D., et al, 2010. A framework for balancing ecosystem and human needs in water management. Ecol. Appl. 20 (6), 1370–1384.

Said, S., Khan, S.A., 2021. Remote sensing-based water quality index estimation using data-driven approaches: a case study of the Kali River in Uttar Pradesh, India. Environment, Development and Sustainability. 1387-585X. 23 (12), 18252–18277. doi:10.1007/s10668-021-01437-6.

Sarker, B., Keya, K.N., Mahir, F.I., Nahiun, K.M., Shahida, S., Khan, R.A., 2021. Surface and Ground Water Pollution: causes and Effects of Urbanization and Industrialization in South Asia. Sci. Rev. 2412-2599. (73, 32–41. doi:10.32861/sr.73.32.41.

Scherman, P.A., Muller, W.J., Palmer, C.G., 2003. Links between ecotoxicology, biomonitoring and water chemistry in the integration of water quality into environmental flow assessments. River Res. Appl. 19 (5-6), 483–493.

Schuler, M.S., Cañedo-Argüelles, M., Hintz, W.D., Dyack, B., Birk, S., Relyea, R.A., 2019. Regulations are needed to protect freshwater ecosystems from salinization. Philos. Trans. R. Soc. B Biol. Sci. 0962-8436. 374 (1764), 20180019. doi:10.1098/rstb.2018.0019.

Smedema, L.K., Shiati, K., 2002. Irrigation and salinity: a perspective review of the salinity hazards of irrigation development in the arid zone. Irrig. Drain. Syst. 01686291. 16 (2), 161–174. doi: 10.1023/A:1016008417327.

Stets, E.G., Sprague, L.A., Oelsner, G.P., Johnson, H.M., Murphy, J.C., Ryberg, K., Vecchia, A.V., Zuellig, R.E., Falcone, J.A., Riskin, M.L., 2020. Landscape Drivers of Dynamic Change in Water Quality of U.S. Rivers. Environ. Sci. Technol. 15205851. 54 (7), 4336–4343. Available from: http://pubs.acs.org/journal/esthag. doi:10.1021/acs.est.9b05344.

Tedeschi, A., Beltran, A., Aragüés, R., 2001. Irrigation management and hydrosalinity balance in a semi-arid area of the middle Ebro river basin (Spain). Agric. Water Manag. 49 (1), 31–50.

Trung, L.V., Lam, L.N., Tam, D.M., 2021. Web Gis and Remote Sensing Solutions for Monitoring and Thematic Mapping in the Mekong Delta Region. IOP Conf. Ser. Earth Environ. Sci. 652 (1), 1–7. doi:https://doi.org/10.1088/1755-1315/652/1/012023.

Williams, W.D., 1987. Salinization of rivers and streams: an important environmental hazard. Ambio. 00447447. 16 (4), 180–185.

Williams, W.D., 2001. Anthropogenic salinisation of inland waters. Hydrobiologia 466, 329–337. doi:10.1023/A:1014598509028.

Williams, W.D., Sherwood, J., 1994. Definition and measurement of salinity in salt lakes. Int. J. Salt Lake Res. 3 (1), 53–63.

Yagmur, N., Bilgilioglu, B.B., Dervisoglu, A., Musaoglu, N., Tanik, A., 2021. Long and short-term assessment of surface area changes in saline and freshwater lakes via remote sensing. Water Environ. J. 17476593. 35 (1), 107–122. Available from: http://onlinelibrary.wiley.com/journal/10.1111/(ISSN)1747-6593. doi:10.1111/wej.12608.

Zaeri, A., Mohammadi, Z., Rezanezhad, F., 2023. Determining the source and mechanism of river salinity: an integrated regional study. J. Hydrol. Reg. Stud. 47, 1–19. doi:https://doi.org/10.1016/j.ejrh.2023.101411.

Zgórska, A., Bondaruk, J., Dudziak, M., Hamerla, A., 2020. Impact of Industrial Discharge on Aquatic Ecosystems of the Klodnica River with Reference to Water Framework Directive Objectives. Pol. J. Environ. Stud. 29 (4), 2945–2953.

Chapter 14

Effect of river pollution on marine ecosystem

Rantim Bhattacharjee[a,b,*], Aanchal Chauhan[a,b,*], Pratika Singh[a], and Abhilash K. Tilak[c]

[a]*National Centre for Polar and Ocean Research (NCPOR), Ministry of Earth Sciences, Vasco-da-Gama, Goa, India*

[b]*Department of Microbiology, School of Biological Sciences and Biotechnology (SBSB), Goa University, Panjim, Goa, India*

[c]*Department of Mechanical and Industrial Engineering, Manipal Institute of Technology Bengaluru, Manipal Academy of Higher Education, Manipal, India*

1 Introduction

In the human body, veins containing blood carry essential nutrients, oxygen, and all important constituents to the cells present throughout the body to sustain life. Similarly, rivers are veins that carry water from the land to the ocean to sustain the marine environment. Just as blood is essential for human life, rivers are highly important for sustaining both terrestrial and marine environments; river water includes vast amounts of essential nutrients, minerals, high organic–inorganic matter, and large quantities of different microbial cells. Weathering processes, soil leaching, and atmospheric deposition are the natural sources of nutrients, minerals, and both dissolved organic carbon (DOC) and particulate organic carbon (POC), which constitute river water (Di Grazia et al., 2023). Since the onset of the Anthropocene, human activities have increased exponentially. These anthropogenic activities include vast levels of industrialization, urbanization, agricultural practices, etc.; these activities are the major drivers of river pollution, which includes agricultural runoff, wastewater discharges, and industrial effluents (Li et al., 2022; Zhu et al., 2022). The increased coastal population in recent years has intensified land use heavily. This has accelerated climate-driven changes in precipitation patterns and hydrological regimes and has also impacted the timing, magnitude, and composition of riverine fluxes (León-Muñoz et al., 2024). Convergence zones where fresh water from rivers mixes with ocean water are called estuaries;

---

* Authors contributed equally.

River Pollution. https://doi.org/10.1016/B978-0-443-45024-2.00018-X

these zones significantly influence the turbidity, salinity, and redox conditions of the ocean, which affect productivity, food web dynamics, and nutrient and carbon cycling across continental shelves (Lønborg et al., 2021). Riverine inputs are heterogeneous in nature and are mostly composed of suspended particles and colloidal substances. They provide suitable surfaces for microbial proliferation, colonization, and biofilm formation, creating microhabitats with physicochemical conditions different from those of the surrounding seawater (Bansal et al., 2023). Seasonal and episodic variations in river flow influence light penetration, change oxygen levels, and modify nutrient composition, all of which impact microbial growth and the microbial community structure that affects the whole marine ecosystem (Dang and Lovell, 2016; Zhang et al., 2023a).

River pollution affects marine ecosystems and the surrounding ocean environment. It influences the structure and function of microbial communities crucial for regulating the biogeochemical cycles of the ocean. Rivers carry polluted water with large amounts of DOC and POC into the ocean. By taking up DOC as an energy source, heterotrophic bacteria (such as Bacteroidetes and Gammaproteobacteria) break down complex organic matter or POC. They release enzymes such as kinases, proteases, and glycosidases to catalyze the mineralization and remineralization of both organic and inorganic carbon, known as carbon fixation and release (Kirchman, 2003). The balance between fixing and releasing determines whether the ocean is acting as a carbon sink or a carbon source (Jiao et al., 2010). As the concentration of pollutants increases, specific bacteria thrive in polluted environments, which decreases the diversity of the microbial community. This significantly affects the efficiency of the "biological carbon pump" of the ocean (Jiao et al., 2010). Rivers also contribute by transporting excess nutrients such as nitrate, nitrite, silicate, and phosphate at elevated concentrations. The presence of these nutrients stimulates the growth of specific algae, e.g., *Pseudonitzschia* sp., *Synechococcus* sp., *Cylindrospermopsin* sp., and *Microcystis* sp., and leads to harmful algal blooms (HABs). As blooms die, bacteria quickly start breaking down POC into DOC. The increased metabolic activity during this process consumes a significant portion of the dissolved oxygen (DO), leading to hypoxic or low-oxygen zones (Diaz and Rosenberg, 2008). Anaerobic microbes thrive in oxygen-limited zones and change the structure and function of marine ecosystems (Cloern et al., 2016). In this chapter, we discuss the influence of river pollutants carried to marine ecosystems. Research has focused on the roles of nutrients, organic matter, and particular changes that affect the structure of microbial communities along with their metabolic functions. Our goal is to understand the broader implications of pollutants exchanged between rivers and the ocean, which will provide solid ground for taking necessary steps toward the sustainability and preservation of delicate ecosystems amidst rapidly changing global climates.

2 Mechanism of river pollution

River pollution is a well-known environmental problem caused by both natural processes and anthropogenic discharges of a complex mixture of contaminants into water bodies. Anthropogenic activities are the primary drivers of rising pollutants in river systems. Industrialization and urban growth release toxic substances and emerging pollutants such as pharmaceuticals and microplastics (MPs), whereas agriculture and improper waste management contribute excess nutrients, pesticides, and organic microcontaminants to rivers. The mechanisms of river pollution are summarized in Fig. 14.1.

2.1 Acidification

Acidification is largely caused by anthropogenic pollutants, including sulfur oxides (SOx) and nitrogen oxides (NOx), which are released through combustion, fossil fuel consumption, and agricultural runoff (Doney et al., 2007). These chemicals are deposited from the air or leach directly into marine ecosystems, where they transform into corrosive acids such as sulfuric and nitric acids, reducing the pH of streams and rivers (Doney et al., 2007). This increase in acidity tips the geochemical balance, increasing the solubility of heavy metals such as mercury and cadmium. The elevated solubility enhances their bioavailability and capacity to be toxic to freshwater and marine organisms (Eklöf et al., 2024). When acidified river water enters marine habitats, it affects the natural buffering capacity of ocean water, resulting in a localized reduction in carbonate ions and interference with essential carbonate chemistry (Gazeau et al.,

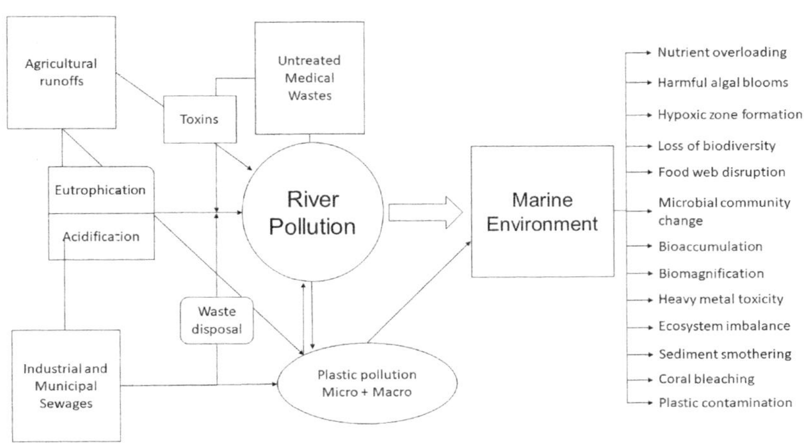

FIG. 14.1 Schematic summary of the possible mechanisms of river pollution and its impact on marine ecosystems.

2013). These changes interfere with biomineralization, the process by which corals, mollusks, and certain plankton construct skeletons and shells. The consequence is decreased skeletal integrity, which makes organisms more vulnerable to disease and predation and ultimately contributes to increased mortality (Gazeau et al., 2013; Osborne et al., 2022). At the cellular level, acidification compromises proton gradients across membranes in microbes and larger life forms. These gradients are critical for energy transduction and nutrient uptake. For marine microorganisms such as phytoplankton and bacteria, low pH affects enzyme activity, induces oxidative stress, and disrupts major nitrogen and carbon cycling pathways. Such physiological disruption can inhibit growth, rearrange community composition, and influence primary production, with these ripple effects propagating upward through the food web (Lund et al., 2020).

2.2 Eutrophication

Eutrophication is an anthropogenic consequence that results from overenrichment due to nutrient surplus, mostly nitrogen (N) and phosphorus (P), which enter aquatic ecosystems via runoff from agriculture, untreated urban wastewater, and industrial effluents. Excessive availability of nutrients changes natural nutrient limitations, causing disproportionate stimulation of primary production (Riza et al., 2023). Phytoplankton and opportunistic macrophytes react with increased growth, tending to produce HABs. These blooms reduce light penetration and outcompete other photosynthetic microbes while emitting an array of secondary metabolites and toxins that are harmful to the physiology of fish, invertebrates, and humans. The decomposition of bloom biomass drives dense heterotrophic microbial activity (Lan et al., 2024). Aerobic breakdown of organic material lowers DO levels, establishing hypoxic or anoxic conditions that are fatal to numerous benthic and pelagic organisms (Jessen et al., 2017). Microbial communities adapt to anaerobic processes, such as denitrification, sulfate reduction, and methanogenesis, which further alter redox balances and favor the release of reduced substances such as hydrogen sulfide and methane (Bastviken et al., 2004). These processes escalate habitat deterioration and are responsible for the development of widespread "dead zones" in coastal areas (Altieri and Diaz, 2019). In addition to oxygen depletion, eutrophication restructures species interactions and ecosystem composition. High levels of nutrients promote rapid-growing, opportunistic phytoplankton (e.g., cyanobacteria and dinoflagellates) over more sensitive organisms, decreasing biodiversity and changing trophic transfer efficiency. At the microbial level, nutrient enhancement modifies bacterial community composition, including structure and metabolic activity, influencing the rates of nutrient recycling and biogeochemical feedback cycles (Brush et al., 2020). These alterations lead to the destabilization of food webs and a decline in ecological resilience, thereby impairing essential ecosystem functions, such as fisheries productivity, nutrient control, and carbon cycling.

2.3 Toxins

Industrial effluents, agricultural chemicals, and untreated sewage discharge a wide variety of toxins, including high concentrations of heavy metals, polychlorinated biphenyls (PCBs), polycyclic aromatic hydrocarbons (PAHs), pesticides, etc., into rivers (Ebele et al., 2017). Once these pollutants enter the marine environment, they accumulate in sediments and aquatic organisms, leading to bioaccumulation and subsequent biomagnification through the food chain. The presence of these toxic substances interferes with crucial biological functions, such as photosynthesis, respiration, and reproduction, in aquatic organisms (Dey et al., 2024). High exposure to elevated concentrations of heavy metals and synthetic chemicals can result in immunosuppression, endocrine disruption, and a high rate of disease occurrence in fish populations (Breitmeyer et al., 2022). This not only influences the individual organisms themselves but also alters species-to-species interactions and is potentially responsible for the total reorganization of the marine community.

3 Impact of MPs on marine ecosystems

3.1 Abundance and distribution of MP pollution

MPs are minute plastic particles with sizes ranging from 1 μm to 5 mm (<5 mm) and are emerging as critical pollutants. MPs are of global concern, as they have emerged as one of the major pollutants identified in oceanic environments, including polar regions, in recent years (Hidalgo-Ruz et al., 2012). MPs form from the breakdown of larger plastic particles, which are known as macroplastics, through the processes of mechanical fragmentation, photo degradation, thermal stress, and biological and chemical degradation (Yee et al., 2021; Zhang et al., 2021). MPs are distributed across ocean surfaces, in water columns, as well as in deep-sea sediments (Auta et al., 2017), as shown in Table 14.1. Rivers act as major conduits that carry MPs (beads, fragments, pellets, and synthetic fibers) into marine ecosystems, such as estuaries and coastal environments. The abundance and distribution of MPs are essential for understanding and estimating marine MP pollution. Global estimates suggest that 236,000 metric tons of MPs in oceans (Van Sebille et al., 2015) occur in various forms, ranging from large macroplastics to MPs, as well as nanoplastics, are distributed throughout the water column, are found in surface waters, and even reach pristine polar regions.

3.2 Rivers as transport pathways for MPs

Rivers constitute one of the main transport pathways for plastic pollution in marine ecosystems. Industrial effluents, urban wastewater, and storm water introduce MPs into rivers, which, in turn, discharge them into oceans and estuaries. Global plastic emissions in aquatic ecosystems have been estimated by

TABLE 14.1 Microplastic abundance in marine ecosystems.

Sl. no.	Location	Types of MPs in the marine ecosystem	References
1.	Bohai Sea	Surface, water columns, and sediments (fibers 75%–96%)	(Dai et al., 2018)
2.	Northwest Pacific, Bering and Chukchi Sea	0.018–0.31 items/m^3	(Mu et al., 2019)
3.	Swedish sea	0.10 MPs/m^3	(Schönlau et al., 2020)
4.	Tropical Eastern Pacific and Galapagos	100% occurrence of MPs ranging in size between 150 and 5000 μm	(Alfaro-Núñez et al., 2021)
5.	Shoreside waters of South Georgia	1.75 MPs/L of seawater	(Buckingham et al., 2022)
6.	Ross Sea (Antarctica)	PE and PP (57.1%), PL (28.6%), PTFE (5.7%), PMMA (5.7%), and PA (2.9%)	(Cincinelli et al., 2017)
7.	Jinhae Bay, South Korea	PP, PE, PS, and PVC	(Song et al., 2015)
8.	North Western Mediterranean Sea	PS	(Collignon et al., 2012)
9.	North Sea	PET, PU, PC, and PVC	(Löder and Gerdts, 2015)
10.	Southern Ocean	PS, PVC, PP, and PE	(Isobe et al., 2017)

PE, polyethylene; *PP*, polypropylene; *PL*, polylactic acid; *PTFE*, polytetrafluoroethylene; *PMMA*, polymethyl methacrylate; *PA*, polyamide; *PS*, polystyrene; *PVC*, polyvinyl chloride; *PET*, polyethylene terephthalate; *PU*, polyurethane; *PC*, polycarbonate.

Jambeck et al. (2015) and Lebreton et al. (2017) on the basis of the amount of mismanaged plastic waste and its likelihood of entering oceans. Meijer et al. (2021) improved the estimation by including geographical factors and climate and reported that over 1000 rivers contribute 0.8–2.7 million metric tons of plastic per year into oceans. Zhang et al. (2023b) used a top-down approach with ocean transport models and sea surface plastic concentration data and estimated global plastic emissions at 0.13–3.8 million metric tons per year. In European rivers, primary MPs from industry, personal care items, and cleaning products, and secondary MPs, largely fibers from synthetic textiles, have been found (Gao et al., 2024). In Asian rivers, higher MP concentrations have been observed near large cities; however, not all studies have shown a clear link between MPs and population density (Lin et al., 2024).

3.3 Wastewater and industrial effluents

Wastewater treatment plants (WWTPs) constitute a major pathway for the discharge of MPs into rivers, seas, estuaries, and eventually into the ocean. Although WWTPs can remove a large fraction of this waste, large amounts of untreated industrial effluents are still discharged into rivers and oceans. Sludge from WWTPs is another indirect route for MPs to enter the ecosystem. Industrial effluents from industries such as plastic manufacturing, paints, cosmetics, and textiles are known to release primary MPs directly into wastewater. Most plastics are not removed during sewage treatment because they tend to accumulate in the sludge. According to Iyare et al. (2020), the MPs concentration of sewage sludge varies considerably, ranging from 720 to 14,900 particles/kg (wet weight) and 1000 to 170,900 particles/kg (dry weight), the majority of which are fibers, accounting for almost 63–90% of the total. A study conducted at 10 sites along a 2575-km stretch of the Ganges River (Napper et al., 2021) revealed that combined flows of the Ganges, Brahmaputra, and Meghna rivers could release 1–3 billion MPs per day into the Bay of Bengal.

4 Impact of MP pollution on marine ecosystems

4.1 Physical and biological interaction ingestion by marine organisms

Plastic debris, which includes MPs and nanoplastics, is widespread in oceans and sediments, where concentrations can reach up to 100,000 particles/m^3. Plastic debris is found from beaches to deep-sea sediments, even in remote regions such as the Arctic and Antarctic (Wright et al., 2013). Marine pollution has major impacts on marine organisms, such as zooplankton and filter feeders (copepods, mussels), which ingest MPs and mistake them for food. A study conducted by Cole et al. (2015) on the copepod *Calanus helgolandicus* revealed that exposure to polystyrene MPs reduced feeding efficiency, with individuals ingesting 11% fewer algal cells and 40% less carbon biomass, while shifting their diet toward smaller prey. Longer exposure periods significantly decreased reproductive output, suggesting that the ingestion of MPs can impair feeding and energy intake, leading to long-term carbon depletion. MPs can leach toxic additives, which can adsorb persistent organic pollutants (POPs), increasing the risk for marine organisms and potentially transferring pollutants to higher trophic levels via biomagnification and bioaccumulation. The abundance of MPs is increasing globally, highlighting the urgent need for research on the ecological impacts on marine invertebrates.

4.2 Biofilm formation and microbial transport

MPs can act as novel substrates in marine ecosystems, serving as surfaces for microbial colonization and the establishment of diverse biofilms, also known as plastispheres, which can harbor potentially pathogenic microbes such as *Vibrio* spp. These microbes can pose risks to marine organisms and human health (Zettler et al., 2013). MPs are highly persistent and mobile, which allows microbial communities to attach and be transported across vast marine regions. This, in turn, facilitates the spread of pathogens, antibiotic resistance genes, and nonnative microorganisms (Amaral-Zettler et al., 2020). They also act as hotspots for horizontal gene transfer and the exchange of antibiotic resistance genes within plastisphere communities. Unlike natural substrates such as wood or algae, MPs enable the long-distance dispersal of opportunistic and invasive microbes across the ocean due to their durability, which allows microbes to remain attached for extended periods and disrupts ecological balance. Therefore MPs act as microbial vectors, functioning as both biological carriers and physical pollutants. The presence of MPs is a concern in aquaculture and coastal zones, where they increase the risk of disease outbreaks and reduce water quality.

4.3 Chemical impacts

MPs act as both carriers and sources of chemical contaminants in marine ecosystems. MPs pose significant chemical risks in marine ecosystems, as their surfaces allow the adsorption of POPs, such as PCBs, PAHs, and pesticides from seawater. These chemicals are concentrated to levels several orders of magnitude higher than those in the surrounding water and leach harmful additives, such as bisphenol A, phthalates, and flame retardants, into seawater (Tanaka et al., 2023), as well as heavy metals, hydrocarbons, and POPs, which increase the mobility and bioavailability of these contaminants. When ingested by marine organisms, these contaminants lead to toxic effects on the organisms, potentially bioaccumulating along the food chain (Wang et al., 2021), as they can be desorbed in their digestive tracts, leading to reduced fertility, endocrine disruption, and various developmental abnormalities (Rochman et al., 2013). The combination of chemical leaching, adsorption, and trophic transfer makes MPs a multifaceted threat, increasing both the ecological and human health risks in marine environments.

5 Impact of river pollution on marine waters

5.1 Impact on macrobenthos

Riverine pollution significantly alters the structure and function of macrobenthic communities in marine ecosystems. Rivers act as the main conduits for marine pollution in the form of heavy metals, nutrients, untreated sewage, and other contaminants, which tend to settle in sediments where most benthic organisms live. Macrobenthos such as mollusks, echinoderms, polychaetes, and crustaceans are deposit feeders that are highly vulnerable to contaminants associated with sediments (Table 14.2). Sediment smothering and oxygen depletion are caused by pollutants and organic-rich river inputs, which lead to a reduction in oxygen penetration and, therefore, create anoxic zones that impact and reduce species richness (Gray et al., 2002). MPs and fine sediments have the ability to adsorb heavy metals and POPs, and when ingested by organisms, they tend to accumulate in benthic tissues, impairing immune responses, enzyme activities, and reproduction (Wright et al., 2013). Long-term exposure to riverine pollution is known to cause a shift in taxa for these sensitive microorganisms, resulting in a decline in biodiversity and altered food web dynamics.

The health of macrobenthic communities is severely impacted by river-led pollution entering marine ecosystems. For example, the Mississippi River discharges its waste into the Gulf of Mexico and carries excess nitrogen and phosphorus from agriculture, creating hypoxia, which smothers benthic habitats by reducing the population and biodiversity of bivalves, worms, and crustaceans present there (Rabalais and Turner, 2019). Similarly, the Ganges delivers untreated sewage, effluents, and heavy metals into the Bay of Bengal. The effluents are bound to the sediments, which, in turn, affect the benthic communities by altering their structure (Siddiqui and Pandey, 2019). Plastic debris from the Yangtze River is among the major contributors to the East China Sea, where it settles in sediments and is ingested by benthic organisms, impairing their growth (Ge et al., 2024).

5.2 Impact on fishes

The impact of river pollution on the marine environment is profound, and the fish population is affected, as rivers act as conduits for excess nutrients, chemical residues, heavy metals, and plastics in marine systems (Fig. 14.2). In South America, the Amazon River transports mercury from gold mining into the Atlantic Ocean, where estuarine fishes bioaccumulate toxic metals in their muscle tissues, which impairs reproductive functions, as well as neurological functions and threatens human consumers through seafood intake (Paiva et al., 2025). Rivers such as the Yangtze River are known to deliver significant MP loads into the East China Sea, where ingestion by fish results in gut blockage, altered feeding behavior, and oxidative stress (Xie et al., 2025). Similarly, in Asia, rivers such as the Ganga and Yamuna discharge major amounts of pesticides and pharmaceutical residues into the Bay of Bengal, thereby leading to reduced fertility, disruption of endocrine regulation in fish, and compromised population sustainability (Singh and Suthar, 2021). Excess amounts of nutrients such as nitrogen and phosphorus from agricultural runoff in the Mississippi Gulf of Mexico are responsible for fueling seasonal hypoxia, forcing fishes to migrate from low-oxygen environments and triggering large-scale mortality among fishes (Breitburg et al., 2018).

TABLE 14.2 Overview of pollutants affecting marine organisms.

Sl. no.	Type of marine pollution source	Transport pathway	Location	Type of microorganism affected	References
1.	Nutrient-rich effluents (N, P) from agricultural runoff	Rivers	Romanian Black Sea coast	Cyanobacteria and dinoflagellates	(Ristea et al., 2025)
2.	Heavy metals from industrial effluents	Rivers	Ganges and Yamuna → Bay of Bengal	*Pseudomonas, Thiobacillus,* and *Rhodopirellula,*	(Chaturvedi et al., 2024)
3.	Microplastics from urban waste and fishing activities	Rivers and coastal activities	Yangtze → East China Sea	*Vibrio* spp., ARGcarrying microbes	(Zettler et al., 2013)
4.	Macroplastics from fishing gear and litter	Coastal runoff	Mediterranean Sea	*Gambierdiscus* spp. and *Prorocentrum lima*	(Do Prado Leite et al., 2022)
5.	Pharmaceuticals and antibiotics	Rivers	Yamuna → Ganges → Bay of Bengal	MDR *E. coli*, antibiotic-resistant bacteria	(Singh and Suthar, 2021)
6.	Thermal pollution from power plants and industries	Coastal discharge	Pearl River Estuary, China	Proteobacteria and cyanobacteria	(Gu et al., 2021)
8.	Oil/petroleum hydrocarbons from oil spills and shipping	Coastal currents	Gulf of Mexico	*Alcanivorax, Alteromonas, Cycloclasticus, Halomonas, Marinobacter,* and *Pseudoalteromonas*	(Gutierrez et al., 2013)
9.	Oil and grease from industrial effluents	River and stormwater	Nile → Mediterranean Sea	*Alcanivorax* and *Marinobacte*	(Chernikova et al., 2020)
10.	Sewage andindustrial effluents	River	Prana River → Río de la Plata Estuary	Antibiotic-resistant *Escherichia coli* and fecal coliforms	(Giowanella et al., 2015)
11.	Nutrient and organic load from WWTP	River	Kadambrayar → Cochin Backwaters, India	*Trachelomonas, Euglena,* and *Phacus*	(Praseetha et al., 2025)

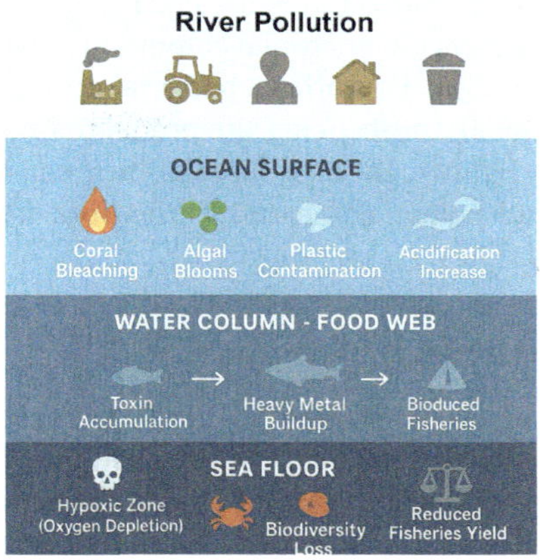

FIG. 14.2 Schematic diagram of river pollution affecting marine ecosystems.

5.3 Impact on microbes

River pollutants are known to strongly influence the microbial diversity present in estuarine and marine ecosystems. The Danube River transports large amounts of agricultural nutrients and untreated wastewater into the Black Sea, which causes eutrophication, HABs, and a shift in microbial communities toward cyanobacteria and toxin-producing dinoflagellates (Zaitsev, 1992). Similarly, the Nile River delivers untreated urban effluents and hydrocarbons into the Mediterranean Sea, thereby stimulating the hydrocarbonoclastic bacteria *Alcanivorax* and *Marinobacter*, which are indicators of chronic pollution (Chernikova et al., 2020). In Asia, the Mekong River carries pesticides and herbicides into the South China Sea from rice agriculture, which suppresses sensitive microbial taxa, thereby promoting pesticide-degrading bacteria and altering carbon turnover and nutrient cycling. The Paraná River in South America discharges sewage and industrial effluents into the Río de la Plata estuary, resulting in an increased abundance of antibiotic-resistant *Escherichia coli* and fecal coliforms in coastal waters, with public health implications (Giowanella et al., 2015). In the Yamuna River, untreated wastewater inputs foster multidrug-resistant *E. coli,* which highlights rivers as hotspots for the spread of antibiotic resistance genes (Singh et al., 2021).

MPs have a significant impact on microbial life in marine environments. Microbes attach themselves to small plastic particles and form complex biofilms known as the plastisphere, which are very different from the microbial communities found in surrounding waters (Zhai et al., 2023; Du et al., 2022). Rivers often carry MPs into estuaries and coastal zones, resulting in microbial communities that have already changed due to pollution. River mouths and estuaries serve as hotspots where MPs are colonized rapidly by dense microbial biofilms, comprising species such as *Vibrio* spp. and *Pseudomonas* spp., which carry antibiotic resistance genes (Liu et al., 2021). Hence, the accumulation of these biofilms on MPs alters the microbial community composition and disrupts normal ecosystem functions. This affects bacterial growth and disrupts biogeochemical cycles, ultimately influencing the productivity of the ocean environment. Since MPs are lightweight and durable, they can travel long distances, spreading invasive and potentially harmful microbes from polluted rivers to cleaner ocean regions.

5.4 Impact on coral reefs

River pollution has a major impact on coral reef ecosystems worldwide. High amounts of inorganic nitrogen from agricultural runoff are being discharged into the Great Barrier Reef, Australia, from the Burdekin, Fitzroy, Tully, and Daintree Rivers. This promotes algal blooms, which outcompete corals and reduce their biodiversity. Similarly, the Kadambrayar River in Kerala, India, is affected by the Brahmapuram waste treatment plant, which has increased the levels of pollution-tolerant algae and water hyacinth, thereby increasing the biological oxygen demand and lowering the DO level, which in turn harms coastal coral communities (Praseetha et al., 2025). The St. Lucie River in Florida contains nutrient-rich freshwater plumes in nearby reefs, which alter carbonate chemistry and support phytoplankton blooms, negatively affecting coral health. In addition, in the Red Sea, the sinking of the Rubymar vessel, which carried ammonium phosphate sulfate fertilizer, has threatened coral reefs by promoting algal blooms and creating oxygen-depleted dead zones. Therefore river pollution severely compromises coral reef health globally by altering coral reef community composition and reducing coral reef biodiversity.

6 Conclusion

River pollution poses a significant threat to marine ecosystems in modern countries. Rivers are continuously becoming overloaded with nutrients, heavy metals, plastics, and untreated sewage, and these contaminants are discharged into ocean waters. It affects fish, coral reefs, and countless microorganisms that are vital to the marine food web. Moreover, the proliferation of MPs worsens the situation, as these particles accumulate, provide surfaces for pathogenic microbes, and contribute to the spread of antibiotic-resistant diseases. To address this escalating issue, it is essential to implement and enforce measures that prevent pollutant dumping into rivers while adopting a broader commitment toward sustainability, recycling, and reuse policies. Although some developed nations have initiated sustainable practices, these efforts must become widespread and standardized globally to effectively combat river pollution. Vigilance, community engagement, and long-term dedication are crucial to ensuring the ongoing health and stability of our oceans in the future.

References

Alfaro-Núñez, A., Astorga, D., Cáceres-Farías, L., Bastidas, L., Soto Villegas, C., Macay, K., Christensen, J.H., 2021. Microplastic pollution in seawater and marine organisms across the Tropical Eastern Pacific and Galápagos. Sci. Rep. 20452322. 11 (1). Available from: www.nature.com/srep/index.html. doi:10.1038/s41598-021-85939-3.

Altieri, A.H., Diaz, R.J., 2019. Dead zones: oxygen depletion in coastal ecosystems. In: Sheppard, C. (Ed.), World Seas: An Environmental Evaluation. Academic Press, pp. 453–473.

Amaral-Zettler, L.A., Zettler, E.R., Mincer, T.J., 2020. Ecology of the plastisphere. Nat. Rev. Microbiol. 17401534. 18 (3), 139–151. Available from: http://www.nature.com/nrmicro/index.html. doi:10.1038/s41579-019-0308-0.

Auta, H.S., Emenike, C.U., Fauziah, S.H., Auta, H., Emenike, C., Fauziah, S., 2017. Distribution and importance of microplastics in the marine environment: a review of the sources, fate, effects, and potential solutions. Environ. Int. 102, 165–176. doi:10.1016/j.envint.2017.02.013.

Bansal, S., Creed, I.F., Tangen, B.A., Bridgham, S.D., Desai, A.R., Krauss, K.W., Neubauer, S.C., Noe, G.B., Rosenberry, D.O., Trettin, C., Wickland, K.P., Allen, S.T., Arias-Ortiz, A., Armitage, A.R., Baldocchi, D., Banerjee, K., Bastviken, D., Berg, P., Bogard, M.J., Chow, A.T., Conner, W.H., Craft, C., Creamer, C., DelSontro, T., Duberstein, J.A., Eagle, M., Fennessy, M.S., Finkelstein, S.A., Göckede, M., Grunwald, S., Halabisky, M., Herbert, E., Jahangir, M.M.R., Johnson, O.F., Jones, M.C., Kelleway, J.J., Knox, S., Kroeger, K.D., Kuehn, K.A., Lobb, D., Loder, A.L., Ma, S., Maher, D.T., McNicol, G., Meier, J., Middleton, B.A., Mills, C., Mistry, P., Mitra, A., Mobilian, C., Nahlik, A.M., Newman, S., O'Connell, J.L., Oikawa, P., van der Burg, M.P., Schutte, C.A., Song, C., Stagg, C.L., Turner, J., Vargas, R., Waldrop, M.P., Wallin, M.B., Wang, Z.A., Ward, E.J., Willard, D.A., Yarwood, S., Zhu, X., 2023. Practical guide to measuring wetland carbon pools and fluxes. Wetlands. 43 (8). doi:10.1007/s13157-023-01722-2.

Bastviken, D., Persson, L., Odham, G., Tranvik, L., 2004. Degradation of dissolved organic matter in oxic and anoxic lake water. Limnol. Oceanogr. 0024-3590. 49 (1), 109–116. doi:10.4319/lo.2004.49.1.0109.

Breitburg, D., Levin, L.A., Oschlies, A., Grégoire, M., Chavez, F.P., Conley, D.J., Garçon, V., Gilbert, D., Gutiérrez, D., Isensee, K., Jacinto, G.S., Limburg, K.E., Montes, I., Naqvi, S.W.A., Pitcher, G.C., Rabalais, N.N., Roman, M.R., Rose, K.A., Seibel, B.A., Telszewski, M., Yasuhara, M., Zhang, J., 2018. Declining oxygen in the global ocean and coastal waters. Science. 10959203. 359 (6371). Available from: http://science.sciencemag.org/content/359/6371/eaam7240/tab-pdf. doi:10.1126/science.aam7240.

Breitmeyer, S.E., Walsh, H.L., Blazer, V.S., Bunnell, J.F., Burritt, P.M., Dragon, J., Hladik, M.L., Bradley, P.M., Romanok, K.M., Smalling, K.L., 2022. Potential health effects of contaminant mixtures from point and nonpoint sources on fish and frogs in the New Jersey Pinelands. Sci. Total Environ. 18791026. 851. Available from: www.elsevier.com/locate/scitotenv. doi:10.1016/j.scitotenv.2022.158205.

Brush, M.J., Giani, M., Totti, C., Testa, J.M., Faganeli, J., Ogrinc, N., Michael Kemp, W., Umani, S.F., 2020. Eutrophication, harmful algae, oxygen depletion, and acidification. In: Malone T.C., Malej, A., Faganeli, J. (Eds.), Coastal Ecosystems in Transition: A Comparative Analysis of the Northern Adriatic and Chesapeake Bay. Wiley, United States, pp. 75–104. Available from: https://www.wiley.com/en-us/Coastal+Ecosystems+in +Transition%3A+A+Comparative+Analysis+of+the+Northern+Adriatic+and +Chesapeake+Bay-p-9781119543589. doi: 10.1002/9781119543626.ch5.

Buckingham, J.W., Manno, C., Waluda, C.M., Waller, C.L., 2022. A record of microplastic in the marine nearshore waters of South Georgia. Environ. Pollut. 02697491. 306, 119379. doi:10.1016/j.envpol.2022.119379.

Chaturvedi, S., Chakraborty, B., Min, L., Kumar, A., Pathak, B., Kumar, R., Yu, Z., 2024. Exploring the dynamic microbial tapestry of South Asian rivers: insights from the Ganges and Yamuna ecosystems. Ecohydrology. 1936-0584. 17 (6). doi:10.1002/eco.2662.

Chernikova, T.N., Bargiela, R., Toshchakov, S.V., Shivaraman, V., Lunev, E.A., Yakimov, M.M., Thomas, D.N., Golyshin, P.N., 2020. Hydrocarbon-degrading bacteria alcanivorax and marinobacter associated with microalgae pavlova lutheri and nannochloropsis oculata. Front. Microbiol. 1664302X. 11. Available from: https://www.frontiersin.org/journals/microbiology#. doi:10.3389/fmicb.2020.572931.

Cincinelli, A., Scopetani, C., Chelazzi, D., Lombardini, E., Martellini, T., Katsoyiannis, A., Fossi, M.C., Corsolini, S., 2017. Microplastic in the surface waters of the Ross Sea (Antarctica): occurrence, distribution and characterization by FTIR. Chemosphere 18791298. 175, 391–400. Available from: www.elsevier.com/locate/chemosphere. doi:10.1016/j.chemosphere.2017.02.024.

Cloern, J.E., Abreu, P.C., Carstensen, J., Chauvaud, L., Elmgren, R., Grall, J., Greening, H., Johansson, J.O.R., Kahru, M., Sherwood, E.T., Xu, J., Yin, K., 2016. Human activities and climate variability drive fast-paced change across the world's estuarine-coastal ecosystems. Glob. Change Biol. 1365-2486. 22 (2), 513–529. Available from: www.blacksci.co.uk/~cgilib/jnlpage.bin?Journal=gcb&File=gcb&Page=aims. doi:10.1111/gcb.13059.

Cole, M., Lindeque, P., Fileman, E., Halsband, C., Galloway, T.S., 2015. The impact of polystyrene microplastics on feeding, function and fecundity in the marine copepod Calanus helgolandicus. Environ. Sci. Technol. 15205851. 49 (2), 1130–1137. Available from: http://pubs.acs.org/journal/esthag. doi:10.1021/es504525u.

Collignon, A., Hecq, J.-H., Glagani, F., Voisin, P., Collard, F., Goffart, A., 2012. Neustonic microplastic and zooplankton in the North Western Mediterranean Sea. Mar. Pollut. Bull. 0025326X. 64 (4), 861–864. doi:10.1016/j.marpolbul.2012.01.011.

Dai, Z., Zhang, H., Zhou, Q., Tian, Y., Chen, T., Tu, C., Fu, C., Luo, Y., 2018. Occurrence of microplastics in the water column and sediment in an inland sea affected by intensive anthropogenic activities. Environ. Pollut. 02697491. 242 (Pt B), 1557–1565. doi:10.1016/j.envpol.2018.07.131.

Dang, H., Lovell, C.R., 2016. Microbial surface colonization and biofilm development in marine environments. Microbiol. Mol. Biol. Rev. 80 (1), 91–138. doi:10.1128/mmbr.00037-15.

Dey, S., Rajak, P., Sen, K., 2024. Bioaccumulation of metals and metalloids in seafood: a comprehensive overview of mobilization, interactive effects in eutrophic environments, and implications for public health risks. J. Trace Elem. Miner. 27730506. 8, 100141. doi:10.1016/j.jtemin.2024.100141.

Di Grazia, F., Garcia, X., Acuña, V., Llanos-Paez, O., Galgani, L., Gumiero, B., Loiselle, S.A., 2023. Modeling dissolved and particulate organic carbon dynamics at basin and sub-basin scales. Sci. Total Environ. 00489697. 884, 163840. doi:10.1016/j.scitotenv.2023.163840.

Diaz, R.J., Rosenberg, R., 2008. Spreading dead zones and consequences for marine ecosystems. Science. 10959203. 321 (5891), 926–929. doi:10.1126/science.1156401.

Do Prado Leite, I., Menegotto, A., da Cunha Lana, P., Júnior, L.L.M., 2022. A new look at the potential role of marine plastic debris as a global vector of toxic benthic algae. Sci. Total Environ. 18791026. 838. Available from: www.elsevier.com/locate/scitotenv. doi:10.1016/j.scitotenv.2022.156262.

Doney, S.C., Mahowald, N., Lima, I., Feely, R.A., Mackenzie, F.T., Lamarque, J.-F., Rasch, P.J., 2007. Impact of anthropogenic atmospheric nitrogen and sulfur deposition on ocean acidification and the inorganic carbon system. Proc. Natl. Acad. Sci. U. S. A. 0027-8424. 104 (37), 14580–14585. doi:10.1073/pnas.0702218104.

Du, Y., Liu, X., Dong, X., Yin, Z., 2022. A review on marine plastisphere: biodiversity, formation, and role in degradation. Comput. Struct. Biotechnol. J. 20010370. 20, 975–988. doi:10.1016/j.csbj.2022.02.008.

Ebele, A.J., Abou-Elwafa Abdallah, M., Harrad, S., 2017. Pharmaceuticals and personal care products (PPCPs) in the freshwater aquatic environment. Emerg. Contam. 24056642. 3 (1), 1–16. Available from: www.keaipublishing.com/en/journals/emerging-contaminants/. doi:10.1016/j.emcon.2016.12.004.

Eklöf, K., von Brömssen, C., Huser, B., Åkerblom, S., Augustaitis, A., Braaten, H.F.V., de Wit, H.A., Dirnböck, T., Elustondo, D., Grandin, U., Holubová, A., Kleemola, S., Krám, P., Lundin, L., Löfgren, S., Markensten, H., Moldan, F., Karlsson, G.P., Rönnback, P., Valinia, S., Vuorenmaa, J., 2024. Trends in mercury, lead and cadmium concentrations in 27 European streams and rivers: 2000–2020. Environ. Pollut. 02697491. 360, 124761. doi:10.1016/j.envpol.2024.124761.

Gao, S., Orlowski, N., Bopf, F.K., Breuer, L., 2024. A review on microplastics in major European rivers. Wiley Interdiscip. Rev. Water 20491948. 11 (3). Available from: http://wires.wiley.com/WileyCDA/WiresJournal/wisId-WAT2.html. doi:10.1002/wat2.1713.

Gazeau, F., Parker, L.M., Comeau, S., Gattuso, J.P., O'Connor, W.A., Martin, S., Pörtner, H.O., Ross, P.M., 2013. Impacts of ocean acidification on marine shelled molluscs. Mar. Biol. 00253162. 160 (8), 2207–2245. doi:10.1007/s00227-013-2219-3.

Ge, X., Xu, F., Li, B., Liu, L., Lu, X., Wang, L., Zhang, Y., Li, J., Li, J., Tang, Y., 2024. Unveiling microplastic distribution and interactions in the benthic layer of the Yangtze River Estuary and East China Sea. Environ. Sci. Ecotechnol. 20. doi:10.1016/j.ese.2023.100340

Giowanella, M., Bozza, A., Do Rocio Dalzoto, P., Dionísio, J.A., Andraus, S., Guimarães, E.L.G., Pimentel, I.C., 2015. Microbiological quality of water from the rivers of Curitiba, Paraná State, Brazil, and the susceptibility to antimicrobial drugs and pathogenicity of *Escherichia coli*. Environ. Monit. Assess. 15732959. 187 (11). Available from: www.wkap.nl/journalhome.htm/0167-6369. doi:10.1007/s10661-015-4896-5.

Gray, J.S., Wu, R.S., Or, Y.Y., 2002. Effects of hypoxia and organic enrichment on the coastal marine environment. Mar. Ecol. Prog. Ser. 0171-8630. 238, 249–279. doi:10.3354/meps238249.

Gu, Q., Wang, W., Li, G., Liu, Y., Wong, P.K., An, T., 2021. Pollution profile of waterborne bacterial and fungal community in urban Rivers of Pearl River estuary: microbial safety assessment. J. Freshw. Ecol. 21566941. 36 (1), 305–322. Available from: http://www.tandfonline.com/loi/tjfe20. doi:10.1080/02705060.2021.1975580.

Gutierrez, T., Singleton, D.R., Berry, D., Yang, T., Aitken, M.D., Teske, A., 2013. Hydrocarbon-degrading bacteria enriched by the Deepwater Horizon oil spill identified by cultivation and DNA-SIP. ISME J. 17517370. 7 (11), 2091–2104. doi:10.1038/ismej.2013.98.

Hidalgo-Ruz, V., Gutow, L., Thompson, R.C., Thiel, M., 2012. Microplastics in the marine environment: a review of the methods used for identification and quantification. Environ. Sci. Technol. 15205851. 46 (6), 3060–3075. doi:10.1021/es2031505.

Isobe, A., Uchiyama-Matsumoto, K., Uchida, K., Tokai, T., 2017. Microplastics in the Southern Ocean. Mar. Pollut. Bull. 18793363. 114 (1), 623–626. Available from: www.elsevier.com/locate/marpolbul. doi:10.1016/j.marpolbul.2016.09.037.

Iyare, P.U., Ouki, S.K., Bond, T., 2020. Microplastics removal in wastewater treatment plants: a critical review. Environ. Sci. Water Res. Technol. 20531419. 6 (10), 2664–2675. Available from: http://pubs.rsc.org/en/journals/journal/ew. doi:10.1039/d0ew00397b.

Jambeck, J.R., Geyer, R., Wilcox, C., Siegler, T.R., Perryman, M., Andrady, A., Narayan, R., Law, K.L., 2015. Plastic waste inputs from land into the ocean. Science. 10959203. 347 (6223), 768–771. Available from: http://www.sciencemag.org/content/347/6223/768.full.pdf. doi:10.1126/science.1260352.

Jessen, G.L., Lichtschlag, A., Ramette, A., Pantoja, S., Rossel, P.E., Schubert, C.J., Struck, U., Boetius, A., 2017. Hypoxia causes preservation of labile organic matter and changes seafloor microbial community composition (Black Sea). Sci. Adv. 23752548. 3 (2). Available from: http://advances.sciencemag.org/content/advances/3/2/e1601897full.pdf. doi:10.1126/sciadv.1601897.

Jiao, N., Herndl, G.J., Hansell, D.A., Benner, R., Kattner, G., Wilhelm, S.W., Kirchman, D.L., Weinbauer, M.G., Luo, T., Chen, F., Azam, F., 2010. Microbial production of recalcitrant dissolved organic matter: long-term carbon storage in the global ocean. Nat. Rev. Microbiol. 8 (8), 593–599. doi:10.1038/nrmicro2386.

Kirchman, D.L., 2003. The contribution of monomers and other low-molecular weight compounds to the flux of dissolved organic material in aquatic ecosystems. In: Findlay, S.E.G., Sinsabaugh, R.L. (Eds.), Aquatic Ecosystems: Interactivity of Dissolved Organic Matter. Academic Press, pp. 217–241.

Lan, J., Liu, P., Hu, X., Zhu, S., 2024. Harmful algal blooms in eutrophic marine environments: causes, monitoring, and treatment. Water. 2073-4441. 16 (17), 2525. doi:10.3390/w16172525.

Lebreton, L.C.M., Van Der Zwet, J., Damsteeg, J.W., Slat, B., Andrady, A., Reisser, J., 2017. River plastic emissions to the world's oceans. Nat. Commun. 20411723. 8. Available from: http://www.nature.com/ncomms/index.html. doi:10.1038/ncomms15611.

León-Muñoz, J., Aguayo, R., Corredor-Acosta, A., Tapia, F.J., Luis Iriarte, J., Reid, B., Soto, D., 2024. Hydrographic shifts in coastal waters reflect climate-driven changes in hydrological regimes across Northwestern Patagonia. Sci. Rep. 2045-2322. 14 (1). doi:10.1038/s41598-024-71008-y.

Li, Y., Fang, L., Yuanzhu, W., Mi, W., Ji, L., Guixiang, Z., Yang, P., Chen, Z., Bi, Y., 2022. Anthropogenic activities accelerated the evolution of river trophic status. Ecol. Indic. 136. 10.1016/j.ecolind.2022.108584

Lin, H.T., Schneider, F., Aziz, M.A., Wong, K.Y., Arunachalam, K.D., Praveena, S.M., Sethupathi, S., Chong, W.C., Nafisyah, A.L., Parthasarathy, P., Chelliapan, S., Kunz, A., 2024. Microplastics in Asian rivers: geographical distribution, most detected types, and inconsistency in methodologies. Environ. Pollut. 18736424. 349. Available from: https://www.sciencedirect.com/science/journal/02697491. doi:10.1016/j.envpol.2024.123985.

Liu, Y., Liu, W., Yang, X., Wang, J., Lin, H., Yang, Y., 2021. Microplastics are a hotspot for antibiotic resistance genes: progress and perspective. Sci. Total Environ. 00489697. 773, 145643. doi:10.1016/j.scitotenv.2021.145643.

Löder, M.G.J., Gerdts, G., 2015. Methodology used for the detection and identification of microplastics—a critical appraisal. In: Bergmann, M., Gutow, L., Klages, M. (Eds.), Marine Anthropogenic Litter. Springer International Publishing, Germany, pp. 201–227. doi:10.1007/978-3-319-16510-3_8.

Lønborg, C., Müller, M., Butler, E.C.V., Jiang, S., Ooi, S.K., Trinh, D.H., Wong, P.Y., Ali, S.M., Cui, C., Siong, W.B., Yando, E.S., Friess, D.A., Rosentreter, J.A., Eyre, B.D., Martin, P., 2021. Nutrient cycling in tropical and temperate coastal waters: is latitude making a difference? Estuar. Coast. Shelf Sci. 02727714. 262. Available from: http://www.elsevier.com/inca/publications/store/6/2/2/8/2/3/index.htt. doi:10.1016/j.ecss.2021.107571.

Lund, P.A., De Biase, D., Liran, O., Scheler, O., Mira, N.P., Cetecioglu, Z., Fernández, E.N., Bover-Cid, S., Hall, R., Sauer, M., O'Byrne, C., 2020. Understanding how microorganisms respond to acid pH is central to their control and successful exploitation. Front. Microbiol. 1664302X. 11. Available from: https://www.frontiersin.org/journals/microbiology#. doi:10.3389/fmicb.2020.556140.

Meijer, L.J.J., van Emmerik, T., van der Ent, R., Schmidt, C., Lebreton, L., 2021. More than 1000 rivers account for 80% of global riverine plastic emissions into the ocean. Sci. Adv. 23752548.

7 (18). Available from: https://advances.sciencemag.org/content/7/18/eaaz5803. doi:10.1126/sciadv.aaz5803.

Mu, J., Zhang, S., Qu, L., Jin, F., Fang, C., Ma, X., Zhang, W., Wang, J., 2019. Microplastics abundance and characteristics in surface waters from the Northwest Pacific, the Bering Sea, and the Chukchi Sea. Mar. Pollut. Bull. 0025326X. 143, 58–65. doi:10.1016/j.marpolbul.2019.04.023.

Napper, I.E., Baroth, A., Barrett, A.C., Bhola, S., Chowdhury, G.W., Davies, B.F.R., Duncan, E.M., Kumar, S., Nelms, S.E., Niloy, M.N.H., Nishat, B., Maddalene, T., Thompson, R.C., Koldewey, H., 2021. The abundance and characteristics of microplastics in surface water in the transboundary Ganges River. Environ. Pollut. 18736424. 274. Available from: https://www.journals.elsevier.com/environmental-pollution. doi:10.1016/j.envpol.2020.116348.

Osborne, E., Hu, X., Hall, E.R., Yates, K., Vreeland-Dawson, J., Shamberger, K., Barbero, L., Hernandez-Ayon, M., Gomez, F.A., Hicks, T., Xu, Y.-Y., McCutcheon, M.R., Acquafredda, M., Chapa-Balcorta, C., Norzagaray, O., Pierrot, D., Munoz-Caravaca, A., Dobson, K.L., Williams, N., Rabalais, N., Dash, P., 2022. Ocean acidification in the Gulf of Mexico: drivers, impacts, and unknowns. Prog. Oceanogr. 209. 10.1016/j.pocean.2022.102882

Paiva, T., Pestana, I., Miranda, L., de Carvalho, G., Bastos, W., Kasper, D., 2025. Mercury scenario in fish from the Amazon Basin: exploring the interplay of social groups and environmental diversity. Toxics. 2305-6304. 13 (7), 580. doi:10.3390/toxics13070580.

Praseetha, A.P., Jayaraj, R., Sreekumar, V.B., 2025. Phytoplankton as environmental indicator for evaluating water pollution in the Kadambrayar River, Ernakulam District, Kerala, India. Vegetos. 2229-4473. doi:10.1007/s42535-025-01229-7.

Rabalais, N.N., Turner, R.E., 2019. Gulf of Mexico hypoxia: past, present, and future. Limnol. Oceanogr. Bull. 15396088. 28 (4), 117–124. Available from: http://onlinelibrary.wiley.com/journal/10.1002/(ISSN)1539-6088;jsessionid=DFD79A0C9BB838AC02BE19E139ADE036.f02t01. doi:10.1002/lob.10351.

Ristea, E., Bisinicu, E., Lavric, V., Parvulescu, O.C., Lazar, L., 2025. A long-term perspective of seasonal shifts in nutrient dynamics and eutrophication in the Romanian Black Sea Coast. Sustainability. 2071-1050. 17 (3), 1090. doi:10.3390/su17031090.

Riza, M., Ehsan, M.N., Pervez, M.N., Khyum, M.M.O., Cai, Y., Naddeo, V., 2023. Control of eutrophication in aquatic ecosystems by sustainable dredging: effectiveness, environmental impacts, and implications. Case Stud. Chem. Environ. Eng. 26660164. 7. Available from: www.journals.elsevier.com/case-studies-in-chemical-and-environmental-engineering/. doi:10.1016/j.cscee.2023.100297.

Rochman, C.M., Hoh, E., Kurobe, T., Teh, S.J., 2013. Ingested plastic transfers hazardous chemicals to fish and induces hepatic stress. Sci. Rep. 20452322. 3. doi:10.1038/srep03263.

Schönlau, C., Karlsson, T.M., Rotander, A., Nilsson, H., Engwall, M., van Bavel, B., Kärrman, A. 2020. Microplastics in sea-surface waters surrounding Sweden sampled by manta trawl and in situ pump. Mar. Pollut. Bull. 0025326X. (153), 111019. doi:10.1016/j.marpolbul.2020.111019.

Siddiqui, E., Pandey, J., 2019. Assessment of heavy metal pollution in water and surface sediment and evaluation of ecological risks associated with sediment contamination in the Ganga River: a basin-scale study. Environ. Sci. Pollut. Res. 26 (11), 10926–10940. doi:10.1007/s11356-019-04495-6.

Singh, N.S., Singhal, N., Kumar, M., Virdi, J.S., 2021. High prevalence of drug resistance and class 1 integrons in *Escherichia coli* isolated from River Yamuna, India: a serious public health risk. Front. Microbiol. 1664302X. 12. Available from: https://www.frontiersin.org/journals/microbiology#. doi:10.3389/fmicb.2021.621564.

Singh, V., Suthar, S., 2021. Occurrence, seasonal variations, and ecological risk of pharmaceuticals and personal care products in River Ganges at two holy cities of India. Chemosphere. 00456535. 268, 129331. doi:10.1016/j.chemosphere.2020.129331.

Song, Y.K., Hong, S.H., Jang, M., Han, G.M., Shim, W.J., 2015. Occurrence and distribution of microplastics in the sea surface microlayer in Jinhae Bay, South Korea. Arch. Environ. Contam. Toxicol. 00904341. 69 (3), 279–287. Available from: http://www.springerlink.com. doi:10.1007/s00244-015-0209-9.

Tanaka, K., Takahashi, Y., Kajiwara, T., Matsukami, H., Kuramochi, H., Osako, M., Suzuki, G. 2023. Identification and quantification of additive-derived chemicals in beached micro–mesoplastics and macroplastics. Mar. Pollut. Bull. 0025326X. (186), 114438. doi:10.1016/j.marpolbul.2022.114438.

Van Sebille, E., Wilcox, C., Lebreton, L., Maximenko, N., Hardesty, B.D., Van Franeker, J.A., Eriksen, M., Siegel, D., Galgani, F., Law, K.L., 2015. A global inventory of small floating plastic debris. Environ. Res. Lett. 17489326. 10 (12). Available from: http://iopscience.iop.org/article/10.1088/1748-9326/10/12/124006/pdf. doi:10.1088/1748-9326/10/12/124006.

Wang, L.-C., Lin, J.C.-T., Dong, C.-D., Chen, C.-W., Liu, T.-K., 2021. The sorption of persistent organic pollutants in microplastics from the coastal environment. J. Hazard. Mater. 03043894. 420, 126658. doi:10.1016/j.jhazmat.2021.126658.

Wright, S.L., Thompson, R.C., Galloway, T.S., 2013. The physical impacts of microplastics on marine organisms: a review. Environ. Pollut. 02697491. 178, 483–492. doi:10.1016/j.envpol.2013.02.031.

Xie, J.-Y., Li, B., Jia, Q.-P., Li, X., Zhao, Q., Fan, P.-D., Wang, C.-Q., Zhang, L.-Y., 2025. Neglected microplastics and their risks in rivers throughout the three gorges reservoir area. Toxics. 2305-6304. 13 (9), 781. doi:10.3390/toxics13090781.

Yee, M.S.-L., Hii, L.-W., Looi, C.K., Lim, W.-M., Wong, S.-F., Kok, Y.-Y., Tan, B.-K., Wong, C.-Y., Leong, C.-O., 2021. Impact of microplastics and nanoplastics on human health. Nanomaterials. 2079-4991. 11 (2), 496. doi:10.3390/nano11020496.

Zaitsev, Y.P., 1992. Recent changes in the trophic structure of the Black Sea. Fish. Oceanogr. 13652419. 1 (2), 180–189. doi:10.1111/j.1365-2419.1992.tb00036.x.

Zettler, E.R., Mincer, T.J., Amaral-Zettler, L.A., 2013. Life in the "plastisphere": microbial communities on plastic marine debris. Environ. Sci. Technol. 15205851. 47 (13), 7137–7146. doi:10.1021/es401288x.

Zhai, X., Zhang, X.-H., Yu, M., 2023. Microbial colonization and degradation of marine microplastics in the plastisphere: a review. Front. Microbiol. 1664-302X. 14. doi:10.3389/fmicb.2023.1127308.

Zhang, K., Hamidian, A.H., Tubić, A., Zhang, Y., Fang, J.K.H., Wu, C., Lam, P.K.S., 2021. Understanding plastic degradation and microplastic formation in the environment: a review. Environ. Pollut. 18736424. 274. Available from: https://www.journals.elsevier.com/environmental-pollution. doi:10.1016/j.envpol.2021.116554.

Zhang, W., Han, S., Zhang, D., Shan, B., Wei, D., 2023a. Variations in dissolved oxygen and aquatic biological responses in China's coastal seas. Environ. Res. 00139351. 223, 115418. doi:10.1016/j.envres.2023.115418.

Zhang, Y., Wu, P., Xu, R., Wang, X., Lei, L., Schartup, A.T., Peng, Y., Pang, Q., Wang, X., Mai, L., Wang, R., Liu, H., Wang, X., Luijendijk, A., Chassignet, E., Xu, X., Shen, H., Zheng, S., Zeng, E.Y., 2023b. Plastic waste discharge to the global ocean constrained by seawater observations. Nat. Commun. 20411723. 14 (1). Available from: https://www.nature.com/ncomms/. doi:10.1038/s41467-023-37108-5.

Zhu, X., Wang, L., Zhang, X., He, M., Wang, D., Ren, Y., Yao, H., Net Victoria Ngegla, J., Pan, H. 2022. Effects of different types of anthropogenic disturbances and natural wetlands on water quality and microbial communities in a typical black-odor river. Ecol. Indic. 1470160X. (136), 108613. doi:10.1016/j.ecolind.2022.108613.

Chapter 15

Human health ailments due to polluted waters

Somvir Singh[a], Rahul Kumar[a], Arti Thakur[b], Vishav Kiran[c], Frank Zakeyo Wandira[d], Arti Sharma[e], Mitanshi Sharma[f], and Nitesh Kumar[g]

[a]*Department of Biosciences, University Institute of Biotechnology, Chandigarh University, Mohali, India*
[b]*Department of Botany, Shoolini Institute of Life Sciences and Business Management, Solan, India*
[c]*LR Institute of Pharmacy Jabil Kyar, Solan, India*
[d]*Department of Agriculture Extension, University of Agriculture and Natural Resources, Lilongwe, Malawi*
[e]*Faculty of Agriculture, Abhilashi University, Mandi, Himachal Pradesh, India*
[f]*Department of Forensic Science, University Institute of Allied Health Sciences, Chandigarh University, Mohali, India*
[g]*Department of Biosciences, Himachal Pradesh University, Shimla, India*

1 Introduction

1.1 Importance of clean water for human health

The introduction and infiltration of substances into water bodies that do not occur naturally in Earth's hydrological cycle is known as water pollution. In addition, by worsening the quality of water (Alrumman et al., 2016) and causing serious risks to the ecosystem, it disrupts the balance of the biosphere's natural processes that make it stable and functional. Consequently, the adverse effects on human health and the environment are unavoidable. Water is the most important natural resource after air, with food ranking third, but its production is entirely dependent on water availability. Most biochemical reactions take place within living organisms, with the most complex life forms originating from microorganisms. Moreover, most ecological and biological processes occur in a multidimensional environment, where water occupies spatial dimensions and interacts with temporal aspects in essential ways. Life originated and is maintained in water, which plays a fundamental role as a medium in evolution, sustaining living systems, and supporting their origin. Water is crucial in biological terms because it aids in the growth and development of living organisms. Therefore water is essential for

River Pollution. https://doi.org/10.1016/B978-0-443-45024-2.00003-8

domestic needs, daily activities, life processes, and overall development (Bibi et al., 2016). Freshwater sources were safe for direct consumption before the industrial revolution. However, the rapid increase in urbanization and industrialization has caused extensive contamination, rendering freshwater no longer drinkable. The term "freshwater" was previously used to differentiate it from marine and saline oceanic water. Water pollution is defined as any alteration in the biological, chemical, and physical characteristics of water that affects the health of humans and aquatic life (Taneja et al., 2019). Water is vital for biogeochemical processes that occur on Earth's surface, supporting household activities such as irrigation, drinking, and bathing (Ullah et al., 2014; Ahmad et al., 2014). The past few decades, natural resources have reached a state of crisis due to the increasing demand for freshwater, rapid population growth, urbanization, industrialization, and intensive agricultural practices (Afzal et al., 2018). Many industries contribute to water pollution through the discharge of their effluents, including cement industries, sugar mills, thermal power plants, fertilizer plants, distilleries, paper industries, pharmaceutical industries, and textile plants that produce dyes and dye intermediates (Edokpayi et al., 2017). These industries discharge untreated or partially treated wastewater into nearby water bodies, sewage systems, or undeveloped lands, causing severe environmental degradation and ecological imbalance (Madhav et al., 2018a; Raju et al., 2011) (Table 15. 1).

TABLE 15.1 Waterborne infectious diseases, causative organism, and major health effects.

Sr. no.	Type of diseases	Name of disease	Causative organism/ contaminant	Major health effects
1.	**Waterborne infectious diseases** (Morris and Levin, 1995)	Cholera	*Vibrio cholerae*	Severe diarrhea, dehydration
		Typhoid fever	*Salmonella typhi*	High fever, abdominal pain
		Dysentery	*Shigella/ Entamoeba histolytica*	Bloody stools, intestinal cramps
		Hepatitis A and E	HAV and HBV	Liver inflammation, jaundice
		Gastroenteritis	Bacteria/virus contamination	Vomiting, diarrhea
		Giardiasis	*Giardia lamblia*	Abdominal pain, fatigue
		Cryptosporidiosis	Cryptosporidium	Watery diarrhea, weakness

Sr. no.	Type of diseases	Name of disease	Causative organism/ contaminant	Major health effects
2.	**Heavy metal-related diseases** (Chisolm, 1980)	Minamata disease	Mercury	Nervous system damage
		Itai-Itai disease	Cadmium	Bone pain, kidney failure
		Lead poisoning	Lead (pipes, industries)	Anemia, brain, and nerve damage
		Arsenicosis	Arsenic in groundwater	Skin lesions, cancer risk
3.	**Chemical/ pesticide-related disorders** (Singh et al., 2025)	Fluorosis	Excess fluoride	Tooth and bone deformities
		Methemo-globinemia (blue baby syndrome)	Nitrate contamination	Reduced oxygen in blood
		Cancer (liver, skin, bladder)	Arsenic, benzene, pesticides	Tumor formation, organ damage
		Endocrine disruption	Industrial chemicals, plasticizers	Hormonal imbalance, fertility issues
4.	**Other disease** (Ali et al., 2021)	Skin diseases	Chemical and microbial pollutants	Dermatitis, rashes, infections
		Eye infections	Poor hygiene, contaminated water	Conjunctivitis, trachoma
		Neurological disorders	Lead, mercury exposure	Cognitive and behavioral problems
		Reproductive issues	Long-term toxin exposure	Birth defects, infertility

1.2 Water pollution and its sources

Water pollution can be classified in several ways based on the nature and origin of the contaminants.

1. **Based on the source of contamination**, water pollution is divided into two main categories, namely, point and nonpoint sources. Point sources are defined as pollutants that originate from a single site, and their location

is also identifiable. A common example of a point source is the release of industrial effluents into water bodies. Nonpoint sources are defined as water sources contaminated by multiple sources. Examples of nonpoint sources include runoff water that flows across urban areas, roads, and agricultural fields after rainfall before entering rivers and lakes.

2. **Based on the nature of the contaminants**, water pollution is divided into natural and anthropogenic/manmade. Natural pollution is defined as pollution caused by geological formations, low-quality surface water, the presence of shallow aquifers, sea water intrusion, and the effects of geothermal fluids, which result in natural ground water pollution. Anthropogenic pollution is mainly caused by human activities such as the discharge of industrial effluents, poorly constructed wells, the excessive use of agricultural fertilizers and pesticides, and mining operations.
3. **Nitrate (NO_3^-)** enters water bodies due to domestic sewage and animal excreta, as microbial activities break down organic nitrogen molecules ($R\text{-}NH_2$). Microbial enzymes convert organic nitrogen into ammonia, which then undergoes oxidation processes. Farmers commonly use nitrogen-based fertilizers because nitrogen helps increase crop yield. However, there is often an imbalance between plant absorption and nitrate availability. When fertilizers are applied during periods of low crop density or at times unfavorable for plant absorption, the excess nitrate is not utilized. This excess nitrate evaporates and later seeps into groundwater, causing contamination (Lisco et al., 2022). Anthropogenic activities exacerbate nitrate pollution, including the discharge of municipal wastewater, overuse of nitrogen-based fertilizers, urban runoff, industrial effluent discharge, landfilling, and the presence of soluble nitrogen compounds in geological formations (Fendall and Sewell, 2009). In addition, the dung and urine of buffaloes and cattle, combined with the decomposition of organic matter (Wright et al., 2013), are considered major sources of groundwater contamination (Ahamad et al., 2018; Taneja et al., 2019).

2 Global and regional impact of water contamination

Based on the UNESCO World Water Development Report (2021), approximately 829,000 deaths occur annually from diarrheal diseases due to the consumption of contaminated water, poor hand hygiene, and inadequate sanitation. Contaminated water has an adverse impact on human health. Over 30,000 deaths occur in children under the age of 5 years, which accounts for 5.3% of all age groups. This is due to the lack of adequate water and sanitation facilities and diseases such as cholera, helminth infections, trachoma, and schistosomiasis. Various health issues, such as respiratory diseases, diarrheal diseases, neurological disorders, and cardiovascular diseases, are caused by the consumption of contaminated water. In both developed and developing countries, the major causes of infectious diseases are heavy rainfall, floods, and adverse climatic conditions. Studies from less economically developed countries show a relationship between the consumption of contaminated water and cholera (Ullah et al., 2014; Ahmad et al., 2014).

3 Major sources of water pollution

3.1 Industrial discharge (heavy metals, chemicals, pharmaceuticals)

Among all industries, manufacturing companies or units play an important role in environmental pollution, as many of them dispose of their untreated effluents directly into lakes and rivers (Bertram et al., 2022). Wastewater contains numerous contaminants, including microbial pathogens, pharmaceutical residues, and toxic metals, which eventually end up in surface waters. These pollutants decrease the recreational quality of the water and also cause long-term ecological harm. Over time, they reduce the availability of freshwater, increase the cost of purification, harm aquatic life, and disrupt food resources (Lisco et al., 2022). Moreover, water quality is degraded by heavy metals, agricultural chemicals, industrial wastes, and dye residues. The continuous discharge of untreated water leads to the loss of aquatic life and an increase in diseases caused by contaminated water, such as cholera and diarrhea, which pose serious risks to human health (Afzal et al., 2018; Edokpayi et al., 2017). Micropollutants and organic contaminants enter the environment through municipal wastewater. A significant number of these substances originate from personal care products and pharmaceuticals, which include sedatives, antibiotics, pain relievers, antiinflammatory drugs, antidepressants, contraceptives, and medicinal formulations. Every year, pharmaceutical industries worldwide introduce new drugs, with more than 300 new products launched annually. Research and development in this sector represent 8% of global investment, reflecting innovation and growth. Globally, the pharmaceutical substance market is valued at approximately US$773 billion, with an annual production volume of 1000 tons. In the United States, per capita pharmaceutical sales average US$676 per person annually (Kummerer, 2013). Even at very low concentrations, lead is highly toxic and causes serious health issues. Human exposure to lead occurs in various ways, such as consuming contaminated food, eating fish from polluted water, and inhaling dust containing lead. Lead exposure is also common through drinking water, particularly in older residences with lead pipes, which increases the risk of metal contamination.

Although the ingested lead is eliminated in urine, the element accumulates in the body over time, which causes risks in children. Prolonged exposure to high amounts of lead can cause severe damage to the kidneys, brain, and central nervous system. In some serious cases, it may even be fatal (USGAO, 2000). Long-term exposure can lead to difficulties in learning, poor memory, high blood pressure, hearing loss, headaches, digestive issues, reproductive disorders, musculoskeletal pain, and slow physical growth. Multiple studies have proven how dangerous lead is, particularly for children, for whom lead poisoning remains one of the biggest health risks. Beyond stunted growth, neurological impairment, and learning disabilities, lead exposure in children has also been associated with increased rates of criminal and antisocial behavior (Kumar et al., 2013). These lasting effects underscore the urgent need for effective measures to prevent and mitigate lead contamination (Kumar et al., 2013).

3.2 Agricultural runoff (pesticides, fertilizers, livestock waste)

There is an urgent need to generate extra agricultural output since everyone needs food to exist, including nonvegetarian items such as chicken, fish, and eggs, as well as vegetables, lentils, beans, and other legumes. Artificial growth-promoting techniques are employed, and as a result of these agricultural endeavors, pollution has escalated at a startling rate in recent years. Water is most negatively impacted by this. Since less than 1% of the world's water is fresh, and if it is polluted, water shortages will arise at the same time that food demands are met. Contamination of agricultural water is a serious issue that affects ecosystems and societies throughout (Afzal et al., 2018). Since over 70% of the freshwater on Earth is used for agriculture, we need to address the origins and sources of this issue. The many sources of agricultural water contamination and the underlying mechanisms that contribute to its persistence will be examined in this essay. Both point and nonpoint sources cause the contamination of the agricultural water supply. Field runoff, concentrated animal feeding operations (CAFOs), and animal livestock are easy to identify because they are part of point sources. Nonpoint sources, such as pesticides and fertilizers from agricultural fields, are not identified easily (Madhav et al., 2018b). Agricultural practices have a tendency to contaminate the water supply through both point and nonpoint sources in both developed and developing countries. In developed countries, large-scale animal livestock and CAFOs release nutrients and chemicals (Akhtar et al., 2022).

3.3 Domestic sewage and wastewater

Nowadays, wastewater and drinking water both cause serious health issues because untreated sewage is released into natural water sources and water supply systems. Serious health problems are increasingly observed due to contaminated water. Moreover, there is little funding and emphasis given to problems related to wastewater treatment, appropriate access to potable water, sanitation, and the implementation of stringent water safety and conservation regulations. These health problems can be mitigated with better sanitation, a safe and sufficient water supply, a sewage line connected to each home, and appropriate waste management and treatment. By 2015, the World Health Organization (WHO) and the United Nations Children's Fund set the Millennium Development Goal (MDG) to assist people in gaining access to basic sanitation facilities and clean drinking water. In addition, since they lack direct economic relevance, sewage treatment and other environmental concerns are not prioritized in developing nations compared to other developmental challenges such as food supply, health, education, and industrial growth. Therefore it is not practical for underdeveloped nations to build extremely expensive and sophisticated treatment technologies. The time has come to create a sewage treatment system that is acceptable, affordable, sustainable, and both technically and financially viable. Determining the best treatment technology for developing nations can be challenging at times because there are significant concerns about selection. This is because low cost is not the only criterion that must be considered; removal efficiency, low operation and maintenance costs, and low sludge withdrawal of the system are also important factors (Schwarzenbach et al., 2010).

3.4 Plastic and microplastic contamination

To learn more about how microplastics (MPs) affect human health, two study papers were reviewed. To determine the Global Average Rate of Microplastic Ingestion, a study (Yu et al., 2024) examined the presence of MPs in a variety of settings, including terrestrial and marine ecosystems, and analyzed diverse organisms. The results of the study demonstrated that the number of MPs consumed depended on several variables, including particle size, shape, type of polymer, and the distribution of particle sizes. The findings showed that, globally, groundwater is the primary source of the 11,845–193,200 MPs that people may ingest each year. The study highlighted the potential risks that the continued usage of MPs might pose to human health. It also revealed that the average weekly consumption of MPs varied from 0.1 to 0.5 g, depending on several factors, such as human age, size, demographics, customs, location, environmental development, and lifestyle choices. The Changjiang Estuary and the nearby East China Sea were the focus of different research (Lin et al., 2022), which assessed the amount and composition of MPs in collected samples. The study concluded that the study location was significantly polluted after evaluating the MPs' pollution load index (Senathirajah et al., 2021).

3.5 Natural sources (arsenic, fluoride, radioactive materials)

Although atomic energy is the foundation of clean air and carbon-free electricity, radioactive ion contamination from sources such as uranium mining tailings and heap leakage remnants (like 226 Ra ions), nuclear fusion derivatives (like 90 Sr), nuclear weapon testing, medical research, and leaks or unintentional explosions in nuclear reactors can have long-term effects that pose a serious risk to human health. Radioactive contamination in water is caused by industrial effluents and nuclear power stations. A tiny amount of the enormous energy created during the fission process can harm human cells, tissues, and organs because it travels in waves. It is impossible to neutralize these dangerous waves chemically, physically, or biologically. Depending on the element's half-life, radioactive impacts might linger in the environment for a long time. All long-term geological deposits, including radioactive minerals, are exposed to geological agents during weathering and erosion, and over time, radioactive elements become mobilized. The mineral composition, ambient factors, and geological agents all affect the rate of weathering. The lowest amount of radioactive elements present in any groundwater is determined by the local geological rock and layers. Water contains both natural and manufactured radionuclides because of their decay products and radius size. Examples of radioactive elements that generate low radiation energy are carbon-14 (14 C), tritium (3 H), potassium-40 (40 K), and rubidium-87 (87 Rb). Radium-226 (226 Ra), the daughters of radium-228 (228 Ra), polonium-210 (210 Po), uranium (U), thorium (Th), radon-220 (220 Rn), and radon-222 (222 Rn) are alpha particles that release high radiation energy and are also found in trace amounts in water. The pertechnetate (TcO 42 −) ion is one of the oxo-anions that has drawn the most interest since the technetium isotope (99 Tc) is a radioactive element with an extremely long half-life (2.1×10^5 years). 99 Tc is a by-product in the fission sector (Samanta et al., 2018; Salbu and Krekling, 1998).

4 Types of waterborne contaminants

4.1 Biological contaminants (bacteria, viruses, protozoa, parasites)

4.1.1 Bacteria that cause disease

By generating poisonous compounds that might harm human tissues, acting as parasites inside human cells, or establishing colonies in the body that could interfere with regular human processes, pathogenic bacteria have the capacity to infect or sicken people. Numerous pathogenic bacteria, such as *Salmonella typhi*, *Escherichia coli*, and *Vibrio cholerae*, may be found in water and can cause cholera, typhoid, diarrhea, and other waterborne illnesses. Water-transmitted viruses include enteroviruses, such as coxsackieviruses and polioviruses, rotavirus, norovirus, hepatitis A and E viruses, adenovirus, and astrovirus. The majority of the time, these viruses can cause gastroenteritis, which can result in diarrhea and other symptoms, including fever, vomiting, and cramping in the abdomen. However, some of the same viruses may cause more serious illnesses, such as hepatitis (hepatitis A and E viruses), cancer (polyomavirus), myocarditis (enteroviruses), encephalitis, and meningitis (WHO, 2011). The majority of these virus-based illnesses occur in developing nations, as many of these nations may be dealing with high rates of HIV infection and widespread malnutrition (Gall et al., 2015).

Humans can contract parasites through a variety of routes, such as by drinking contaminated water. Every year, these parasites cause 842,000 fatalities (Omarova et al., 2018). Giardiasis may be caused by *Giardia intestinalis*, also known as *Giardia duodenalis* and *Giardia lamblia*. In industrialized nations, it affects around 2% of adults and 8% of children, whereas in underdeveloped nations, it affects over 33% of the population. Humans are primarily exposed to it through the fecal–oral pathway, typically via contaminated water (fecal pollution) (Omarova et al., 2018).

4.2 Chemical pollutants (heavy metals, pesticides, endocrine disruptors)

Heavy metals are a class of metals and metalloids that are widely employed in consumer products and industrial production. They are defined as elements with a density greater than 5 g/cm^3. Because of the existing low level of industrial manufacture, poor recycling rates, informal dismantling operations, and insufficient and uneven regional socioeconomic conditions, they are regarded as one of the main environmental pollutants. Stated differently, heavy metals have the ability to persist in environmental media, including soil, water, sediment, dust, and air (Alrumman et al., 2016). Notably, heavy metals are not biodegradable in the natural environment, meaning they often cannot be transformed into less hazardous by-products, in contrast to organic pollutants. In addition, these metals may be biologically amplified, increasing their concentration thousands of times.

Endocrine-disrupting chemicals (EDCs) are exogenous compounds that can affect the production and metabolism of endogenous hormones by mimicking them through similar structures or actions. They can alter the way the endocrine system functions, interacting with hormone receptors such as thyroid hormone receptors and sex hormone receptors (Bertram et al., 2022), primarily estrogen receptors, androgen receptors, and progestational receptors, and can have negative health effects by interfering with or counteracting endocrine hormonal effects (Fig. 15.1) (Taneja et al., 2019; Ullah et al., 2014; Ahmad et al., 2014).

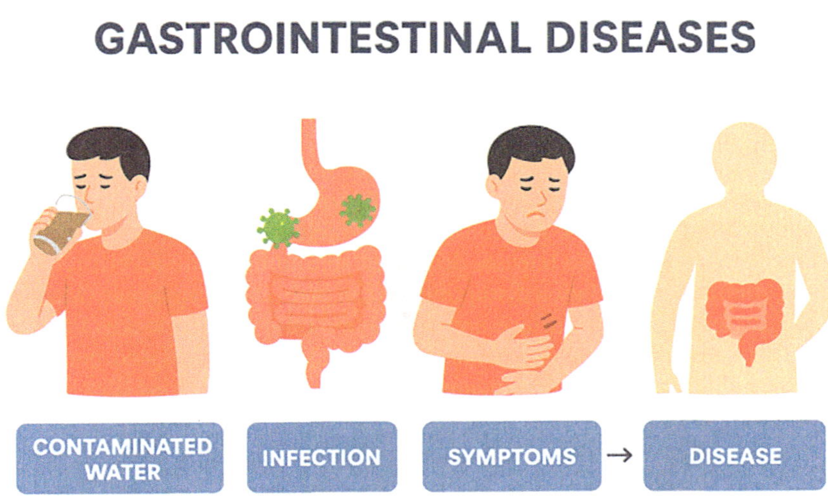

FIG. 15.1 Pathway of gastrointestinal diseases. This figure illustrates the pathways of gastrointestinal disease development.

These chemicals can act as ligands, binding to specific hormone receptors, which then attach to response elements in target genes, regulating gene expression and causing undesirable side effects. In addition, other glands within the endocrine system produce and release hormones. If these exogenous EDCs disrupt the production, transport, or metabolism of natural hormones in the human body, hormone homeostasis cannot be maintained. Furthermore, heavy metals, pesticides, plasticizers, medications, personal care products, food items, and food packaging are just a few examples of common substances that contain EDCs. It is widely recognized that the endocrine system is essential for regulating metabolic processes, maintaining the volume and composition of extracellular fluid, controlling growth and development, modulating reproductive function, and preserving endocrine homeostasis (resistance and adaptability) (Lisco et al., 2022).

4.3 Physical pollutants (suspended solids, MPs)

The quantity of inorganic and organic material held in the water column of a lake, pond, river, canal, or dam due to instability is referred to as suspended solids (SS). Particulate matter with a diameter of <62 μm is typically seen in SS (95). Four important elements determine how SS affects aquatic biota. These include: (1) the amount of SS, (2) the duration of interaction with the amount of SS, (3) the geochemistry of SS, and (4) the distribution of SS particle sizes. When suspended particles are present, they can block sunlight from penetrating the water, which is essential for benthic flora to perform photosynthesis. Fish gill damage caused by finer suspended particles might result in asphyxiation. MPs are microscopic pieces of plastic that are <5 mm long and can harm marine life and our oceans. MPs can be categorized according to their source. Primary and secondary are the two main types. Secondary MPs are tiny fragments of plastic that break off from larger ones, whereas primary MPs are intentionally manufactured to be that small. MPs are problematic because they are so small that water filtration systems typically fail to remove them, allowing them to enter rivers and seas. The fact that fish and other aquatic species consume MPs and either die or suffer from health issues makes them particularly harmful (Wright et al., 2013). If consumed, MPs pose a physical threat to living organisms. The consequences include digestive tract blockage, delayed stomach enzyme release, decreased steroid hormone levels, reduced eating stimulation, ovulation disruption, and an inability to reproduce (Madhav et al., 2018a). The high surface area-to-volume ratio of MPs makes them an effective vehicle for harmful pollutants, such as trace metals and persistent organic pollutants, which also endanger human health and marine life (Fendall and Sewell, 2009).

4.4 Human health ailments linked to polluted water

4.4.1 *Gastrointestinal diseases (cholera, dysentery, typhoid, hepatitis A and E)*

Water contamination poses one of the most pressing public health challenges of our time, affecting millions of individuals worldwide through diverse pathways of exposure. The relationship between contaminated water and human health issues causes multiple diseases, ranging from acute gastrointestinal infections to chronic systemic diseases that appear years after the initial exposure. Long-term exposure to pollutants causes chronic health issues related to water contamination, which have been analyzed through comprehensive studies.

4.4.2 *Gastrointestinal diseases and waterborne pathogens*

The most immediate and widespread health consequences of polluted water consumption manifest through gastrointestinal disorders. Contaminated water serves as a primary transmission vehicle for numerous pathogenic microorganisms, including bacteria, viruses, and parasites that cause severe digestive system infections (Wolf et al., 2023; Morales-Novelo et al., 2024). Conservative estimates place diarrhea among the top five causes of death worldwide, with 3.5–5 million deaths occurring every year, most of which are among young children in nonindustrialized countries (Jain et al., 2013). According to the WHO, diarrheal diseases account for 4.1% of the total disability-adjusted life years in the global burden of disease, 88% of which is attributable to unsafe water supply, sanitation, and hygiene (Wolf et al., 2023).

Fig. 15.1 shows how a person becomes infected with gastrointestinal diseases through polluted water. It begins with a person drinking contaminated water containing harmful microorganisms. The second stage shows the pathogens entering the stomach and intestines. Then, the onset of symptoms occurs in the third stage, such as pain, diarrhea, vomiting, and discomfort. The final stage represents the development of full gastrointestinal disease if the infection is not treated.

Cholera represents one of the most devastating waterborne diseases, caused by the bacterium *V. cholerae,* which thrives in contaminated water supplies. This acute intestinal infection produces severe watery diarrhea, vomiting, and rapid dehydration that can prove fatal within hours if left untreated (Morens et al., 1979; Schroeder et al., 1968). A waterborne outbreak of gastroenteritis analysis revealed that symptoms included vomiting (81%), diarrhea (65%), and fever (49%), with a median duration of illness of 24 h, and attack rates that increased with the consumption of contaminated water (Morens et al., 1979). The disease spreads rapidly in communities with inadequate sanitation infrastructure and limited access to clean water, making it particularly dangerous in developing regions and emergency situations.

Dysentery, another significant gastrointestinal ailment linked to water pollution, results from bacterial infection by Shigella species or parasitic infection by *Entamoeba histolytica* (The Lancet Gastroenterology & Hepatology, 2022). The major cause of waterborne diseases is poor sanitation and hygiene practices, which lead to diseases like cholera, dysentery, hepatitis A and E, polio, and giardiasis (The Lancet Gastroenterology & Hepatology, 2022). Cholera and diarrhea cause severe abdominal cramps, dehydration, bloody diarrhea, and fever. If medication is not provided on time, it may lead to long-term health effects such as intestinal perforation and systemic sepsis (Table 15.2).

TABLE 15.2 Water pollutants and their impacts on human health.

Health impact category	Specific ailments/diseases	Primary pollutants/contaminants	Mechanism of impact	Key human health effects
Gastrointestinal diseases	Cholera, dysentery, typhoid fever, hepatitis A, and hepatitis E	Bacteria, viruses, protozoa (fecal contamination)	Pathogens invade the GI tract/liver after ingestion	Severe diarrhea, vomiting, dehydration, and liver inflammation
Neurological disorders (heavy metals)	Lead poisoning, mercury toxicity, arsenic neurotoxicity, cadmium poisoning	Lead, mercury, arsenic, cadmium	Heavy metals cross the blood–brain barrier and disrupt nerve signaling	Memory loss, cognitive impairment, tremors, and developmental delays
Reproductive and developmental impacts	Infertility, miscarriages, birth defects, low birth weight	Endocrine disruptors, pesticides, nitrates, heavy metals	Hormonal disruption and fetal development interference	Reduced fertility, miscarriages, congenital defects
Cancer risks	Liver, bladder, stomach, skin cancers	Arsenic, benzene, PFAS, pesticides, industrial solvents	Long-term exposure causes DNA mutations	Increased cancer risk
Dermatological and organ system damage	Dermatitis, eczema, fluorosis, liver/kidney toxicity	Fluoride, arsenic, chemicals, industrial pollutants	Chemical absorption and organ accumulation	Skin lesions, bone weakness, liver/kidney failure
Antibiotic resistance	Drug-resistant infections	Antibiotic residues, resistant bacteria	Resistance genes spread through water	Hard-to-treat infections, longer illness
Pharmaceutical contamination	Hormonal imbalance, metabolic disorders	Hormones, antidepressants, painkillers, Antiretroviral Drugs (ARVs)	Drugs disrupt the endocrine and nervous systems	Reproductive issues, behavioral changes

4.5 Neurological disorders from heavy metal contamination

Heavy metal contamination in water sources poses severe neurological health risks, particularly affecting developing nervous systems in children and pregnant women. The most harmful neurotoxic metals found in contaminated water supplies are lead, mercury, and arsenic, each one of them causing different neurological diseases through different mechanisms of action (Tchounwou et al., 2012; Yu et al., 2024). These heavy metals exhibit neurotoxic effects by altering the epigenome and inducing epigenetic changes, which suggests that these heavy metals have the potential to cause permanent neurological damage (Yu et al., 2024).

The exposure to lead in drinking water causes neurological disorders, mostly in developing children, where it leads to behavioral disorders, learning disabilities, and severe cognitive damage (Tchounwou et al., 2012). Chromosomal abnormalities, DNA damage, and alterations in DNA replication and transcription are induced by heavy metals like lead. These metals interfere with synaptic transmission and promote inflammatory responses in the central nervous system. Even low-level chronic exposure to lead can reduce intelligence quotient scores, impair working memory, and compromise executive function in children.

Mercury toxicity through water contamination, historically demonstrated in cases such as Minamata disease, causes severe neurological symptoms, including ataxia, muscle weakness, numbness in extremities, and, in severe cases, coma and death. Mercury is a highly toxic metal found in air, water, and soil, considered to be highly carcinogenic by the EPA, and exposure can result in various health problems, including Alzheimer's disease, lung damage, and skin issues. Mercury compounds readily cross the blood-brain barrier and accumulate in neural tissues, where they disrupt cellular metabolism and promote oxidative stress that damages neurons and glial cells.

4.6 Reproductive and developmental health impact

Water contamination significantly threatens reproductive health through exposure to EDCs and toxic substances that interfere with normal hormonal function and fetal development. Nitrates, commonly found in agricultural runoff and groundwater contamination, represent a major reproductive health hazard with documented associations with adverse birth outcomes (Picetti et al., 2022). Pollution of water sources, largely from wide-scale agricultural fertilizer use, has resulted in nitrate and nitrite contamination of drinking water, with potentially serious effects on human health (Picetti et al., 2022).

The exposure to nitrate in drinking water causes various health issues. According to a systematic review, meta-analysis, and epidemiological research, a relationship was found between stomach cancer and nitrate exposure. The researchers identified this relationship in the meta-analysis, with an odds ratio of 1.91 for every 10 mg/L increase in nitrate ion concentration (Picetti et al., 2022). Nitrate has a carcinogenic risk that is concentration-dependent, which increases significantly when the level of nitrate in drinking water exceeds the specific limit.

During pregnancy, exposure to nitrates causes neural tube defects. The consumption of nitrate-contaminated water increases the risk of spina bifida and abnormalities in the population. During the early stages of pregnancy, during the formation of the neural tube, before a woman is aware that she is pregnant, ensuring water quality improvement is important.

4.7 Cancer risk and carcinogenic exposure

In contaminated water, prolonged exposure to carcinogens increases the risk of cancer in multiple organs. Arsenic represents the most well-documented water-related carcinogen, with research showing that prolonged exposure to certain water contaminants can increase the risk of specific cancers (Lin et al., 2022; Jiang et al., 2024). Water contamination is a critical public health issue that has been linked to various adverse health outcomes, including cancer.

Arsenic exposure through drinking water demonstrates strong dose–response relationships with several cancer types, including skin, lung, bladder, kidney, and liver cancers. According to WHO statistics, the number of cancer patients diagnosed in 2020 reached 19.3 million, while the number of deaths from cancer increased to 10 million, with one-fifth of the global population expected to develop cancer during their lifetime (Lin et al., 2022). Ingestion of arsenic from drinking water can cause skin cancer and kidney and bladder cancer, with the risk of cancer in the population from arsenic in the United States water supply being comparable to the risk from tobacco smoke and radon in the home environment (Lin et al., 2022).

Bladder cancer shows particularly strong associations with arsenic in drinking water, with positive correlations found between bladder cancer incidence and various water contaminants, including arsenic. A high association of arsenic in drinking water with lung cancer was demonstrated in northern Chilean controlled studies, with research showing a synergistic effect of smoking and arsenic intake in drinking water in causing lung cancer (Lin et al., 2022).

4.8 Dermatological and organ system damage

Heavy metal contamination in water sources causes extensive damage to the skin and vital organs through multiple mechanisms of toxicity. Arsenic exposure produces characteristic dermatological effects that often serve as early indicators of chronic poisoning (Singh et al., 2024). Toxic heavy metal ions can be introduced into the water through several point and nonpoint sources, including industrial activity, with these toxic heavy metals posing threats to both humans and animals when they infiltrate water and soil (Singh et al., 2024).

Chronic arsenic exposure typically manifests first through skin changes, including hyper pigmentation, hyperkeratosis (thickening of the skin on palms and soles), and the development of skin lesions that may progress to skin cancer. A groundbreaking study in the Gangetic plains of Bihar, India, elucidated for the first time the catastrophic effect of manganese causing cancer, with cancer patients' blood samples showing high contamination and their household hand pump water samples also containing elevated metal contamination. After approximately five years of exposure, the dermatological effects appear and persist with prolonged exposure. The cell proliferation and differentiation are abnormal because arsenic interferes with cellular metabolism and keratinocyte function, leading to the observed changes in the skin.

The skin damage or systemic organ damage is caused by heavy metals such as lead, mercury, and cadmium. The exposure to mercury causes skin discoloration, contact dermatitis, and acrodynia or pink disease, which is characterized by painful discoloration of the feet and hands. Various health issues are caused by heavy metals, including skin problems, renal failure, liver failure, cancer, and many more (Singh et al., 2024). The exposure to lead causes various skin issues and also worsens dermatological conditions.

4.9 Antibiotic resistance and pharmaceutical contamination

The residue of pharmaceuticals in water systems is a threat to public health, contributing to the global crisis of antibiotic resistance. The major global health concern is antimicrobial resistance (AMR), which arises from the mismanagement and misuse of antibiotics in veterinary and clinical settings, leading to greater expenses, longer treatment durations, and higher death rates (Cheung et al., 2025). The environment plays an important role as a transmitter of AMR, with bacterial antibiotic resistance in recent years being one of the major public health issues.

In water sources, antibiotic compounds enter in various ways, such as wastewater from pharmaceutical industries and hospitals, agricultural runoff from livestock operations, and incomplete removal by traditional water treatment systems (Xi et al., 2009; Liguori et al., 2022). Wastewater treatment plants have been identified as significant contributors, releasing large amounts of AMR bacteria into the environment. Hospital wastewater represents a particularly concentrated source of antibiotic contamination, containing high levels of clinically relevant antibiotics along with antibiotic-resistant bacteria and resistance genes.

5 Case studies and epidemiological evidence

5.1 Notable water contamination incidents and their health impacts

Disease prevalence is classified at several levels in epidemiology; community diseases can be classified as either epidemic or endemic (CDC, 2012). The term "endemic disease" describes a condition that is persistently present or prevalent in a particular demographic and/or geographic area. The comparatively steady and predictable yearly incidence of chickenpox (varicella zoster virus) among children in the United States, UK, and Canada is an example of an endemic disease (Seward et al., 2008). The terms "endemic" and "sporadic" are sometimes used interchangeably to characterize sickness levels; however, this is not always true because "sporadic" usually refers to low infection rates that are neither steady nor predictable. Sudden increases in illness exceeding local endemic levels are referred to as epidemic diseases. Globally, endemic, sporadic, and pandemic levels of disease are linked to groundwater sources, according to a review of the available research. Because outbreak-associated diseases are monitored and reported more often within healthcare systems, they are more commonly connected to the source or organisms that caused them. Since cases of endemic and sporadic diseases linked to groundwater supplies are, by definition, isolated and more likely to be associated with healthy adults (i.e., mild gastrointestinal symptoms with low levels or absence of associated sequelae), they are usually much harder to track, leading to high rates of underreporting and/or underdiagnoses, as well as frequent self-medication (Thomas et al., 2013).

5.2 Long-term studies linking water pollution to chronic diseases

Irregularities in the circadian cycle: As of now, six main groups of environmental substances have been shown to interfere with circadian rhythms: metals, biocides and pesticides, neuroactive medications, polychlorinated biphenyls, steroid hormones, and other substances, including bisphenol A (45) and cyanobacterial toxins. By oxidizing essential thiol groups in the key circadian proteins cryptochrome and period circadian protein homolog 1, as well as by redox-regulating factors that modulate the activity of transcription factors such as CLOCK (circadian locomotor output cycles protein kaput) and BMAL1 (basic helix–loop–helix ARNT-like protein 1), such as sirtuins, heme oxygenase, and hypoxia-inducible factor 1α (46,4), these environmental compounds cause dysregulation of circadian rhythms. This dysregulation can lead to cardiovascular and other health consequences. Disease can result from exposure to inorganic pollutants, such as metalloids and trace metals, which are both natural and man-made. Likewise, natural and artificial organic pollutants, such as volatile organic compounds, polychlorinated biphenyls, and Polycyclic Aromatic Hydrocarbons (PAHs), present a number of health hazards, including cancer, neurological diseases, and reproductive issues. Many of the more than 300,000 synthetic compounds that humans have created in the last 50 years have been shown to have cardiotoxic effects. Numerous of these compounds are present in consumer goods, exposing the general public to them on a large scale (Omarova et al., 2018). This is exacerbated by the absence of strict regulations in the chemical manufacturing sector and the inability of several governments to implement health-protective measures (Drwal et al., 2019).

6 Strategies for prevention and mitigation

6.1 Water purification and treatment technologies (filtration, chlorination, UV treatment, reverse osmosis)

Osmosis: It is the process by which water or a solvent diffuses passively or spontaneously across a semipermeable membrane as a result of osmotic pressure. A semipermeable membrane allows liquid to pass through from a diluted to a more concentrated solution. When osmosis occurs, a lower-concentration solution or water molecules will "filter" or tend to move toward the higher-concentration solution without applying pressure across a membrane, diluting it until equilibrium is reached. By creating equal osmotic pressure and balancing the solute concentrations on both sides of the membrane, solvent mobility lowers the system's free energy. Reverse osmosis (RO): In the RO process, the osmotic flow is inverted. Solvents are forced to filter through the membrane against a pressure gradient into the lower-concentration solution by applying hydraulic pressure to the solution's high-concentration side. In RO, the intrinsic osmotic pressure, a thermodynamic parameter, is overcome by applying pressure to a solution via one side of the semipermeable membrane using a mechanical pump. In addition, the procedure eliminates soluble and particulate matter from the solution of interest, including salt from seawater in desalination. There are still many conventional methods for purifying water. Boiling and chlorination are two of the numerous methods that are well-known and in use around the world. Boiling water eliminates pathogens, iron, turbidity, taste, odors, and other harmful substances. It guarantees the water's safety from a microbiological perspective. In addition, boiling ensures pathogen-free water, which increases the water's microbiological safety. In many areas, boiling is considered the industry standard due to its ease of use and 100% effectiveness. Long-term boiling of water can lower its iron content, which will improve its taste and make it less turbid. When using a fabric-based model, solar energy is used to purify the water through a cotton fabric towel (Genter et al., 2022).

6.2 Policy and regulatory frameworks (WHO guidelines, national water safety laws)

The provinces and territories of Canada control the quality of drinking water for the majority of their communities. Since the Government of Canada is the only body with the power to enact laws for First Nations communities, provincial laws and regulations do not apply to them. As of right now, First Nations communities are not subject to any laws controlling the purity of their drinking water. The Office of the Auditor General of Canada (OAG) (2011) notes in audits conducted in 2005 and 2011 that, because provincial laws and regulations are not implemented on reserves, First Nations communities do not enjoy the same level of drinking water protection as those who live off of them (p. 15).

The absence of drinking water regulations for First Nations communities will be the policy issue this project identifies.

To address drinking water issues in First Nations communities, the Minister of Aboriginal Affairs and Northern Development introduced the Safe Drinking Water for First Nations Act (Bill S-8). In addition to sharing joint authority with the Minister of Aboriginal Affairs and Northern Development over monitoring, reporting, testing frequencies, drinking water advisories, and specific emergency measures in the event of water contamination on First Nations lands, the Act, which was passed into law on June 19, 2013, grants the Minister of Health the sole authority to create regulations pertaining to drinking water quality standards. According to Aboriginal Affairs and Northern Development Canada (AANDC, 2013), the Act is intended to serve as enabling legislation for a regulatory framework (Morales-Novelo et al., 2024).

6.3 Public awareness and community-driven initiatives

In Pakistan, community-driven development is not a novel strategy. In the Karakoram Mountain Range in Northern Pakistan, the Aga Khan Rural Support Program (AKRSP) encouraged the formation of community organizations that promote saving in 1982. This program addresses infrastructure, savings mobilization, skill development, and institutional development. In general, the participatory method has proven effective in South Asia for managing water resources, particularly for rural irrigation (Shah, 2009). Governments and nongovernmental organizations were drawn to self-governing "irrigation communities" that embraced the communitarian ideal, particularly when public tube well projects began to wane in the 1970s and 1980s. A digital version is accessible at: https://ssrn.com/abstract=1464690. An alternative to the unreliability of government institutions and politics, which frequently resulted in the failure of infrastructure construction and maintenance, was the participatory development strategy. Based on their experiences in rural Pakistan (Schwarzenbach et al., 2010), the World Bank recommended in July 1994 that community resources be used for financing schemes and user management prior to the Asian Development Bank (ADB)-funded Punjab Community Water Supply and Sanitation Project. According to case studies in Pakistan, women are crucial in determining the success of projects, and in Punjab, there must be adequate training for Sanitation Promoters (16), local involvement, and reasonably priced technologies (Samanta et al., 2018). This supported the conclusion drawn from the ADB's initial phase project in 2002, which was that the project needed to include efficient training initiatives for community organizations. However, the focus of these discussions remained on promoting more user involvement to effectively meet community needs.

7 Conclusion

Polluted water poses a serious threat to human health, contributing to a wide range of ailments such as gastrointestinal infections, neurological disorders, reproductive issues, and cancers. Heavy metals, pathogens, pesticides, and industrial chemicals are contaminants that disrupt vital biological functions, especially in susceptible populations. Addressing these issues requires efforts that include public health awareness and education, wastewater treatment, and strict environmental regulations. Access to safe and clean water is not only a fundamental human right but also essential for global welfare, sustainable development, and protecting future generations from diseases caused by waterborne contaminants.

References

Afzal, M.S., Ashraf, A., Nabeel, M., 2018. Characterization of industrial effluents and groundwater of Hattar industrial estate, Haripur. Adv. Agric. Environ. Sci. Open Access. 1 (2), 70–77.

Ahamad, A., Madhav, S., Singh, P., Pandey, J., Khan, A.H., 2018. Assessment of groundwater quality with special emphasis on nitrate contamination in parts of Varanasi City, Uttar Pradesh, India. Appl. Water Sci. 8 (4), 115.

Ahmad, S.M., Yusafzai, F., Bari, T., 2014. Assessment of heavy metals in surface water of River Panjkora Dir Lower, KPK Pakistan. J. Biol. Environ. Sci. 5, 144–152.

Akhtar, A., Hanif, M.A., Rashid, U., Bhatti, I.A., Alharthi, F.A., Kazerooni, E.A., 2022. Advanced treatment of direct dye wastewater using novel composites produced from hoshanar and sunny grey waste. Separations. 22978739. 9 (12). Available from: www.mdpi.com/journal/separations. doi:10.3390/separations9120425.

Ali, S., Amir, S., Ali, S., Rehman, M.U., Majid, S., Yatoo, A.M., 2021. Water pollution: diseases and health impacts. In: Dar, G.H., Hakeem, K.R., Mehmood, H., Qadri, M.A. (Eds.), Freshwater Pollution and Aquatic Ecosystems. Apple Academic Press, Boca Raton, FL, pp. 1–23.

Alrumman, S.A., El-kott, A.F., Kehsk, M.A., 2016. Water pollution: source and treatment. Am. J. Environ. Eng. 6 (3), 88–98.

Bertram, M.G., Gore, A.C., Tyler, C.R., Brodin, T., 2022. Endocrine-disrupting chemicals. Curr. Biol. 32, R727–R730.

Bibi, S., Khan, R.L., Nazir, R., et al., 2016. Heavy metals in drinking water of LakkiMarwat District, KPK, Pakistan. World Appl. Sci. J. 34 (1), 15–19.

CDC (Centers for Disease Control and Prevention), 2012. Principles of Epidemiology in Public Health Practice: An Introduction to Applied Epidemiology and Biostatistics, third ed. Self-Study Course SS1978. Centers for Disease Control and Prevention, Atlanta, GA, USA.

Cheung, C., Naughton, P.J., Dooley, J.S.G., Corcionivoschi, N., Brooks, C., 2025. The spread of antimicrobial resistance in the aquatic environment from faecal pollution: a scoping review of a multifaceted issue. Environ. Monit. Assess. 197 (4), 467.

Chisolm, Jr., J.J., 1980. Poisoning from heavy metals (mercury, lead, and cadmium). Pediatr. Ann. 9 (12), 458–468.

Drwal, E., Rak, A., Gregoraszczuk, E.L., 2019. Review: polycyclic aromatic hydrocarbons (PAHs)—action on placental function and health risks in future life of newborns. Toxicology. 411, 133–142.

Edokpayi, J.N., Odiyo, J.O., Durowoju, O.S., 2017. Impact of wastewater on surface water quality in developing countries: a case study of South Africa. In: Tutu, H. (Ed.), Water Quality. IntechOpen, London, pp. 401–416.

Fendall, L.S., Sewell, M.A., 2009. Contributing to marine pollution by washing your face: microplastics in facial cleansers. Mar. Pollut. Bull. 58 (8), 1225–1228.

Gall, A.M., Mariñas, B.J., Lu, Y., Shisler, J.L., 2015. Waterborne viruses: a barrier to safe drinking water. PLoS Pathog. 15537374. 11 (6). Available from: https://journals.plos.org/plospathogens/. doi:10.1371/journal.ppat.1004867.

Genter, F., Putri, G.L., Pratama, M.A., Priadi, C., Willetts, J., Foster, T., 2022. Microbial contamination of groundwater self-supply in urban Indonesia: assessment of sanitary and socio-economic risk factors. Water Resour. Res. 0043-1397. 58 (10). doi:10.1029/2021wr031843.

Jiang, J., Zhang, L., Wang, Z., Gu, W., Yang, C., Shen, Y., Zhao, J., Han, W., Hu, Y., Xue, F., Chen, W., 2024. Spatial consistency of co-exposure to air and surface water pollution and cancer in China. Nat. Commun. 15 (1), 7813.

Kumar, R., Gupta, A.K., Chattree, A., Tripathi, R.M., 2013. A review on the detection of heavy metals in water bodies, fish organs, sediment river beds. Int. J. Curr. Res. Rev. 5 (18), 41.

Kummerer, K., 2013. Pharmaceuticals in the Environment: Sources, Fate, Effects and Risks. Springer, Berlin, Heidelberg.

Liguori, R., Mesto, E., Di Onofrio, V., Masi, S., Cesaro, A., 2022. Antibiotic contaminants in the aquatic environment: occurrence, sources, fate, and remediation strategies. Environ. Sci. Pollut. Res. 29, 65718–65743.

Lin, S., Wang, Z., Li, Y., Chen, J., Xu, Q., 2022. Arsenic in drinking water and urinary tract cancers: a systematic review of 30 years of epidemiological evidence. Environ. Health 13, 44. doi:10.1186/1476-069X-13-44.

Lisco, G., Giagulli, V.A., Iovino, M., Guastamacchia, E., De Pergola, G., Triggiani, V., 2022. Endocrine-disrupting chemicals: introduction to the theme. Endocr. Metab. Immune Disord. Drug Targets. 22123873. 22 (7), 677–685. Available from: http://www.eurekaselect.com/636/journal/endocrine-metabolic-amp-immune-disorders-drug-targets. doi:10.2174/1871530321666210413124425.

Madhav, S., Ahamad, A., Kumar, A., Kushawaha, J., Singh, P., Mishra, P.K., 2018a. Geochemical assessment of groundwater quality for its suitability for drinking and irrigation purpose in rural areas of SantRavidas Nagar (Bhadohi), Uttar Pradesh. Geol. Ecol. Landsc. 2 (2), 127–136.

Madhav, S., Ahamad, A., Singh, P., Mishra, P.K., 2018b. A review of textile industry: wet processing, environmental impacts, and effluent treatment methods. Environ. Qual. Manag. 15206483. 27 (3), 31–41. Available from: http://onlinelibrary.wiley.com/journal/10.1002/(ISSN)1520-6483. doi:10.1002/tqem.21538.

Morales-Novelo, J.A., Rodríguez-Tapia, L., Medina-Rivas, C.M., Revollo-Fernández, D.A., 2024. Waterborne gastrointestinal diseases and child mortality: a study of socioeconomic inequality in Mexico. Int. J. Environ. Res. Public Health. 16604601. 21 (11). Available from: http://www.mdpi.com/journal/ijerph. doi:10.3390/ijerph21111399.

Morens, D.M., Holman, R.C., Louria, D.B., 1979. Cholera deaths in the United States, 1965–1978. Ann. Intern. Med. 91 (6), 827–831.

Morris, R.D., Levin, R. (1995). Estimating the incidence of waterborne infectious disease related to drinking water in the United States. In: IAHS Publications-Series of Proceedings and Reports-Intern Assoc Hydrological Sciences, vol. 233. pp. 75–88.

Omarova, A., Tussupova, K., Berndtsson, R., Kalishev, M., Sharapatova, K., 2018. Protozoan parasites in drinking water: a system approach for improved water, sanitation and hygiene in developing countries. Int. J. Environ. Res. Public Health. 1660-4601. 15 (3), 495. doi:10.3390/ijerph15030495.

Picetti, R., De Roda Husman, A.M., Posthuma, L., Evers, E.G., van der Hoek, J.P., 2022. Nitrate and nitrite contamination in drinking water and cancer risk: a systematic review with meta-analysis. Environ. Res. 210, 112988. doi:10.1016/j.envres.2022.112988.

Raju, N.J., Shukla, U.K., Ram, P., 2011. Hydrogeochemistry for the assessment of ground water quality in Varanasi: a fast-urbanizing center in Uttar Pradesh, India. Environ. Monit. Assess. 173 (1–4), 279–300.

Salbu, B., Krekling, T., 1998. Characterisation of radioactive particles in theenvironment. Analyst. 123 (5), 843–850.

Samanta, P., Chandra, P., Dutta, S., Desai, A.V., Ghosh, S.K., 2018. Chemically stable ionic viologen-organic network: an efficient scavenger of toxic oxo-anions from water. Chem. Sci. 20416539. 9 (40), 7874–7881. Available from: http://pubs.rsc.org/en/journals/journal/sc. doi:10.1039/C8SC02456A.

Schroeder, S.A., Woodward, W.E., Gangarosa, E.J., 1968. Clinical and epidemiological features of cholera. J. Am. Med. Assoc. 206 (5), 949–953.

Schwarzenbach, R.P., Egli, T., Hofstetter, T.B., von Gunten, U., Wehrli, B., 2010. Global water pollution and human health. Annu. Rev. Environ. Resour. 35, 109–136.

Senathirajah, K., Attwood, S., Bhagwat, G., Carbery, M., Wilson, S., Palanisami, T., 2021. Estimation of the mass of microplastics ingested—a pivotal first step towards human health risk assessment. J. Hazard. Mater. 03043894. 404 (Pt B), 124004. doi:10.1016/j.jhazmat.2020.124004.

Seward, J.F., Watson, B.M., Peterson, C.L., Mascola, L., Pelosi, J.W., Zhang, J.X., Maupin, T.J., Goldman, G.S., Wharton, M., 2008. Varicella disease after introduction of varicella vaccine in the United States, 1995–2000. JAMA 287 (5), 606–611.

Shah, T., 2009. Taming the Anarchy: Groundwater Governance in South Asia. Resources for the Future Press, Washington, DC; Resources for the Future Press, Washington, DC.

Singh, V., Ahmed, G., Vedika, S., Kumar, P., Chaturvedi, S.K., Rai, S.N., Vamanu, E., Kumar, A., 2024. Toxic heavy metal ions contamination in water and their sustainable reduction by eco-friendly methods: isotherms, thermodynamics and kinetics study. Sci. Rep. 14, 7595.

Singh, V., Srivastava, R.K., Bhatt, A.K., 2025. Effects of water pollution. In: Singh, V., Srivastava, R.K., Bhatt, A.K. (Eds.), Battling Air and Water Pollution: Protecting Our Planet's Vital Resources. Springer Nature Singapore, Singapore, pp. 145–164.

Taneja, P., Labhasetwar, P., Nagarnaik, P., 2019. Nitrate in drinking water and vegetables: intake and risk assessment in rural and urban areas of Nagpur and Bhandara districts of India. Environ. Sci. Pollut. Res. 26 (3), 2026–2037.

Tchounwou, P.B., Yedjou, C.G., Patlolla, A.K., Sutton, D.J., 2012. Heavy metal toxicity and the environment. In: Luch, A. (Ed.), Molecular, Clinical and Environmental Toxicology. Springer, Basel, pp. 133–164.

The Lancet Gastroenterology & Hepatology, 2022. Dysentery and global gastrointestinal infections: continuing challenges in prevention and treatment. Lancet gastroenterol. Lancet Gastroenterol. Hepatol. 7, 1012–1014.

Thomas, M.K., Murray, R., Flockhart, L., Pintar, K., Pollari, F., Fazil, A., et al., 2013. Estimates of the burden of foodborne illness in Canada for 30 specified pathogens and unspecified agents, circa 2006. Foodborne Pathog. Dis. 10 (7), 639–648.

Ullah, S., Javed, M.W., Shafique, M., Khan, S.F., 2014. An integrated approach for quality assessment of drinking water using GIS: a case study of lower dir. J. Himalayan Earth Sci. 23056959. 47 (2), 163–174. Available from: http://nceg.upesh.edu.pk/GeologicalBulletin/Vol-47(2)-2014/Vol-47(2)-2014-Paper12.pdf.

USGAO (United States General Accounting Office), 2000. Lead Contamination: Actions Are Needed to Reduce Environmental Sources of Lead to Protect Children. GAO/HEHS-00-18. United States General Accounting Office, Washington, DC.

WHO (World Health Organization), 2011. Guidelines for Drinking-water Quality, fourth ed. World Health Organization, Geneva, Switzerland.

Wolf, J., Johnston, R., Freeman, M.C., Ram, P.K., Slaymaker, T., Laurenz, E., Prüss-Ustün, A., 2023. Burden of disease attributable to unsafe drinking water, sanitation, and hygiene in domestic settings: a global analysis for selected adverse health outcomes. Lancet 401 (10393), 1851–1867.

Wright, S.L., Thompson, R.C., Galloway, T.S., 2013. The physical impacts of microplastics on marine organisms: a review. Environ. Pollut. 178, 483–492.

Xi, C., Zhang, Y., Marrs, C.F., Ye, W., Simon, C., Foxman, B., Nriagu, J., 2009. Prevalence of antibiotic resistance in drinking water treatment and distribution systems. Appl. Environ. Microbiol. 75 (17), 5714–5718.

Yu, H., Wang, L., Zhang, X., Chen, J., 2024. Neurotoxic effects of heavy metal pollutants in the environment: focusing on epigenetic mechanisms. Clin. Epigenetics Available at:. doi:10.1016/j.envpol.2024.123563.

Chapter 16

Impact of water pollution: Disturbance in the food chain

Sahana Savappla Anand[a], **and Shivaraju Harikaranahalli Puttaiah**[a,b]
[a]*Department of Environmental Sciences, School of Life Sciences, JSS Academy of Higher Education and Research, Mysuru, India*
[b]*Centre for Water, Food and Energy, GREENS TRUST, Turuvekere, Karnataka, India*

1 Introduction

Water is a major natural resource we obtain from our environment. Even though a large portion of the Earth is covered by water, <0.1% of it is available to us as potable water. Unfortunately, with the rise in pollution, water pollution is also increasing. Water pollution refers to the presence of chemical, physical, and biological pollutants in water bodies such as oceans, rivers, lakes, and groundwater (Puckett, 2008). This contamination makes water unfit for its respective uses (Wang et al., 2008). Currently, over 70% of freshwater sources are contaminated, leading to numerous severe health problems and fatalities worldwide. Reports indicate that approximately 14,000 people die each day globally due to water-related pollution (Bian et al., 2018; Fn and Mf, 2017).

Agricultural surface runoff, improper disposal of sewage, pipeline leakage, oil spills, and urban stormwater are the primary causes of river pollution (Sikder et al., 2019; Jayaraj et al., 2016; Global organochlorine contamination trends: an overview on JSTOR, n.d.). This leads to the release of toxic metals, detergents, soaps, infectious agents, and other nonbiodegradable materials into water, causing disturbances in the food chain. As they move from one trophic level to another trophic level, their concentration increases through bioaccumulation and biomagnification (Mahboob et al., 2019).

For example, mercury released from the burning of fossil fuels, combustion, weathering of rocks, incineration, and other human activities leads to the release of mercury into the environment (Walker et al., 2019; Warren, 1971). Through surface runoff, Hg is released into water and gets converted into methylmercury, which is one of the organic forms of mercury (Jayaswal et al., 2018). Mercury-contaminated drinking water, fish, seafood, and inhalation of ambient

River Pollution. https://doi.org/10.1016/B978-0-443-45024-2.00013-0

mercury are the major sources of mercury. Consumption of fish leads to the accumulation of mercury in cells and tissues, especially adipose tissue, as it is lipophilic in humans and other wildlife, and its concentration increases as they reach higher trophic levels (Younger, 2000). The animals present at the higher trophic levels face more lethal effects due to bioaccumulation. The energy of each organism is completely dependent on the organisms present below its trophic level (Fritz and Schiefer, 2008).

Therefore any disruption affecting one organism can directly impact others within the chain. This chapter mainly focuses on the different substances or pollutants responsible for water pollution and how they affect human beings, the environment, and the ecosystem through the food chain. This chapter encompasses the major pollutants responsible for water pollution, including their types and sources. It also explains their harmful effects on aquatic life, human health, and the overall ecosystem. Special attention is given to how these pollutants impact the food chain through the processes of bioaccumulation and biomagnification.

2 Types of water pollutants and sources

Substances or organisms that cause water pollution when they enter water bodies, such as rivers, oceans, and seas, are called water pollutants (Kuchangi et al., 2023). They are categorized into different groups based on their residence time and nature.

Based on residence time, they are classified into biodegradable and nonbiodegradable pollutants (Abdallah and Cobbina, 2022). Biodegradable pollutants are those that are easily degraded by natural biological processes and have less residence time, for example, food waste, agricultural waste, and domestic waste (Mudgal et al., 2010; Ali et al., 2019). Nonbiodegradable pollutants, including pesticides, fertilizers, and plastic waste, have lower absorptive properties and a high residence time. These pollutants are mainly responsible for much of the water pollution by accumulating in the food chain and the environment, including heavy metals, pesticides, etc. (Velvizhi et al., 2020).

Solid waste can be classified into hazardous and nonhazardous types. Hazardous waste includes materials such as medical waste and radioactive substances, which may be flammable, corrosive, infectious, or explosive (Wootton and Power, 1993). These wastes require intensive chemical treatment and high-temperature processing for safe disposal, whereas nonhazardous waste includes single-use plastic bottles, cans, and plastic covers. Compared to hazardous chemicals, these are less harmful and can be recycled (Freedman et al., 2024).

Detection of sources from which pollutants are coming is a necessary step to prevent pollution. Based on the location, pollutant sources are divided into point sources and nonpoint sources. Pollutants coming from an identifiable, single location are referred to as point sources. These include emissions from industries, pipelines, oil spills, etc. Pollutants coming from various locations are referred to as nonpoint sources. These include pollutants originating from surface runoff, sewage, etc. (Chaudhry and Ecosyst, 2017).

Surface runoff, improper disposal of sewage, and emissions from industries are the primary contributors to water pollution.

Industrial discharges: The release of effluents from industries and power plants directly into water bodies reduces the qualitative properties of water and contributes to water pollution. Effluents contain various toxic organic and inorganic compounds, as well as heavy metals (Kılıç, 2021).

Surface runoff: Remnants of fertilizers and pesticides are carried to water bodies through surface runoff. The release of nitrate, phosphate, and other nutrients causes the excessive growth of algae and fungi, forming an algal bloom. This is called nutrient pollution. This reduces the amount of sunlight entering the water and decreases the level of dissolved oxygen. It creates hypoxia, reduces photosynthesis and productivity, and disrupts the entire food chain (Goel, 2006) (Fig. 16.1).

Based on their nature, pollutants are classified into physical, chemical, and biological (Pandit and Kumar, 2019). Among these, chemical pollutants are the primary contributors to water contamination, encompassing both natural and xenobiotic compounds, such as (1) inorganic nutrients: nitrate and phosphate; (2) organic materials; (3) synthetic pesticides and synthetic detergents; (4) heavy metals: arsenic and mercury; and (5) chlorinated substances such as dichlorodiphenyltrichloroethanes (DDTs), dichlorodiphenyldichloroethanes (DDEs), and various others (Pandit and Kumar, 2019; Fn and Mf, 2017; U. E. N. C. for E. Assessment, 2009).

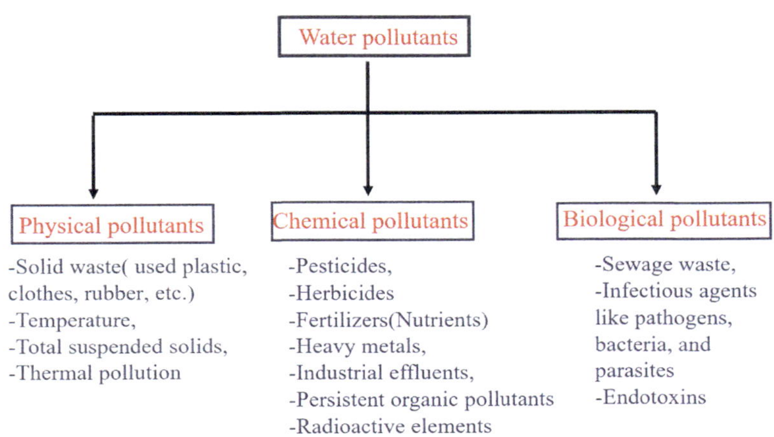

FIG. 16.1 The classification of water pollutants.

2.1 Physical pollutants

Physical pollutants include factors like temperature, pressure, salinity, turbidity, suspended solids, dissolved solids, and floating large objects like plastic bottles, covers, rubber, metal sheets, textiles, wood, and bags. These pollutants typically affect the health of aquatic life, including both flora and fauna, and reduce the esthetic value of water (Wine, 2010). Compared to chemical and biological pollutants, these are not hazardous to humans, but they affect the metabolism of aquatic animals (U. E. N. C. for E. Assessment, 2009).

Turbidity (suspended solids) measures the cloudiness or haziness of water caused by the presence of dust, construction residues, sediments, industrial waste, and floating materials from erosion and surface runoff. The presence of suspended solids decreases water transparency and limits sunlight penetration and oxygen into the water. This reduces the oxygen availability to aquatic animals and disrupts productivity (Davies-Colley and Smith, 2001).

Disposal of used plastic bottles, plastic covers, and food packets in the water irritates aquatic animals (MacLeod et al., 2021).

2.2 Chemical pollutants

2.2.1 Heavy metals

Mercury: Mercury is released from both manmade and natural activities. Natural sources like the weathering of rocks, forest fires, and volcanic eruptions release low levels of mercury into the environment. However, anthropogenic activities such as the combustion of fossil fuels, mining of ore (especially cinnabar), incineration, and sewage discharge are leading to an increase in mercury levels in the environment. Mercury is highly volatile in nature, remains in the environment for a longer time, and can reach various water bodies such as seepages and rivers. Once mercury enters a water body, it is converted into its organic forms by sulfate and iron-reducing bacteria (Davidson and Bradshaw, 1967). Methyl mercury is one such organic form produced after methylation. It is bioavailable and toxic to humans. A notable example is Minamata disease, which killed 1043 people in Japan. Methyl mercury is accumulated in the cells and tissues of fish when they consume plankton that grow in mercury-contaminated water. As large predatory fish feed on smaller fish, the concentration of mercury increases within their bodies, a process known as biomagnification. Consumption of mercury-contaminated food and water and inhalation are the main routes of exposure (Ozuah, 2000).

Arsenic: Arsenic is a toxic metalloid present in low amounts in rocks, soils, and water. For centuries, it has been used in the manufacture of poisons, herbicides, fertilizers, insecticides, paints, and wood preservatives. According to the World Health Organization (WHO), a permissible level of arsenic is 10 μg/L, beyond which it causes lethal effects to animals (I. P. on C. Safety and W. H. Organization, 1986). Ingestion of arsenic-contaminated food and water leads to nausea, vomiting, gut problems, and coma, while prolonged exposure causes cancer and death. A high concentration of arsenic is toxic to plants; it also leads to enzyme disruption and retards overall growth (Wang and Mulligan, 2006). Inhalation, drinking, and consumption of arsenic-contaminated air, water, and food grown with arsenic-contaminated water, respectively, lead to the accumulation of arsenic in tissues and cells (Ngegba et al., 2022; Kordbacheh and Heidari, 2023). Arsenate, the inorganic form of arsenic, enters the cell through the phosphate transport pathway, as it is structurally similar to phosphate. In *E. coli*, phosphate uptake occurs via two main transport systems: Pst (phosphate-specific transport) and Pit (phosphate inorganic transport), both of which can also facilitate the uptake of arsenate (Susan Murcott, 2012).

Selenium: An optimal level of selenium is needed by the body to maintain proper brain function in humans and the function of certain enzymes. Primary sources include emissions from petroleum refineries, mining, agricultural runoff (from pesticides, sewage, and fertilizers), and the combustion of fossil fuels. Prolonged exposure to selenium causes selenosis, hair loss, brittleness of skin, and other neurological disorders (Rayman et al., 2008). Like other elements, selenium also builds up and magnifies as it reaches higher trophic levels and poses severe health effects. The effect of selenium on the food chain depends on the geographical region, pH, and type of soil present. The presence of clay and Fe hydroxides reduces the uptake of selenium by plants, even though selenium is abundant.

2.2.2 Organic nutrients

Aquatic organisms are facing severe health risks due to an increase in nitrate and phosphate-containing compounds in water without treatment. Even though the optimal level of nitrate and phosphate is required for the physiological processes of many microorganisms, increasing above the recommended levels is fatal. Surface runoff of pesticides, fertilizers, animal waste, and dead and decaying matter is the major source of nitrate. Surface runoff from fertilizers and pesticides, weathering of phosphate-containing rocks, and detergent emissions are the major phosphate contributors (Isiuku and Enyoh, 2020). Upon release into the water, compounds favor the excessive growth of algae and other aquatic plants, leading to an algal bloom or eutrophication. This results in the formation of a dense covering over the water surface, which limits the penetration of sunlight and oxygen. Consequently, it reduces oxygen availability for aquatic animals and decreases photosynthesis and productivity (Travis and Hester, 1991). It also changes the color and gives a pungent odor to the water, thereby decreasing its esthetic property and disrupting the entire food chain.

Lead is a toxic heavy metal, second only to arsenic, and is usually present at lower levels in the Earth's crust (Travis and Hester, 1991). The permissive level in drinking water as per the WHO is 0.01 mg/L. Gasoline, lead pipes, old paints, vehicle emissions, and industrial waste are the primary sources of lead. Plants absorb the lead present in the soil and transfer it to their aerial parts. Ingestion of these plants by humans and other animals leads to the absorption of lead through the gastrointestinal epithelium, which is then transported to other organs, a process known as bioaccumulation. The concentration of lead increases as it moves from one trophic level to another, and animals at higher trophic levels experience adverse health effects (Kumar et al., 2020). Lead is classified as a Group 2B carcinogen for humans. Prolonged exposure to excess lead can result in cognitive behavior, delayed puberty, kidney dysfunction, fertility problems, and stunted fetal development during pregnancy. The absorption of lead is higher in children compared to adults (Kundu et al., 2016).

2.2.3 Synthetic pesticides and chemical detergents

Synthetic pesticides encompass organochlorines, organophosphates, and carbamates.

Organochlorines: These are chlorinated, persistent compounds. Endosulfan, cyclodienes, lindane, DDT, DDE, and dieldrin are widely used organochlorines. They can be easily absorbed through the skin and stored in fat tissues as they are lipophilic (Dhankhar and Kumar, 2023). Excess amounts also reach the kidneys and are excreted through urine and feces. Prolonged exposure causes ataxia, hyperreflexia, lack of coordination, vomiting, and nausea and affects the central nervous system, and some act as endocrine disruptors, altering hormone functions (Ensley, 2018).

DDT was used to kill insects and pests in the 20th century. It was later found that it started accumulating and magnifying in different animals through the food chain via bioaccumulation and biomagnification (Tripathi et al., 2020). Thinning of eggshells, early hatching, death of premature embryos, and early death of fetus were observed upon DDT exposure. It is lipophilic and binds to adipose tissue upon exposure, causing harmful effects (Beard, 2006). DDT is considered a potent carcinogen. Studies showed that bioaccumulated dieldrin and lindane in rats expressed high levels of intracellular reactive oxygen species by reducing mitochondrial membrane potential and causing neurotoxic effects. Prolonged exposure to organochlorines leads to Parkinson's disease (Turusov et al., 2002). DDT, upon dehydrohalogenation, converts into DDE. Both are lipophilic and accumulate in the tissues of animals in the food chain (Whitacre, 2008; Ahmed, 2024).

Organophosphates: These are a class of synthetic pesticides produced by the esterification of acids (phosphoric acid and alcohol). They are degradable and have less residence time. This class includes parathion, dimethoate, and malathion, which are used to kill insects (Curl et al., 2020). Recent studies show that organophosphates affect the activity of cholinesterase enzymes, leading to the disruption of normal cellular metabolism of proteins, carbohydrates, and fats (Pooja et al., 2020). They have been linked to reduced insulin secretion, genotoxic effects, and impaired mitochondrial function (Rusyniak and Nañagas, 2004). These disruptions can induce anxiety-related symptoms, oxidative stress at the cellular level, and disturbances in the endocrine system, especially with prolonged or high exposure.

Carbamates: These are esters of carbamic acid and are structurally and mechanistically similar to organophosphates (Ahsan and Valls, 2011). Carbamic acid and its derivatives, thiocarbamic acid and dithiocarbamic acid, are used as herbicides (I. P. on C. Safety and W. H. Organization, 1986; Dias et al., 2015; Mdeni et al., 2022). They belong to a class of insecticides that includes carbaryl, aldicarb, and carbofuran. Recent studies have shown that exposure to carbamates can induce necrosis and apoptosis in human immune cells, including T lymphocytes and natural killer cells, as observed in in vitro studies (Dias et al., 2015; I. P. on C. Safety and W. H. Organization, 1986).

2.3 Biological pollutants

Bacteria, viruses, protozoa, and helminths are common biological pollutants. Their sources include open defecation, poor sanitation, unscientific sewage disposal, leakage from septic tanks, and conventional sanitation practices (Binnie, 2021). These pathogens enter water bodies and cause various waterborne diseases such as diarrhea, cholera, ascariasis, and amoebiasis. Although about 75% of the Earth's surface is covered with water, $<0.1\%$ is available for drinking, highlighting the scarcity of microbe-free water (Akpor et al., 2014; Segerson and Walker, 2002).

Studies report that rural areas are more prone to biological pollutants due to the usage of conventional and open toilets, improper sewage disposal, and poor maintenance of hygiene. Eighty percent of the biological pollution is contributed by domestic and human nitrogenous waste (Wang et al., 2013).

2.4 Thermal pollution

Release of hot water from industries directly into water bodies causes thermal pollution (Råman Vinnå et al., 2017). The direct release of hot water increases the temperature of the water. High temperatures favor microorganism's growth, reduce the dissolved oxygen level, and alter the physiology of aquatic organisms, including both flora and fauna (Susan. Murcott, 2012). The contaminants present in the discharged hot water also pose lethal effects on fish and other organisms in the food chain (Råman Vinnå et al., 2017; Nordell, 2003).

2.5 Contaminants of emerging concern

These are new, frequently detected compounds that surpass the removal treatment methods. Their long-term exposure is yet to be fully understood, but they have the potential to cause harmful effects to both humans and the ecosystem. Nanomaterials, antibiotics, birth control pills, cosmetic products, pharmaceuticals, microplastics, etc., are some common emerging concerns (Yadav et al., 2021; Vandermeersch et al., 2015). As they escape from the treatment methods, there is a quest for the development of new techniques to detect and degrade these contaminants (Kim et al., 2018).

3 Structure of the food chain

A food chain is a visual representation of the energy flow from one organism to another in an ecosystem. It is characterised by a linear, unbranched chain, with each organism feeding on and deriving energy from the preceding ones (González, 2021). Each level in a food chain is known as a trophic level. A trophic level refers to the sequential stages in a food chain, starting with producers at the bottom, followed by primary, secondary, and tertiary consumers (Pimm, 1979). The food chain also explains the feeding pattern or relationship between living organisms (Fig. 16.2).

3.1 The food chain contains four components

Sunlight serves as a basic and indispensable source of energy. Without it, the food chain would not begin (Fn and Mf, 2017).

Producers are the starting organisms in a food chain and occupy the lowest trophic level (Water pollution policy, 2016). They are mainly autotrophs, such as phytoplankton, cyanobacteria, and certain types of algae. These organisms produce their food using sunlight through a process called photosynthesis (Sintes et al., 2004).

Consumers are organisms that rely on producers, or plants, for their food. They form the largest part of the food chain, as this group includes almost all living organisms, such as herbivores, carnivores, scavengers, and parasites. Herbivores feed on producers, occupy the second trophic level, and are known as primary consumers, whereas carnivores occupy the third trophic level and are called secondary consumers (Schmitz, 1992).

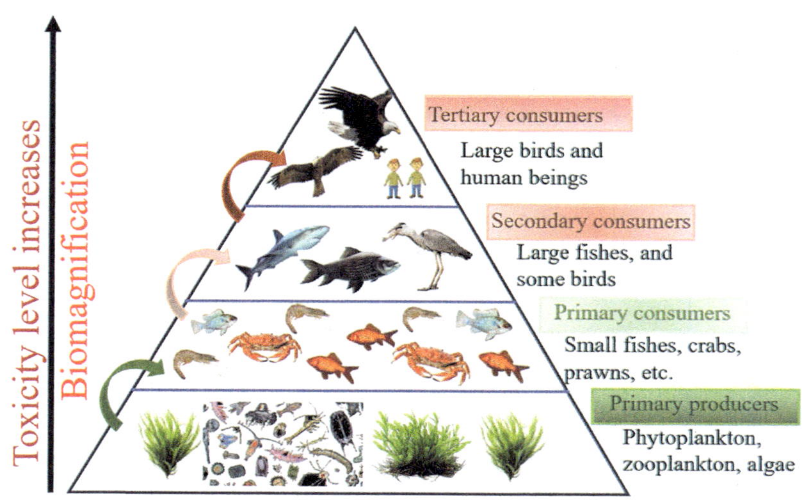

FIG. 16.2 The diagrammatic representation of the food chain.

Decomposers are organisms that convert organic waste into inorganic materials. They obtain energy from dead and decaying organic matter. They occupy the last level of the food chain and play a crucial role in recycling nutrients, providing them to oceans and soils. These inorganic nutrients are again utilized by producers (Pimm, 1979; Post, 2002).

This is how, through the food chain, energy transfers from producers to consumers, and the stability and strength of each organism mainly depend on the organisms at the lower trophic levels.

4 Mechanism of the effect of pollutants on the food chain components

4.1 Bioaccumulation and biomagnification

As these pollutants are ingested by primary producers such as phytoplankton, they enter the food chain and are subsequently transferred to organisms like zooplankton, small fish, and then larger fish, finally to humans at higher trophic levels, a process referred to as bioaccumulation (Beek et al., 2000). Lipophilic pollutants, such as organochlorines (e.g., DDT and DDE) and organophosphates, tend to accumulate in the adipose tissue of organisms (Lotufo et al., 2001). In contrast, heavy metals like mercury bind to proteins and accumulate in vital organs (Morel et al., 1998). Over time, the concentration of these pollutants increases through bioaccumulation and biomagnification, posing serious threats to aquatic life and, ultimately, to human health (Streit, 1992) (Fig. 16.3).

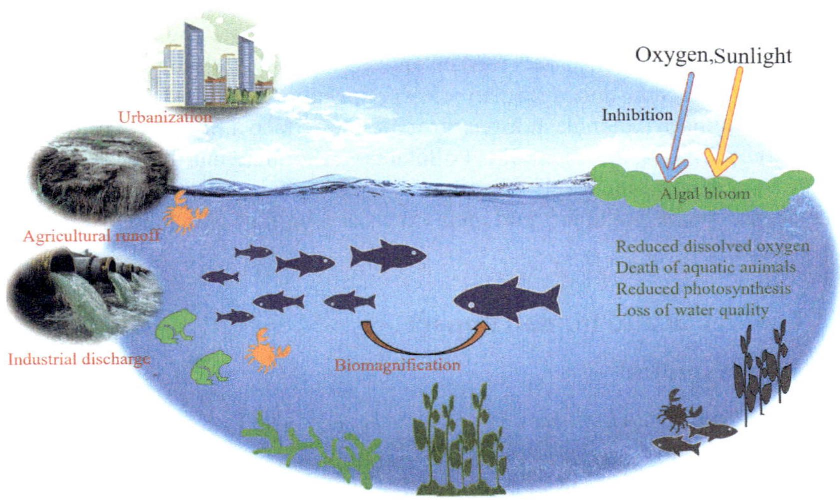

FIG. 16.3 Schematic representation of water pollution.

4.2 Decline in primary producers

The natural food chain is disrupted by pollution. Small animals consume pollutants like lead and cadmium present in water, and these small animals are ingested by larger animals, continuing the disruption along the chain until it reaches higher levels (Rossi et al., 2023). Since producers depend entirely on inorganic materials and sunlight for the preparation of food, any disturbance in this process can lead to their death (Allan et al., 2021).

Primary producers begin to decline due to increasing water pollution. Inorganic chemical pollutants, such as nitrates and phosphates, promote the excessive growth of algae and fungi, leading to nutrient pollution or eutrophication (Nfon et al., 2008). As these organisms die, decomposers consume the available oxygen during decomposition, resulting in hypoxia, which reduces the availability of dissolved oxygen for aquatic life (Howarth, 2000). In addition, pesticides and herbicides from agricultural runoff enter water bodies, killing or inhibiting the growth of phytoplankton and other aquatic organisms. Oxides of sulfur and nitrogen released from industries can lower the pH of water, further hindering the growth of certain organisms and disrupting photosynthesis (Waters et al., 2013).

4.3 Toxicity in consumers

Herbivorous animals that feed on plants or plankton, carnivorous animals that feed on herbivorous animals, and omnivorous animals that feed on both come under consumers. Fishes and other aquatic animals exposed to contaminated water bodies take up various pollutants (Dórea, 2008). The pollutants accumulate in the cells, tissues, and organs based on their nature. Seafood and fish are consumed by many people around the world for their taste and essential nutrients. When primary consumers feed on autotrophs, they show high toxicity, and when secondary consumers feed on primary consumers, they exhibit the highest toxicity. Similarly, animals at higher trophic levels face a high level of toxicity (Wang et al., 2024; Moss, 2008). Pollutants accumulate and build up as they move to higher trophic levels. A well-known example is the bioaccumulation of persistent organic pollutants and methylmercury, both of which are neurotoxic to animals.

5 Ecological and human impact

5.1 Case studies

5.1.1 Minamata disease

Minamata disease, also known as mercury poisoning, was first reported in Minamata Bay, Japan, due to the ingestion of fish contaminated with methylmercury, formed after the methylation of mercury released from the vinyl chloride industry (Eto, 2000). Methyl mercury is highly toxic, stable, and persistent, and it accumulates and is magnified as it reaches higher trophic levels. Between 1953 and 1956, Minamata disease was first diagnosed, and a total of 2263 cases were found by 1911, out of which 1368 people were affected due to high levels of toxicity. Prolonged exposure to methylmercury causes ataxia, affects the central nervous system, causes birth defects, and leads to brain damage (Segerson and Walker, 2002).

5.1.2 Methemoglobinemia or blue baby syndrome

Prolonged exposure to nitrate beyond the recommended levels leads to methemoglobinemia. Consumption of nitrate-contaminated food and water results in its accumulation in blood cells (Curry, 1982). Hemoglobin is converted to methemoglobin by nitrate, reducing the oxygen-carrying capacity. Compared to children and adults, infants are more prone to methemoglobinemia as they have a low level of cytochrome b5 reductase, an enzyme involved in the conversion of methemoglobin into hemoglobin (Glibert et al., 2005). The presence of 2% methemoglobin is normal; however, if it exceeds 2%, it leads to the discoloration of skin to blue (Harada, 1995; Wright et al., 1999).

Similarly, the release of nitrate and phosphate above the standard level causes the excessive growth of algae and other aquatic plants, forming a thick, dense cover called an algal bloom or eutrophication that acts as a cover on the water surface (Rabalais et al., 2002). A well-noted example is that the Gulf of Mexico is known as the second-largest oxygen-depleted zone in the world. Light penetration and dissolved oxygen are reduced due to an algal bloom, which causes suffocation for bottom-dwelling aquatic animals. With the passage of time, the algal bloom dies, further decreasing the oxygen and causing hypoxia (Dodds, 2006; Rabotyagov et al., 2014). The lack of dissolved oxygen and sunlight inhibits photosynthesis and reduces productivity, leading to a lower number of producers, such as plankton. This reduces the food availability for the animals present at the higher trophic levels. Finally, it leads to a food chain imbalance and disrupts the overall ecosystem (Schwarzenbach et al., 2010; Kim et al., 2018).

5.1.3 Polychlorinated biphenyls

Polychlorinated biphenyls (PCBs) are toxic, stable xenobiotic compounds that have been used in the manufacture of heat exchange transformers, plastics, and paint additives (Ross, 2004; Safe, 1994). Consumption and inhalation of PCBs, as well as PCB-contaminated food, water, and air, lead to several health issues. PCBs can cross the plasma membrane by binding to various receptors and altering the biological response. PCBs were first found in the Great Lakes and North America, alongside microplastics (Safe, 1994). Like other pollutants, PCBs also undergo bioaccumulation and biomagnification, causing hormonal disruption, dysfunction in the reproductive system, and other liver-related problems.

6 Methods to prevent pollution and to protect the food chain and ecosystem

6.1 Reduce agricultural runoff

Surface runoff is the major contributor to water pollution. It washes away the remnants of pesticides, herbicides, and fertilizers from agricultural land, which are carried into nearby water bodies (Krutz et al., 2005; Cooper et al., 2004). The construction of vegetative barriers involves planting shrubs and larger trees across borders, planting members of the Fabaceae family as they absorb many nutrients, and practicing the use of conventional organic, neem-based fertilizers. The incorporation of Integrated Pest Management practices can reduce surface runoff and its effects (Torres and de F. Bueno, 2018).

6.2 Regulate industrial discharge

To reduce pollution and prevent the toxic effects of emissions (heavy metals, dyes, organic pollutants, etc.) from industries, following the standards of the WHO is crucial. Checking the toxic levels of compounds, constructing effluent treatment plants, and monitoring emissions before discharge may reduce the harm (Magat and Viscusi, 1990; Rajaram and Das, 2008). Converting sewage (domestic and human nitrogenous waste) into useful products like biogas and biofuel not only helps reduce the toxicity of certain pollutants but is also beneficial to the ecosystem. The usage of treated water for nonpotable purposes, such as animal washing and cloth washing, can save water (Salgot and Folch, 2018; Sonune and Ghate, 2004).

6.3 Continuous monitoring

Continuous monitoring to check the quality of water bodies is crucial to understanding their status. The usage of sensors to detect the presence of certain pollutants and to check physical parameters (like pH, temperature, total suspended solids, etc.) is essential (Usali and Ismail, 2010). In addition, checking for biological indicators like cyanobacteria, sludge worms, and *E. coli* helps determine the pollution level, while pollution-sensitive organisms like penny beetles, stonefly nymphs, and mayflies, which thrive only in clean water bodies, provide further insights. Remote sensing and GIS assist in identifying pollution sources and eutrophicated water bodies (Yaroshenko et al., 2020).

6.4 Public awareness and education

Creating awareness about the current water status, the effects of pollution being faced at present, and the need to conserve water among people and organizations through campaigns and advertisements may reduce the further effects of pollution. The practice of environmental education in schools and colleges helps create awareness at an early stage.

7 Conclusion

Water pollution is a growing global concern. Improper sewage disposal, surface runoff, oil spills, leakage of pipelines, and emissions from industries are the primary sources of water pollution. The concentration of these pollutants increases as they move up the food chain, a process known as biomagnification. One of the major effects of water pollution is its disruption of the food chain. These pollutants lead to a decline in primary producers and cause ecosystem imbalance. Biomagnification affects top predators at higher trophic levels, such as large fish and birds, and eventually impacts human health. Prevention and control of pollutants are essential to maintaining ecosystem sustainability. In addition to current methods, new and innovative sewage treatment technologies, along with the development of real-time sensors for pollution monitoring, are necessary to effectively combat water pollution.

References

Abdallah, C.K., Cobbina, S.J., 2022. Advanced Water Pollution Control Strategies: A Systematic Review. Springer Science and Business Media LLC. doi:10.21203/rs.3.rs-1315815/v1.

Ahmed, A.S.O., 2024. Environmental toxicology. In: Sulieman, A.M.E., Alshammari, N.I. (Eds.), Microbial Toxins in Food Systems: Causes, Mechanisms, Complications, and Metabolism. Springer Nature, Switzerland, pp. 81–95. doi: 10.1007/978-3-031-62839-9_7.

Ahsan, D.A., Valls, D., 2011. Impact of arsenic contaminated irrigation water in food chain: an overview from Bangladesh. Int. J. Environ. Res. 5 (3), 627–638.

Akpor, O.B., Otohinoyi, D.A., Olaolu, D.T., Aderiye, B.I., 2014. Pollutants in wastewater effluents: impacts and remediation processes. Int. J. Environ. Res. Earth Sci. 3 (3).

Ali, H., Khan, E., Ilahi, I., 2019. Environmental chemistry and ecotoxicology of hazardous heavy metals: environmental persistence, toxicity, and bioaccumulation. J. Chem. 2019 (1), 6730305. doi:10.1155/2019/6730305.

Allan, J.D., Castillo, M.M., Capps, K.A., 2021. Primary producers. Stream Ecol. 141–176. doi:10.1007/978-3-030-61286-3_6.

Beard, J., 2006. DDT and human health. Sci. Total Environ. 355 (1–3), 78–89. doi:10.1016/J.SCITOTENV.2005.02.022.

Beek, B., Böhling, S., Bruckmann, U., Franke, C., Jöhncke, U., Studinger, G., 2000. The assessment of bioaccumulation. In: Beek, B (Ed.), Bioaccumulation—New Aspects and Developments. Springer-Verlag, pp. 235–276. doi: 10.1007/10503050_4.

Bian, Y., Xiong, N., Zhu, G., 2018. Technology for the remediation of water pollution: a review on the fabrication of metal organic frameworks. Processes. 6 (8), 122. doi:10.3390/PR6080122.

Binnie, P.W.H., 2021. Biological pollutants in the indoor environment. Indoor Air Pollut. 13–24. doi:10.1201/9781003210566-4.

Chaudhry, F., Ecosyst, M.M.-J., 2017. Ecography, and undefined 2017, "Factors affecting water pollution: a review". J. Ecosyst. Ecogr. 7 (1). doi:10.4172/2157-7625.1000225.

Cooper, C.M., et al., 2004. Innovative uses of vegetated drainage ditches for reducing agricultural runoff. Water Sci. Technol. 49 (3), 117–123. doi:10.2166/WST.2004.0176.

Curl, C.L., Spivak, M., Phinney, R., Montrose, L., 2020. Synthetic pesticides and health in vulnerable populations: agricultural workers. Curr. Environ. Health Rep. 7 (1), 13–29. doi:10.1007/S40572-020-00266-5.

Curry, S., 1982. Methemoglobinemia. Ann. Emerg. Med. 11 (4), 214–221. doi:10.1016/S0196-0644(82)80502-7.

Davidson, B., Bradshaw, R.W., 1967. Thermal pollution of water systems. Environ. Sci. Technol. 1 (8), 618–630. doi:10.1021/ES60008A606.

Davies-Colley, R.J., Smith, D.G., 2001. Turbidity, suspended sediment, and water clarity: a review. J. Am. Water Resour. Assoc. 37 (5), 1085–1101. doi:10.1111/J.17521688.2001.

Dhankhar, N., Kumar, J., 2023. Impact of increasing pesticides and fertilizers on human health: areview. Mater. Today Proc. doi:10.1016/J.MATPR.2023.03.766.

Dias, E., Garcia E Costa, F., Morais, S., de Lourdes Pereira, M., 2015. A review on the assessment of the potential adverse health impacts of carbamate pesticides. Top. Public Health. doi:10.5772/59613.

Dodds, W.K., 2006. Nutrients and the ‘dead zone’: the link between nutrient ratios and dissolved oxygen in the northern Gulf of Mexico. Front. Ecol. Environ. 4 (4), 211–217. doi:10.1890/1540-9295(2006)004[211:NATDZT]2.0.CO;2.

Dórea, J.G., 2008. Persistent, bioaccumulative and toxic substances in fish: human health considerations. Sci. Total Environ. 400 (1–3), 93–114. doi:10.1016/J.SCITOTENV.2008.06.017.

Ensley, S.M., 2018. Organochlorines. In: Gupta, R.C. (Ed.), Veterinary Toxicology: Basic and Clinical Principles, third ed. Elsevier, pp. 509–513. doi: 10.1016/B978-0-12-811410-0.00038-6.

Eto, K., 2000. Minamata disease. Neuropathology. 20 (SUPPL), 14–19. doi:10.1046/J.1440-1789.2000.00295.X.

Fn, C., Mf, M., 2017. Factors affecting water pollution: a review. J. Ecosyst. Ecogr. 7 (1). doi:10.4172/2157-7625.1000225.

Freedman, B., Dorsey, W., Frazier, A., Kambhampati, M., Galiotos, J., Mukherjee, S., 2024. Biodegradable and non-biodegradable waste (Chapter 5). Available from: https://louis.pressbooks.pub/environmentalscience/chapter/chapter-5/.

Fritz, M., Schiefer, G., 2008. Food chain management for sustainable food system development: a European research agenda. Agribusiness. 24 (4), 440–452. doi:10.1002/AGR.20172.

Glibert, P.M., et al., 2005. Harmful algal blooms eutrophication th e role of in the global proliferation of harmful algal blooms new perspectives and new approaches. 18 (2) Available from: www.ut.ee/~olli/eutr/. (Accessed 14 May 2025).

Global organochlorine contamination trends: an overview on JSTOR, n.d. Available from: https://www.jstor.org/stable/4314197. (Accessed 11 November 2025).

Goel, P.K., 2006. Water pollution—causes, effects & control. 418. Available from: https://books.google.com/books/about/Water_Pollution.html?id=4R9CYYoiFCcC. (Accessed 14 May 2025).

González, R., et al., 2021. Effects of pollution in aquatic food chains. In: Hader, D.P., Helbling, E.W., Villafane, V.E. (Eds.), Anthropogenic Pollution of Aquatic Ecosystems. Springer International Publishing, pp. 61–89. doi: 10.1007/978-3-030-75602-4_4.

Harada, M., 1995. Minamata disease: methylmercury poisoning in Japan caused by environmental pollution. Crit. Rev. Toxicol. 25 (1), 1–24. doi:10.3109/10408449509089885.

Howarth, R.W., et al., 2000. Nutrient pollution of coastal rivers, bays, and seas. Issues Ecol. 7 (7), 1–16.

I. P. on C. Safety and W. H. Organization, 1986. Carbamate pesticides: a general introduction. Available from: https://iris.who.int/handle/10665/38687. (Accessed 14 May 2025).

Isiuku, B.O., Enyoh, C.E., 2020. Pollution and health risks assessment of nitrate and phosphate concentrations in water bodies in South Eastern, Nigeria. Environ. Adv. 2, 100018. doi:10.1016/J.ENVADV.2020.100018.

Jayaraj, R., Megha, P., Sreedev, P., 2016. Organochlorine pesticides, their toxic effects on living organisms and their fate in the environment. Interdiscip. Toxicol. 9 (3–4), 90–100. doi:10.1515/intox-2016-0012.

Jayaswal, K., Sahu, V., Gurjar, B.R., Dec 2018. Water pollution, human health and remediation. In: Energy, Environment, and Sustainability. Singapore, pp. 11–27. doi:10.1007/978-981-10-7551-3_2 (Accessed 23 December 2018).

Kim, S., et al., 2018. Removal of contaminants of emerging concern by membranes in water and wastewater: a review. Chem. Eng. J. 335, 896–914. doi:10.1016/J.CEJ.2017.11.044.

Kılıç, Z., 2021. Water pollution: causes, negative effects and prevention methods. İstanbul Sabahattin Zaim Üniversitesi Fen Bilimleri Enstitüsü Dergisi. 3 (2), 129–132. doi:10.47769/IZUFBED.862679.

Kordbacheh, F., Heidari, G., 2023. Water pollutants and approaches for their removal. Mater. Chem. Horiz. 2 (2), 139–153. doi:10.22128/MCH.2023.684.1039.

Krutz, L.J., Senseman, S.A., Zablotowicz, R.M., Matocha, M.A., 2005. Reducing herbicide runoff from agricultural fields with vegetative filter strips: a review. Weed Sci. 53 (3), 353–367. doi:10.1614/WS-03-079R2.

Kuchangi, S.N., Mruthunjayappa, M.H., Sanna Kotrappanavar, N., 2023. An overview of water pollutants in present scenario. 3D Printing Technology for Water Treatment Applications. Elsevier. pp. 83–105. doi:10.1016/B978-0-323-99861-1.00001-1.

Kumar, A., et al., 2020. Lead toxicity: health hazards, influence on food chain, and sustainable remediation approaches. Int. J. Environ. Res. Public Health 17 (7), 2179. doi:10.3390/IJERPH17072179.

Kundu, D., Mondal, S., Dutta, D., Haque, S., Ghosh, A.R., 2016. Accumulation and contamination of lead in different trophic levels of food chain in sewage-fed East Kolkata Wetland, West Bengal, India. Int. J. Environ. Technol. Sci. 2, 61–68. Available from: www.journalijets.org. Accessed 29 October 2025.

Lotufo, G.R., Landrum, P.F., Gedeon, M.L., 2001. Toxicity and bioaccumulation of ddt in freshwater amphipods in exposures to spiked sediments. Environ. Toxicol. Chem. 20 (4), 810–825. doi:10.1002/ETC.5620200417.

MacLeod, M., Arp, H.P.H., Tekman, M.B., Jahnke, A., 2021. The global threat from plastic pollution. Science. 373 (6550), 61–65. doi:10.1126/SCIENCE.ABG5433.

Magat, W.A., Viscusi, W.K., 1990. Effectiveness of the EPA's regulatory enforcement: the case of industrial effluent standards. J. Law Econ. 33 (2), 331–360. doi:10.1086/467208.

Mahboob, S., et al., 2019. Impact of water pollution on trophic transfer of fatty acids in fish, microalgae, and zoobenthos in the food web of a freshwater ecosystem. Biomolecules. 9 (6), 231. doi:10.3390/BIOM9060231.

Mdeni, N.L., Adeniji, A.O., Okoh, A.I., Okoh, O.O., 2022. Analytical evaluation of carbamate and organophosphate pesticides in human and environmental matrices: a review. Molecules. 27 (3), 618. doi:10.3390/MOLECULES27030618.

Morel, F.M.M., Kraepiel, A.M.L., Amyot, M., 1998. The chemical cycle and bioaccumulation of mercury. Annu. Rev. Ecol. Syst. 29, 543–566. doi:10.1146/ANNUREV.ECOLSYS.29.1.543.

Moss, B., 2008. Water pollution by agriculture. Philos. Trans. R. Soc. Lond. B Biol. Sci. 363 (1491), 659–666. doi:10.1098/RSTB.2007.2176.

Mudgal, V., Madaan, N., Mudgal, A., Singh, R.B., Mishra, S., 2010. Effect of toxic metals on human health. Open Nutraceuticals J. 3, 94–99.

Murcott, S., 2012. Arsenic contamination in the world : an international sourcebook 2012. Available from: https://books.google.com/books/about/Arsenic_Contamination_in_the_World.html?id=CHKNoVj0t_cC. (Accessed 14 May 2025).

Murcott, S., 2012. Arsenic contamination in the world : an international sourcebook 2012. Available from: https://books.google.com/books/about/Arsenic_Contamination_in_the_World.html?id=CHKNoVj0t_cC. (Accessed 29 October 2025).

Nfon, E., Cousins, I.T., Broman, D., 2008. Biomagnification of organic pollutants in benthic and pelagic marine food chains from the Baltic Sea. Sci. Total Environ. 397 (1–3), 190–204. doi:10.1016/J.SCITOTENV.2008.02.029.

Ngegba, P.M., Cui, G., Khalid, M.Z., Zhong, G., 2022. Use of botanical pesticides in agriculture as an alternative to synthetic pesticides. Agriculture. 12 (5), 600. doi:10.3390/AGRICULTURE12050600.

Nordell, B., 2003. Thermal pollution causes global warming. Glob. Planet. Change. 38 (3–4), 305–312. doi:10.1016/S0921-8181(03)00113-9.

Ozuah, P.O., 2000. Mercury poisoning. Curr. Probl. Pediatr. 30 (3), 91–99. doi:10.1067/MPS.2000.104054.

Pandit, A.B., Kumar, J.K., 2019. Drinking water treatment for developing countries : physical, chemical and biological pollutants. Available from: https://books.google.com/books/about/Drinking_Water_Treatment_for_Developing.html?id=1hSMDwAAQBAJ. (Accessed 14 May 2025).

Pimm, S.L., 1979. The structure of food webs. Theor. Popul. Biol. 16 (2), 144–158. doi:10.1016/0040-5809(79)90010-8.

Pooja, D., Kumar, P., Singh, P., Patil, S. (Eds.), 2020. Sensors in water pollutants monitoring: role of material. In: Advanced Functional Materials and Sensors. Springer Singapore. doi:10.1007/978-981-15-0671-0.

Post, D.M., 2002. The long and short of food-chain length. Trends Ecol. Evol. 17 (6), 269–277. doi:10.1016/S0169-5347(02)02455-2.

Puckett, L.J., 2008. Identifying the major sources of nutrient water pollution. Environ. Sci. Technol. 29 (9), 408A–414A. doi:10.1021/ES00009A001.

Rabalais, N.N., Turner, R.E., Wiseman, W.J., 2002. Gulf of Mexico hypoxia, a.k.a. 'The dead zone,'. Annu. Rev. Ecol. Syst. 33, 235–263. doi:10.1146/ANNUREV.ECOLSYS.33.010802.150513.

Rabotyagov, S.S., Kling, C.L., Gassman, P.W., Rabalais, N.N., Turner, R.E., 2014. The economics of dead zones: causes, impacts, policy challenges, and a model of the Gulf of Mexico hypoxic zone. Rev. Environ. Econ. Policy. 8 (1), 58–79. doi:10.1093/REEP/RET024.

Rajaram, T., Das, A., 2008. Water pollution by industrial effluents in India: discharge scenarios and case for participatory ecosystem specific local regulation. Futures. 40 (1), 56–69. doi:10.1016/J.FUTURES.2007.06.002.

Råman Vinnå, L., Wüest, A., Bouffard, D., 2017. Physical effects of thermal pollution in lakes. Water Resour. Res. 53 (5), 3968–3987. doi:10.1002/2016WR019686.

Rayman, M.P., Infante, H.G., Sargent, M., 2008. Food-chain selenium and human health: spotlight on speciation. Br. J. Nutr. 100 (2), 238–253. doi:10.1017/S0007114508922522.

Ross, G., 2004. The public health implications of polychlorinated biphenyls (PCBs) in the environment. Ecotoxicol. Environ. Saf. 59 (3), 275–291. doi:10.1016/J.ECOENV.2004.06.003.

Rossi, G.D., Hoque, M.M., Ji, Y., Kling, C.L., 2023. The economics of nutrient pollution from agriculture. Annu. Rev. Resour. Econ. 15, 105–130. doi:10.1146/ANNUREV-RESOURCE-111820-021317.

Rusyniak, D.E., Nañagas, K.A., 2004. Organophosphate poisoning. Semin. Neurol. 24 (2), 197–204. doi:10.1055/S-2004-830907/ID/23/BIB.

Safe, S.H., 1994. Polychlorinated bphenyls (PCBs): environmental impact, biochemical and toxic responses, and implications for risk assessment. Crit. Rev. Toxicol. 24 (2), 87–149. doi:10.3109/10408449409049308.

Salgot, M., Folch, M., 2018. Wastewater treatment and water reuse. Curr. Opin. Environ. Sci. Health 2, 64–74. doi:10.1016/J.COESH.2018.03.005.

Schmitz, O.J., 1992. Exploitation in model food chains with mechanistic consumer-resource dynamics. Theor. Popul. Biol. 41 (2), 161–183. doi:10.1016/0040-5809(92)90042-R.

Schwarzenbach, R.P., Egli, T., Hofstetter, T.B., Gunten, U.V., Wehrli, B., 2010. Global water pollution and human health. Annu. Rev. Environ. Resour. 35, 109–136. doi:10.1146/ANNUREV-ENVIRON-100809-125342.

Segerson, K., Walker, D., 2002. Nutrient pollution: an economic perspective. Estuaries. 25 (4), 797–808. doi:10.1007/BF02804906.

Sikder, M.T., Rahman, M.M., Jakariya, M., Hosokawa, T., Kurasaki, M., Saito, T., 2019. Remediation of water pollution with native cyclodextrins and modified cyclodextrins: a comparative overview and perspectives. Chem. Eng. J. 355, 920–941. doi:10.1016/J.CEJ.2018.08.218.

Sintes, E., Martínez-Taberner, A., Moyà, G., Ramon, G., 2004. Dissecting the microbial food web: structure and function in the absence of autotrophs. Aquat. Microb. Ecol. 37 (3), 283–293. doi:10.3354/AME037283.

Sonune, A., Ghate, R., 2004. Developments in wastewater treatment methods. Desalination. 167 (1–3), 55–63. doi:10.1016/J.DESAL.2004.06.113.

Streit, B., 1992. Bioaccumulation processes in ecosystems. Experientia. 48 (10), 955–970. doi:10.1007/BF01919142.

Torres, J.B., Bueno, A.de F., 2018. Conservation biological control using selective insecticides—a valuable tool for IPM. Biol. Control. 126, 53–64. doi:10.1016/J.BIOCONTROL.2018.07.012.

Travis, C.C., Hester, S.T., 1991. Global chemical pollution. Environ. Sci. Technol. 25 (5), 814–819. doi:10.1021/ES00017A001.

Tripathi, G., Yadav, V.K., Singh, J., Mishra, V., 2020. Analytical methods of water pollutants detection. In: Pooja, D., Kumar, P., Singh, P., Patil, S. (Eds.), Advanced Functional Materials and Sensors. Springer, Singapore, pp. 63–78. doi: 10.1007/978-981-15-0671-0_5.

Turusov, V., Rakitsky, V., Tomatis, L., 2002. Dichlorodiphenyltrichloroethane (DDT): ubiquity, persistence, and risks. Environ. Health Perspect. 110 (2), 125–128. doi:10.1289/EHP.02110125.

U. E. N. C. for E. Assessment, 2009. Carbamate pesticides. In: Encyclopedia of Toxicology: Third Edition. Elsevier, pp. 661–664. doi:10.1016/B978-0-12-386454-3.00106-8.

Usali, N., Ismail, M.H., 2010. Use of remote sensing and gis in monitoring water quality. J. Sustain. Dev. 3 (3). Available from: www.ccsenet.org/jsd. (Accessed 14 May 2025).

Vandermeersch, G., et al., 2015. Environmental contaminants of emerging concern in seafood—European database on contaminant levels. Environ. Res. 143, 29–45. doi:10.1016/J.ENVRES.2015.06.011.

Velvizhi, G., et al., 2020. Biodegradable and non-biodegradable fraction of municipal solid waste for multifaceted applications through a closed loop integrated refinery platform: paving a path towards circular economy. Sci. Total Environ. 731, 138049. doi:10.1016/J.SCITOTENV.2020.138049.

Walker, D.B., Baumgartner, D.J., Gerba, C.P., Fitzsimmons, K., 2019. Surface water pollution. Environ. Pollut. Sci. 261–292. doi:10.1016/B978-0-12-814719-1.00016-1.

Wang, H., Zhang, W., Chen, L., Wang, J., Liu, T., 2013. The contamination and control of biological pollutants in mass cultivation of microalgae. Bioresour. Technol. 128, 745–750. doi:10.1016/J.BIORTECH.2012.10.158.

Wang, M., Webber, M., Finlayson, B., Barnett, J., 2008. Rural industries and water pollution in China. J. Environ. Manage. 86 (4), 648–659. doi:10.1016/J.JENVMAN.2006.12.019.

Wang, S., Mulligan, C.N., 2006. Occurrence of arsenic contamination in Canada: sources, behavior and distribution. Sci. Total Environ. 366 (2–3), 701–721. doi:10.1016/J.SCITOTENV.2005.09.005.

Wang, Y., Javeed, A., Jian, C., Zeng, Q., Han, B., 2024. Precautions for seafood consumers: an updated review of toxicity, bioaccumulation, and rapid detection methods of marine biotoxins. Ecotoxicol. Environ. Saf. 274, 116201. doi:10.1016/J.ECOENV.2024.116201.

Warren, C.E., 1971. Biology and Water Pollution Control. W. B. Saunders Company, West Washington Square, Philadelphia, PA. 19105.

Water pollution policy, 2016. 97–149, doi: 10.4324/9781315651477-4.

Waters, M.N., Smoak, J.M., Saunders, C.J., 2013. Historic primary producer communities linked to water quality and hydrologic changes in the Northern Everglades. J. Paleolimnol. 49 (1), 67–81. doi:10.1007/S10933-011-9569-Y/METRICS.

Whitacre, D.M. (Ed.), 2008. Reviews of environmental contamination and toxicology. In: Reviews of Environmental Contamination and Toxicology, vol. 194. Springer, New York. doi:10.1007/978-0-387-74816-0.

Wine, P.H., 2010. Editorial: atmospheric and environmental physical chemistry: pollutants without borders. J. Phys. Chem. Lett. 1 (11), 1749–1751. doi:10.1021/JZ1006252.

Wootton, J.T., Power, M.E., 1993. Productivity, consumers, and the structure of a river food chain. Proc. Natl. Acad. Sci. U. S. A. 90 (4), 1384–1387. doi:10.1073/PNAS.90.4.1384.

Wright, R.O., Lewander, W.J., Woolf, A.D., 1999. Methemoglobinemia: etiology, pharmacology, and clinical management. Ann. Emerg. Med. 34 (5), 646–656. doi:10.1016/S0196-0644(99)70167-8.

Yaroshenko, I., et al., 2020. Real-time water quality monitoring with chemical sensors. Sensors. 20 (12), 3432. doi:10.3390/S20123432.

Yadav, D., et al., 2021. Environmental and health impacts of contaminants of emerging concerns: recent treatment challenges and approaches. Chemosphere. 272, 129492. doi:10.1016/J.CHEMOSPHERE.2020.129492.

Younger, P.L., 2000. Holistic remedial strategies for short- and long-term water pollution from abandoned mines. Min. Technol. 109, 210–218. doi:10.1179/MNT.2000.109.3.210.

Chapter 17

Socioeconomic impacts of river pollution

Amit Kumar Singh and Ekta Verma
Post Graduate Department of Botany, Magadh University, Bodh-Gaya, India

1 Introduction

Rivers have always been essential to human civilization, supporting agriculture, fueling industry, and influencing societies. Approximately 2.5%–2.75% of the Earth's water is freshwater. Freshwater is primarily stored in glaciers, ice caps, and groundwater, with only a small portion being accessible in lakes, rivers, and swamps. Rivers and lakes collectively constitute approximately 0.3% of the world's freshwater. Consequently, rivers provide just a small percentage of worldwide freshwater; however, this portion is significant due to its accessibility for human, agricultural, and industrial use. Rivers are crucial resources for billions globally, supplying needed drinking water, agricultural needs, industrial use, and transportation. Nonetheless, rapid urbanization, industrialization, and intensified agriculture have resulted in the considerable deterioration of river ecosystems, especially in developing nations such as India. This deterioration emerges in multiple forms, including the accumulation of pollutants such as heavy metals, nutrients, plastics, and new contaminants, which negatively impact both environmental and human health (Micella et al., 2024; United Nations Environment Programme (UNEP), 2021a, 2021b). Seventy percent of water in developing nations is progressively contaminated due to the release of pollutants, as streams transform into channels for untreated sewage, industrial effluent, and agricultural runoff, thereby undermining the same currents that formerly supported societies. In addition to polluted fish and turbid waterways, river pollution substantially negatively affects socioeconomic conditions by disrupting livelihoods, jeopardizing public health, and exacerbating wealth disparities. As toxic pollutants devastate aquatic ecosystems, fishing families in numerous low-income areas experience a reduction in their catches. Boatmen who once guided tourists down tranquil riverbanks are now struggling to survive through low-wage employment; their traditional vocation has nearly vanished. Crop yields diminish as

River Pollution. https://doi.org/10.1016/B978-0-443-45024-2.00017-8

farmers contend with contaminated irrigation water, forcing them to choose between reducing usage or incurring losses through sales. As foot traffic diminishes, small enterprises dependent on clean water, such as ritual ashram vendors, launderettes, and lakeside cafés, face challenges in maintaining viability. In India, rivers such as the Yamuna, Ganga, and Gomati have become significantly contaminated due to the discharge of untreated sewage, industrial effluents, and agricultural runoff. A study evaluating human impacts on the water quality of the Gomati River identified substantial pollution levels linked to anthropogenic activities, underscoring the critical necessity for efficient water quality management and public health measures (Singh et al., 2024). The Yamuna River continues to confront ongoing pollution issues, with studies from 2014 to 2024 revealing that, despite heightened awareness and corrective actions, the river remains severely contaminated, necessitating immediate and sustained interventions (Srivastava et al., 2025). The economic effects of river pollution are significant and complex. Contaminated water supplies result in numerous health problems, including waterborne diseases like cholera, typhoid, and diarrhea, which disproportionately impact vulnerable communities, especially in rural and underdeveloped regions (Babuji et al., 2023). The concealed expense of contaminated waterways is also evident in human health. Waterborne illnesses, such as diarrhea and cholera, increase significantly. Healthcare expenditures surge, depleting limited household earnings into perpetual medical expenses. In India, polluted water results in an annual economic loss of over $6 billion due to health and productivity losses, while global assessments of river-related economic harm reach hundreds of billions of dollars annually. The economic repercussions are extensive, including heightened healthcare expenses, decreased labor productivity, and lower agricultural output resulting from the use of contaminated water for irrigation (Greenstone and Jack, 2022; Asian Development Bank (ADB), 2020). Moreover, river pollution intensifies social disparities, as disadvantaged groups frequently endure the most severe consequences. However, the burden of pollution is not distributed equitably. Marginalized groups—the urban poor, transient workers, women, and children—suffer the most. Without access to dependable piped water or cost-effective filtration, people face regular exposure to heavy metals and pathogens. As disease increases, school attendance declines and household incomes diminish, ensnaring families in a cycle of poverty and environmental degradation. A study examining human–river interactions and perceived health issues among the riverbank population in Kerala revealed that individuals from low socioeconomic backgrounds faced increased rates of skin infections and fungal diseases, highlighting the nexus between environmental degradation and social vulnerability (Rani et al., 2024). Despite the growing recognition of these issues, there remains a significant shortfall in comprehensive assessments that integrate the ecological, health, and economic dimensions of river pollution. This chapter seeks to address this gap by integrating new studies from India and beyond, offering a comprehensive picture of the socioeconomic effects of river pollution (Fig. 17.1).

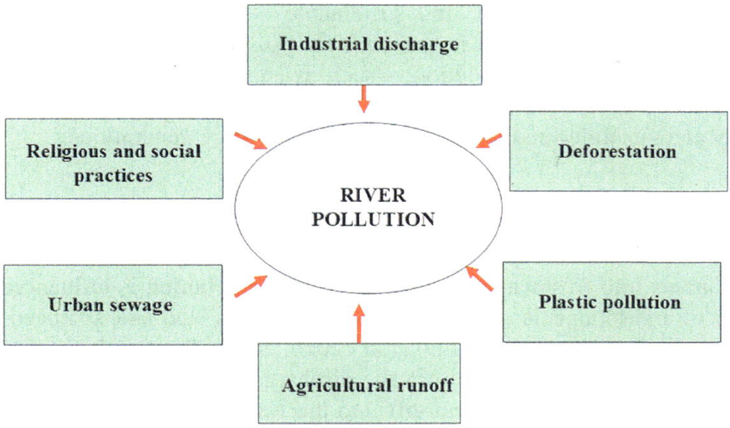

FIG. 17.1 Sources of river pollution. This figure illustrates the various sources of river pollution, including industrial discharge, deforestation, religious and social practices, urban sewage, plastic pollutants, and agricultural runoff.

This chapter analyzes the mechanisms through which pollution impacts human welfare and economic stability, aiming to guide policy responses and promote sustainable river management methods that emphasize environmental health and social fairness. The subsequent chapters delineate how contaminated rivers disrupt social cohesion and weaken regional economies. We examine case studies from the chemically contaminated banks of the Yamuna to the transboundary cleanup challenges of the Danube. This chapter ultimately promotes comprehensive solutions: enhancing wastewater treatment, enforcing industrial regulations, and enabling communities to assume stewardship of their waterways. Restoring these waterways is essential for safeguarding economic growth and human well-being. The subject of "Socioeconomic Impact of River Pollution" is particularly pertinent now, as river contamination poses a significant challenge to both environmental sustainability and human growth. As populations expand and economies industrialize, rivers globally are encountering unparalleled levels of contamination from sewage, industrial effluent, plastic garbage, and agricultural pesticides. This issue impacts ecosystems, and millions of individuals rely on rivers for potable water, agriculture, fishing, and economic endeavors. Contaminated rivers result in hazardous drinking water, diminished livelihoods for agricultural laborers and fishing communities, increased healthcare expenses due to waterborne illnesses, and the deterioration of local enterprises dependent on river tourism or a clean water supply. The resultant economic losses and health crises exacerbate social inequities, imposing a greater burden on marginalized and impoverished communities, particularly those in rapidly emerging nations and urban slums. Moreover, there is a connection between river pollution and broader concerns like food security, public health, social justice, and sustainable development. Comprehending these socioeconomic effects is essential for politicians, researchers, and the public to devise effective solutions—such as enhanced pollution regulations, infrastructure investment, and community involvement—to rehabilitate river quality and guarantee equitable development for future generations.

2 Pollution pathways and exposure dynamics

River pollution arises from many human activities that introduce a wide range of pollutants into freshwater environments. River pollution is influenced by a variety of anthropogenic and natural factors, such as land use, socioeconomic status, and climate variations (Luo et al., 2019, 2020; Chen et al., 2016). River pollution generally originates from the discharge of domestic sewage, industrial effluents, or agricultural run-off into the waters of rivers. The assault on Indian rivers from a growing population, agricultural modernization, unplanned urbanization, and industrialization is enormous and growing day by day (Birnie and Boyle, 2004). Broadly, the human causes responsible for the pollution of river water are growing population, poverty, urbanization, industrialization, agricultural run-off and improper agricultural practices, and religious and social practices.

Proper water resources management is essential to achieving an ecologically balanced environment since access to water is a fundamental right for a healthy quality of life (da Silva Antunes de Souza and de Souza, 2019). Human, animal, and plant life all depend on water. It is a component of vital ecological functions like respiration, chemosynthesis, and photosynthesis. For many plant and animal species, it serves as an ecological niche and habitat. Because they supply us with water for drinking, farming, and industry, rivers are the lifelines of our world (https://testbook.com/geography/river-water-pollution). Rivers are multifaceted, multifunctional, and intricate structures (Willett et al., 2014; Wang and He, 2022). Their development, circulation, and applications are all greatly impacted by human activity and the effects of climate change (Thorslund et al., 2021; Chung et al., 2021).

Principal river pollutants encompass biological agents (pathogenic bacteria, viruses), chemical pollutants (heavy metals, pesticides, industrial effluents), physical contaminants (plastics, sediments), and emerging pollutants, including pharmaceuticals and microplastics (Micella et al. 2024; Sharma et al., 2022). In India, untreated household sewage constitutes a primary source of microbiological contamination in rivers, with over 80% of sewage released directly into water bodies without sufficient treatment (Central Pollution Control Board (CPCB), 2020). A report submitted to the Yamuna Pollution Control Committee by the Delhi Pollution Control Committee and the Central Pollution Control Board (CPCB) indicates that the Yamuna receives a minimum of 90% of the city's domestic wastewater. The predominant source of wastewater is domestic activities, which account for the substantial quantities of laundry detergents and phosphate compounds. The analyzed samples revealed a phosphate concentration of 0.51 mg/L, beyond the standard range of 0.005–0.05 mg/L. The rivers were enveloped in strata of toxic froth as a result of the phosphate concentration. Industrial effluents are liquid waste generated during industrial activities. The swift industrial growth in India has resulted in a considerable waste management problem. The primary contributor to natural water pollution is the substantial leakage of contaminants into nearby river systems. This increases the burden on wastewater treatment and may ultimately lead to point source pollution, which not only intensifies environmental contamination but also poses significant health risks. The sources of industrial effluents include the sugar industry, paper and pulp industry, chemical fertilizer industry, textile industry, and pharmaceutical sector, which present significant health risks, such as antibiotic resistance (ABR) and many health-related issues, including endocrine disruption.

The main contaminants in industrial wastewater are nitrogen, phosphorus, hydrocarbons, heavy metals, and microbes. Industrial effluents introduce heavy metals such as lead, mercury, and cadmium, which accumulate in sediments and infiltrate the food chain, presenting chronic health hazards (Azam and Tripathi, 2025; Naqash et al., 2023). Delhi's densely urbanized watershed area of the Yamuna River is home to numerous drains. The primary conduits discharging significant pollutants into the river are the Najafgarh and Shahdara drains. Industrialization, driven by rapid urbanization and population growth, presents a significant threat of heavy metal contamination in nearby water bodies. A multitude of heavy metals are substantially present in the water of the River Yamuna. Iron (Fe) has been identified as the most abundant heavy metal in the water of the Yamuna River, exceeding permissible limits. Elevated concentrations of heavy metals in water can lead to many health consequences, including stunted growth and development, cancer, organ damage, and neurological impairment. The Hooghly Estuary in the Calcutta-Asansol region receives industrial effluents with diverse properties from various enterprises, including tanneries, paper mills, textile mills, jute mills, and large distilleries. Industrial effluents have consistently been discharged into the Damodar River from the heavily industrialized Durgapur-Asansol region of West Bengal, encompassing steel mills, coke oven facilities, distilleries, pulp and paper manufacturing, coal-based chemical industries, a fertilizer factory, and various other enterprises. The River Damodar is now one of the most polluted rivers in the country due to rampant pollution. The water is excessively polluted, rendering it unfit for human consumption. Urban industrial effluents are an increasing concern and present significant environmental challenges. Recent global studies have been undertaken to quantify wastewater and offer future estimates. Qadir et al. (2020) report that 380 billion m^3 of wastewater is generated globally each year. According to the rates of urbanization and population expansion, it is anticipated that daily wastewater production will rise by 51% (574 billion m^3) by 2050 and by 24% (470 billion m^3) by the end of the century, in 2030. Asia was the largest producer of wastewater, making up 42% (159 billion m^3) of the world's total. Projections indicate that wastewater output will increase by 44% by 2030, necessitating urgent attention.

Agricultural runoff and urban stormwater are significant contributors to the contamination of surface water (such as lakes, rivers, reservoirs, estuaries, and coastal ocean waters) and groundwater globally, underscoring the increasing human influence on natural ecosystems. Nitrogen and phosphorus are primary nutrients of concern in urban stormwater and agricultural runoff (USEPA, 2000). Agriculture, through the application of pesticides, fertilizers, and other chemicals, significantly contributes to water pollution in developed countries. These pollutants may infiltrate neighboring lakes and rivers, leading to eutrophication and jeopardizing aquatic organisms. Agricultural runoff contributes excessive nutrients (nitrogen, phosphorus), resulting in eutrophication and detrimental algal blooms that impair water quality and aquatic ecosystems (India Water Portal, 2021). An increasing body of literature associates nitrogen pollution in water with a rising incidence of health impacts. Nitrogen in its oxidized forms (nitrates and nitrites) in water, when ingested in high concentrations by newborns, has long been recognized as boosting the risk of methemoglobinemia, commonly referred to as blue baby syndrome, which can be fatal. Furthermore, runoff from agricultural areas may pollute groundwater, an essential source of potable water for numerous towns. In impoverished nations, the circumstances are often more dire. Agriculture serves as the primary source of income for millions, rendering it a vital element of numerous economies. Nevertheless, insufficient resources and technology hinder numerous farmers in these countries from employing ecologically advantageous and sustainable practices. Such behavior leads to significant contamination and degradation of water sources. Pollutants infiltrate the human system via various exposure routes: direct intake of contaminated water, skin contact during household or occupational activities, and consumption of contaminated food, especially fish and crops irrigated with dirty water (Babuji et al., 2023; Rani et al., 2024). River-dependent populations are particularly vulnerable, as their limited access to clean water forces them to rely on contaminated sources (Dekaraja and Mahanta, 2022). Sophisticated modeling approaches in recent studies have demonstrated seasonal fluctuations in pollution loads, influenced by monsoon cycles and human activity patterns. Singh et al. (2024) used a weighted fuzzy soft set model to evaluate the water quality of the Gomati River, demonstrating that pollution surges correlate with heightened human activity and insufficient waste treatment at specific times. Climate change exacerbates exposure dynamics by modifying hydrological patterns, increasing pollutant concentration during droughts, and promoting extensive dispersal during floods (van Vliet et al., 2023). This fluctuation poses problems to current monitoring methods and necessitates dynamic risk assessments to adequately protect vulnerable groups. Comprehending these pollution channels and exposure mechanisms is essential for formulating tailored interventions that mitigate health risks and socioeconomic burdens linked to river pollution.

3 Human health burden

River pollution adversely affects human health by contaminating drinking water, residential supplies, and food sources, leading to various waterborne infections and chronic health issues. Pathogenic microbial contamination, including *Escherichia coli*, *Salmonella*, and *Vibrio cholerae*, continues to be a primary cause of diarrheal diseases, disproportionately impacting children under five and resulting in considerable morbidity and mortality, particularly in low-income communities dependent on contaminated river water (Babuji et al., 2023; WHO, 2023). Cholera, an acute diarrheal sickness, is estimated to cause the deaths of 21,000 to 143,000 individuals per year, with the potential to claim up to 4 million lives if untreated (Brainerd and Menon, 2014). It is linked to water polluted with feces. The 2017 research by the Lancet Commission on Population and Health linked 1.8 million mortalities to water pollution. Heavy metals such as arsenic, lead, mercury, and cadmium, often found in numerous Indian rivers due to industrial discharges and mining operations, bioaccumulate in human tissues and result in neurological, renal, and cardiovascular disorders (Azam and Tripathi, 2025; Naqash et al., 2023). A study in the Harike wetland region of Punjab recorded heightened heavy metal levels associated with chronic health risks for local populations consuming contaminated water and fish (Naqash et al., 2023). Furthermore, exposure to organic pollutants and pesticides present in agricultural runoff has been linked to heightened occurrences of cancer, reproductive problems, and endocrine disruption (Dasgupta et al., 2022). These health burdens result in substantial economic costs associated with healthcare expenses, diminished labor productivity, and decreased quality of life, placing considerable pressure on households and public health systems (Greenstone and Jack, 2022). Research on riverfront villages in India indicates that marginalized groups encounter elevated exposure risks due to insufficient access to clean water and sanitation facilities. Rani et al. (2024) reported increased incidences of dermatological and respiratory conditions among communities residing along the rivers in Kerala, linking these health issues to persistent exposure to contaminated river water. Worldwide, analogous health effects have been noted in areas experiencing significant river pollution. Investigations of the Citarum River in Indonesia, recognized as one of the most contaminated rivers globally, revealed heightened concentrations of heavy metals in both water and sediments, correlating with detrimental health effects on surrounding populations (Asian Development Bank (ADB), 2020). This highlights the global significance of river contamination as a critical public health issue. Emerging pollutants, including medicines and microplastics, are attracting scrutiny due to their possible long-term health implications, despite the limited availability of evidence. A major concern with water contamination is the rise of ABR. When pharmaceuticals contaminate aquatic environments, particularly rivers and other water sources, they exert selection pressure on microorganisms, leading to the emergence of antibiotic-resistant bacteria (ARB). These bacteria can induce more severe and persistent diseases and diminish the effectiveness of conventional treatments when they infect patients through contaminated food, beverages, or direct contact. The evolution of germs and their development of resistance to antibiotics pose a significant threat to human health and survival. ABR, or more broadly, antimicrobial resistance (AMR), refers to the occurrence of microbes acquiring resistance to antibiotics. The veterinary and agricultural sectors significantly contribute to antibiotic contamination, with manure and wastewater irrigation serving as the primary sources of antibiotics in agricultural land. Antibiotic accumulation in East Asian agricultural soils has been documented, with concentrations ranging from 5000 to 15,000 μg/kg. The median antibiotic concentrations in agricultural soils varied from 0.008 to 160.0 μg/kg, but the median concentrations in sewage sludge and cattle dung were 100–1000 μg/kg and 10–100,000 μg/kg, respectively. Fang et al. (2023) identified 56 antibiotic compounds across five therapeutic categories—tetracyclines, macrolides, sulfonamides, quinolones, and chloramphenicols—in agricultural soils on at least one occasion. These data confirm that humans frequently misuse antibiotics, sometimes for inappropriate reasons, without understanding the repercussions, necessitating focused public health education and awareness. The increase in bacteria harboring heavy metal genes (HMGs), mobile genetic elements (MGEs), and antibiotic resistance genes (ARGs) is a prominent consequence of antibiotic contamination (Kraemer et al., 2019). These contaminants endure in aquatic systems, bioaccumulate in marine animals, and may interfere with human hormonal and immunological functioning, warranting immediate study attention (Micella et al., 2024; Sharma et al., 2022). In summary, river pollution represents a complex health risk that necessitates comprehensive water quality management and effective public health strategies to reduce disease prevalence, safeguard at-risk populations, and enhance socioeconomic welfare.

4 Economic productivity and growth impacts

This section aims to clarify the fragmented understanding of the impact of water quality degradation on economic growth and sectoral productivity. The chapter attempts to examine macroeconomic effects on GDP, evaluate sectoral impacts, define empirical approaches, compare regional and temporal changes, and analyze policy efficacy. A comprehensive examination of empirical, theoretical, and policy research utilizing econometric models, input–output analysis, and scenario simulations was conducted, focusing on data from global and Indian contexts in recent decades. Research indicates that water contamination results in substantial GDP reductions ranging from 0.8% to over 2% at the regional level, significantly impacting agricultural production, health-related efficiency, and industrial performance. Regional inequalities and temporal trends reveal deteriorating pollution levels despite economic growth in numerous Indian states, thereby undermining the Environmental Kuznets Curve argument. Policy evaluations underscore inadequate pollution control methods, recommending integrated management and financial mechanisms to mitigate economic risks. The summary highlights the intricate interaction of sectoral, regional, and governance elements influencing the economic consequences of water contamination. These insights guide focused responses and emphasize the need for strong, context-specific regulations to protect sustainable economic development in the face of increasing water quality concerns worldwide and in India. Water pollution and economic growth are fundamentally interconnected and share an unfavorable relationship. All forms of industry generate pollution as a by-product. Pollution produced by upstream sources can impact downstream users, as clean water is essential for life, health, and economic productivity. Regulators and policymakers must assess the benefits of polluting activities against the costs of pollution, using the principle of sustainability. A simplistic method to evaluate the "economic cost of pollution" is to correlate water contamination with GDP. However, while assessing indicators of water pollution, one encounters the curse of dimensionality. A plethora of water pollutants and factors exist that may adversely affect the economy, and these parameters are constantly expanding. To mitigate the curse of dimensionality, this chapter utilizes the water quality parameter "biological oxygen demand" (BOD). It is sometimes regarded as an "umbrella pollutant" because of its capacity to represent various water quality metrics. Jason Russ et al. report that when pollution is mild (i.e., BOD ranges from 2 to 8 mg/L), downstream regions experience a 1.4% decline in GDP growth. When BOD surpasses 8 mg/L, indicating a seriously polluted condition, GDP growth declines by 2%. The impacts are somewhat more pronounced in middle-income nations, with moderate pollution resulting in a growth decline of 1.8% and heavy pollution causing a growth decline of 2.5%. The impact dissipates in high-income countries.

River pollution incurs substantial economic burdens at several levels, including personal livelihoods, community well-being, and national economic development. Contaminated water diminishes labor productivity by elevating disease prevalence, resulting in absenteeism and reduced work efficiency, particularly in sectors reliant on manual labor, such as agriculture, fisheries, and informal industries (Greenstone and Jack, 2022; Azam and Tripathi, 2025). Agricultural productivity declines when contaminated river water is utilized for irrigation, leading to diminished crop yields, soil degradation, and the accumulation of hazardous compounds in food products, which adversely impact marketability and food security (India Water Portal, 2021). In the Ganga basin, contamination by heavy metals and nutritional imbalances has diminished the quality of arable land, thereby affecting farmer incomes and rural economies (Naqash et al., 2023). The quality of irrigation water is essential for assessing agricultural productivity and food security. Russ et al. (2020, 2022) estimated that irrigation with saline water diminishes worldwide agricultural productivity by approximately 124 trillion kilocalories annually. Fisheries, an essential source of nourishment and revenue for millions, confront significant challenges due to declining water quality. Pollution diminishes fish populations through habitat destruction and the bioaccumulation of toxins, consequently jeopardizing food security and the livelihoods of communities reliant on rivers (Dekaraja and Mahanta, 2022; Asian Development Bank (ADB), 2020). The Citarum River in Indonesia exemplifies the detrimental impact of industrial pollution on fisheries, resulting in significant income reductions and driving communities into poverty (Asian Development Bank (ADB), 2020). Industrial sectors incur fees for treating contaminated water used in production processes, elevating operational costs and occasionally necessitating the transfer or closure of firms, hence impacting employment and economic growth (CPCB, 2020). Moreover, the tourism and recreation sectors are adversely affected in areas where river pollution reduces esthetic and recreational appeal, hence constraining potential for economic diversification (WWF, 2023). Degradation of water quality can lead to diminished land and property values, indicating that water quality adversely affects development and prosperity beyond its influence on health, labor productivity, agricultural output, recreation, or tourism. At the macroeconomic level, health expenses resulting from pollution and the depletion of human capital impede GDP growth. Greenstone and Jack (2022) predict that pollution-induced productivity losses may reduce economic growth rates by as much as 1.5% per annum in impacted locations. Furthermore, the unrecognized losses in ecosystem services from damaged rivers, estimated globally at over $58 trillion per year, represent a significant concealed expense for economies dependent on natural capital (Micella et al., 2024; World Wildlife Fund (WWF), 2023). The World Bank asserts that upstream pollution significantly hinders downstream economic development. A World Economic Forum analysis indicates that contaminated water can lead to a 33% reduction in economic output (GDP) in certain regions (World Economic Forum, 2019). This is a substantial obstacle to the overall economic progress of those regions. In middle-income countries such as India, where water pollution is a significant concern, the situation deteriorates. In this context, GDP growth may be reduced by 50%, while economic losses could increase roughly fourfold. Studies indicate a significant 16% reduction in crop production and a 9% decline in agricultural revenue in India for those located downstream of contaminated regions (World Economic Forum, 2019). The economic burden of environmental degradation caused by industrial water pollution in India is substantial. Estimates indicate that the annual expenditure totals an impressive INR 3.75 trillion, approximately $80 billion. Estimates indicate that the annual health costs attributable to water pollution range from INR 470–610 billion ($6.7–8.7 billion) (World Economic Forum, 2019). Waterborne infections, particularly diarrheal diseases that predominantly impact children under 5 years, account for a significant share of these costs. The repercussions of polluted water extend far beyond monetary considerations. Approximately 400,000 Indians are said to perish annually due to insufficient access to clean water, sanitation, and proper hygiene. This situation is considerably more dire globally, since water-related diseases result in around 200 million lost workdays annually and the fatalities of 1.5 million children under the age of 5 years. The cascading economic effects underscore the pressing necessity for investments in pollution control and integrated water resource management, with the formulation of economic policies that internalize the environmental costs associated with river pollution.

5 Livelihood vulnerability in agriculture and fisheries

River pollution significantly impacts livelihoods reliant on freshwater ecosystems, particularly agriculture and fisheries, which sustain hundreds of millions globally, especially in low- and middle-income nations such as India. The degradation of water quality due to chemical, microbiological, and physical contaminants disrupts ecological equilibrium and the accessibility of potable water, significantly jeopardizing economic stability in riparian communities (Babuji et al., 2023; Dekaraja and Mahanta, 2022).

5.1 Agriculture

Irrigated agriculture constitutes around 80% of total water consumption in India, with a significant portion derived directly from rivers (Central Pollution Control Board (CPCB), 2020). Polluted water utilized for irrigation can contaminate soils with heavy metals and persistent organic contaminants, diminish soil fertility, and hinder crop growth. Research in Punjab and Uttar Pradesh indicates that the utilization of water tainted with cadmium, lead, and arsenic has resulted in the bioaccumulation of hazardous elements in food crops such as rice and wheat, endangering both farmers and consumers (Sharma et al., 2022; Naqash et al., 2023). Farmers often lack understanding or alternatives, resulting in the continued use of contaminated river water, which diminishes productivity and crop quality while also posing long-term health hazards. This establishes a feedback cycle in which agricultural loss exacerbates rural poverty, heightens indebtedness, and prompts migration, especially in climate-affected areas (India Water Portal 2021; Rani et al., 2024).

5.2 Fisheries

Riverine fisheries are also susceptible to river pollution. Contaminated water results in hypoxia, the destruction of breeding habitats, and fish mortality owing to toxins, thereby diminishing fish availability, biomass, and biodiversity (Asian Development Bank (ADB), 2020; World Wildlife Fund (WWF), 2023). Fisherfolk communities, particularly in the Ganga, Yamuna, and Brahmaputra basins, have shown diminishing catches and revenue, compelling some to forsake traditional livelihoods (Dekaraja and Mahanta, 2022). Naqash et al. (2023) conducted a study on the Harike Wetland (Punjab) that revealed elevated concentrations of mercury and lead in both water and fish tissue, posing direct health hazards for consumers and indirect economic repercussions on fish marketing and exportation. Pollution diminishes yields and tarnishes the reputation of fisheries and related sectors, thereby lowering prices and income. Furthermore, gender and caste contribute to vulnerability, as women and marginalized communities frequently engage in small-scale fishing, processing, and vending. These people experience diminished access to increasingly scarce resources and frequently lack political representation in water governance frameworks (Bullard, 2007; FAO, 2021). The degradation of river systems results in livelihood loss, economic marginalization, and heightened food insecurity, particularly for those least capable of adapting or relocating. This underscores the pressing necessity of amalgamating pollution management with livelihood restoration, skill diversification, and environmental justice frameworks.

6 Impacts of river pollution on microbial assemblages and aquatic animals

6.1 Microbial assemblages: Community structure and functional shifts

River pollution dramatically modifies the composition and functional capabilities of microbial communities in both water and sediment. Major Indian rivers, including the Ganga and Yamuna, contain sediments that harbor a diverse bacterial community characterized by increased levels of pollution-tolerant taxa and ARGs, suggesting pollution-induced selective pressure (Rai et al., 2023; Kumari et al., 2023). Metagenomic investigations indicate a heightened prevalence of Proteobacteria and species such as *Pseudomonas* and *Acinetobacter*, which are linked to the breakdown of xenobiotic chemicals and resistance to heavy metals (Kumari et al., 2023; Rai et al., 2023). Seasonal fluctuations additionally influence microbial communities, with increased antibiotic residues and ARG prevalence noted during the monsoon and postmonsoon periods in rivers such as Kshipra, highlighting temporal pollution dynamics (Singh et al., 2022; Kaur et al., 2018). Contaminated river sediments serve as reservoirs of ABR, containing a broad resistome with genes that impart resistance to several classes of antibiotics, including essential therapeutic medications (Singh et al., 2022; Rai et al., 2023). Environmental resistomes carry a risk of horizontal gene transfer to pathogenic bacteria, becoming a notable public health issue (Rai et al., 2023; Kaur et al., 2018). In addition, microbial functional pathways transition toward increased nitrogen and sulfur metabolism and the breakdown of aromatic hydrocarbons in contaminated waters, signifying adaptability to human-made pollutants (Kumari et al., 2023). This functional reorganization affects biogeochemical cycling and ecological services typically offered by microbial communities (Kumari et al., 2023).

6.2 ABR proliferation in riverine environments

Indian rivers, including the Yamuna, Ghaghara, and Ganga, exhibit heightened levels of antibiotic residues and multidrug-resistant bacteria in both water and sediments (Singh et al., 2022; Rai et al., 2023; Kumar et al., 2024). Quantifiable amounts of antibiotics like ofloxacin, amoxicillin, and penicillin have been found, which have led to the rise of ABR in microbial communities (Kumar et al., 2024). Environmental *E. coli* isolates from these rivers exhibit resistance to "reserve" antibiotics, underscoring the environmental aspect of AMR (Kaur et al., 2018). The dissemination of antibiotics into the environment is facilitated by urban aquatic ecosystems across heavily populated regions. The principal sources of antibiotic contamination in global aquatic ecosystems include wastewater treatment facilities (Arun et al., 2022), healthcare institutions (Zhang et al., 2020), agricultural methods (Kim et al., 2023), and anthropogenic activities (Jiang et al., 2022). Aquaculture employs antibiotics such as oxytetracycline, florfenicol, and ormetoprim extensively to treat infections and regulate the growth of Chilean salmon and Nile tilapia (*Oreochromis niloticus*) (de Freitas et al., 2023; Leal et al., 2019; Avendaño-Herrera et al., 2023). Pharmaceutical pollutants such as carbamazepine, ofloxacin, and paracetamol are detected in the effluents of wastewater treatment facilities, thereby further polluting urban rivers (Shigei et al., 2021). Surface runoffs are significant contributors to antibiotic pollution in aquatic environments, with ARB, ARGs, HMGs, and MGEs as key causes of this contamination. The findings highlight polluted rivers as critical sites for the establishment and spread of ARGs, raising concerns over the contamination of drinking water supplies and the transmission of resistant bacteria to humans and animals (Rai et al., 2023; Kumar et al., 2024).

6.3 Effects on aquatic animals and biodiversity

Pollution-driven changes in microbial communities adversely affect aquatic organisms, including fish, macroinvertebrates, and zooplankton, influencing their health, diversity, and population dynamics (Singh et al., 2023; Sharma et al., 2022). Alterations in fish gut microbiomes resulting from water pollution can influence digestion, immunological responses, and growth rates, as evidenced in Cyprinus carpio in the Yamuna River (Sharma et al., 2022). Changes in macroinvertebrate and plankton communities in polluted rivers indicate reductions in biodiversity and functional diversity, as sensitive species are supplanted by tolerant taxa (Singh et al., 2023). Research on Indian river systems indicates that heightened levels of heavy metals, organic pollutants, and nutrient loading negatively impact sediment-dwelling organisms and fish, resulting in physiological stress, increased vulnerability to diseases, and modified reproductive success (Naqash et al., 2023; Rani et al., 2024). Changes in parasite dynamics in fish subjected to pollution have been documented, indicating intricate ecological disturbances (Naqash et al., 2023).

7 Unequal burden and environmental injustice

Not all populations experience the effects of river contamination equally. Instead, they mirror and perpetuate preexisting social and economic disparities, particularly impacting impoverished, marginalized castes, Indigenous populations, women, and individuals residing in informal settlements along contaminated riverbanks. This situation corresponds with the tenets of environmental injustice, wherein marginalized populations disproportionately endure the consequences of environmental degradation while possessing restricted access to mitigating resources and decision-making processes (Bullard, 2007; Dekaraja and Mahanta, 2022). In India, various research indicates that low-income communities frequently inhabit regions with the most exposure to contaminated water sources. These communities rely significantly on rivers for potable water, sanitation, fishing, and irrigation owing to the lack of piped water infrastructure and sewage systems (Rani et al., 2024; Central Pollution Control Board (CPCB), 2020). In certain regions of Assam and Kerala, riverbank communities have documented elevated occurrences of gastrointestinal diseases, dermatological infections, and reproductive health complications associated with extended exposure to contaminated rivers (Dekaraja and Mahanta, 2022; Rani et al., 2024). Marginalized groups, such as Dalits and tribal populations, frequently have limited power over local water governance. Although they endure the most severe repercussions of pollution, they are infrequently consulted in water management strategies or incorporated into river restoration initiatives such as the Namami Gange program (India Water Portal, 2021; Srivastava et al., 2025). This structural exclusion exacerbates socioeconomic inequities, resulting in what scholars term a "double vulnerability," exposure to both environmental risks and institutional neglect. Gender introduces an additional vital feature. Women, especially in rural regions, serve as the principal water collectors and carers. They are adversely impacted by contaminated rivers, since substandard water quality amplifies their daily burdens, exposes them to health hazards, and restricts time for schooling or economic pursuits (FAO, 2021; Babuji et al., 2023). However, policy determinations and scientific evaluations concerning water quality predominantly exclude women's viewpoints. Globally, this pattern remains consistent. The Citarum River Basin in Indonesia and river systems in sub-Saharan Africa exhibit analogous patterns, wherein impoverished, river-dependent people are subjected to industrial pollutants yet lack the political agency to advocate for regulatory enforcement (Asian Development Bank (ADB), 2020; United Nations Environment Programme (UNEP), 2021a, 2021b). Consequently, river pollution transcends mere technical or ecological concerns; it is fundamentally a political and ethical matter. Addressing this issue necessitates environmental justice frameworks that prioritize the voices and rights of the most impacted populations while ensuring equal access to clean water, healthcare services, and compensation procedures. The Indian community of Bichhri in Rajasthan offers an alternative perspective on the problem.

The struggle here is not for scarcity but for the intrinsic right to clean water. For >35 years, the inhabitants of Bichhri have been involved in a clandestine struggle against industrial pollution. The arrival of five chemical facilities in 1987 heralded a promise of wealth. However, the reality was far from idyllic. These companies, particularly those producing H-acid, a dye precursor, discharged their hazardous waste into open sewers and unlined pits. The toxic, acidic water infiltrated the soil, polluting the village's sole source of sustenance, the groundwater. It was a rapid and vicious strike. The previously clear water transformed into a turbid brown, rendering it unsuitable for consumption or irrigation. Agricultural yields decreased by >75% on certain farms. This poisonous influx primarily affected livestock. The inhabitants of Bichhri declined to remain passive, while they fought for their survival. Protests commenced, petitions were submitted, and voices were articulated as early as 1988. In 1996 their unwavering quest for justice culminated in a landmark Supreme Court decision. This decision was a symbol of hope. The Supreme Court's ruling mandated the closure of polluting facilities and implemented the "polluter pay principle," which assigns accountability to firms for environmental damage and associated remediation costs (Mongabay-India, 2023). The calamities of Bichhri and Kanpur exemplify the grave human toll of industrial negligence. The livelihoods and health of the villagers are at risk as they await justice. These occurrences act as a clarion call. These instances constitute a summons to action: (1) Enhanced enforcement: Environmental standards require rigor. Robust enforcement procedures are essential to ensure accountability for polluters. (2) Support for affected communities: Villages like Bichhri require financial and technical support to restore their water supplies and rebuild their lives.

8 River ecosystems and unpriced losses

River pollution poses significant risks to human health and livelihoods while inflicting severe, often irreversible harm to freshwater ecosystems, impacts that are frequently overlooked in conventional economic evaluations. Unpriced ecological losses encompass declines in biodiversity, disruption of ecological functions, wetland degradation, and the collapse of services, including water purification, carbon sequestration, and climate regulation (Micella et al., 2024; World Wildlife Fund (WWF), 2023). Freshwater ecosystems represent significant biological diversity; however, they face considerable global threats. The Living Planet Report by WWF (2023) indicates that freshwater species populations have experienced an average decline of 83% since 1970, primarily attributed to pollution, overextraction, and habitat modification. Rivers serve as essential migration corridors and breeding habitats for numerous species. Contamination by heavy metals, pesticides, and organic pollutants disrupts reproductive processes, modifies behavior, and contributes to species extinction (FAO, 2021; Naqash et al., 2023). In India, rivers such as the Ganga and Yamuna have experienced considerable ecological decline. Excess nutrient loading has resulted in algal blooms, which create hypoxic zones detrimental to aquatic life (Central Pollution Control Board (CPCB), 2020). Endocrine-disrupting chemicals originating from pharmaceuticals and industrial effluents have been demonstrated to impact fish populations in the Ganga basin, posing a threat to long-term ecosystem resilience (Sharma et al., 2022). Pollution diminishes the ability of rivers to deliver ecosystem services. Clean river systems provide natural filtration, recharge groundwater, support local microclimates, and assist in flood regulation. Upon degradation, these services either fail or require artificial replacement, incurring significant financial and ecological costs (Micella et al., 2024). Restoring polluted rivers to achieve basic water quality standards necessitates substantial financial investment; however, few evaluations account for the opportunity cost associated with the degradation of ecological functions over time. Rivers frequently possess significant cultural, spiritual, and esthetic importance in various societies. In India, rivers such as the Ganga serve as both ecological systems and sacred sites. Pollution diminishes these intangible benefits, impacting mental well-being, community identity, and traditional practices (Srivastava et al., 2025; India Water Portal, 2021). The inability to quantify and integrate ecological and cultural losses into development planning and environmental policy results in ongoing underinvestment in river conservation. Emerging frameworks such as Natural Capital Accounting and ecosystem service valuation serve as tools to address this gap and enhance holistic water governance (United Nations Environment Programme (UNEP), 2021a, 2021b; World Wildlife Fund (WWF), 2023).

9 Policy response and governance gaps

Notwithstanding the increasing recognition of the gravity of river pollution and its socioeconomic ramifications, governmental responses continue to be disjointed, inadequately financed, and ineffectively enforced throughout most of the Global South, including India. Efficient river governance necessitates collaboration among several sectors (water, agriculture, industry, health, and urban development); nevertheless, institutional silos, regulatory gaps, and political stagnation persistently obstruct advancement (Central Pollution Control Board (CPCB), 2020; United Nations Environment Programme (UNEP), 2021a, 2021b). In India, prominent initiatives such as the Namami Gange Mission and the National River Conservation Plan have undertaken ambitious pledges to rehabilitate contaminated rivers, focusing on investments in sewage infrastructure, surface cleaning, and public awareness campaigns. However, inadequate monitoring systems, implementation delays, and insufficient community involvement constrain their efficacy (Srivastava et al., 2025; India Water Portal, 2021). An incisive evaluation of the Yamuna Action Plan indicated that, despite prolonged financial commitment, pollution levels in numerous segments persist at dangerously elevated rates due to the ongoing discharge of untreated sewage and industrial effluents (Sharma et al., 2022). Global experiences reflect these challenges. In Indonesia's Citarum River restoration initiative, supported by the Asian Development Bank, initial top-down clean-up efforts neglected local livelihood requirements and community involvement, resulting in opposition and minimal compliance (Asian Development Bank (ADB), 2020). In Africa and Latin America, river restoration initiatives frequently fail due to inadequate enforcement, corruption, and the lack of baseline environmental data (United Nations Environment Programme (UNEP), 2021a, 2021b). A significant deficiency in governance exists regarding oversight and data openness. Water quality data frequently suffers from obsolescence, inaccessibility, or incompleteness, hindering effective policy evaluation and public scrutiny (Micella et al., 2024). Moreover, the funding of enforcement authorities is inadequate, and they lack the technical expertise necessary to supervise the numerous small and medium enterprises that discharge waste into waterways. Environmental restrictions, exemplified by the Water (Prevention and Control of Pollution) Act, 1974 in India, are frequently compromised by inadequate fines and political meddling. Industrial polluters often circumvent compliance by taking advantage of regulatory ambiguities or by paying nominal fines (Babuji et al., 2023; Sharma et al., 2022). Local administrations, responsible for implementing water regulations, lack sufficient resources, competence, and authority. A significant disjunction exists between environmental and public health policy. Health departments are infrequently engaged in water administration, despite compelling evidence of the burdens posed by waterborne diseases. River management strategies still inadequately employ integrated methodologies like the One Health framework, which links environmental and human health hazards (FAO, 2021). The ongoing prevalence of river contamination, despite several interventions over decades, highlights fundamental governance flaws. Addressing these deficiencies necessitates robust regulatory enforcement, real-time monitoring systems, interagency collaboration, and community-oriented strategies that integrate environmental justice into river restoration plans.

10 Future directions and policy recommendations

Addressing the socioeconomic consequences of river pollution necessitates a transition from disjointed, reactive approaches to cohesive, proactive, and equity-centered initiatives. Future policy solutions must integrate environmental, health, and livelihood considerations into a unified river basin management framework, drawing on lessons from India and global case studies. **1. Enhance integrated river basin governance**: An integrated, river-basin-level strategy that encompasses water quality, land use, industrial development, and public health is important. Enhancing intersectoral cooperation among pollution control boards, municipal authorities, agricultural departments, and health agencies can augment coherence and enforcement (Micella et al., 2024; United Nations Environment Programme (UNEP), 2021a, 2021b). **2. Shift to real-time surveillance and public data transparency**: Implementing affordable, sensor-driven water quality monitoring systems at key river sites—connected to open-access platforms—can improve accountability and facilitate early intervention (Sharma et al., 2022). Citizen science and community monitoring initiatives, when bolstered by training and institutional support, can enhance state-led surveillance (Srivastava et al., 2025). **3. Implement the polluter pays principle and enhance industrial compliance**: In a landmark 1996 judgment regarding the "Bichhari village" case, the Supreme Court affirmed that the right to safe and healthy drinking water is a basic right of all citizens and mandated the closure of all companies, establishing the "polluter pays principle." Rigorous implementation of the "polluter pays" idea is essential. Industries polluting rivers must encounter more stringent legal repercussions and be encouraged to implement zero-liquid-discharge or circular water utilization technologies. Enhancing regulatory clearance procedures through digital compliance monitoring can mitigate corruption and augment transparency (Asian Development Bank (ADB), 2020; Central Pollution Control Board (CPCB), 2020). **4. Enhance wastewater infrastructure and implement decentralized treatment**: Urban and rural regions require enhanced sewage infrastructure and the promotion of decentralized wastewater treatment systems, particularly in periurban areas. Investments in nature-based solutions, such as created wetlands, provide economical and sustainable alternatives (India Water Portal, 2021; World Wildlife Fund (WWF), 2023). Integrate Livelihood and Health Considerations into River Restoration Strategies Restoration initiatives must prioritize the needs of communities dependent on rivers. The integration of health risk assessments, economic rehabilitation, and livelihood diversification initiatives (such as training in sustainable aquaculture and ecotourism) might enhance social support and mitigate vulnerability (Naqash et al., 2023; Rani et al., 2024). **5. Acknowledge rivers as sociocultural–ecological systems**: Policy frameworks must extend beyond biophysical remediation to acknowledge the sociocultural significance of rivers. This involves engaging religious leaders, traditional water users, and Indigenous communities in stewardship duties, as well as safeguarding sacred river segments from industrial encroachment (Srivastava et al., 2025; Bullard, 2007). **6. Institutionalize environmental justice and inclusive governance**: Establishing institutional frameworks that facilitate the involvement of marginalized groups, women, and local people in water governance can increase the legitimacy and efficacy of river management programs. Instruments such as environmental impact assessments, public grievance resolution mechanisms, and local river parliaments (as tested in Uttarakhand) might facilitate the democratization of decision-making (FAO, 2021; United Nations Environment Programme (UNEP), 2021a, 2021b). **7. Conform to international frameworks and financial structures**: India and other nations can synchronize their national river policies with international frameworks, such as the UN Sustainable Development Goals (SDG 6 concerning clean water), and secure green financing via mechanisms like the Green Climate Fund or the Global Environment Facility to facilitate sustainable river pollution management (United Nations Environment Programme (UNEP), 2021a, 2021b).

11 Conclusion

River pollution is a multifaceted developmental challenge that adversely affects health, livelihoods, food security, ecosystems, and economic growth, disproportionately harming marginalized and river-dependent populations in India and beyond. Diminished agriculture, failing fisheries, heightened disease prevalence, and insecurity in livelihoods manifest these effects. Notwithstanding policy initiatives, deficiencies in governance and insufficient investment impede successful rehabilitation. Rivers should be governed as socioecological systems using integrated, participatory, and rights-based methodologies that merge environmental restoration with social equity and inclusive growth. Pollution fundamentally modifies microbial populations and aquatic animals, transforming sediments into reservoirs of ARB and affecting ecosystem functioning. These biological alterations intensify health risks and diminish biodiversity, jeopardizing the resilience of freshwater ecosystems and the populations relying on them. Effective river management must incorporate microbiological and biodiversity monitoring alongside chemical evaluations to inform targeted treatments. Only through comprehensive policies can the socioeconomic impacts of river pollution be alleviated and sustainable futures for riverine communities secured.

References

Arun, S., Xin, L., Gaonkar, O., Neppolian, B., Zhang, G., Chakraborty, P., 2022. Antibiotics in sewage treatment plants, receiving water bodies and groundwater of Chennai city and the suburb, South India: occurrence, removal efficiencies, and risk assessment. Sci. Total Environ. 0048-9697. 851 (Pt 2), 158195. doi:10.1016/j.scitotenv.2022.158195.

Asian Development Bank (ADB), 2020. Reviving the Citarum River. ADB Publications.

Avendaño-Herrera, R., Mancilla, M., Miranda, C.D., 2023. Use of antimicrobials in Chilean Salmon farming: facts, myths and perspectives. Rev. Aquac. 15 (1), 89–111.

Azam, M.M., Tripathi, J.K., 2025. Health and ecological risk assessment of metals in surface water from the Himalayan tributaries of the Ganga river, India. Geochem. Trans. 1467-4866. 26 (1). Available from: https://geochemicaltransactions.biomedcentral.com/. doi:10.1186/s12932-025-00100-7.

Babuji, P., Thirumalaisamy, S., Duraisamy, K., Periyasamy, G., 2023. Human health risks due to exposure to water pollution: a review. Water. 2073-4441. 15 (14), 2532. doi:10.3390/w15142532.

Birnie, P.W., Boyle, A.E., 2004. "International Law And The Environment", Second ed. Oxford University Press, New Delhi, p. 306.

Brainerd, E., Menon, N., 2014. Seasonal effects of water quality: the hidden costs of the Green Revolution to infant and child health in India. J. Dev. Econ. 107, 49–64.

Bullard, R.D., 2007. Dumping in Dixie: Race, Class, and Environmental Quality, Westview Press.

Central Pollution Control Board (CPCB). (2020). Status of water quality in India. Government of India.

Chen, Q., Mei, K., Dahlgren, R.A., Wang, T., Gong, J., Zhang, M.H., 2016. Impacts of land use and population density on seasonal surface water quality using a modified geographically weighted regression. Sci. Total Environ. 5724, 50–66.

Chung, M.G., Frank, K.A., Pokhrel, Y., Dietz, T., Liu, J., 2021. Natural infrastructure in sustaining global urban freshwater ecosystem services. Nat. Sustain. 2398-9629. 4 (12), 1068–1075. Available from: www.nature.com/natsustain/. doi:10.1038/s41893-021-00786-4.

da Silva Antunes de Souza, M.C., Antunes de Souza, G.K., 2019. Invisible pollutants: environmental, economic and social impacts as threats to water quality. Revista Jurídicas 16 (2), 95–107.

Dasgupta, S., et al, 2022. Pesticide contamination and health risks in Indian river systems. Environ. Toxicol. Chem. 41 (5), 1115–1127.

de Freitas, L.V.P., da Mota Júnior, C.A.A., Damaceno, M.A., Silveira, J.G.F., Portela, A.C.V., Campanharo, S.C., da Silva, A.F.B., Assane, I.M., Pilarski, F., Sasanya, J.J., Paschoal, J.A.R., 2023. Residue depletion profile and estimation of withdrawal period for sulfadimethoxine and ormetoprim in edible tissues of Nile Tilapia (*Oreochromis* sp.) on medicated feed. Animals. 2076-2615. 13 (15). Available from: http://www.mdpi.com/journal/animals. doi:10.3390/ani13152499.

Dekaraja, S., Mahanta, P., 2022. Exploring the socioeconomic impact of riverbank erosion in the Brahmaputra Valley of Assam, India: a case study. Landsc. Online 97, 1105.

Fang, L., Chen, C., Li, S., Ye, P., Shi, Y., Sharma, G., Sarkar, B., Shaheen, S.M., Lee, S.S., Xiao, R., Chen, X., 2023. A comprehensive and global evaluation of residual antibiotics in agricultural soils: accumulation, potential ecological risks, and attenuation strategies. Ecotoxicol. Environ. Saf. 0147-6513. 262, 115175. doi:10.1016/j.ecoenv.2023.115175.

FAO, 2021. The State of World Fisheries and Aquaculture 2020. Food and Agriculture Organization of the United Nations.

Greenstone, M., Jack, B., 2022. Pollution and economic growth. J. Environ. Econ. 45 (3), 215–230.

India Water Portal. (2021). Impacts of river pollution on health and livelihoods. Retrieved from https://www.indiawaterportal.org.

Jiang, Q., Feng, M., Ye, C., Yu, X., 2022. Effects and relevant mechanisms of non-antibiotic factors on the horizontal transfer of antibiotic resistance genes in water environments: a review. Sci. Total Environ. 0048-9697. 806 (Pt 3), 150568. doi:10.1016/j.scitotenv.2021.150568.

Kaur, H., Singh, J., Nagpal, A., 2018. Antibiotic resistance characterization of environmental *Escherichia coli* isolated from River Mula-Mutha, Pune District, India. Int. J. Environ. Res. Public Health 15 (6), 1247.

Kim, J.J., Seong, H.J., Johnson, T.A., Cha, C.-J., Sul, W.J., Chae, J.-C., 2023. Persistence of antibiotic resistance from animal agricultural effluents to surface water revealed by genome-centric metagenomics. J. Hazard. Mater. 457, 131761.

Kraemer, S.A., Ramchandran, A., Perron, G.G., 2019. Antibiotic pollution in the environment: from microbial ecology to public policy. Microorganisms. 7 (6), 180. 22.

Kumar, S., Sharma, A., Singh, P., 2024. Profiling of antibiotic residues in surface water of River Yamuna stretch passing through Delhi, India. Water 15 (3), 527.

Kumari, R., Rai, V., Singh, N., 2023. Unveiling the microbiome landscape: a metagenomic study of bacterial diversity, antibiotic resistance, and virulence factors in the sediments of the River Ganga, India. Antibiotics 12 (12), 1735.

Leal, J.F., Santos, E.B.H., Esteves, V.I., 2019. Oxytetracycline in intensive aquaculture: water quality during and after its administration, environmental fate, toxicity and bacterial resistance. Rev. Aquac. 11 (4), 1176–1194.

Luo, Z., Shao, Q., Zuo, Q., Cui, Y., 2020. Impact of land use and urbanization on river water quality and ecology in a dam dominated basin. J. Hydrol. 584, 124655.

Luo, Z., Zuo, Q., Shao, Q., Ding, X., 2019. The impact of socioeconomic system on the river system in a heavily disturbed basin. Sci. Total Environ. 1879-1026. 660, 851–864. Available from: www.elsevier.com/locate/scitotenv. doi:10.1016/j.scitotenv.2019.01.075.

Micella, F., et al, 2024. Future scenarios for river exports of multiple pollutants worldwide. Earth Future 12 (4), e2024EF004712.

Mongabay-India. (2023). A village contaminated by industrial waste has been waiting for justice for over three decades.

Naqash, N., Jamal, M.T., Singh, R., 2023. Heavy metal contamination in surface water of Harike Wetland, India: source and health risk assessment. Water. 2073-4441. 15 (18), 3287. doi:10.3390/w15183287.

Qadir, M., Drechsel, P., Jiménez Cisneros, B., Kim, Y., Pramanik, A., Mehta, P., Olaniyan, O., 2020. Global and regional potential of wastewater as a water, nutrient and energy source. Nat. Resour. Forum. 1477-8947. 44 (1), 40–51. Available from: http://onlinelibrary.wiley.com/journal/10.1111/(ISSN)1477-8947. doi:10.1111/1477-8947.12187.

Rai, V., Singh, S., Pandey, A., 2023. Comparative bacteriome and antibiotic resistome analysis of water and sediment of the Ganga River of India. Microorganisms 11 (3), 657. doi:10.3390/microorganisms11030657.

Rani, P., et al., 2024. Human river interaction and perceived morbidities among the riverbank population in Kerala. J. Integr. Med. Public Health 6 (1), 1–9.

Russ, J., Zaveri, E., Damania, R., Desbureaux, S., Escurra, J., Rodella, A.S., 2020. Salt of the Earth: quantifying the impact of water salinity on global agricultural productivity. Water Global Practice, Policy Research Working Paper, 9144.

Russ, J., Zaveri, E., Desbureaux, S., Damania, R., Rodella, A.S., 2022. The impact of water quality on GDP growth: evidence from around the world. Water Secur. 2468-3124. 17, 100130.

Sharma, A., Nagpal, A., Kaur, H., 2022. Potentially toxic elements in river water and associated health risks in Ropar Wetland, India and its vicinity. Environ. Sci. Pollut. Res. 29, 21444–21459.

Shigei, M., Assayed, A., Hazaymeh, A., Dalahmeh, S.S., 2021. Pharmaceutical and antibiotic pollutant levels in wastewater and the waters of the zarqa river, Jordan. Appl. Sci. (Switz. 2076-3417. 11 (18). Available from: https://www.mdpi.com/2076-3417/11/18/8638/pdf. doi:10.3390/app11188638.

Singh, D., et al., 2022. Seasonal variations in water quality, antibiotic residues, resistant bacteria and antibiotic resistance genes of *Escherichia coli* isolates from water and sediments of the Kshipra River in Central India. Environ. Sci. Pollut. Res. 29 (42), 62635–62650. doi:10.1007/s11356-022-18958-x.

Singh, M., Singh, S., Kumar, A., 2024. Assessing anthropogenic influences on the water quality of Gomati River using an innovative weighted fuzzy soft set based water pollution rating system. Environ. Sci. Pollut. Res. 4, 73.

Singh, S., et al, (2023). Aquatic biodiversity loss in macroinvertebrates and plankton communities due to urban river pollution in India. Environ. Monit. Assess. 195, 212. doi:10.1007/s10661-023-11092-4.

Srivastava, A., Singhal, S., Zehra, R., Verma, J., 2025. Rising tides of contamination from source to sink: the Yamuna's struggle with pollution (2014–2024). Pollution Study. 2811-0153. 6 (2), 3199. doi:10.54517/ps3199.

Thorslund, J., Bierkens, M.F.P., Oude Essink, G.H.P., Sutanudjaja, E.H., van Vliet, M.T.H., 2021. Common irrigation drivers of freshwater salinisation in river basins worldwide. Nat. Commun. 2041-1723. 12 (1). Available from: http://www.nature.com/ncomms/index.html. doi:10.1038/s41467-021-24281-8.

United Nations Environment Programme (UNEP). (2021a). *Measuring Ecosystem Services and Natural Capital: a Guide for Policy Makers*. UNEP.

United Nations Environment Programme (UNEP). (2021b). *State of the world's rivers report*. UNEP.

USEPA (U.S. Environmental Protection Agency), 2000. Nutrient criteria technical guidance manual —rivers and streams (EPA 822B-00-002). Office of Water, Washington, D.C.

van Vliet, M.T.H., Thorslund, J., Strokal, M., Hofstra, N., Flörke, M., Macedo, H.E., Nkwasa, A., Tang, T., Kaushal, S.S., Kumar, R., van Griensven, A., Bouwman, L., Mosley, L.M., 2023. Global river water quality under climate change and hydroclimatic extremes. Nat. Rev. Earth Environ. 2662-138X. 4 (10), 687–702. Available from: https://www.nature.com/natreveearthenviron/. doi:10.1038/s43017-023-00472-3.

Wang, H., He, G., 2022. Rivers: linking nature, life, and civilization. River. 2750-4867. 1 (1), 25–36. doi:10.1002/rvr2.7.

Willett, S.D., McCoy, S.W., Taylor Perron, J., Goren, L., Chen, C.Y., 2014. Dynamic reorganization of River Basins. Science. 1095-9203. 343 (6175). Available from: http://www.sciencemag.org/content/343/6175/1248765.full.pdf. doi:10.1126/science.1248765.

World Economic Forum. (2019). Data and technology can provide solutions to India's water pollution problem.

World Health Organization (WHO). (2023). *Global water, sanitation and hygiene report*. WHO.

World Wildlife Fund (WWF). (2023). *Living Planet Report 2022 – freshwater Focus*. WWF International.

Zhang, S., Huang, J., Zhao, Z., Cao, Y., Li, B., 2020. Hospital wastewater as a reservoir for antibiotic resistance genes: a meta-analysis. Front. Public Health. 2296-2565. 8. Available from: http://journal.frontiersin.org/journal/public-health/section/public-health-education-and-promotion#archive. doi:10.3389/fpubh.2020.574968.

Section III

Assessment of pollution and remedial measures

Chapter 18

Assessing water quality with advanced instrumental methods: Concepts, methods, and new approaches

Ashutosh Kumar, Niranjan Kumar, Prakash Kumar, and Santosh Dixit
Amity School of Engineering & Technology, Amity University Patna, Patna, Bihar, India

1 Introduction

1.1 The need for water quality evaluation

Water controls the biochemical cycles that support civilization, maintains life, and regulates planetary thermodynamics. However, growing development, intensified agriculture, and urban effluents are gradually compromising its cleanliness. In addition to impairing ecosystem function, declining water quality has a direct impact on human health, agriculture, and the economy (Russ et al., 2022; Okorogbona, 2018). As a result, the need for sensitive and precise water quality monitoring has evolved from a local issue to a global scientific and policy imperative.

Classical volumetric, gravimetric, and colorimetric techniques—valuable for their simplicity but constrained in scope and sensitivity—were historically a major component of traditional water analysis. The sub-ppm (parts per million) or ppb (parts per billion) levels of harmful metals, pharmaceutical residues, and organic micropollutants that are currently common in water systems cannot be addressed by these methods, even though they are appropriate for evaluating gross contamination (Guo et al., 2019). Analytical chemistry in the present day requires advanced equipment that combines automation, sensitivity, and selectivity.

River Pollution. https://doi.org/10.1016/B978-0-443-45024-2.00008-7

1.2 The development of sophisticated instrumentation

A range of analytical tools with high sensitivity, multielement capacity, and computational interfaces are referred to as sophisticated instruments. From manual titrations to multidimensional chromatography and high-resolution spectrometry, their evolution mirrors that of analytical chemistry. Water analysis has been transformed by devices like ion-selective electrodes (ISEs), high-performance liquid chromatographs (HPLC), and inductively coupled plasma–mass spectrometers (ICP–MS), which provide precise, real-time quantification of various analytes within complex matrices (Sengupta et al., 2025).

The frontier of water quality evaluation now includes sensor-based, field-deployable devices that can monitor continuously in situ, outside of the laboratory. These systems, which are frequently connected by Internet-of-Things (IoT) infrastructures (Manjakkal et al., 2023), represent a paradigm change in which data collection, signal processing, and decision-making take place in a nearly autonomous manner. These shifts from bench-scale to field-scale investigation represent a redefining of scientific observation in environmental chemistry, as well as a technical improvement.

1.3 The chapter's goals and scope

This chapter aims to:

1. Classify and define essential parameters to establish the conceptual framework for water quality evaluation.
2. Examine the fundamentals and uses of advanced instruments, such as electrochemical, chromatographic, spectroscopic, and hybrid ones.
3. Group instruments according to the development of technology (primary, secondary, and advanced sensor systems).
4. Talk about data analytics, quality control, and upcoming developments in water quality equipment. By doing this, the conversation emphasizes that instrumentation is a way to strengthen scientific temper and ecological accountability—values essential to sustainable environmental governance—rather than an objective in and of itself.

2 Conceptual framework of water quality assessment

2.1 Defining water quality

The physical, chemical, and biological properties of water that define its fitness for a particular use—be it drinking, irrigation, industrial, or ecosystem support—are referred to as water quality (Boyd, 2015). This quality is defined by a variety of criteria, including dissolved gases, particulate matter, inorganic ions, organic contaminants, and microorganisms. Each of these makes a distinct contribution to the overall comprehension of the functional condition of water.

Water quality can be analytically represented as a multidimensional dataset, with each parameter reflecting the interplay between anthropogenic influences and natural geochemical processes. For example, lithological interactions are revealed by calcium and bicarbonate concentrations, whereas agricultural runoff is indicated by nitrate and phosphate levels. Because it determines not only what is present but also why and from where, instrumental analysis in water chemistry provides both diagnostic and predictive functions (Durmishi et al., 2025).

2.2 Classification of parameters

Parameters are conventionally classified as **physical, chemical, and biological**, though their interactions often transcend disciplinary boundaries.

- **Physical parameters:** Temperature, turbidity, electrical conductivity (EC), total dissolved solids (TDS), and color. These govern light penetration, thermal stratification, and the ionic character of water.
- **Chemical parameters:** pH, dissolved oxygen (DO), biochemical oxygen demand, chemical oxygen demand (COD), alkalinity, hardness, nutrients (NO_3^-, PO_4^{3-}, NH_4^+), and heavy metals (Pb, Cd, Cr, As, Hg).
- **Biological parameters:** Microbial counts, coliform indices, and enzymatic activity, serving as indicators of pathogenic or eutrophic conditions.

Each parameter demands an analytical method tailored to its chemical nature and concentration range. For example, DO can be estimated by Winkler titration or electrochemical sensors, while trace arsenic requires ICP–MS or hydride generation [atom absorption spectroscopy (AAS)].

2.3 Instrumental role in quantitative analysis

Sophisticated instruments function on the principle of transducing a chemical property into a measurable physical signal: absorbance, emission, mass-to-charge ratio, or potential difference. The accuracy of such quantification depends on three pillars:

1. **Sensitivity:** The lowest detectable limit of an analyte, often determined by the signal-to-noise ratio.
2. **Selectivity:** The instrument's ability to discriminate against the target analyte from interfering species.
3. **Precision and accuracy:** Repeatability of measurements and proximity to true values, respectively (Alfonso, 2024).

Modern instruments incorporate automated calibration, self-diagnostics, and digital error correction (Taymanov et al., 2021), thereby minimizing human error and enhancing reproducibility, an indispensable requirement for regulatory reporting.

2.4 Instrumental classification: Primary, secondary, and advanced systems

The technological evolution of analytical instrumentation can be classified into three tiers, each reflecting the degree of automation, resolution, and integration.

1. **Primary instruments:** These include standard electrochemical and optical instruments, including turbidimeters, conductivity meters, and pH meters. They serve as the foundation for standard water testing and provide single-parameter measurements, frequently under laboratory settings.
2. **Secondary instruments:** This intermediate category includes devices such as ion chromatographs, ultraviolet–visible (UV–Vis) spectrophotometers, and AAS. They provide enhanced sensitivity, the capacity to identify trace pollutants, and multianalyte capabilities. For compliance testing, the majority of regulatory laboratories use this class.
3. **Advanced and sensor-based systems:** Inductively coupled plasma–optical emission spectroscopy (ICP–OES), ICP–MS, gas chromatography–mass spectrometry (GC–MS), HPLC, and field-deployable biosensors fall under this category. These systems combine software-driven analytics, microprocessor control, and signal digitalization. They can frequently identify many elements and compounds at ppb levels at the same time. IoT-enabled sensor networks and remote telemetry are increasingly being used to adapt advanced systems for real-time, in situ monitoring (Khati et al., 2025).

Such categorization represents the philosophical shift from measurement to monitoring—where data streams are continuous, integrative, and dynamic—as well as a technological advancement.

2.5 Regulatory and index frameworks

While there are differences in water quality regulations around the world, ecological safety and health preservation are at their core. National organizations like the Bureau of Indian Standards (BIS) and the US Environmental Protection Agency (EPA) establish parameter-wise restrictions, while the World Health Organization (WHO) sets the global standards for drinking water (BIS, 2012; WHO, 2022). For instance, the WHO has set a maximum of 0.01 mg/L for arsenic in drinking water and 500 mg/L for TDSs in drinkable water.

Multidimensional data is simplified into a single numerical value that represents overall water quality using composite indices like the Water Quality Index (WQI). By reducing analytical uncertainty, sophisticated tools improve the accuracy of WQI computations, making index-based decision-making a more robust scientific endeavor.

2.6 Data integrity and analytical ethics

Data integrity becomes crucial as reliance on automated tools grows. Traceability, standard reference materials, and calibration procedures need to be strictly adhered to. These procedures are followed by laboratories accredited under ISO/IEC 17025 (Bekmirzayevich et al., 2024), guaranteeing reproducibility and cross-regional comparability.

3 Instrumental techniques for water quality analysis

3.1 Overview

The physicochemical properties, concentration range, and matrix complexity of the analyte all influence the selection of the instrumental technique. By exposing elemental speciation, molecular structures, and trace-level dynamics, advanced instrumentation not only improves analytical accuracy but also expands the interpretive potential of water chemistry. With new developments in sensor-based automation and miniaturization, instruments can be generically classified as spectroscopic, chromatographic, electrochemical, and hybrid systems, as shown in Fig. 18.1.

3.2 Spectroscopic techniques

The foundation of spectroscopy is the interaction of electromagnetic radiation with matter, each of which produces a distinct analytical signature. Spectroscopic instruments make up the most varied and popular category for water analysis (Gauglitz and Dakin, 2017).

3.2.1 UV–Vis spectrophotometry

The amount of UV or Vis light that a sample absorbs or transmits at particular wavelengths is measured quantitatively using UV–Vis spectrophotometry (Passos et al., 2019). By examining a compound's distinct UV–Vis spectrum—which is produced when electrons absorb energy and transition to a higher energy state—it is possible to ascertain its concentration and characteristics.

Principle: Absorbance of light in the UV–Vis region corresponds to electronic transitions in atoms or molecules, which are described quantitatively by the Beer–Lambert law ($A = \varepsilon bc$).

Applications: Determination of nitrate, phosphate, iron, manganese, and organic carbon via colorimetric reagents.

Advantages: Simplicity, low cost, and rapid operation.

Limitations: Interference from turbidity and matrix coloration; not ideal for trace-level quantification.

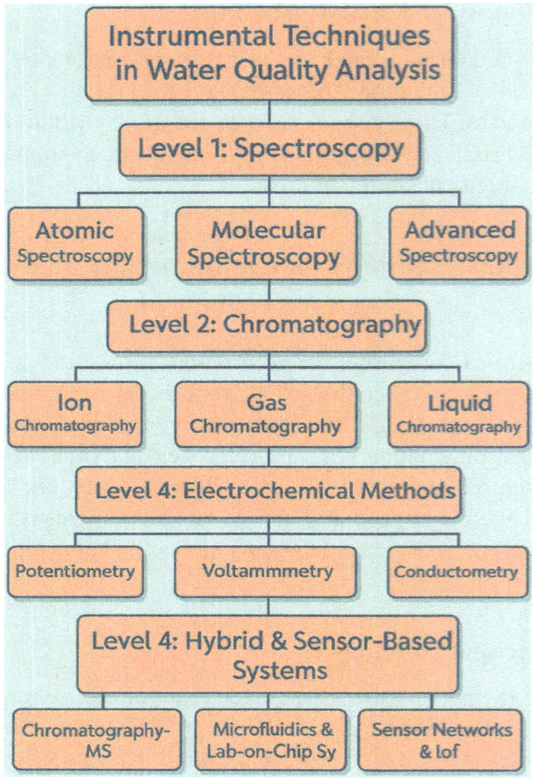

FIG. 18.1 Hierarchy of instruments. Classification of instrumental techniques used in water quality analysis, showing hierarchical relationships among spectroscopy, chromatography, electrochemistry, and hybrid/sensor domains.

3.2.2 *Atom absorption spectroscopy*

AAS quantifies metals based on the absorption of element-specific radiation by free atoms generated in a flame or graphite furnace (Ritgen, 2023). Each element exhibits a characteristic wavelength, e.g., Pb at 217 nm and Cd at 228.8 nm. A brief schematic of the instrument is shown in Fig. 18.2.

Utility: Accurate for Pb, Cd, Zn, Cu, Fe, and Mn at μg/L concentrations.

Variants: Flame AAS for major elements; graphite furnace AAS for trace and ultra-trace detection.

Limitations: Single-element analysis; time-intensive for multiple metals.

3.2.3 *Inductively coupled plasma–optical emission spectroscopy*

ICP–OES excites atoms and ions in an argon plasma (~10,000 K), causing the emission of element-specific photons detected by optical spectrometers (Senila, 2024).

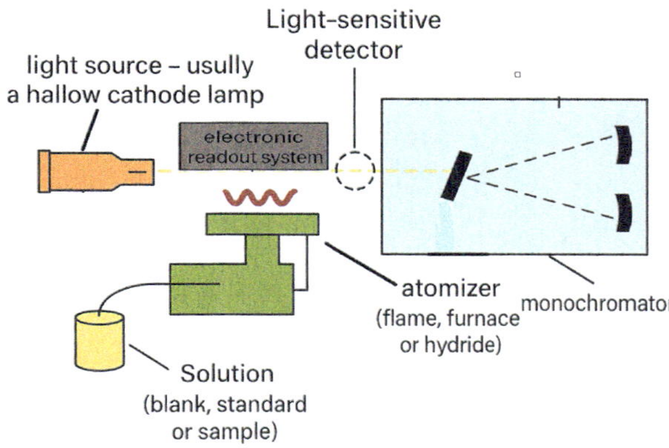

FIG. 18.2 Atomic absorption spectroscopy. Schematic diagram of an AAS system indicating the light source, atomizer, monochromator, detector, and data processor. *Ritgen, U., 2023. Atomic absorption spectroscopy (AAS). Analytical Chemistry I. Berlin, Heidelberg: Springer Berlin Heidelberg, pp. 247-253.*

Strengths: Multielement capability, high throughput, and wide dynamic range.

Applications: Simultaneous quantification of up to 70 elements, including toxic metals and micronutrients.

Detection limit: Typically, 0.1–10 μg/L, depending on the matrix.

3.2.4 Inductively coupled plasma–mass spectrometers

ICP–MS represents the apex of elemental analysis. The argon plasma atomizes and ionizes the sample, and ions are separated by mass-to-charge ratios (Klika et al., 2024). Table 18.1 shows a comparative summary of ICP–OES and ICP–MS performance parameters.

Advantages: Ultra-trace detection (ppt–ppb), isotopic analysis, and rapid multielement detection.

Challenges: Matrix suppression, high cost, and the need for clean-room operation. ICP–MS has become indispensable for evaluating contamination in groundwater, wastewater, and desalinated supplies, where precision is crucial for regulatory compliance.

3.2.5 Molecular and advanced spectroscopy

To investigate the molecular structure of complex dissolved materials and organic contaminants, a distinct set of sophisticated optical techniques is needed. Molecular spectroscopy provides distinctive "fingerprints" that aid in characterizing the compositional, structural, and functional characteristics of pollutants present even at trace levels by capturing vibrational, rotational, and electronic changes within molecules. These methods are particularly useful for

TABLE 18.1 Comparison between two instrument parameter detection limit.

Feature	ICP–OES	ICP–MS
Detection limit	0.1–10 μg/L	0.001–1 μg/L
Elements analyzed	60–70	>80
Throughput	High	Very high
Interferences	Spectral	Matrix, polyatomic ions
Cost and maintenance	Moderate	High

Klika, K.D., Han, J., Busse, M.S., Soloshonok, V.A., Javahershenas, R., Vanhaecke, F., Makarem, A., 2024. Inductively coupled plasma-mass spectrometry (ICP-MS): an emerging tool in radiopharmaceutical science. J. Am. Chem. Soc. 0002-7863. 146 (45), 30717–30727. doi:10.1021/jacs.4c12254.

evaluating microplastics, natural organic matter (NOM), emerging organic contaminants, and biomolecular markers of ecological stress.

3.2.6 Fourier transform infrared spectroscopy

Fourier transform infrared (FTIR) spectroscopy measures the absorption of mid-infrared radiation. These radiations are sufficient to cause molecular vibrations: stretching, bending, and deformation of chemical bonds.

Analytical significance:

- Each functional group (e.g., C=O, N–H, O–H, C–H) exhibits characteristic absorption bands. These are unique because each bond differs in its polarity and strength. This enables qualitative and semi-quantitative evaluation of organic molecules.
- To identify and categorize microplastics in water systems, FTIR has become essential. In the 400–4000 cm^{-1} range, polymer fragments (such as polyethylene, polypropylene, and polystyrene) exhibit unique fingerprints that enable researchers not only to detect but also occasionally to track the stages of plastic residue decomposition (Yang et al., 2023).
- By examining changes in spectral signatures during treatment stages, FTIR is employed in wastewater characterization to evaluate COD surrogates, surfactants, phenolic chemicals, and industrial dyes.

Attenuated total reflectance accessories for modern FTIR systems enable the direct measurement of aqueous samples with little pretreatment, increasing their application in environmental monitoring.

3.2.7 Raman spectroscopy

Raman spectroscopy is a nondestructive analytical method that determines a substance's molecular and chemical structure by using light scattering. It creates a distinct "fingerprint" for every molecule by shining a high-intensity laser on a sample and examining the scattered light for wavelength shifts (Orlando et al., 2021).

Environmental applications:

- Using diagnostic vibrational bands, organic contaminants such as industrial solvents, insecticides, and polycyclic aromatic hydrocarbons (PAHs) can be identified.
- Characterization of microplastics, such as colored or dark polymers, where FTIR performance might be compromised.
- Determination of dissolved organic matter (DOM) quality in rivers and lakes, where subtle changes in Raman bands reveal shifts in aromaticity or microbial processing of organic substrates (Almaviva et al., 2022).

The main advantage of Raman is that it can be used for in situ analysis. The development of portable Raman spectrometers has been made possible by recent technological advancements, such as tiny diode lasers, enhanced notch filters, and high-sensitivity CCD detectors. For the immediate detection of organic pollutants in groundwater, industrial discharge sites, and coastal sediments, these portable devices can be set up at field locations. By enabling real-time measurements inside pipes, wells, and aquatic chambers, fiber-optic Raman probes further expand this capability.

3.2.8 Fluorescence spectroscopy

It is an analytical method that employs a laser beam to excite molecules in a sample, causing them to emit light (fluoresce) at a particular wavelength. The characteristics of the sample are examined by measuring the intensity and wavelength of the light that is released, which provides details about its composition, concentration, and structure.

Key attributes:

- The emission profiles of fluorescent substances have extremely unique molecular signatures because they absorb photons at a specific wavelength and emit at a longer wavelength.
- Fluorescence is especially sensitive to PAHs and other aromatic hydrocarbons, detecting them at sub-ppb concentrations; DOM components like humic-like, fulvic-like, and protein-like substances exhibit distinct fluorescence patterns, providing insights into anthropogenic pollution, wastewater fingerprinting, and biogeochemical processing (Zacharioudaki et al., 2022).

When combined with parallel factor analysis (PARAFAC), excitation–emission matrix (EEM) fluorescence has become one of the most effective chemometric techniques for characterizing dissolved organic materials in environmental labs. A three-dimensional dataset with overlapping spectral signatures from several fluorophores is produced by an EEM spectrum, which records fluorescence intensity as a function of both excitation and emission wavelengths. Precise source apportionment is made possible by the combination of EEM and PARAFAC. For example, it is possible to separate treated effluent signatures from upstream NOM or to differentiate terrestrial humic inputs from autochthonous algal products. Tracing pollutant routes in rivers and estuaries, assessing the effectiveness of wastewater treatment, and tracking seasonal or man-made changes in DOM composition have all benefited greatly from this approach.

3.2.9 Emerging trends and technological innovations

Molecular spectroscopy is rapidly evolving toward portability, sensitivity, and multidimensional analytical capacity.

Recent innovations include:

- **Portable Raman analyzers** for field detection of hazardous organics and microplastics.
- **Fiber-optic IR and Raman probes,** enabling deep-water, subsurface, and confined-space measurements.
- **Time-resolved fluorescence** for differentiating DOM fractions based on excited-state lifetimes.
- **Surface-enhanced Raman spectroscopy** utilizing metallic nanostructures to amplify weak Raman signals, enabling ultra-trace detection of pesticides, endocrine disruptors, and pharmaceuticals.
- **Quantum cascade lasers** revolutionizing mid-IR absorption measurements with high spectral brightness and tunability (Apostolakis et al., 2024).

Collectively, these advancements have transformed molecular spectroscopy into a cornerstone of modern water quality science, capable not only of detecting contaminants but also of deciphering their sources, transformation pathways, and ecological implications.

3.3 Chromatographic techniques

Chromatography separates analytes based on their differential partitioning between mobile and stationary phases, achieving unparalleled resolution for complex mixtures.

3.3.1 Ion chromatography

Ion chromatography (IC) is a type of liquid chromatography that uses charge to separate, identify, and quantify ions and polar compounds. It operates by passing a sample through a stationary phase (an ion exchange resin) in a column using a mobile phase (eluent) (Yelampalli et al., 2023).

Principle: Ionic species are separated on resin columns and detected by conductivity or spectrophotometry. **Applications:** Major cations (Na^+, K^+, Ca^{2+}, Mg^{2+}) and anions (Cl^-, NO_3^-, SO_4^{2-}, F^-). **Advantages:** Automation, precision, and minimal sample handling. **Emerging trends:** Capillary IC and gradient elution for ultra-trace ions in desalinated or ultrapure water.

3.3.2 Gas chromatography

An analytical method for separating, identifying, and measuring the constituents of a volatile mixture is gas chromatography. The sample is transported through a column (the stationary phase) by a gas (the mobile phase), and the components segregate according to their varying volatilities and interactions with the stationary phase.

Injecting a sample into a hot port causes it to evaporate. A carrier gas is then used to sweep the vaporized sample into the column. The mixture passes through a stationary phase-containing column. As components pass through the column at varying speeds, they separate: less volatile components interact more with the stationary phase and move more slowly, eluting later, while more volatile components—those that evaporate more readily—travel quicker and elute first. A detector (such as a flame ionization detector) detects components as they leave the column and determines their concentration and presence. After processing the detector's signal, a computer creates a chromatogram—a graph showing the detector's response over time (Hites, 2024).

By comparing it with established criteria, the peak's location on the time axis aids in the compound's identification. The component's concentration can be inferred from the area under the peak. GC is employed for volatile organic compounds (VOCs), chlorinated solvents, and hydrocarbons. Coupled with flame ionization or electron-capture detectors, GC achieves sub-ppb sensitivity. Sample preparation often involves headspace analysis or solid-phase microextraction (SPME).

Applications of gas chromatography in water quality analysis

Please see Table 18.2.

3.3.3 High-performance liquid chromatograph

HPLC is an analytical chemistry technique that separates, identifies, and quantifies components in a liquid mixture. It works by using high-pressure pumps to force a liquid mobile phase through a column packed with a stationary phase. Different components of the mixture interact differently with the two phases, causing them to separate as they move through the column. Schematics of the instrumentation are shown in Fig. 18.3.

Pesticides, medications, and phenols are examples of nonvolatile and thermolabile substances that are separated using HPLC. UV–Vis, fluorescence, and diode-array are among the detection modes. Environmental applications are dominated by reverse-phase columns because of their polarity-based retention mechanisms.

TABLE 18.2 Applications of GC.

Application	Purpose	Typical GC technique	Outcome
Detection of volatile organic compounds (VOCs)	Detects VOCs, such as benzene, toluene, ethylbenzene, and xylene (BTEX), in industrial or groundwater samples.	Purge-and-trap GC or headspace GC	Quantifies VOCs at ppb levels and assesses pollution sources.
Analysis of pesticides and herbicides	Monitors pesticide residues from agricultural runoff, including organochlorine and organophosphorus pesticides.	GC–ECD (electron capture detector)	Ensures trace-level monitoring of persistent organic pollutants (POPs).
Determination of disinfection by-products (DBPs)	Determines toxic by-products like trihalomethanes (THMs) and haloacetic acids formed during chlorination.	Liquid–liquid extraction followed by GC–MS or GC–ECD	Checks compliance with regulatory limits for disinfection by-products in drinking water.
Identification of industrial pollutants	Identifies phenols, halogenated solvents, and hydrocarbons from industrial effluents.	Direct aqueous injection GC or SPME–GC–MS	Provides chemical fingerprinting of pollutant sources and contamination tracking.
Analysis of petroleum hydrocarbons	Assesses petroleum hydrocarbon contamination from oil spills or leakages in aquatic systems.	GC-FID for total petroleum hydrocarbons (TPH)	Quantifies aliphatic and aromatic fractions to evaluate biodegradation and cleanup efficiency.
Monitoring of odorous compounds	Detects sulfur- or nitrogen-containing odorous compounds (e.g., mercaptans, amines) that affect water taste and odor.	GC coupled with sulfur/nitrogen-specific detectors (SCD/NPD)	Identifies and quantifies odor-causing agents at very low concentrations.
Micropollutant and emerging contaminant analysis	Analyzes pharmaceuticals, personal care products.	Derivatization followed by GC–MS or GC–MS/MS	Detects and quantifies complex, emerging contaminants with high sensitivity and selectivity.

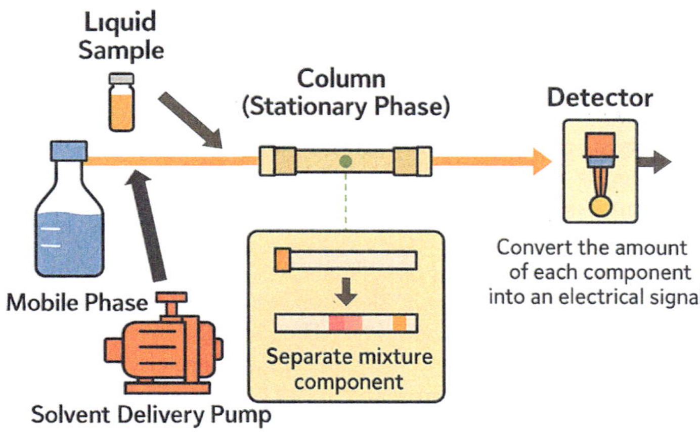

FIG. 18.3 Schematics of HPLC. A simplified schematic of an HPLC system illustrating solvent delivery, injector, column, detector, and data acquisition unit.

Ultra-fast and highly sensitive analysis is made possible by developments in liquid chromatography–tandem mass spectrometry (LC–MS)/MS and ultra-performance liquid chromatography (Ali, 2022).

3.4 Electrochemical and sensor-based methods

Techniques that employ electrochemical reactions to identify and gauge the concentration of compounds are known as electrochemical and sensor-based methods. Electrochemical techniques are based on redox reactions occurring at electrode surfaces. They function by transforming a chemical reaction into an electrical signal that can be measured, such as a potential difference or current. The quantifiable electrical characteristics of analyte–electrode interactions are exploited by electrochemical instruments. They meet the needs of contemporary field monitoring by providing portability, low power consumption, and real-time detection.

3.4.1 Potentiometry and ISEs

An ISE is a specialized sensor used in potentiometry to measure the concentration of a particular ion, while potentiometry measures the potential difference of an electrochemical cell under static conditions. In ISEs, a single-type ion-permeable membrane is used, and the potentiometric system measures the potential difference across the membrane, which is directly proportional to the activity or concentration of the ion. Applications include the measurement of ions such as potassium, nitrate, and fluoride, as well as pH, in industrial, environmental, and clinical contexts (Dauzvardis and Lisensky, 2021).

Applications: ISEs are frequently used in environmental analysis to measure important parameters like pH, salinity-related ions (Na^+ and Cl^-), hardness ions (Ca^{2+} and Mg^{2+}), and environmentally sensitive species like ammonium, nitrate, nitrite, fluoride, and cyanide. The quality of drinking water, industrial pollution detection, agricultural runoff assessment, eutrophication prevention, and wastewater treatment plant process monitoring all depend on these measurements. Because of their portability, quick reaction times, and capacity to deliver continuous, real-time data without the need for complex sample preparation, ISEs are particularly appreciated in online sensor networks and field monitoring.

Advantages: Potentiometric techniques have several benefits, including high target ion selectivity, affordability, nondestructiveness, and the ability to measure a broad range of concentrations. They are appropriate for both in-situ and laboratory applications due to their speed and simplicity.

Limitations: ISE-based measurements do have some drawbacks, though. Accuracy may be impacted by interference from chemically similar ions, and signal drift frequently necessitates frequent electrode calibration. While temperature changes can alter electrode response as determined by the Nernst equation, membrane fouling by organic matter or suspended solids may decrease sensitivity. Furthermore, some electrodes are not suitable for trace-level (ultra-low concentration) analysis or may have a short lifespan.

3.4.2 Conductometry

Voltammetry measures the current generated when a voltage is applied to ascertain the concentration of electroactive substances, whereas conductometry measures the EC of a solution to ascertain the concentration of ions. While voltammetry uses a controlled, variable voltage to induce a reaction and then measures the current that results, conductometry uses a constant voltage to measure how well ions conduct electricity. Voltammetry is a very sensitive technique for quantifying particular analytes, and conductometry is helpful for monitoring the progress of reactions (Majeed et al., 2022).

Advantages: Conductometry offers several advantages: it is simple, rapid, highly reproducible, and requires minimal sample preparation. Moreover, the instruments are robust and well-suited for field measurements.

Limitations: However, the technique measures only total ionic strength and does not identify individual ions, which limits its specificity. Conductivity readings are also influenced by temperature, requiring compensation, and very low or very high ionic strengths can reduce accuracy.

3.4.3 Voltammetry

The current generated by the oxidation or reduction of analytes at an electrode surface is measured as a function of applied potential in this extremely sensitive technique. Voltammetric techniques are frequently employed in water quality monitoring to identify trace and ultra-trace levels of heavy metals, including lead, cadmium, mercury, copper, arsenic, and zinc, all of which present serious health and environmental hazards. Even at parts-per-billion levels, these metals

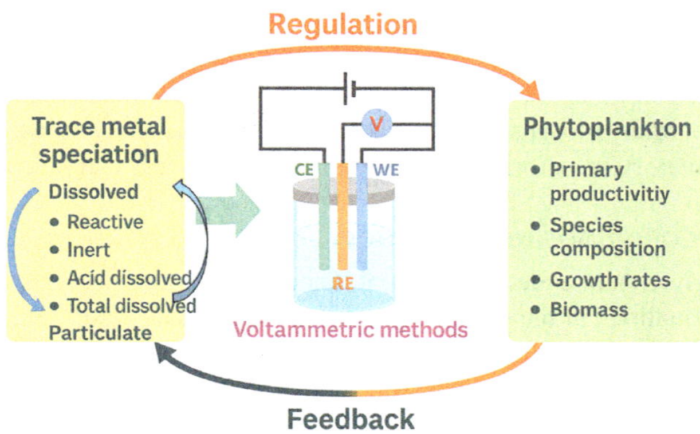

FIG. 18.4 Voltammetry feedback–regulation loop. A representative voltammogram showing the feedback-regulation relationship for trace metal determination.

can be precisely quantified using methods like differential pulse voltammetry and anodic stripping voltammetry (Antonietta Baldo et al., 2024). Fig. 18.4 shows feedback–regulation relationships for trace metal detection.

Advantages: Excellent sensitivity, selectivity (often made possible by altering the electrode surfaces), and suitability for on-site analysis using portable instruments are all provided by voltammetry.

Limitations: However, organic matter that can contaminate the electrode surface may have an impact, and it necessitates careful control of experimental parameters like pH and DO. Furthermore, the technique requires expert calibration and operation, and pretreatment might be necessary for some complex matrices.

3.4.4 Biosensors

Because they combine biological recognition elements—such as enzymes, antibodies, whole cells, DNA, or aptamers—with transducers that translate biological interactions into quantifiable signals, biosensors have become extremely effective tools in the monitoring of water quality. These instruments are especially useful for identifying contaminants such as pesticides, phenols, endocrine disruptors, pathogens, and some heavy metals that are hard to quantify using conventional physicochemical techniques (Hui et al., 2022). Biosensors provide high specificity because the biological component selectively binds to the target contaminant, enabling rapid and often real-time detection even at very low concentrations. Field-based microbial contamination studies, wastewater treatment facilities, and environmental surveillance are increasingly using them.

Advantages: High sensitivity, rapid reaction, portability, and the capacity to identify toxic effects that are biologically significant rather than merely chemical concentrations are some of their benefits.

Limitations: Biosensors do have certain drawbacks, though, such as the potential for biological components to lose their activity over time, the need for controlled storage, and potential sensitivity to matrix interferences, pH, and temperature. Many biosensors still lack long-term stability or standardization for regular regulatory monitoring, and the fabrication process can be complicated.

3.4.5 *Nanocomposite electrodes*

These greatly enhance electrochemical detection in water analysis by utilizing the special qualities of nanomaterials, such as high surface area, enhanced conductivity, catalytic activity, and tunable surface chemistry. To create sensitive, reliable, and selective sensing platforms, these electrodes frequently use materials such as metal nanoparticles, metal oxides, graphene, carbon nanotubes, or polymer–nanoparticle hybrids. Nanocomposite electrodes are used in water quality monitoring to detect traces of pesticides, dyes, heavy metals, medications, and other emerging contaminants (Kanoun et al., 2021).

Advantages: Even at ultra-trace concentrations, accurate analysis is made possible by their large active surface area, which improves electron transfer and lowers detection limits. In addition, they enable quicker reaction times and can be modified on the surface to enable selective recognition.

Limitations: However, there are some drawbacks to their use: some nanomaterials can aggregate or lose their activity, fabrication can be expensive or complicated, and long-term stability in actual environmental conditions can be unpredictable. Furthermore, reproducibility in electrode preparation is still a problem for widespread use.

3.5 Hybrid instrumentation and emerging interfaces

It describes how various measurement technologies or control signals are combined to produce more reliable, effective, and adaptable systems, frequently utilizing cutting-edge techniques for human-machine interaction. It entails integrating various sensor types or data processing techniques to improve functionality, performance, and dependability.

3.5.1 *Hybrid instrumentation*

By combining optical, spectroscopic, chromatographic, or electrochemical modules, these hybrid systems can take advantage of each technique's strengths while compensating for its weaknesses. For instance, devices that combine voltammetry and microfluidic separation or potentiometry and optical sensors enable the simultaneous detection of several contaminants, such as organic pollutants, heavy metals, and nutrients, in a single, compact device (Alves et al., 2025). For complex environmental samples, where no single technique can provide complete reliability, hybrid instrumentation is especially useful.

Advantages: Improved sensitivity, multiparameter capability, lower sample consumption, and suitability for continuous monitoring in automated stations are some of its main benefits.

Limitations: Hybrid systems, however, can be costly to develop and maintain, necessitate complex calibration, and require more intricate data integration and interpretation.

3.5.2 Emerging interfaces

These are the newest generation of sophisticated water monitoring technologies, emphasizing the use of contemporary engineering advancements to strengthen the link between the sensing element, data processing unit, and end user (Miller et al., 2025). These interfaces include wearable or flexible sensors, wireless IoT modules, smartphone-connected electrodes, microfluidic lab-on-a-chip systems, and platforms for AI-assisted interpretation. They seek to facilitate almost instantaneous data sharing and decision-making while improving the portability, automation, and usability of water analysis. Emerging interfaces improve detection accuracy, enable predictive analysis, and facilitate extensive environmental surveillance by seamlessly integrating sensors with cloud databases or machine learning algorithms.

Advantages: Miniaturization, low power consumption, remote accessibility, and appropriateness for point-of-use monitoring in industrial or rural environments are some of its main benefits.

Limitations: These technologies might, however, have drawbacks, such as poor resilience in challenging field settings, reliance on dependable power or connectivity, and the requirement for standardized testing prior to regulatory approval.

3.5.3 Gas chromatography–mass spectrometry

Among the most effective analytical methods for monitoring water quality, GC–MS is particularly useful for identifying and measuring organic contaminants, even at trace or ultra-trace levels. Based on their interactions with the chromatographic column and boiling points, it separates complex mixtures of volatile and semivolatile compounds. Following separation, the compounds go through a mass spectrometer, where their distinct mass-to-charge ratios are used to identify, ionize, and fragment them. GC–MS is particularly well-suited for the analysis of pesticides, industrial solvents, petroleum hydrocarbons, VOCs, PAHs, phenols, and disinfection by-products in water because of its dual capability of high-resolution separation followed by molecular-level identification (Nguyen et al., 2025). It is essential for environmental surveillance, forensic pollution investigations, drinking water standard compliance testing, and industrial effluent monitoring. Fig. 18.5 shows a block diagram of a GC–MS system, illustrating injection, separation, ionization, mass filtering, and detection.

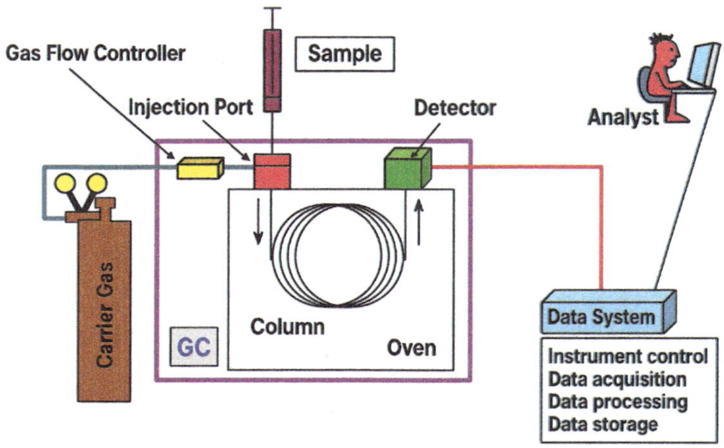

FIG. 18.5 Block diagram of a GC–MS system. The block diagram of a GC–MS system showing the processes of injection, separation, ionization, mass filtering, and detection.

Advantages: Excellent sensitivity, high selectivity, and the capacity to offer conclusive structural information regarding unknown contaminants are just a few benefits of GC–MS. It is perfect for early toxic pollutant detection because its detection limits frequently approach the nanogram-per-liter level. Furthermore, when combined with sophisticated sample preparation methods like headspace analysis, purge-and-trap, or solid-phase extraction (SPE), GC–MS can accurately analyze complex samples.

Limitations: Notwithstanding its advantages, the technique has drawbacks. For example, it is impossible to analyze many nonvolatile or thermally unstable compounds without derivatization, and the equipment is costly, requires frequent maintenance, and skilled operators. The method is generally not appropriate for quick on-site measurements, and sample preparation can be time-consuming.

3.5.4 Liquid chromatography–tandem mass spectrometry/mass spectrometry

LC–MS/MS is a sophisticated analytical method that combines tandem mass spectrometry for compound identification and quantification with liquid chromatography for compound separation. The detection of low concentrations of substances in complex mixtures is made possible by this method's high sensitivity and selectivity, which is essential in domains like clinical diagnostics, environmental monitoring, and drug development. Peptides, metabolites, and hormones are examples of nonvolatile or thermally unstable samples that are particularly well-suited for LC–MS/MS analysis (Liu et al., 2025).

Advantages: The method has a number of significant benefits. It provides very low detection limits—often at ng/L to pg/L levels—making it suitable for trace and ultra-trace analysis. Even in complex samples like wastewater or industrial effluents, its tandem mass filtering steps guarantee exceptional specificity by reducing matrix interference. Compounds that cannot be investigated with GC-based techniques can be directly analyzed thanks to LC–MS/MS, which also requires minimal sample derivatization.

Limitations: Notwithstanding its advantages, LC–MS/MS has some drawbacks, including expensive equipment that needs to be operated by professionals and requires frequent maintenance. Quantification may be impacted by ion suppression or enhancement brought on by matrix effects, requiring careful method development and calibration. For some matrices, sample preparation—including solid-phase extraction or filtration—can be time-consuming.

3.5.5 Sample preparation innovations

Modern analytical techniques used in water quality monitoring, such as GC–MS, LC–MS/MS, voltammetry, biosensors, and nanocomposite-based electrochemical methods, depend on developments in sample preparation to increase their efficiency, sensitivity, and dependability. Advanced preparation techniques are necessary to isolate, concentrate, and purify analytes prior to instrumental analysis because environmental water samples frequently contain complex matrices, including organic matter, suspended solids, salts, microorganisms, and nanoparticles. To support quick, high-quality results, recent innovations have focused on streamlining processes, reducing solvent usage, and enabling automation.

Innovations in sample preparation for electrochemical methods like voltammetry, biosensors, and nanocomposite electrodes focus on reducing fouling, eliminating interfering ions, and stabilizing target analytes. This includes preconcentration with ion-exchange resins, membrane-based separation, and microfiltration, all of which enhance electrode performance and prolong sensor life. Sample pretreatment techniques for biosensors, such as magnetic nanoparticle capture, enzyme-assisted digestion, and microfluidic purification modules, ensure the removal of biological contaminants and debris that could hinder biomolecular recognition. Automated miniaturized modules for on-chip filtration, dilution, and extraction are also integrated into hybrid instrumentation and emerging interfaces, enabling near real-time monitoring with minimal human assistance.

- **SPE:** It integrates preconcentration, extraction, and sampling into a single solvent-free step and is one significant development.
- **SPME:** These fibers or coatings enable direct uptake of volatile and semi-volatile compounds, significantly increasing sensitivity and streamlining field sampling.
- **Dispersive liquid–liquid microextraction (DLLME):** It is a sophisticated sample preparation method that is frequently employed in water quality analysis due to its remarkable enrichment capability, speed, and ease of use

(Amaral et al., 2025). In DLLME, a disperser solvent is used to quickly disperse a small volume of an extraction solvent—typically one that has a high affinity for the target analytes—into the aqueous sample. The contact surface area between the sample and the extractant is significantly increased as a result of the formation of a fine, hazy emulsion of microdroplets. Analytes consequently partition into the extraction solvent rapidly, reaching high preconcentration in a matter of seconds or minutes. After centrifugation, the enriched extract is collected from the bottom of the tube and can be directly analyzed using techniques such as GC–MS or LC–MS/MS.

4 Data analytics, automation, and quality assurance

4.1 Calibration and standardization protocols

Instrument calibration ensures a linear correlation between signal intensity and analyte concentration. Common practices include external calibration curves, standard addition, and internal standard methods (Zero, 2025). Sophisticated instruments maintain built-in calibration routines and self-check algorithms. Fig. 18.6 shows a typical calibration curve.

4.2 Method validation parameters

Analytical validity encompasses:

- **Precision (repeatability)**
- **Accuracy (trueness)**
- **Limit of detection (LOD), limit of linearity (LOL), and limit of quantification (LOQ)**

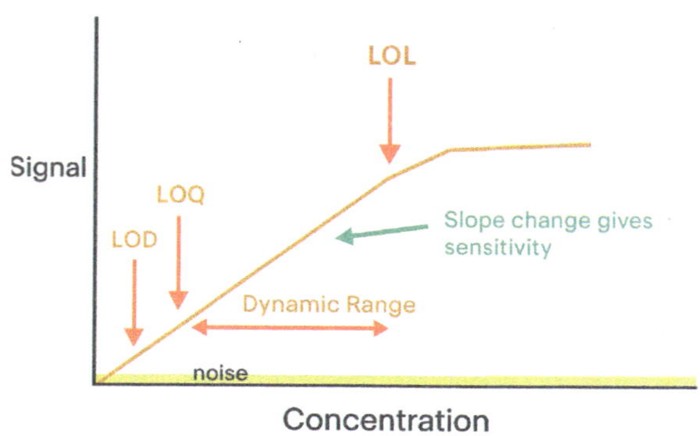

FIG. 18.6 Calibration curve. Example of a calibration curve illustrating the linear dynamic range and coefficient of determination (R^2).

- **Selectivity and specificity**
- **Robustness and ruggedness**

Regulatory frameworks (e.g., ISO 17025, EPA 600 Series) mandate the documentation of these metrics before data reporting.

4.3 Automation and digital integration

The latest generation of analytical instruments is microprocessor-driven, featuring:

- **Autosamplers** reducing human handling.
- **Programmable methods** ensuring reproducibility.
- **Remote monitoring** via IoT interfaces.
- **Real-time data logging** integrated with cloud analytics.

For instance, modern spectrometers allow wireless transmission of spectral data to centralized servers for instantaneous quality flagging, a transition from reactive to proactive monitoring.

4.4 Chemometrics and multivariate analysis

Chemometrics applies statistical and mathematical tools to extract meaningful patterns from large analytical datasets. Principal component analysis (PCA), cluster analysis (CA), and partial least squares regression are used to:

- Differentiate between pollution sources.
- Detect temporal or spatial trends.
- Validate analytical consistency.

When coupled with spectral or chromatographic data, chemometrics enhances interpretative accuracy and reduces redundancy in monitoring programs (Kalivas and Spiers, 2025).

4.5 Quality assurance and laboratory accreditation

A sound quality assurance program includes:

- **Standard operating procedures** for every analytical step.
- **Interlaboratory comparisons** and proficiency testing.
- **Control charts (Shewhart or CUSUM)** for long-term precision tracking.

ISO/IEC 17025 accreditation serves as the global benchmark for testing laboratories, mandating documentation of traceability, competence, and impartiality.

4.6 Ethical dimensions of analytical practice

Instrumental sophistication confers responsibility. Data fabrication, selective reporting, or omission of uncertainty undermines public trust and environmental governance. Adherence to Good Laboratory Practice and transparent disclosure of analytical uncertainties are, therefore, integral to the scientific process.

5 Case studies and applied scenarios

5.1 Overview

The true measure of analytical sophistication lies not merely in laboratory precision but in its capacity to interpret and improve environmental realities. The following section highlights real-world applications of advanced instrumentation in assessing water quality at diverse spatial and temporal scales. Each example underscores the synergy between methodological rigor, environmental insight, and scientific integrity.

5.2 Case study: Multielemental analysis of the ganga river using ICP–OES and ICP–MS

Title: Spatio-temporal variation of trace elements distributed over surface water of Upper Ganga River Basin in Western Himalayan Region (Dimri et al., 2023).

Reference: Dimri, Deepika, et al., (2023). Spatio-temporal variation of trace elements distributed over surface water of Upper Ganga River Basin in Western Himalayan Region. J. Mt. Sci. 20(1): 145–162.

Background: The River Ganga, supporting over 400 million people, faces severe anthropogenic pressure from industrial effluents, agricultural runoff, and urban sewage. Monitoring its trace metal profile is essential for water resource management.

Methodology: Water samples were collected from nineteen monitoring locations during three hydrological seasons: premonsoon (May–June), monsoon (August–September), and postmonsoon (December). Samples underwent acid digestion (HNO_3–$HClO_4$) and were analyzed to detect traces of 20 elements: Ag, Al, Ba, Cd, Ca, Cr, Cu, Fe, Ga, K, Mn, Mg, Na, Ni, Pb, Sr, Th, U, Zn, and Zr, using ICP–OES.

Findings:

- The study reports that in the "hyporhithron" (lower-Himalayan/lower-elevation) stretch of the river, the computed Metal Index (MI) often exceeded 1.0, which the authors interpret as a *warning level of contamination.* PCA distinguished industrial clusters from agricultural catchments.
- The multivariate statistical (PCA/CA) analysis suggests that, while many metals are of lithogenic/geogenic origin, some trace metals reflect anthropogenic input, implying possible exceedances relative to baseline or background levels.

Significance: The study demonstrated the necessity of integrating multi-instrumental strategies for a holistic picture. ICP–OES provided macroscale patterning, while ICP–MS offered trace-level confirmation. Together, they formed a dual-tiered analytical framework, a model for national water surveillance systems.

5.3 Application 2: Groundwater characterization in the Indo-Gangetic plains

Title: Groundwater quality appraisal and its hydrochemical characterization in Ghaziabad (a region of Indo-Gangetic plain), Uttar Pradesh, India (Singh et al., 2014).

Reference: Singh, Uday Veer, et al. "Groundwater quality appraisal and its hydrochemical characterization in Ghaziabad (a region of Indo-Gangetic plain), Uttar Pradesh, India." Applied Water Science 4.2 (2014): 145–157.

Background: Ghaziabad, part of the rapidly urbanizing Indo-Gangetic plains, experiences intense industrial, agricultural, and domestic pressure on its groundwater reserves. The region's aquifers are highly exploited and vulnerable to degradation due to wastewater infiltration, industrial discharge, and accelerated water–rock interaction. Assessing groundwater quality is essential for safeguarding drinking-water security and understanding the geochemical evolution of the aquifer under anthropogenic influence.

Methodology:

- **Sample size:** 250 groundwater samples collected across premonsoon and postmonsoon seasons.
- **Analyses conducted:**
- Physicochemical parameters: pH, EC, TDS, hardness, alkalinity.
- Major ions: Ca^{2+}, Mg^{2+}, Na^{+}, K^{+}, HCO_3^{-}, Cl^{-}, SO_4^{2-}.
- Heavy metals: Fe, Cu, Pb, Cr (and others).
- **Analytical tools:** Standard APHA methods, ICP–OES, or AAS for metals.
- **Hydrochemical evaluation:**
- Piper diagrams and facies classification.
- Base-exchange indices, ionic ratios.
- Seasonal comparison to identify temporal shifts.

Findings:

- **Hydrochemical facies:**
- 76% of samples belonged to *Na–HCO₃ facies*, indicating significant ion exchange and anthropogenic influence.
- **Heavy metals detected:**
- **Fe (iron):** Most abundant heavy metal; elevated in several locations.
- **Cr (chromium):** "Exceptionally high" in some samples; attributed to industrial discharge (chrome-plating, wood-preservatives).
- **Cu and Pb:** Present in lower, but notable, concentrations.

- **Water quality concerns:**
- Many samples exhibited high TDS and EC values, making the water unsuitable for drinking without treatment.
- Broad variance and high standard deviation indicated strong spatial heterogeneity and multisource contamination.

Significance:

This study highlights the dual origin of groundwater chemistry, natural aquifer lithology and significant anthropogenic modification. It emphasizes the presence of metal contamination hotspots (notably Fe and Cr) and the importance of geochemical facies in diagnosing hydrochemical evolution. For groundwater monitoring across the Indo-Gangetic plains, the study provides a template for integrating physicochemical analysis, trace-metal screening, and hydrochemical modeling, supporting policy frameworks for sustainable groundwater use and contamination mitigation. This study also highlights the significance of policy shifting. The push toward sophisticated instrumentation is not technological vanity; it is a scientific necessity shaped by the complexity of modern water contaminants. Water quality threats today arise from sources that were unimaginable a century ago: nanomaterials, personal care residues, engineered chemicals, pharmaceutical metabolites, microplastic sorption complexes, antibiotic resistance markers, and more.

5.4 Synthesis

These studies collectively demonstrate that water quality assessment is evolving toward multimodal integration, a convergence of precision (spectroscopy), resolution (chromatography), and continuity (sensors). Analytical outcomes are no longer static measurements but dynamic indicators of anthropogenic impact and ecological resilience. In addition, hybrid analytical systems and sophisticated instrumentation are the modern pillars of water quality science. They ensure that monitoring is sensitive enough (to detect trace pollutants), specific enough (to identify their sources), fast enough (for real-time alerts), and reliable enough (to inform policy and public health).

Together, they transform water monitoring from a reactive exercise into a predictive, preventative, and scientifically rigorous discipline, one aligned with the demands of a rapidly evolving environmental landscape.

6 Challenges and future directions

6.1 Analytical and technical constraints

Despite their precision, sophisticated instruments face intrinsic limitations:

- **Matrix interference:** High ionic strength or organic load can suppress signals, particularly in ICP–MS.
- **Sample integrity:** Preservation and transportation often introduce variability, especially for volatile or redox-sensitive species.

- **Calibration drift:** Long-term stability of electrodes and light sources remains a concern in field sensors.
- **Operational cost:** The acquisition and maintenance of advanced instruments like ICP–MS or LC–MS/MS require skilled personnel and significant capital investment.

Such constraints necessitate methodological prudence and contextual adaptation, choosing instruments not merely for their sophistication but for their *fitness for the analytical purpose*.

6.2 Miniaturization and portability

Technological advances in microfabrication and nanomaterials have fostered the development of lab-on-chip devices capable of performing multiparameter analyses on micro-volumes of water. Microfluidic sensors integrate fluid channels (Chen et al., 2023), electrodes, and optical detectors within compact platforms, enabling rapid, on-site diagnostics. The movement toward portable spectrometers and handheld XRF analyzers promises democratization of water testing, bridging the gap between high-end laboratories and community-level monitoring. Fig. 18.7 shows a conceptual illustration of a lab-on-chip microfluidic water quality sensor.

6.3 Green analytical chemistry

In alignment with sustainable science, analytical chemists are adopting green principles: reducing solvent use, energy consumption, and waste. Techniques such as microwave-assisted extraction, solvent-free SPME, and aqueous–organic mobile phases in chromatography exemplify eco-conscious innovation. Furthermore, the design of biobased sensor membranes and recyclable nanocomposites reflects an ethical reorientation of instrumentation, science in service of both truth and sustainability.

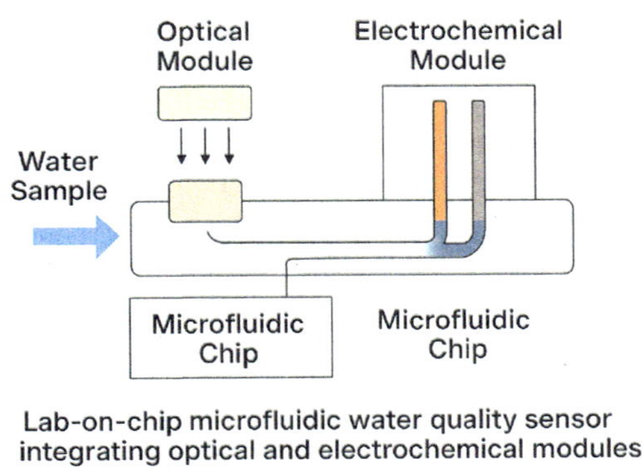

FIG. 18.7 Illustration of a lab-on-chip microfluidic water quality sensor integrating optical and electrochemical modules.

6.4 Artificial intelligence and predictive analytics

Machine learning algorithms are increasingly integrated into analytical workflows for predictive diagnostics. Neural networks can model nonlinear correlations between spectral signatures and contaminant levels, while clustering algorithms facilitate anomaly detection in sensor networks (Vaikunta Rao et al., 2025). AI-assisted calibration can self-correct instrumental drift and optimize analytical conditions dynamically, a frontier where instrumentation and computation converge to emulate cognitive precision.

6.5 Ethical and institutional imperatives

Sophisticated instruments alone cannot guarantee sound science. As analytical ecosystems become data-rich, the ethical dimension—*what we measure, why, and for whom*—becomes ever more critical. Ensuring open data policies, transparent methodology, and public engagement must accompany technological advancement. Scientific temper, in its truest sense, entails both precision and conscience, a principle central to your own academic philosophy.

7 Conclusion

Water quality assessment through sophisticated instrumentation represents the intersection of chemistry, technology, and societal responsibility. The instruments discussed—from AAS and ICP–MS to biosensors and IoT platforms—embody the evolutionary trajectory of analytical science: from descriptive observation to predictive understanding.

The transition from classical to advanced techniques has not only expanded our capacity to detect contaminants but also deepened our grasp of hydrogeochemical interactions. Yet, as this chapter underscores, instrumentation is not merely a tool, but it is a philosophical extension of human curiosity and discipline.

Future directions point toward miniaturized, AI-integrated, and sustainable analytical systems, capable of autonomous operation and real-time intelligence. However, the enduring challenge remains the same: to align technological sophistication with ecological compassion and ethical transparency.

Thus in the pursuit of accurate measurement, we reaffirm a larger scientific ideal, the cultivation of reasoned observation, the hallmark of both scientific temper and environmental stewardship.

References

admin_zero. "Differences between Verification, Calibration, Inspection, and Standardization in Instrument Measurement: definitions and Scope." Just Measure It, 2 Mar. 2025, https://zeroinstrument.com/differences-between-verification-calibration-inspection-and-standardization-in-instrument-measurement-definitions-and-scope/.

Alfonso, C., 2024. Quantitative and qualitative methods, and primary methods. In: Environmental Toxicology: Non-bacterial Toxins. Laboratorio CIFGA SA, Avenida Benigno Rivera 56, pp. 65–101.

Ali, A.H., 2022. High-performance liquid chromatography (HPLC): a review. Ann. Adv. Chem 6, 010–020.

Almaviva, S., Artuso, F., Giardina, I., Lai, A., Pasquo, A., 2022. Fast detection of different water contaminants by Raman spectroscopy and surface-enhanced Raman spectroscopy. Sensors. 1424-8220. 22 (21), 8338. doi:10.3390/s22218338.

Alves, E., Gurupadayya, B.M., Prabhakaran, P., 2025. Next-generation miniaturized separation platforms: converging detection, automation, and sustainable design for intelligent analytical science. Crit. Rev. Anal. Chem. 55 , 1–50 doi:10.1080/10408347.2025.2553122.

Amaral, L.R., Gama, M.R., Pizzolato, T.M., 2025. Ionic liquids in dispersive liquid–liquid microextraction for environmental aqueous samples. Microchem. J. 208, 112254.

Antonietta Baldo, M., Fabris, S., Stortini, A.M., Daniele, S., 2024. Protolysis studies and quantification of acids and bases in aqueous solutions by microelectrode voltammetry. J. Solid State Electrochem. 1432-8488. 28 (3-4), 1049–1068. doi:10.1007/s10008-023-05675-8.

Apostolakis, A., Aoust, G., Maisons, G., Laurent, L., Pereira, M.F., 2024. Photoacoustic spectroscopy using a quantum Cascade Laser for analysis of ammonia in water solutions. ACS Omega. 2470-1343. 9 (17), 19127–19135. doi:10.1021/acsomega.3c10175.

Bekmirzayevich, A.N., Bakhodirovich, K.D., 2024. A process approach to ISO/IEC 17025 in the implementation of a quality management system in testing laboratories. Web Discov. J. Anal. Invent. 2 (12), 62–70.

BIS, (2012). Drinking Water Specification (IS 10500:2012). Bureau of Indian Standards, New Delhi.

Boyd, C.E., 2015. Water Quality, Springer International Publishing AG.

Chen, S., Qiao, Z., Niu, Y., Yeo, J.C., Liu, Y., Jiaming, Q., Fan, S., Liu, X., Lee, J.Y., Lim, C.T., 2023. Wearable flexible microfluidic sensing technologies. Nat. Rev. Bioeng. 2731-6092. 1 (12), 950–971. doi:10.1038/s44222-023-00094-w.

Dauzvardis, F., Lisensky, G., 2021. Assembling and testing a series of ion-selective electrodes in analytical chemistry. J. Chem. Educ. 98 (10), 3312–3318.

Dimri, D., Kumar, A., Mishra, D.K., Sharma, A., 2023. Spatio-temporal variation of trace elements distributed over surface water of upper Ganga River Basin in Western Himalayan Region. J. Mountain Sci. 1672-6316. 20 (1), 145–162. doi:10.1007/s11629-021-6703-9.

Durmishi, B.H., Durmishi, A., Shabani, A., 2025. The role of instrumental methods in chemical analysis for environmental protection. Int. J. Chem. Mater. Sci. 10 (1), 1–22.

Gauglitz, G., Dakin, J.P., 2017. Spectroscopic analysis. In: Dakin, J.P., Brown, R.G.W. (Eds.), Handbook of Optoelectronics, second ed. CRC Press, Boca Raton, pp. 569–600, 9781315156996. doi:10.1201/9781315156996.

Guo, W., Pan, B., Sakkiah, S., Yavas, G., Ge, W., Zou, W., Tong, W., Hong, H., 2019. Persistent organic pollutants in food: contamination sources, health effects and detection methods. Int. J. Environ. Res. Public Health. 1660-4601. 16 (22), 4361. doi:10.3390/ijerph16224361.

Hites, R.A., 2024. The analysis of organic water pollutants by gas chromatography and gas chromatography-mass spectrometry. In: Advances in Chromatography. Informa UK Limited, Vol. 15, pp. 69–112 doi:10.1201/9781003573579-3.

Hui, Y., Huang, Z., Alahi, M.E.E., Nag, A., Feng, S., Mukhopadhyay, S.C., 2022. Recent advancements in electrochemical biosensors for monitoring the water quality. Biosensors. 2079-6374. 12 (7), 551. doi:10.3390/bios12070551.

Kalivas, J.H., Spiers, R.C., 2025. Expanding multivariate analysis principles in conventional chemometric processes. J. Chemom. 39(9), e70067.

Kanoun, O., Lazarević-Pašti, T., Pašti, I., Nasraoui, S., Talbi, M., Brahem, A., Adiraju, A., Sheremet, E., Rodriguez, R.D., Ali, M.B., Al-Hamry, A., 2021. A review of nanocomposite-modified electrochemical sensors for water quality monitoring. Sensors. 1424-8220. 21 (12), 4131. doi:10.3390/s21124131.

Khati, H.S., Vishwakarma, K., Kumar, S., Bahuguna, A.S., Pundir, Y.P., Kumar, K., Kala, S. 2025 IoT-based water quality monitoring system. In: 2025 IEEE International Conference on Computer, Electronics, Electrical Engineering and their Applications, IC2E3 2025. 9798331524395. Institute of Electrical and Electronics Engineers Inc., India, http://ieeexplore.ieee.org/xpl/mostRecentIssue.jsp?punumber=11166672. doi:10.1109/IC2E365635.2025.11167598.

Klika, K.D., Han, J., Busse, M.S., Soloshonok, V.A., Javahershenas, R., Vanhaecke, F., Makarem, A., 2024. Inductively coupled plasma-mass spectrometry (ICP-MS): an emerging tool in radiopharmaceutical science. J. Am. Chem. Soc. 0002-7863. 146 (45), 30717–30727. doi:10.1021/jacs.4c12254.

Liu, Y., Xu, X., Wang, Y., Zhang, Y., Lu, J., Liu, C., Duan, J., Sun, H., 2025. Effectiveness of different organic solvent additions to water samples for reducing the adsorption effects of organic pesticides using ultra-high-performance liquid chromatography–tandem mass spectrometry. Molecules. 1420-3049. 30 (1), 200. doi:10.3390/molecules30010200.

Majeed, S., Naqvi, S.T.R., Ul Haq, M.N., Ashiq, M.N., 2022. Electroanalytical techniques in biosciences: conductometry, coulometry, voltammetry, and electrochemical sensors. In: Analytical Techniques in Biosciences. Elsevier BV, pp. 157–178. doi:10.1016/b978-0-12-822654-4.00004-x.

Manjakkal, L., Lorenzelli, L., Willander, M., eds. 2023. Sensing Technologies for Real Time Monitoring of Water Quality. John Wiley & Sons.

Miller, T., Durlik, I., Kostecka, E., Kozlovska, P., Łobodzińska, A., Sokołowska, S., Nowy, A., 2025. Integrating artificial intelligence agents with the internet of things for enhanced environmental monitoring: applications in water quality and climate data. Electron. (Switzerland). 2079-9292. 14 (4). Available from: www.mdpi.com/journal/electronics. doi:10.3390/electronics14040696.

Nguyen, M.K., Ahn, J., Han, T.G., Tung, N.K.T., Lee, J.-Y., Kang, J.S., Kim, H.M., 2025. Determination of organophosphate flame retardants in bottled drinking water via solid-phase extraction and gas chromatography-mass spectrometry. J. Food Compos. Anal. 0889-1575. 141, 107322. doi:10.1016/j.jfca.2025.107322.

Okorogbona, A.O.M., et al., 2018. Water quality impacts on agricultural productivity and environment. In: Lichtfouse, E. (Ed.), Sustainable Agriculture Reviews, Vol. 27. Springer International Publishing, Cham, pp. 1–35, 978-3-31-975189-4, 978-3-31-975190-0. doi:10.1007/978-3-319-75190-0_1.

Orlando, A., Franceschini, F., Muscas, C., Pidkova, S., Bartoli, M., Rovere, M., Tagliaferro, A., 2021. A comprehensive review on raman spectroscopy applications. Chemosensors. 2227-9040. 9 (9), 262. doi:10.3390/chemosensors9090262.

Passos, M.L.C., Lúcia MFS Saraiva, M., 2019. Detection in UV–Visible spectrophotometry: detectors, detection systems, and detection strategies. Measurement 135, 896–904.

Ritgen, U., 2023. Atomic absorption spectroscopy (AAS). In: Ritgen, U. (Ed.), Analytical Chemistry I. Springer, Berlin, Heidelberg, pp. 247–253, 978-3-662-66335-6. doi:10.1007/978-3-662-66336-3_21

Russ, J., Zaveri, E., Desbureaux, S., Damania, R., Rodella, A.-S., 2022. The impact of water quality on GDP growth: evidence from around the world. Water Secur. 2468-3124. 17, 100130. doi:10.1016/j.wasec.2022.100130.

Sengupta, A., Jasmine, J., Chaudhary, B., Singh, K., Sharma, S., Sable, H., Sachan, R.S.K., 2025. Advanced monitoring and detection techniques for emerging contaminants in aquatic environments. In: Jain, H., Rath, P., Bhardwaj, L.K., Shah, M.P. (Eds.), Emerging Contaminants in the Aquatic Environment: Global Overview, Advances in Wastewater Research, first ed. Springer Nature Singapore, Singapore, pp. 129–156, 978-981-96-8547-9, doi:10.1007/978-981-96-8548-6.

Senila, M., 2024. Recent advances in the determination of major and trace elements in plants using inductively coupled plasma optical emission spectrometry. Molecules. 1420-3049. 29 (13), 3169. doi:10.3390/molecules29133169.

Singh, U.V., Abhishek, A., Singh, K.P., Dhakate, R., Singh, N.P., 2014. Groundwater quality appraisal and its hydrochemical characterization in Ghaziabad (a region of indo-gangetic plain). Appl. Water Sci. 2190-5487. 4 (2), 145–157. doi:10.1007/s13201-013-0137-7.

Taymanov, R., Sapozhnikova, K., Baksheeva, I., 2021. Trends in the development of metrological self-check and automatic technical diagnostics. In: 2021 XXXI International Scientific Symposium Metrology and Metrology Assurance (MMA). IEEE.

Vaikunta Rao, L., Shirsagar, P.R.K., Tan, K.T., Krishnan, S.B., Sonekar, S.V., Zade, S., 2025. The assessment of water quality forecasting using AI-based ML algorithms. Metall. Mater. Eng. 2812-9105. 31 (4), 942–961. doi:10.63278/1539.

WHO, 2022. Guidelines for Drinking-Water Quality, 4th ed. World Health Organization.

Yang, J., et al., 2023. Microplastics in different water samples (seawater, freshwater, and wastewater): methodology approach for characterization using micro-FTIR spectroscopy. Water Res. 232, 119711.

Yelampalli, S.R., Gandla, K., Reddy, K.T.K., Ibrahim, A.E., Deeb, S.E., 2023. Determination of sodium, potassium, and magnesium as sulfate salts in oral preparations using ion chromatography and conductivity detection. Separations. 2297-8739. 10 (2), 99. doi:10.3390/separations10020099.

Zacharioudaki, D.-E., Fitilis, I., Kotti, M., 2022. Review of fluorescence spectroscopy in environmental quality applications. Molecules 27 (15), 4801.

Chapter 19

Wastewater treatment

Subhadwip Ghorai, Soham Bachaspati, Soham Hazra, Sourav Roy, Rajdeep Mohanta, Parijat De, Sk Md Asif, Sanjay Mochary, and Pabitra Kumar Ghosh

Department of Agriculture, Brainware University, Kolkata, West Bengal, India

1 Introduction

1.1 Background

Wastewater management has become a significant environmental and public health issue in the wake of increasing urbanization, industrialization, and population growth. It is estimated that global municipal wastewater production amounts to around 380 billion m^3 per annum and is expected to rise by 24% by 2030 and 51% by 2050. Nevertheless, only around 48% of it is treated prior to discharge. These are the estimates relayed by the European Investment Bank (2022) and summarized in Tariq and Mushtaq (2023). The World Health Organization reports that over 80% of wastewater worldwide is released untreated, polluting freshwater sources and creating dire ecological hazards (Tariq and Mushtaq, 2023).

The efficiency of wastewater treatment varies across economic groups: high-income economies treat about 70% of wastewater generated, upper-middle-income economies treat 38%, lower-middle-income economies treat 28%, and low-income economies treat merely 8%. Tariq and Mushtaq (2023) and Kesari et al. (2021) pointed out that nearly 40% of the world's population lives in water-scarce basins, which makes it crucial to manage wastewater to avoid surface water degradation.

In India, almost 70% of surface water is unsafe for consumption due to the direct discharge of wastewater, and roughly half of the human waste is discharged without proper treatment, contaminating the Ganges, Yamuna, and other rivers (Anawar and Chowdhury, 2020). The same conditions are found in most developing countries, where untreated municipal, industrial, and agricultural effluents directly flow into river systems (Tariq and Mushtaq, 2023). Inadequate wastewater management reduces riverine ecological resilience, accelerates eutrophication, and increases the occurrence of waterborne illnesses like typhoid and cholera, as explained by Kesari et al. (2021).

River Pollution. https://doi.org/10.1016/B978-0-443-45024-2.00004-X

1.2 Definition and character of wastewater

Wastewater is considered used water that is polluted by domestic, industrial, commercial, or agricultural processes, as well as stormwater and drainage. It is usually made up of >90% water and <10% solids, which consist of 40%–50% organic compounds, 30%–40% inert materials, 10%–15% nondegradable residues, and 5%–8% trace constituents (Tariq and Mushtaq, 2023).

It can be classified mainly into four categories: domestic sewage, industrial wastewater, agricultural runoff, and urban or combined wastewater. Domestic wastewater is primarily blackwater and greywater, both of which are high in organic matter, detergents, and microbial pollutants. Industrial wastewater is generally composed of toxic substances like heavy metals, acids, dyes, and solvents. Agricultural wastewater is high in fertilizers, nitrates, phosphates, and pesticide residues, whereas urban runoff is high in hydrocarbons, plastics, heavy metals, and suspended solids. These categories and their contents are explained in Kesari et al. (2021) and Anawar and Chowdhury (2020).

The heterogeneity of these pollutants dictates the choice of treatment processes and technologies to achieve environmental discharge limits. Treatment of municipal wastewater is frequently characterized by the need to address complex mixtures of organic compounds, nutrients, suspended solids, and toxic pollutants, as highlighted by Kesari et al. (2021).

1.3 Significance of wastewater treatment

Wastewater treatment is essential for maintaining environmental quality, protecting public health, and supporting economic growth. Untreated wastewater degrades aquatic habitats, leads to oxygen depletion, eutrophication, and biodiversity loss, and contributes to the spread of infectious diseases (Kesari et al., 2021). Contaminated water bodies also accumulate heavy metals, pharmaceuticals, and antibiotic-resistant bacteria (ARB), which disrupt ecosystems and pose long-term risks to human health.

Economically, the World Bank (2020) cited that the lack of proper sanitation and wastewater management causes developing countries to lose 1%–7% of Gross Domestic Product (GDP) every year. On the contrary, efficient wastewater management converts waste into a precious commodity. Treated effluents are used for agriculture, industrial cooling, or groundwater recharge, and recovered nutrients and biosolids can feed circular bioeconomy models (Kesari et al., 2021).

Wastewater treatment also supports the realization of the United Nations Sustainable Development Goal 6 (SDG 6) (Clean Water and Sanitation), which indirectly contributes to other goals such as SDG 3 (Good Health and Well-being), SDG 11 (Sustainable Cities and Communities), and SDG 14 (Life Below Water). Full-integration wastewater management, according to

Anawar and Chowdhury (2020), is a major driver of sustainable river-basin restoration, connecting environmental conservation with social and economic development.

2 Wastewater sources and their characteristics

Wastewater is a heterogeneous compound that results from various human and natural activities. It is composed of water (~90%–99%) and suspended, dissolved, and colloidal solids consisting of organic and inorganic compounds, nutrients, heavy metals, and pathogens. The composition and pollutant load are highly influenced by the source—domestic, industrial, agricultural, or urban runoff—and the degree of treatment prior to discharge (Akpor et al., 2014; Kesari et al., 2021; Tariq and Mushtaq, 2023).

2.1 Domestic or municipal sources

Domestic or municipal wastewater is mostly produced from residential areas, institutions, and small businesses. It comprises blackwater (human urine and excreta) and graywater (bathing, washing, and cooking) (Akpor et al., 2014). Its composition includes organic matter from human waste, food residues, and detergents, as well as inorganic salts and pathogens.

Typical composition:

- **Biochemical oxygen demand (BOD_5):** 200–400 mg/L
- **Chemical oxygen demand (COD):** 400–800 mg/L
- **Total suspended solids (TSSs):** 250–400 mg/L
- **Total nitrogen (N):** 20–85 mg/L
- **Total phosphorus (P):** 4–15 mg/L
- **Fecal coliforms:** up to 10^6–10^7 CFU/100 mL

High COD and BOD suggest a large biodegradable organic load, while fecal coliforms and enteric viruses are serious microbial threats (Akpor et al., 2014).

The volume and intensity of wastewater within homes grow with urbanization, poor sanitation, and aging sewerage infrastructure. In developing countries, high rates of population growth and inadequate sewage systems result in direct discharge into natural water bodies, promoting eutrophication and disease transmission (cholera, typhoid, and gastroenteritis) (Tariq and Mushtaq, 2023).

2.2 Industrial wastewater

Industrial effluents depend enormously on the process category and raw material. They are usually characterized by the presence of high concentrations of inorganic chemicals, heavy metals, and persistent organic pollutants.

TABLE 19.1 Classification of industries and major pollutants.

Industry type	Typical pollutants
Textile and dyeing	Synthetic dyes (azo, anthraquinone), surfactants, salts, chromium
Leather tanning	Chromium (Cr^{3+}/Cr^{6+}), sulfides, fats, organic nitrogen
Chemical and pharmaceutical	Acids, solvents, phenols, nitrates, antibiotics
Food processing and distillery	Sugars, proteins, oils, high BOD (1000–6000 mg/L)
Electroplating and metal finishing	Cyanides, cadmium, nickel, copper, lead
Paper and pulp mills	Lignin, chlorinated organics, resins, suspended solids

Industrial effluents depend enormously on the process category and raw material. They are usually characterized by the presence of high concentrations of inorganic chemicals, heavy metals, and persistent organic pollutants (Table 19.1).

Industrial wastewater is usually acidic in pH, is highly conductive, and contains heavy metals like Cu, Cr, Mn, Fe, Pb, Zn, and Ni, which are nonbiodegradable and bioaccumulate in the food chain (Kesari et al., 2021). Discharging it without pretreatment causes ecological toxicity, carcinogenic hazards, and the spread of antimicrobial resistance.

2.3 Agricultural runoff

Agricultural return flows and drainage are sources of diffuse, nonpoint source pollution. The main constituents are as follows:

- **Fertilizers:** nitrates, phosphates, ammonium salts
- **Pesticides:** organochlorine Dichlorodiphenyltrichloroethane, Hexachlorocyclohexane (DDT, HCH), organophosphate, and carbamate residues
- Erosion sediments and suspended solids
- Animal waste and pathogenic microorganisms

Excess nutrients (N, P, K) cause eutrophication and algal blooms, leading to oxygen consumption and fish kills. Monsoon rainfall enhances runoff, resulting in high loads during the monsoon and low dilution during dry periods (Hossain et al., 2021). Fertilizer applications and irrigation return flows are major sources of nitrate and phosphate contamination in groundwater in countries such as India and China (Tariq and Mushtaq, 2023).

2.4 Stormwater and urban runoff

Stormwater comes from rain over hard urban surfaces such as roads, parking lots, and rooftops that prohibit infiltration and transport pollutants directly to storm drains. Common pollutants include:

- Hydrocarbons and motor oil from vehicles
- Heavy metals (Zn, Cu, Pb) from tire and brake wear
- Litter, plastics, and sediment
- Microbial pollutants from animal and pet waste
- Washed lawn and fertilizers' nutrients

The initial runoff of stormwater may carry the greatest load of pollutants. Combined sewers receiving both domestic sewage and stormwater commonly overflow in spontaneous cities and exacerbate urban water pollution (Hossain and Chowdhury, 2020).

2.5 General composition and pollutant categories

The chemical and biological structure of wastewater is highly variable, but general ranges for important parameters are given in Table 19.2.

TABLE 19.2 General composition and pollutant load ranges for major wastewater sources.

Parameter	Domestic	Industrial (avg.)	Agricultural runoff	Typical range (mg/L)
BOD_5	200–400	500–5000	10–30	100–5000
COD	400–800	1000–10,000	30–100	200–10,000
TSS	250–400	300–1000	50–200	50–1000
Total nitrogen (N)	20–85	10–200	5–50	5–200
Total phosphorus (P)	4–15	5–100	0.5–10	0.5–100
Heavy metals (Zn, Pb, Cd, Cr, Ni)	<1	1–100	0.1–5	0.1–100
Pathogens (fecal coliforms)	10^6–10^7 CFU/100 mL	Variable	Variable	–

Source: Naidoo and Olaniran, 2014, Kesari et al. (2021), and Tariq and Mushtaq (2023).

The heterogeneity of these parameters is a function of the source, collection time, and extent of dilution with stormwater or other effluents. On the whole, untreated wastewater from domestic and industrial sources remains a primary cause of surface water degradation, groundwater pollution, and public health hazards worldwide.

3 Environmental and ecological impacts of untreated wastewater

3.1 Degradation of river water quality

Untreated municipal and industrial effluents impact the quality of river water in an intense and permanent way. BOD and COD are elevated, while dissolved oxygen (DO) is decreased. Suspended solids rise, turbidity increases, light penetration decreases, and photosynthesis processes are lowered. When DO falls below $4\,mg\,L^{-1}$, aquatic organisms like fish and invertebrates start to die due to suffocation. This condition alters community composition and results in decreased biodiversity.

In the Ganga–Yamuna system (India) and Haihe–Huaihe basins (China), BOD concentrations have been recorded to be greater than $300\,mg\,L^{-1}$ and DO was reported as low as $0–2\,mg\,L^{-1}$ during the dry season (Tariq and Mushtaq, 2023; Zhang et al., 2021). Tannery, dyes, and industrial effluents' heavy metals also prevent rivers' natural ability to self-clean, leading to anoxia, odors, and blackened water bodies that become permanent (Sharma et al., 2020).

3.2 Eutrophication

Eutrophication, resulting from the high levels of N and P from fertilizers and detergents, leads to nutrient-rich conditions characteristic of untreated wastewater inflow and is further augmented by sewage. Due to the degradation processes of algae, these nutrients can also cause DO depletion as a result of the loss of light necessary for photosynthesis. Cyanobacterial blooms in Taihu, Dianchi, and Chaohu Lakes (China) and Lake Victoria (East Africa) are classic examples of nutrient-rich conditions caused by untreated wastewater inflow (Li et al., 2021; Bruce and Limin, 2021).

The oxygen sag curves show that DO concentrations frequently fall below $1\,mg\,L^{-1}$, which can result in significant fish kills and the production of foul odors. The transition from oligotrophic water to eutrophic water typically causes a loss of biodiversity, an increase in opportunistic species, and the danger of poisoning the source of drinking water due to the production of cyanotoxins by *Microcystis aeruginosa* (Paerl and Otten, 2013).

3.3 Bioaccumulation and toxicity

Industrial effluents commonly comprise persistent substances (e.g., cadmium (Cd), chromium (Cr), lead (Pb), mercury (Hg), and nickel (Ni)) and/or organic xenobiotics (e.g., chlorophenols, DDTs, HCHs, PAHs) that not only accumulate in aquatic organisms but also resist degradation. Such contaminants frequently bioconcentrate as they move along the food chain, eventually accumulating to toxic levels in top predators and, ultimately, in human beings.

Heavy levels of both cadmium and chromium have been found in the Ravi and Kabul Rivers in Pakistan, and also in the Ganga Basin (Afzaal et al., 2022; Patel et al., 2021; Tariq and Mushtaq, 2023). Similarly, in the Niger Delta province of Nigeria, hydrocarbons and metals are deposited in fish tissues and exceed safety limits set by the FAO guidelines (Anele and Okorite, 2025). Chronic exposure to contaminant mixtures has the potential to result in higher incidences of hepatic, renal, and neurological disorders, as well as compromised reproductive fitness in aquatic organisms (Ali et al., 2019).

3.4 Public health implications

Untreated sewage is a reservoir and a transmission route for pathogens such as *Vibrio cholerae*, *Salmonella typhi*, *Giardia lamblia*, *Escherichia coli*, and multiple enteric viruses that induce infectious disease outbreaks of cholera, typhoid, dysentery, and hepatitis A/E. This is a problem in developing countries, where it is estimated that 80% of all infectious diseases are waterborne.

According to Tariq and Mushtaq (2023), unsafe water and sanitation cause about 3.1% of total deaths globally, accounting for approximately 297,000 child deaths annually.

In addition, ARB and antibiotic resistance genes are present in sewage that finds its way into aquatic ecosystems and promote antimicrobial resistance (Landrigan et al., 2024; Zhang et al., 2021). The recycling of untreated wastewater for irrigation also compounds the problem by offering an avenue for enteric pathogens to infect vegetables intended for human consumption, with potential direct effects on food chains (Prasad et al., 2020).

3.5 Socioeconomic implications

Impacts of untreated wastewater degrading water quality are complex and encompass extensive socioeconomic consequences. Fishing industries can collapse due to decreased levels of oxygen and toxins from heavy metals. This denies communities that depend on those resources the opportunity to earn an income (Javed and Usmani, 2019). Similarly, rivers that are highly polluted, like the Ganga and Yamuna, also lose all recreational, cultural, or religious importance, with losses in pilgrimage tourism as a consequence (Bhatt and Singh, 2022). In rural settings, watering crops with polluted water can also lead to salinization problems in the soil, causing lower yields. Increased healthcare costs as a result of issues linked to waterborne diseases and the rehabilitation of ecosystems also exert adverse economic pressure on national economies. As the WHO (2022) estimates, each USD 1 invested in wastewater management saves USD 5 in health and environmental impact costs. Ultimately, the total lost ecosystem services can be linked to lost GDP and even local displacement in very polluted basins.

4 Overview of wastewater treatment processes

4.1 Treatment goals

Treatment of wastewater is primarily aimed at eliminating contaminants and pathogens to suitable levels before discharge or reuse, in a way that safeguards public health, aquatic environments, and water resources. Raw municipal and industrial wastewater typically contains suspended solids, organic matter, nutrients (N and P), heavy metals, pathogens, and toxic substances that must be eliminated to prevent environmental harm (Seow et al., 2016).

Wastewater treatment plants (WWTPs) are planned to:

- For the removal of suspended and dissolved solids, oil, and greases;
- For BOD and COD reduction;
- For pathogen and nutrient (primarily N and P) removal, which causes eutrophication; and
- For stabilization and enabling safe reuse or disposal of produced sludge (Denysiuk et al., 2018).

Moreover, treated wastewater also needs to comply with national and international discharge standards to safeguard downstream uses such as irrigation, aquaculture, and drinking water supply.

4.2 Stages of treatment classification

Typically, wastewater treatment takes place in three sequentially progressive stages, each associated with distinct groups of contaminants. These individual operations are addressed in greater detail in subsequent sections, but the brief overview is as follows:

4.2.1 Primary or physical treatment

The primary level employs mechanical and hydraulic operations: screening, grit removal, sedimentation, and flow equalization, to take out solids and floating material, which are easy to remove. This stage will typically remove 30%–40% of BOD and 50%–60% of suspended solids and conditions the influent for biological treatment. Removed solids will amalgamate to become "primary sludge," which will need stabilization prior to disposal or secondary treatment.

4.2.2 Secondary or biological treatment

At this process phase, microorganisms reduce soluble and colloidal organics to stable end products (CO_2, H_2O, and minor amounts of biomass).

This phase can occur with suspended-growth systems [e.g., activated sludge, sequencing batch reactors (SBRs)] or attached-growth systems [e.g., trickling filters, rotating biological contactors (RBCs)]. This process can remove 80%–95% of BOD and COD and reduce pathogens to safe levels. The secondary sludge obtained in this step will be used to mix with the primary sludge for digestion.

4.2.3 Tertiary or advanced treatment

The last step in the process gives wastewater an additional polish so that effluent analysis can meet stringent standards for disposal or reuse. The final stage of treatment may involve filtration, adsorption (pollutant removal), nutrient removal, disinfection (e.g., chlorine), and advanced oxidation or membrane technologies. Tertiary treatment may eliminate remaining BOD $< 10\ mg\ L^{-1}$, total nitrogen $< 10\ mg\ L^{-1}$, and total phosphorus $< 1\ mg\ L^{-1}$ (Gerba and Pepper, 2019).

Sludge treatment or solids handling is maintained through all stages and involves thickening, digestion, dewatering, and sludge disposal from the process.

4.3 Important parameters for treatment efficiency evaluation

The overall system performance of the processes for treating lunch water, or a WWTP efficiency, will be systematically evaluated using numerous different physicochemical and biological parameters (Table 19.3). Ongoing monitoring will enable an assumed system control—not regulated—but measured to ensure safety for disposal or reuse of the effluent.

5 Primary (physical) treatment

5.1 Screening and grit removal

Large floatable and settleable solids are removed in the initial step of wastewater treatment, i.e., primary treatment, prior to biological treatment. Screening is performed by passing influent through bar racks or fine screens to prevent rags, plastics, leaves, and other wastes from blocking pumps or pipes. Microscreens (<1 mm) and mechanically cleaned bar screens (6–25 mm spacing) are used by municipal plants.

Following screening, grit chambers eliminate sand, gravel, silt, and other heavier inorganic particles that can cause damage to downstream equipment. The regulated hydraulic velocity (approximately $0.3\ m\ s^{-1}$) in grit removal allows heavier particles to settle while organic matter remains buoyant. After being separated, the grit is washed, dried, and finally disposed of in a landfill.

The Kolkata Municipal Corporation sewage treatment plant has a grit removal efficiency of 85%–95% for particles larger than 0.2 mm in size (Metcalf and Eddy, 2003).

5.2 Sedimentation

Gravity settling is employed by sedimentation tanks, or primary clarifiers, to remove suspended particles and organic waste. After a residence time of one

TABLE 19.3 Key parameters for evaluating wastewater treatment performance and their environmental implications.

Parameter	Description/significance	Typical influent range	Treated effluent target	Principal environmental effect (if untreated)	Key reference
BOD (mg L^{-1})	Indicates biodegradable organic load and microbial oxygen demand.	200–600	<30	Oxygen depletion and aquatic mortality	(Ramzanipour et al., 2023; Tariq and Mushtaq, 2023)
COD (mg L^{-1})	Represents total oxidizable organic matter (biodegradable and nonbiodegradable).	300–1000	<50	High organic pollution load	(Seow et al., 2016; Tariq and Mushtaq, 2023)
pH	Reflects acidity/alkalinity; the optimal microbial range is 6.5–8.5.	6–9	6.5–8.5	Extremes suppress microbial activity and treatment efficiency.	(Metcalf and Eddy, 2003)
TSS (mg L^{-1})	Indicates particulate removal efficiency.	100–500	<20	Turbidity, sedimentation, reduced light penetration	(Gerba and Pepper, 2019)
Total N (mg L^{-1})	Includes organic, ammoniacal, and nitrate-N; excessive levels cause eutrophication.	20–80	<10	Eutrophication and algal blooms	(Li et al., 2021)
Total P (mg L^{-1})	Key nutrient promoting algal blooms and eutrophication.	5–20	<1	Nutrient enrichment and oxygen depletion	(Paerl and Otten, 2013)
Heavy metals (Pb, Cd, Cr, Hg) (μg L^{-1})	Persistent, bioaccumulative, toxic elements.	10–500	<100	Bioaccumulation and aquatic toxicity	(Ali et al., 2019)
Pathogens (CFU 100 mL^{-1})	Disease-causing microbes (bacteria, viruses, protozoa).	10^5–10^7	Nil detectable	Waterborne disease outbreaks	(WHO, 2022)
Oil and Grease (mg L^{-1})	Hydrocarbons forming surface films on water.	10–50	<10	Oxygen barrier and esthetic pollution	(Aransiola et al., 2024)

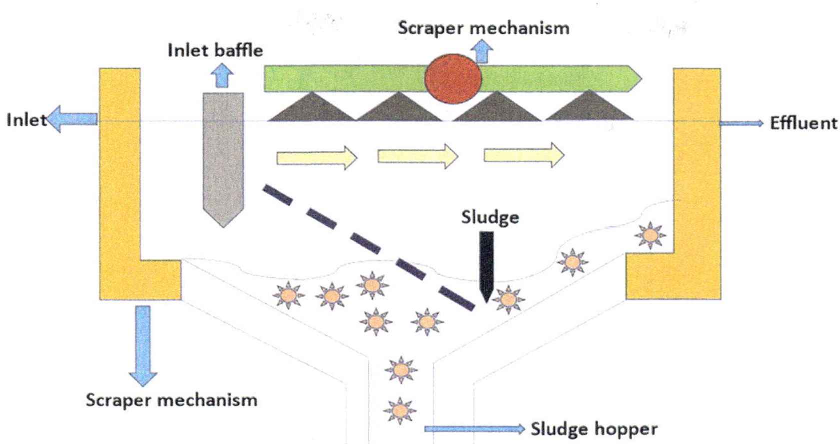

FIG. 19.1 Schematic of a rectangular sedimentation tank showing the inlet baffle, sludge hopper, and scraper mechanism (Mandal et al., 2023).

and a half hours to two hours, settleable solids become sludge. The sludge is then mechanically scraped off and periodically pumped to digesters. Secondary treatment is then applied to the clarified effluent (Fig. 19.1).

The typical effluent removal efficiency is between 25% and 40% for biochemical oxygen demand (BOD) and 50% and 70% for TSSs. For example, Delhi and Chennai municipal plants reduce TSS and BOD in primary clarifiers by 60% and 35%, respectively, on average.

5.3 Flotation and equalization

Flotation is created by utilizing solids and oils lighter than water. Air-diffused tiny bubbles are formed; these bubbles adhere to oil, grease, or suspended particles and float, allowing them to be skimmed off [dissolved air flotation (DAF)]. Grease and oil separation will typically be >90% effective. Equalization tanks balance flow and load fluctuations on an hourly or daily basis to prevent shock loads on the downstream biological system. By aeration or mixing, odor formation or settling is eliminated, and a consistent concentration is achieved (Tchobanoglouset al., 2003) (Fig. 19.2).

6 Secondary (biological) treatment

6.1 Principle and objectives

Secondary treatment biologically breaks down dissolved and colloidal organic matter using aerobic or anaerobic bacteria. The primary objective is to lower BOD and COD to satisfactory discharge values, usually below 30 mg L^{-1} BOD.

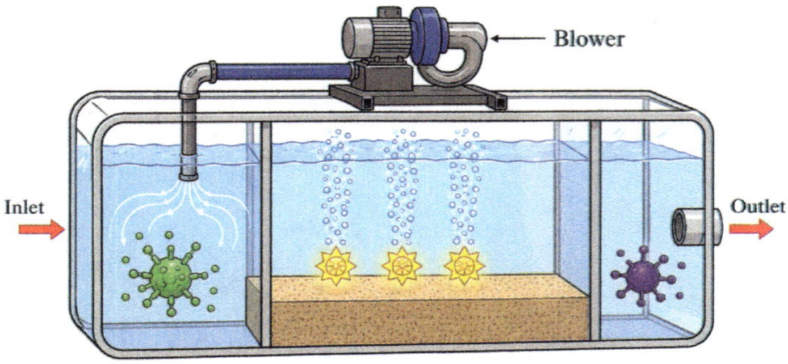

FIG. 19.2 Equalization basin layout with a diffused aeration system.

The bacteria, at the same time, create biomass that is separated into sludge by oxidizing organic carbon to CO_2 and water.

6.2 Suspended growth systems

6.2.1 Activated sludge process

Wastewater flows through an aeration tank in an activated sludge process (ASP), where microorganisms (mixed liquor) degrade the organic matter. Air is supplied to maintain the DO concentration at a level of approximately 2 mg L^{-1}. The mixture flows through a secondary clarifier, where solids accumulate, and part of the biomass is returned (return activated sludge) in a bid to maintain the microbial population.

Key parameters:

- **Hydraulic retention time (HRT):** 4–8 h
- **Sludge age (mean cell residence time):** 5–15 days
- **F/M ratio:** 0.2–0.5 kg BOD kg^{-1} MLSS d^{-1}
- **Mixed Liquor Suspended Solids (MLSS) concentration:** 2000–4000 mg L^{-1}
- **Strengths:** High efficiency (85%–95% BOD removal), flexible operation.
- **Weaknesses:** energy-intensive, requires skilled control, bulking/sludge-settling problems.

6.2.2 Sequencing batch reactors

SBRs obtain aeration and clarity within a single reactor through an automated fill, react, settle, decant, and idle cycle. They are space-saving, efficiently treat variable flows, and deplete BOD/COD by 90%–95%. Extensively used in decentralized plants, like the Surat and Nagpur Sewage Treatment Plants (STPs).

6.3 Attached growth systems

In trickling filters, wastewater is distributed over stone or plastic media beds. Biofilm breaks down organic matter that has adhered to the surface and is seeping down. Wastewater is filtered before being discharged or further treated.

RBCs possess rotating disks that are kept partially immersed in wastewater. The rotating motion achieves effective aeration with minimal energy input by alternating the microbial contact between air and wastewater. Both systems provide 60%–90% BOD elimination, low sludge generation, and ease of operation.

6.4 Anaerobic systems

Anaerobic treatment is most suitable for wastewater that is high in strength. Upflow Anaerobic Sludge Blankets (UASBs) utilize an influent that moves upward through a dense bed of anaerobic granules to decompose organics into biogas ($CH_4 + CO_2$). Advantages include minimal sludge, lower energy usage, and the recovery of biogas. The removal of COD from industrial effluents ranges between 70% and 90%. Stabilization of primary and secondary sludge by anaerobic digesters produces methane-rich biogas (60%–70% CH_4). Running plant operations with the gas is one way to achieve energy neutrality (Table 19.4).

7 Tertiary and advanced treatment processes

7.1 Filtration and adsorption

Removal of turbidity and any residual suspended matter is done by filtration. Rapid sand filters can treat water with a turbidity as low as 5 NTU (Nephelometric Turbidity Unit). Activated carbon filters adsorb color, odor, and chemicals. New materials, like biochar-based filters, are cost-effective and have a high capacity for adsorption (50–200 $mg g^{-1}$ for metals and colors).

TABLE 19.4 Typical performance comparison of aerobic and anaerobic systems.

Parameter	Aerobic (ASP/SBR)	Anaerobic (UASB/digester)
BOD removal (%)	85–95	70–90
COD removal (%)	80–95	70–90
Energy requirement	High (aeration)	Low
Sludge yield	High	Low
Biogas production	Negligible	0.2–0.35 m^3 CH_4 kg^{-1} COD removed

7.2 Membrane technologies

Membrane processes separate contaminants physically by pore size:

- **Microfiltration (MF):** suspended particles and bacteria are removed.
- **Ultrafiltration (UF):** macromolecules and viruses are retained.
- **Nanofiltration:** multivalent ions and organic molecules are removed.
- **Reverse osmosis:** almost all dissolved salts and organics are removed.

While membrane fouling and high capital/operating costs remain problems, compactness and high-quality effluents [total dissolved solid (TDS) $< 50\,mg\,L^{-1}$] are benefits. Hybrid systems [such as membrane bioreactors (MBRs)] use biological treatment and UF membranes together to produce >95% removal of BOD and >99% reduction of pathogens.

7.3 Advanced oxidation processes

Advanced oxidation processes (AOPs) utilize highly reactive hydroxyl radicals (•OH) for the degradation of refractory pollutants such as pharmaceuticals, pesticides, and endocrine disruptors. Typical AOPs include:

- Ozone oxidation (O_3)
- UV/H_2O_2 photolysis
- Fenton reaction (Fe^{2+}/H_2O_2)
- Photocatalysis (TiO_2-based)

AOPs have the capability of achieving >90% micropollutant removal and are generally applied after biological treatment.

7.4 Ion exchange and chemical precipitation

Ion exchange is used to remove heavy metals (Pb^{2+}, Cr^{3+}) and nutritional ions (NH_4^+, NO_3^-) through synthetic resins. Lime, alum, or ferric salts are introduced during chemical precipitation to form insoluble hydroxides or phosphates, which are then filtered or sedimented. Typical applications include phosphate and metal removal, and their comparative performance is summarized in Table 19.5. The combination of tertiary treatment processes is illustrated in Fig. 19.3.

8 Nature-based and emerging technologies

8.1 Constructed wetlands

Constructed wetlands (CWs) are artificial systems created to mimic the natural treatment processes of swamps and marshes through the concerted action of vegetation, microorganisms, and substrate. CWs are low-energy, low-capital-cost alternatives to traditional treatment plants that are very well adapted for decentralized wastewater treatment (Rosendo et al., 2022).

TABLE 19.5 Comparative efficiency and cost of tertiary treatment methods.

Method	Target pollutants	Typical removal (%)	Relative cost	Remarks	Reference
Sand filtration	TSS, turbidity	70–90	Low	Simple, effective polishing	Khelladi et al. (2020)
Activated carbon	Organics, odor	80–95	Medium	Regeneration needed	Torres-Lozada et al. (2018)
Membrane filtration	Salts, pathogens	95–99	High	Fouling issues	Zhang et al. (2023)
Advanced Oxidation Processes (AOPs)	Micropollutants	85–95	High	High energy requirement	Morin-Crini et al. (2022)
Ion exchange	Nutrients, metals	80–98	Medium	Periodic regeneration	Marin et al. (2024)
Chemical precipitation	Phosphorus (P), metals	70–90	Low–Medium	Sludge disposal needed	Abdoli et al. (2024)

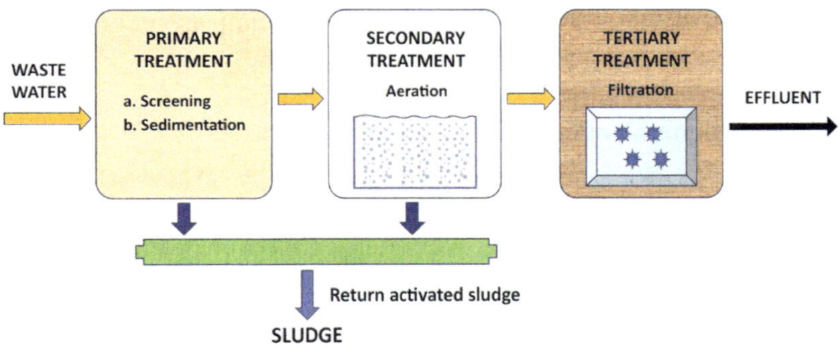

FIG. 19.3 Integrated wastewater treatment chain combining primary, secondary, and tertiary processes (Maxcy Brown et al., 2023).

CWs are generally classified into surface-flow (SF) and subsurface-flow (SSF) systems. In SF wetlands, wastewater runs over a layer of vegetation on soil, where sedimentation and microbial decomposition occur at the water–air interface (Malyan et al., 2021). SSF wetlands, on the contrary, direct wastewater horizontally or vertically through gravel or sand beds with planted macrophytes like *Phragmites australis*, *Typha latifolia*, and *Cyperus papyrus* (Vásquez et al., 2022).

The prevailing mechanisms of pollutant removal are sedimentation, filtration, adsorption, uptake by plants, nitrification, denitrification, and microbial decomposition (James and Vijayanandan, 2023). Normal removal efficiencies are around 70%–90% for BOD, 60%–80% for COD, 40%–70% for total nitrogen, and 30%–60% for total phosphorus, based on HRT and loading rate (Munavalli et al., 2022; Awad et al., 2023). In India, large-scale examples include the East Kolkata Wetlands (EKW), which treat millions of liters of municipal sewage daily while simultaneously supporting aquaculture and agriculture, exemplifying a circular bioeconomic model.

8.2 Algal-based treatment systems

Algae, especially microalgae like Chlorella, Scenedesmus, and Spirulina, have a dual function in wastewater treatment: absorbing nutrients (N and P) and producing oxygen through photosynthesis, which aids aerobic bacterial oxidation of organic compounds (Oruganti et al., 2022; Wang and Hong, 2022). The produced algal biomass contains proteins and lipids, which can be valorized to produce biofertilizer, animal feed, or biofuels.

High-rate algal ponds (HRAPs) are mostly employed because of their shallow depth, constant mixing, and exposure to sunlight, allowing for efficient uptake of nutrients at a high rate (Zhao et al., 2024). Algal–bacterial symbiosis increases the removal of ammonia, nitrates, phosphates, and heavy metals. In river restoration situations, HRAPs coupled with oxidation ponds can diminish the threat of eutrophication by stripping off excess nutrients prior to the discharge of effluent (Aparicio et al., 2024; Kaur, 2025).

8.3 Electrochemical and photocatalytic processes

Electrochemical treatment involves the utilization of oxidation and reduction processes at electrode surfaces for breaking down organic pollutants or settling heavy metals. Applications like electrocoagulation, electrooxidation, and electro-Fenton have recorded high removal rates with low chemical input (Yang et al., 2023).

Photocatalytic oxidation, based on semiconductors such as TiO_2, ZnO, or doped graphene composites, uses UV or sunlight to produce reactive oxygen species (•OH, $O_2•^-$) that mineralize persistent pollutants. Such advanced systems hold potential for addressing emerging pollutants, drugs, endocrine disruptors, and dyes, which evade traditional treatment.

8.4 Biochar and nanomaterial-based adsorbents

Biochar, a carbonaceous porous material that is pyrolyzed biomass, has a high surface area and functional groups with adsorptive properties for organic and inorganic contaminants (Ambaye et al., 2021). Inclusions of metal oxides

(Fe, Zn, Mn) or chitosan improve its arsenic, lead, and phosphate sorption capacity.

Nanomaterial-derived adsorbents, such as graphene oxide, carbon nanotubes, and magnetic nanoparticles, provide enhanced reactivity and renderability. Yet, possible ecotoxicological impacts and the cost of production persist as limitations. These materials are being considered for hybrid systems (biochar + wetland + electrochemical cell) for low-energy, sustainable treatment.

At large, nature-based and hybrid eco-technologies are a paradigm shift toward decentralized, climate-resilient, and low-carbon wastewater management, concordant with India's National Water Mission and global SDG 6 targets.

9 Treatment and recovery of sludge resources

9.1 Wastewater sludge generation and composition

Sludge refers to the concentrated semisolid waste generated from the physical, chemical, and biological treatment of wastewater (Panda et al., 2025). It is mainly generated from three sources:

- Primary sludge, from sedimentation tanks, that consists of settleable solids, grit, and organic detritus;
- Secondary (biological) sludge, made up of microbial flocs and metabolic by-products of activated sludge or biofilm systems;
- Tertiary sludge, due to chemical precipitation or higher treatment stages, usually enriched with metal salts or adsorbents.

The physicochemical makeup of sludge depends on the influent type, treatment arrangement, and seasonal pattern. Raw sewage sludge normally consists of a high concentration of water, fewer solids, and a particular proportion of volatile (organic) to fixed (inorganic) solids (Panda et al., 2025). The inorganic fraction may differ according to the treatment process, but here is the general breakdown: the inorganic fraction is made up of silicates, phosphates, and heavy metals like Cd, Pb, and Cr (Genua et al., 2025).

The intense concentration of pathogenic microbes, parasitic eggs, and toxic substances creates a significant environmental risk when not handled appropriately. Thus stabilization and safe disposal, or, even better, valorization, are necessary to prevent secondary pollution and make valuable resources available.

9.2 Process line of sludge management

Sludge treatment is a process involving a series of volume reduction, stabilization, dewatering, and disposal or recycling, all designed to enhance handling characteristics and minimize environmental hazards. The most important steps are discussed in the following.

9.2.1 Thickening

Thickening concentrates solids to 4%–6% by removing some of the free water (Chaedir et al., 2021). Typical technologies include gravity thickeners, DAF, and centrifugal thickeners. Polymer conditioning tends to enhance floc development and settleability. The process has a profound impact on the footprint reduction of downstream units.

9.2.2 Stabilization

Stabilization converts putrescible organic material into biologically stable material with low odor and pathogen content. This is done via aerobic or anaerobic digestion (AD):

- Aerobic digestion is based on endogenous respiration under aerated conditions, appropriate for small-scale plants but energy-intensive.
- AD, carried out in closed digesters at mesophilic (35°C) or thermophilic (55°C) temperatures (Lanko et al., 2021), breaks down volatile solids through hydrolysis, acidogenesis, acetogenesis, and methanogenesis. AD produces a mixture of biogas, which can be used for heat and electricity, thus balancing plant energy requirements (Inayat et al., 2025).

9.2.3 Dewatering and drying

After digestion, sludge usually has 94%–96% water content and must be dewatered to enable transportation and disposal (Qrenawi and Rabah, 2021). Centrifuges, belt-filter presses, screw presses, and drying beds are commonly used. Chemical conditioners (lime, ferric chloride, or cationic polymers) enhance separation efficiency. Thermal drying (150°C–250°C) removes further moisture to <10%, producing a granular biosolid suitable for incineration or coprocessing (Ling et al., 2024).

9.2.4 Thermal treatment and incineration

In land-scarce or pathogen-sensitive environments, incineration achieves close-to-total volume reduction and pathogen elimination. Energy recovered from combustion gases can be utilized for sludge drying or grid sale. Pyrolysis and gasification are nascent alternatives, producing biochar, syngas, and heat with smaller greenhouse gas footprints (Matarru and Shin, 2025).

9.3 Resource recovery pathways

Modern sludge management goes beyond traditional disposal and fits the paradigm of a circular economy, where wastewater solids are considered secondary resources, not waste. Major recovery pathways are given in the following.

9.3.1 Energy recovery

- **Biogas utilization:** Biogas from AD can power combined heat and power plants, generating 1.5–2.5 kWh m^{-3} of sludge, thus achieving partial or total plant energy neutrality (Salano et al., 2024).
- **Thermal valorization:** Pyrolysis produces energy-dense syngas and biooil. Gasification residues contain high fixed-carbon content and can be used as soil conditioners or activated-carbon precursors (Hoang et al., 2024).

The installation of such energy recovery systems, particularly in large-scale Indian STPs like Okhla (Delhi) and Keshopur, demonstrates 15%–25% savings in operational energy expenditure (Dubey et al., 2024).

9.3.2 Nutrient recovery

Recovery of phosphorus and nitrogen is essential in closing the nutrient loop and breaking nonrenewable phosphate rock dependence. Controlled precipitation of struvite ($MgNH_4PO_4 \cdot 6\ H_2O$) in digester effluent sequesters orthophosphate and ammonium ions (Wagner and Karthikeyan, 2022). Recovered struvite has slow-release fertilizer characteristics akin to superphosphate, with high recovery yields.

9.3.3 Material and product reuse

Organic matter-rich stabilized sludge (biosolids) may be composted and used as a soil amendment to enhance soil structure, water retention, and fertility. Cocomposting with municipal solid waste improves the C/N ratio and decreases heavy metal bioavailability through humification (Hassanzadeh Moghimi et al., 2023).

Alternatively, incineration ash of sludge contains 8%–12% P_2O_5 and silica, which can be reused in the production of bricks, cement, and ceramics. Sludge-pyrolysis-derived biochar is used as a sorbent for heavy metal remediation of river sediments, thereby integrating sludge recycling with river pollution abatement (Meghea et al., 2025).

9.4 The circular economy perspective

Sludge treatment and resource recovery represent the shift from linear "take–use–dispose" systems toward closed-loop resource cycles. Contemporary wastewater utilities are increasingly conceived as Water Resource Recovery Facilities (WRRFs), with the integration of water reclamation, energy generation, and material recycling (Varma et al., 2024).

Combining anaerobic digesters with solar-powered drying, membrane-based dewatering, and nutrient capture reactors is likely to produce self-sustaining treatment centers with very small environmental footprints. The secondary benefits reach beyond pollution control to encompass renewable energy generation, replacement of agricultural inputs, and mitigation of climate change.

10 Case studies of wastewater treatment and river restoration

10.1 Indian context

Rivers in India suffer from severe deterioration owing to high flows of untreated sewage and industrial effluents. The document reports >72,000 Million Litres per Day (MLD) of sewage production against a treatment capacity of a mere 40%, leading to recurring organic and microbial pollution in rivers like the Ganga and Yamuna (CPCB, 2024). The Ganga Action Plan (1985) initiated wastewater interception and treatment through ASPs, oxidation ponds, and UASB reactors with 70%–85% removal of BOD but faced constrained long-term efficiency due to inadequate maintenance and operation. The later Namami Gange Program (2014 onward) implemented basin-scale technology combining SBRs, MBRs, bioremediation, and decentralized wetland systems. More than 160 STPs (≈6000 MLD) now operate in the basin with real-time monitoring of effluent, resulting in enhanced DO and lowered BOD in various reaches (NMCG, 2023). The Yamuna Action Plan, aided by JICA, renovated STPs in Delhi, Agra, and Mathura with tertiary filtration and public-participation modules, lowering effluent BOD to <10 mg L^{-1} (Srivastava et al., 2023). A nature-based hybrid system functions in the EKW, a Ramsar wetland processing ≈ 900 million L/day of municipal wastewater by sedimentation, algal–bacterial symbiosis, and macrophyte absorption, with >85% and achieving >90% TSS removal, as well as sustaining aquaculture and agriculture (Biswas et al., 2023). Together, these projects demonstrate India's evolution from traditional, centralized treatment to comprehensive and sustainable wastewater management systems connecting pollution control with livelihood and ecosystem revival.

10.2 Global examples

Global lessons point to the potential of technology, policy, and governance convergence in bringing about river-basin rejuvenation. Singapore's NEWater project is an exemplary closed-loop reclamation system in which treated effluent undergoes MF, reverse osmosis, and UV disinfection to yield ultra-pure water that adheres to WHO standards and meets approximately 40% of national demand (Wang, 2022). In Japan, the widespread application of MBRs facilitates high-density, automated city treatment of BOD < 5 mg L^{-1} and TN < 10 mg L^{-1}, with effluent used for irrigation and recharge (Ferri and Bolelli, 2025). The Netherlands' "Room for the River" program incorporates CWs and downstream riparian buffer strips at treatment plants to mitigate nutrient load in the Rhine–Meuse delta, promoting aquatic biodiversity and flood resistance (Havinga, 2020). These global examples validate that effective wastewater management relies on technological development facilitated by enduring governance, stakeholder engagement, and circular economy approaches.

11 Challenges in wastewater management

Wastewater management is critical for water security, conservation of the environment, and public health. The sector globally faces a number of challenges, including high capital and operating costs, technological constraints, pollution from so-called "emerging pollutants," and low public involvement. This section of the book addresses these challenges from an international perspective, with examples from India and beyond. It summarizes the impacts of each challenge and proposes solutions directed toward policymakers and scholars. Information is drawn from recent review papers and studies, along with policy documents (Abreham Awono and Adam, 2025).

11.1 Economic constraints and maintenance problems

Wastewater treatment facilities (WWTFs) are expensive, with significant costs associated with operation. WWTFs in North America and Europe consume 1%–4% of overall energy. The biological process aeration systems require about 58% of total plant energy, and costly technologies are needed for nutrient removal (Hernández-Chover et al., 2023). Interviews with stakeholders in India indicated that high operation and maintenance (O&M) costs, inadequate funding, and a lack of qualified employees prevent most municipal authorities from keeping treatment plants operational. A Portuguese study identified energy as the primary cost of operations, noting that natural options like CWs still require energy to reuse water. In Dakar, Senegal, an activated-sludge sewer-based system was costing ten times more per capita in capitalized annual costs and five times more in overall capital and O&M costs than a fecal sludge management system. Similarly, municipal utility audits in Tamil Nadu and Jharkhand found underutilized treatment plants due to incomplete sewerage networks, poor financial management, and exorbitant energy levies (Pegu and Samal, 2025).

11.2 Technological and infrastructural impediments

Technological shortcomings reduce the efficiency of existing treatment plants. Membrane processes can be applied to remove almost all dissolved solids but are marred by fouling from colloids, organics, and biofilms, which reduces water flux, increases transmembrane pressure, and necessitates costly cleaning and replacement (Pathak et al., 2020). Emerging technologies such as MBRs and AOPs facilitate nutrient and energy recovery but are hindered by high capital costs and energy consumption. The majority of Indian WWTPs perform suboptimally because the sewerage system is incomplete or inadequately maintained (Abreham Awono and Adam, 2025). Operational audits of Addis Ababa plants identified equipment malfunctions, unsatisfactory sludge disposal, odor release, and low hydraulic capacity; 70% of plant staff described cost as the main hindrance to operation. These technological and infrastructural issues are compounded in low-income countries where utilities cannot afford to fund maintenance or modernization (Sha et al., 2024).

11.3 Emerging contaminants: Microplastics and antibiotics

New pollutants such as microplastics, pharmaceuticals, and antibiotics pose new challenges. Microplastics (microfibers, microfragments, and micropellets < 5 mm) are emitted to wastewater via personal care products, textiles, and industrial effluents. Removal efficiency is reported to be highly variable: in Iran and Finland, removal of microplastics is 84%–99%, whereas removal of microplastics is 48%–78% in Turkey and Lithuania (Bodzek et al., 2024). Even advanced multistaged systems retain residual microplastics in effluent, and the majority of trapped particles accumulate in sludge. Antibiotics and personal care products are also an issue. They reach the environment through human excretion, disposal, and pharmaceutical industries' effluents. Treatment facilities most likely do not eliminate them; studies show concentrations in effluent near pharmaceutical plants are 10–1000 times greater than in regular plants (Sharma et al., 2023). Wastewater is a good environment to facilitate horizontal gene transfer and may serve as a reservoir of resistant antibiotic genes and bacteria. Unnecessary application of antibiotics, uncontrolled sales, and inadequate sanitation contribute to the issue.

11.4 Social barriers: Public participation and enforcement lack

Public participation influences the success of wastewater projects. The NITI Aayog report on India's urban wastewater sector revealed that sometimes citizens themselves remain unaware of water shortages and the benefits of reuse. Negative attitudes, health risk concerns, and fear of color or odor reduce acceptance of reclaimed water. Public opinion studies imply that acceptance increases when human contact is minimal, treatment quality is reassuring, costs are reasonable, and environmental benefits are communicated (Hartley, 2006). Without regulation, many industries and municipalities discharge untreated wastewater. Regulatory penalties under India's Water Act are negligible, state pollution control boards (SPCBs) are undermanned and lack laboratories, and monitoring networks are thin. Stakeholders demand the imposition of mandatory recycling, subsidization, training, and penalties for noncompliance (Balkrishna et al., 2024).

11.5 Impact on policy and practice

The financial and technological challenges lead to subpar performance of WWTPs, wasting energy and incurring unsustainable expenses. Ineffective treatment of microplastics, antibiotics, and other contaminants poses threats to aquatic environments and human health. Institutional and social barriers limit investment in infrastructure, the installation of new technology, and water reuse practices. The overview of these hurdles, their effects, and solutions for researchers and policymakers is provided in Table 19.6.

12 Future outlook and sustainable methods

Wastewater infrastructure plays a vital role in sustainable development, but traditional centralized treatment facilities are not capable of dealing with the rapid expansion of cities, climate variability, and the need for resource recovery. Contemporary research confirms that wastewater may be viewed as a renewable resource for water, energy, and nutrient recovery rather than being treated and discharged. In the following sections, future-looking strategies like decentralized systems, renewable energy integration, smart monitoring, and circular economy principles are introduced to inform policymakers and academics (Sha et al., 2024; Hernández-Chover et al., 2023).

12.1 Modular and decentralized systems

Decentralized systems of sewers have high up-front expenses and are shock-sensitive. Summaries of review studies suggest decentralized treatment systems are flexible, resilient, and effective in operation, particularly for small towns and rural communities where the construction of long sewer lines is not feasible. Individual on-site systems (e.g., septic tanks) treat oils and solids at the household level and discharge the effluent to the soil or compact treatment plants. Cluster systems collect wastewater from several households through small-diameter pipes and treat it in a local plant, offering a compromise between on-site and central plants (Sha et al., 2024; Strande and Brdjanovic, 2014).

12.2 Integration of renewable energy

12.2.1 *Solar-powered aeration*

Wastewater treatment works are energy-consuming, and the integration of renewable energy can reduce operating costs and greenhouse gas emissions. Solar photovoltaic (PV) systems convert direct sunlight into electricity, and a review of renewable energy integration in wastewater treatment identifies PV modules as being widely applied due to declining costs and the simplicity of installation; stand-alone or grid-connected PV can power pumps, aerators, and sensors in small plants (Hernández-Chover et al., 2023). A pilot project in Thailand demonstrated a solar-powered decentralized treatment system for canal water: the system utilized mechanical aeration and PV-powered biofilters to reduce oil and grease by 94%, biological oxygen demand by 71.9%, and TDSs by 54%, and increase DO levels (Tharasawatpipat et al., 2025). The

TABLE 19.6 The overview of these hurdles, their effects, and solutions for researchers and policymakers.

Challenge	Impact	Potential solutions
High operating and capital costs	Energy-consuming processes make wastewater treatment too expensive; underutilization of plants; and limited nutrient recovery.	Implement energy-efficient technologies (e.g., high-rate algal ponds, anaerobic digestion); subsidize renewable energy; pursue decentralized fecal sludge management (FSM) that distributes costs.
Maintenance and manpower shortages	Equipment breakdowns, inadequate sludge management, odor, and downtime problems.	Offer training programs; create separate O&M funds; establish performance-based contracts for private operators; and employ resilient natural treatment systems, such as constructed wetlands.
Technological limitations (fouling, incomplete sewer networks)	Membrane fouling raises energy consumption and time offline; partial sewer networks result in untreated releases.	Invest in research to create antifouling membranes and low-cost pretreatment; emphasize sewer network extension; install decentralized treatment systems where central sewers are unfeasible.
Microplastics pollution	Microplastics remain in effluent; fragments and fibers predominate; removal efficiencies are highly variable by country.	Introduce source reduction (e.g., microbead bans); upgrade treatment operations through finer filtration; formulate sludge management practices to avoid re-release; promote international norms for microplastic discharge.
Pharmaceuticals and antibiotic resistance	Antibiotics in effluent contribute to resistant bacteria and genes; high concentrations are found in effluents around pharmaceutical facilities.	Strengthen regulations on disposing of pharmaceuticals; introduce advanced oxidation and biological treatment, focusing on antibiotics; enforce effluent regulations; track antimicrobial resistance in WWTPs.
Weak enforcement and institutional overlap	Low fines and underresourced regulators enable illicit releases; duplicated institutional responsibilities slow projects.	Raise the penalty for noncompliance; rationalize the government with well-defined responsibilities; finance laboratories and monitoring networks; incorporate wastewater planning into urban planning.
Public skepticism and NIMBYism	Reuse initiatives are resisted on the basis of the psychological "yuck" factor; esthetic objections undermine acceptance.	Conduct public education campaigns highlighting the protection of health and the environment; engage communities in decision-making; implement decentralized or subsurface WWTPs to alleviate land-use conflicts; and showcase replicable projects (e.g., Singapore's NEWater).

study highlights that energy from the sun is universal and scalable, as a result of which it can be easily utilized in disaster-affected areas and off-grid villages (Tharasawatpipat et al., 2025). The state governments in India have begun funding solar aeration systems for drainage channels and community tanks, highlighting alignment on an international scale at the policy front.

12.2.2 Biogas application

AD of sewage sludge is a proven technology in which the sludge is stabilized by releasing biogas, a mixture of carbon dioxide and methane, which can be used for electricity or heat generation. Technical reports published by IEA Bioenergy state that AD is common in Europe and India; methane yield from primary and secondary sludge is around 0.7–1.0 m^3 per kg of volatile solids, and codigestion of sludge with organic waste (e.g., food waste) may result in enhanced methane yield (Bachmann et al., 2015). Yet, with full-scale plants being crucial in the United States, a survey showed that fewer than 10% of the facilities use biogas effectively, with the majority of the gas being flared. Slow biogas production, low energy value, and the upgrading cost to pipeline quality are the primary inhibitions (Shen et al., 2015). Policymakers will ensure the promotion of biogas capture and utilization, linking biogas with renewable fuel targets, and incentivizing research on small-scale digesters as a viable option for rural WWTPs. Coupling PV with biogas will help achieve energy neutrality through day-time aeration and using biogas for night-time or peak loads (Bachmann et al., 2015; Shen et al., 2015).

12.3 Internet of things and artificial intelligence in wastewater monitoring

The future of wastewater management lies in smart sensors and artificial intelligence (AI) to sense in real time and control processes automatically. Comprehensive reviews emphasize that IoT-based systems constantly monitor water-quality parameters such as pH, DO, turbidity, TDSs, electrical conductivity (EC), temperature, COD, biochemical oxygen demand (BOD), oxidation–reduction potential, and calculate the water quality index. Commercial sensors, such as multiparameter spectrometer probes (spectro:lyser) and fluorescence spectroscopy instruments, measure TSS, turbidity, color, TOC, DOC, BOD, COD, and nitrate; such equipment is fitted with onboard web servers, dataloggers, and autocleaning mechanisms (Essamlali et al., 2024; Alprol et al., 2024). A LoRaWAN and NB-IoT network was utilized in a case study for monitoring pH, BOD, COD, ammoniacal nitrogen, suspended solids, and DO in real time over vast geographic areas. Machine learning models also enhance predictive control: support vector regression has been employed to forecast BOD and COD from MBR parameters so that aeration or chemical dosing can be automatically adjusted (Essamlali et al., 2024). The integration of Internet of Things (IoT) and Artificial Intelligence (AI) thereby offers early warning of failure, reduces the amount of chemicals used, and decreases operating costs, but requires investment in digital infrastructure and human capacity development. Low-cost sensors and open-source platforms need to be considered by developing nations to democratize intelligent monitoring (Alprol et al., 2024; Essamlali et al., 2024; Abdulameer et al., 2025).

12.4 Circular economy and zero-liquid discharge

Circular economy strategies aim to close material and water loops by recovering resources from wastewater. Zero-liquid discharge (ZLD) technologies focus on avoiding liquid effluent and achieving full water reuse, but their application requires significant amounts of energy and chemical inputs. A life-cycle assessment of ZLD for thermal power plants notes that newer membrane technologies, such as low-salt-rejection reverse osmosis, forward osmosis, bipolar membrane electrodialysis, and membrane distillation, can recover more than 95% of liquid waste and dry the remainder into solid form (Hernández-Chover et al., 2023). Industrial research from Indian industries identifies that pretreatment steps (coagulation, neutralization, and biological treatment) remove oils, organics, and suspended solids, thereby reducing COD and minimizing fouling of the membrane. The core dilemma lies in balancing energy efficiency and water recovery; guidelines suggest combining closed-loop resource management with waste-heat recovery and renewable energy to save costs and protect the environment (Hernández-Chover et al., 2023). ZLD has increasingly been adopted in textile and dye units along India's Noyyal and Sabarmati rivers due to strict discharge standards, and such adoption is also growing in China and the Middle East. Policymakers must promote energy-efficient ZLD technology research, support industries in embracing circular economy models, and ensure accountability for concentrate disposal or crystallized salts (Hernández-Chover et al., 2023).

13 Policy, regulation, and governance frameworks

13.1 International policies

Global guidelines such as the UN SDG 6 and the Water Framework Directive (WFD) of the European Union shape national and subnational policy by setting targets, indicators, and a rights-based strategy for water quality and wastewater management. SDG 6 officially links wastewater treatment to public health and ecosystem objectives and offers internationally comparable indicators to monitor progress (Bilalova et al., 2025). The WFD has established a bold "good status" target for surface water and groundwater and, importantly, illustrates how cross-sector integration, legally binding river-basin management plans, and basin-scale planning can drive water quality improvements. However, there are implementation gaps in all regions due to uneven capacity, finance, and cross-sector coordination (Copetti and Erba, 2024).

13.2 Indian regulatory framework

India's legal foundation for wastewater and pollution control remains the Water (Prevention and Control of Pollution) Act, 1974, which created the Central Pollution Control Board (CPCB) and SPCBs and delegated discharge standards and enforcement power. National programs such as the National River Conservation Plan and, more recently, the Namami Gange initiative have spurred huge investments in sewage interception, treatment, and riverfront works and introduced hybrid financing/public–private partnership (PPP) structures for the rapid development of infrastructure. Despite significant capital expenditure, operating performance, equitable sewer coverage, and sustainable O&M finance remain critical bottlenecks in the majority of towns and cities (Balkrishna et al., 2024).

13.3 Institutional roles and PPP

Overseeing primarily falls on the CPCB (policy, standards, national monitoring) and SPCBs (permitting and enforcement). Institutional weaknesses, personnel shortages, limited laboratory capacity, and fragmented mandates among urban local governments, water utilities, and pollution regulators hamper compliance and plant performance over the long term. PPPs have been widely used to mobilize technical capability and capital for STPs and Common Effluent Treatment Plants. Evidence from around the world and experience confirm that PPPs can increase efficiency and build capacity where contracts have balanced incentives, included robust monitoring (e.g., OCEMS/SCADA), and had clear risk allocation and government control (Lima et al., 2021; Cheng et al., 2023). However, weakly designed contracts, poor enforcement, and fiscal guarantees can transfer long-term risks to governments and erode value for money unless treated with care (Ruan et al., 2024).

14 Conclusion

Wastewater treatment is the cornerstone of sustainable, long-term environmental management and river restoration. As discussed in this chapter, raw effluent from households, industries, and farming severely deteriorates river water quality and creates eutrophication, loss of biocomplexity, and increasing public-health crises. A hierarchical treatment train, consisting of primary, secondary, and tertiary stages, is still indispensable for the efficient removal of organic load, nutrients, heavy metals, and pathogenic microorganisms. However, traditional centralized systems alone cannot confront the rising challenges of urbanization, climate change, and emerging contaminants such as microplastics and pharmaceuticals. The new developments integrating nature-based solutions, bioelectrochemical and photocatalytic processes, and AI-driven smart monitoring represent a paradigm shift toward adaptive, energy-saving, and low-carbon wastewater management.

Wetlands, algal–bacterial consortia, and biochar filters are some of the eco-technologies that combine pollutant removal and resource recovery according to circular economy principles. Valorization of sludge through biogas production, nutrient recovery, and material recycling also transforms WWTPs into WRRFs, concurrently providing advantages in energy neutrality, food security, and climate resilience. However, their sustainability also depends on robust governance, financial provisions, and public engagement. Robust PPPs, decentralized treatment infrastructure, and strict enforcement of discharge norms are all imperative in bridging the functional gaps that still prevail in the majority of developing nations, like India. Moving forward, there has to be an integrated policy mechanism that brings together technological innovation, socioeconomic justice, and environmental sustainability.

In reality, wastewater should no longer be viewed as waste but as a renewable resource, a water, energy, and nutrient recovery nexus. Embracing this circular perspective will not only serve to safeguard riverine ecosystems but also drive the global pursuit of SDG 6 (Clean Water and Sanitation) and associated goals of environmental resilience, economic efficiency, and public health protection.

References

Abdoli, S., Asgari Lajayer, B., Dehghanian, Z., Bagheri, N., Vafaei, A.H., Chamani, M., Rani, S., Lin, Z., Shu, W., Price, G.W., 2024. A review of the efficiency of phosphorus removal and recovery from wastewater by physicochemical and biological processes: challenges and opportunities. Water 16 (17), 2507. doi:10.3390/w16172507.

Abdulameer, L., Al Maimuri, N.M.L., Nama, A.H., Rashid, F.L., Al-Dujail, A.N., 2025. The role of artificial intelligence in managing sustainable water resources: a review of smart solution implementations. Water Conserv. Manage. 25235672. 9 (2), 281–291. Available from: https://www.watconman.org/download/9525/. doi:10.26480/wcm.02.2025.281.291.

Abreham Awono, Y., Adam, D.B., 2025. Evaluation of operational challenges and technological performance of wastewater treatment plants in Addis Ababa. Sci. Rep. 2045-2322. 15 (1). doi:10.1038/s41598-025-17945-8.

Afzaal, M., Hameed, S., Liaqat, I., Khan, A.A.A., Abdul Manan, H., Shahid, R., Altaf, M., 2022. Heavy metals contamination in water, sediments and fish of freshwater ecosystems in Pakistan. Water Pract. Technol. 1751-231X. 17 (5), 1253–1272. doi:10.2166/wpt.2022.039.

Akpor, O.B., Otohinoyi, D.A., Olaolu, D.T., Aderiye, B.I., 2014. Pollutants in wastewater effluents: impacts and remediation processes. Int. J. Environ. Res. Earth Sci. 3, 50–59.

Ali, H., Khan, E., Ilahi, I., 2019. Environmental chemistry and ecotoxicology of hazardous heavy metals: environmental persistence, toxicity, and bioaccumulation. J. Chem. 2090-9063. 2019, 1–14. doi:10.1155/2019/6730305.

Alprol, A.E., Mansour, A.T., Ibrahim, M.E.E.-D., Ashour, M., 2024. Artificial intelligence technologies revolutionizing wastewater treatment: current trends and future prospective. Water. 2073-4441. 16 (2), 314. doi:10.3390/w16020314.

Ambaye, T.G., Vaccari, M., van Hullebusch, E.D., Amrane, A., Rtimi, S., 2021. Mechanisms and adsorption capacities of biochar for the removal of organic and inorganic pollutants from industrial wastewater. Int. J. Environ. Sci. Technol. 17352630. 18 (10), 3273–3294. Available from: http://www.springerlink.com/content/1735-1472. doi:10.1007/s13762-020-03060-w.

Anawar, H.M., Chowdhury, R., 2020. Remediation of polluted river water by biological, chemical, ecological and engineering processes. Sustainability 12 (17), 7017. doi:10.3390/su12177017.

Anele, B.C., Okorite, G.W., 2025. Assessment of pathogenic microbial and heavy metal contamination in stream water at Omuhuhechi, Aluu, Nigeria: health implications. SSR J. Med. Sci. 2 (4), 1–6. doi:10.5281/zenodo.16417383.

Aparicio, S., Borrás-Falomir, L., Jiménez-Benítez, A., Seco, A., Robles, Á., 2024. Urban wastewater treatment at ambient conditions using microalgae-bacteria consortia in a membrane high-rate algal pond (MHRAP): the effect of hydraulic retention time and influent strength. Environ. Technol. Innov. 23521864. 36, 103846. doi:10.1016/j.eti.2024.103846.

Aransiola, S.A., Zobeashia, S.L.T., Ikhumetse, A.A., Musa, O.I., Abioye, O.P., Ijah, U.J.J., Maddela, N.R., 2024. Niger Delta mangrove ecosystem: biodiversity, pollution, threats and mitigation. Reg. Stud. Mar. Sci. 75, 103568. doi:10.1016/j.rsma.2024.103568.

Awad, A.M., Ibrahim, H.T., Al-Aboodi, A.H., 2023. Performance of compact bio-contact oxidation reactors for municipal wastewater treatment under different hydraulic retention time. J. Ecol. Eng. 22998993. 24 (12), 304–318. Available from: http://www.jeeng.net. doi:10.12911/22998993/173579.

Bachmann, N., la Cour Jansen, J., Bochmann, G., Montpart, N., 2015. Sustainable biogas production in municipal wastewater treatment plants (Vol. 20). Massongex, Switzerland: IEA bioenergy.

Balkrishna, Singh, S.K., Ghosh, S., Banerjee, S., Verma, S., Arya, V., 2024. An analytical review on the integrated management of river resources through Namami Gange. Water Policy. 19969759. 26 (5), 462–479. Available from: https://iwaponline.com/wp/article-pdf/26/5/462/1427096/026050462.pdf. doi:10.2166/wp.2024.209.

Bhatt, A., Singh, S., 2022. An analysis of repercussions of religious waste. Indian J. L. Legal Res 4, 1.

Bilalova, S., Newig, J., Villamayor-Tomas, S., 2025. Toward sustainable water governance? taking stock of paradigms, practices, and sustainability outcomes. WIREs Water. 2049-1948. 12 (1). doi:10.1002/wat2.1762.

Biswas, J.K., Prasad, M.N., 2023. Wastewater-based circular economy operations in East Kolkata Wetlands (largest Ramsar site in South Asia), India. In: Smol, M., Prasad, M.N.V., Stefanakis, A.I. (Eds.), Water in Circular Economy. Advances in Science, Technology & Innovation, Springer, Cham. doi:10.1007/978-3-031-18165-8_19.

Bodzek, M., Pohl, A., Rosik-Dulewska, C., 2024. Microplastics in wastewater treatment plants: characteristics, occurrence and removal technologies. Water. 2073-4441. 16 (24), 3574. doi:10.3390/w16243574.

Bruce, M., Limin, M., 2021. Recent advances on water pollution research in Africa: a critical review. Int. J. Sci. Adv. 2708-7972. 2 (3). doi:10.51542/ijscia.v2i3.23.

Chaedir, B.A., Kurnia, J.C., Sasmito, A.P., Mujumdar, A.S., 2021. Advances in dewatering and drying in mineral processing. Dry. Technol. 15322300. 39 (11), 1667–1684. Available from: https://www.tandf.co.uk/journals/titles/07373937.asp. doi:10.1080/07373937.2021.1907754.

Cheng, S., Yu, Y., Meng, F., Chen, J., Chen, Y., Liu, G., Fan, W., 2023. Potential benefits of public–private partnerships to improve the efficiency of urban wastewater treatment. Npj Clean Water. 20597037. 6 (1). Available from: https://www.nature.com/npjcleanwater/. doi:10.1038/s41545-023-00232-2.

Copetti, D., Erba, S., 2024. A bibliometric review on the water framework directive twenty years after its birth. Ambio. 16547209. 53 (1), 95–108. Available from: https://www.springer.com/journal/13280. doi:10.1007/s13280-023-01918-0.

CPCB, 2024. Ambient air quality monitoring network (manual) under NAMP as on 19.11.2024. Central Pollution Control Board, https://cpcb.nic.in/uploads/Stations_NAMP.pdf.

Denysiuk, R., Santo, I.E., Costa, L. 2018. Portugal wastewater treatment plant design: optimizing multiple objectives. In: ICORES 2018 - Proceedings of the 7th International Conference on Operations Research and Enterprise Systems, 9789897582851. SciTe Press, pp. 327–334. http://www.scitepress.org/DigitalLibrary/HomePage.aspx. doi: 10.5220/0006660303270334

Dubey, D., Kumar, S., Dutta, V., 2024. A circular economy approach for sustainable water reuses in India: policies, practices and future prospects. Environ. Sustain. 2523-8922. 7 (3), 303–323. doi:10.1007/s42398-024-00321-z.

Essamlali, I., Nhaila, H., El Khaili, M., 2024. Advances in machine learning and IoT for water quality monitoring: a comprehensive review. Heliyon. 24058440. 10 (6), e27920. doi:10.1016/j.heliyon.2024.e27920.

Ferri, E.N., Bolelli, L., 2025. Wastewater remediation treatments aimed at water reuse: recent outcomes from pilot- and full-scale tests. Appl. Sci. 2076-3417. 15 (5), 2448. doi:10.3390/app15052448.

Genua, F., Lancellotti, I., Leonelli, C., 2025. Geopolymer-based stabilization of heavy metals, the role of chemical agents in encapsulation and adsorption: review. Polymers. 2073-4360. 17 (5), 670. doi:10.3390/polym17050670.

Gerba, C.P., Pepper, I.L., 2019. Municipal wastewater treatment. In: Environmental and Pollution Science. Elsevier BV, 393–418. doi:10.1016/b978-0-12-814719-1.00022-7.

Hartley, T.W., 2006. Public perception and participation in water reuse. Desalination. 00119164. 187 (1-3), 115–126. doi:10.1016/j.desal.2005.04.072.

Hassanzadeh Moghimi, O., Nabi Bidhendi, G., Daryabeigi Zand, A., Rabiee Abyaneh, M., Nabi Bidhendi, A., 2023. Effect of forest-based biochar on maturity indices and bio-availability of heavy metals during the composting process of organic fraction of municipal solid waste (OFMSW). Sci. Rep. 20452322. 13 (1). Available from: https://www.nature.com/srep/. doi:10.1038/s41598-023-42835-2.

Havinga, H. 2020. Towards sustainable river management of the Dutch Rhine River. Water 12 (6), 1827. doi:10.3390/w12061827.

Hernández-Chover, V., Castellet-Viciano, L., Fuentes, R., Hernández-Sancho, F., 2023. Circular economy and efficiency to ensure the sustainability in the wastewater treatment plants. J. Clean. Prod. 09596526. 384. Available from: https://www.journals.elsevier.com/journal-of-cleaner-production. doi:10.1016/j.jclepro.2022.135563.

Hoang, T.D., Van Anh, N., Yusuf, M., Ali, S.A.M., Subramanian, Y., Hoang Nam, N., Minh Ky, N., Le, V.G., Huyen, N.T.T., Bianasari, A.A., Azad, K.A., 2024. Valorization of agricultural residues into value-added products: a comprehensive review. Chem. Rec. 24 (8), e202300333. doi:10.1016/j.jclepro.2022.135563.

Hossain, M.I., Soliman, M.M., El-Naggar, M.E., Sultan, M.Z., Kechi, A., Abdelsalam, N.R., Abu-Saied, M.A., Chowdhury, M., 2021. Synthesis and characterization of graphene oxide-ammonium ferric sulfate composite for the removal of dyes from tannery wastewater. J. Mater. Res. Technol. 12, 1715-1727. doi:10.1016/j.jmrt.2021.03.097.

Inayat, A., Dafalla, M., Asaad, S., Jamil, F., Al-Haj, L., Shah, F.M., Ghenai, C., Shanableh, A., 2025. Sustainable energy production from waste: a review of hybrid approaches combining anaerobic digestion and gasification. Int. J. Energy Res. 1099114X. 2025 (1). Available from: https://onlinelibrary.wiley.com/journal/ijer. doi:10.1155/er/6644084.

James, S.N., Vijayanandan, A., 2023. Recent advances in simultaneous nitrification and denitrification for nitrogen and micropollutant removal: a review. Biodegradation. 15729729. 34 (2), 103–123. Available from: https://www.springer.com/journal/10532. doi:10.1007/s10532-023-10015-8.

Javed, M., Usmani, N., 2019. An overview of the adverse effects of heavy metal contamination on fish health. Proc. Natl. Acad. Sci. India Sect. B Biol. Sci. 03698211. 89 (2), 389–403. Available from: http://www.springerlink.com/content/0369-8211. doi:10.1007/s40011-017-0875-7.

Kaur, B., 2025. Environmental and economic benefits of algal-microbial remediation. In: Algal Bioengineering and Microbial Synergy to Green Remediation. Springer Science and Business Media LLC, 257–295. doi:10.1007/978-981-96-8054-2_11.

Kesari, K.K., Soni, R., Jamal, Q.M.S., Tripathi, P., Lal, J.A., Jha, N.K., Siddiqui, M.H., Kumar, P., Tripathi, V., Ruokolainen, J., 2021. Wastewater treatment and reuse: a review of its applications and health implications. Water, Air, and Soil Pollution. 15732932. 232 (5). Available from: http://www.kluweronline.com/issn/0049-6979/. doi:10.1007/s11270-021-05154-8.

Khelladi, R.M.B., Fellah, A.C., Pontié, M., Shabani, M., Guellil, F.Z., 2020. Performance of sand and granular activated carbon filtration coupling in tertiary urban wastewater treatment in Algeria. Desalin. Water Treat. 205, 111–123. doi:10.5004/dwt.2020.26330.

Landrigan, P.J., Britt, M., Fisher, S., Holmes, A., Kumar, M., Mu, J., Rizzo, I., Sather, A., Yousuf, A., Kumar, P., 2024. Assessing the human health benefits of climate mitigation, pollution prevention, and biodiversity preservation. Ann. Glob. Health. 22149996. 90 (1). Available from: https://storage.googleapis.com/jnl-up-j-agh-files/journals/1/articles/4161/659808212e84e.pdf. doi:10.5334/aogh.4161.

Lanko, I., Hejnic, J., Říhová-Ambrožová, J., Ferrer, I., Jenicek, P., 2021. Digested sludge quality in mesophilic, thermophilic and temperature-phased anaerobic digestion systems. Water. 2073-4441. 13 (20), 2839. doi:10.3390/w13202839.

Li, Q., Zhu, Y., Chen, Q., Li, Y., Chen, J., Gao, Y., Lin, P., 2021. Spatio-temporal dynamics of water quality and eutrophication in Lake Taihu, China. Ecohydrology. 1936-0584. 14 (5). doi:10.1002/eco.2291.

Lima, S., Brochado, A., Marques, R.C., 2021. Public-private partnerships in the water sector: a review. Util. Policy. 09571787. 69 (101182). doi:10.1016/j.jup.2021.101182.

Ling, W., Xing, Y., Zhao, C., Hong, C., Feng, W., Wang, Y., Hu, J., Hou, J., Feng, L., 2024. Research progress on exhaust gas emissions and control technology during sewage sludge thermal drying: a review. Dry. Technol. 15322300. 42 (1), 1–18. Available from: https://www.tandf.co.uk/journals/titles/07373937.asp. doi:10.1080/07373937.2023.2260895.

Malyan, S.K., Yadav, S., Sonkar, V., Goyal, V.C., Singh, O., Singh, R., 2021. Mechanistic understanding of the pollutant removal and transformation processes in the constructed wetland system. Water Environ. Res. 15547531. 93 (10), 1882–1909. Available from: https://onlinelibrary.wiley.com/loi/15547531. doi:10.1002/wer.1599.

Mandal, D., Mondal, S., Biswas, S., Seal, S., Das, S., Bagchi, S., Mandal, R., Fulchand, S., Mondal, A., Basu, A., 2023. Wastewater Treatment in India—A New Perspective. In: Das, J., Bhattacharya, S.K., (Eds.), Monitoring and Managing Multi-hazards. GIScience and Geo-environmental Modelling. Springer, Cham.

Marin, N.M., Nita Lazar, M., Popa, M., Galaon, T., Pascu, L.F., 2024. Current trends in development and use of polymeric ion-exchange resins in wastewater treatment. Materials 17 (23), 5994. doi:10.3390/ma17235994.

Matarru, A.A., Shin, D., 2025. Biochar production technology as a negative emission strategy: a review. Energies. 1996-1073. 18 (18), 4898. doi:10.3390/en18184898.

Maxcy Brown, J., Elliott, M.A., Bearden, B., 2023. Household-level wastewater management and disposal data in the US: history, shortcomings, and policy implications. Water Policy 25 (9), 927–947. doi:10.2166/wp.2023.147.

Meghea, I., Stefan, D.S., Ioniţă, F., Lesnic, M., Manea-Saghin, A.M., 2025. An integrative approach to hazardous effects caused by pharmaceutical contaminants on aquatic effluents. Molecules.

14203049. 30 (17). Available from: http://www.mdpi.com/journal/molecules. doi:10.3390/molecules30173483.

Metcalf, L., Eddy, H.P., 2003. Wastewater engineering treatment and reuse. McGraw-Hill, New York.

Morin-Crini, N., Lichtfouse, E., Fourmentin, M., Ribeiro, A.R.L., Noutsopoulos, C., Mapelli, F., Fenyvesi, É., Vieira, M.G.A., Picos-Corrales, L.A., Moreno-Piraján, J.C., Giraldo, L., 2022. Removal of emerging contaminants from wastewater using advanced treatments. A review. Environ. Chem. Lett. 20 (2), 1333–1375. doi:10.1007/s10311-021-01379-5.

Munavalli, G.R., Sonavane, P.G., Koli, M.M., Dhamangaokar, B.S., 2022. Field-scale decentralized domestic wastewater treatment system: effect of dynamic loading conditions on the removal of organic carbon and nitrogen. J. Environ. Manage. 10958630. 302. Available from: https://www.sciencedirect.com/journal/journal-of-environmental-management. doi:10.1016/j.jenvman.2021.114014.

Naidoo, S., Olaniran, A.O., 2014. Treated wastewater effluent as a source of microbial pollution of surface water resources. Int. J. Environ. Res. Public Health 1 (1), 249-270. doi:10.3390/ijerph110100249.

NMCG, 2023. National Framework on Safe Reuse of Treated Water, National Mission for Clean Ganga, Govt. of India.

Oruganti, R.K., Katam, K., Show, P.L., Gadhamshetty, V., Upadhyayula, V.K.K., Bhattacharyya, D., 2022. A comprehensive review on the use of algal-bacterial systems for wastewater treatment with emphasis on nutrient and micropollutant removal. Bioengineered. 21655987. 13 (4), 10412–10453. Available from: http://www.tandfonline.com/loi/kbie20. doi:10.1080/21655979.2022.2056823.

Paerl, H.W., Otten, T.G., 2013. Harmful cyanobacterial blooms: causes, consequences, and controls. Microb. Ecol. 00953628. 65 (4), 995–1010. doi:10.1007/s00248-012-0159-y.

Panda, D., Kumar, A., Gangawane, K.M., 2025. Principles of Wastewater Treatment and Sludge Generation. Springer Science and Business Media LLC, pp. 1–12. doi:10.1007/978-3-031-45758-6_1.

Patel, A., Chaudhary, V.K., Singh, A., Rai, D., Patel, N., 2021. Heavy-metal pollution in river Ganga and mitigation plans: a review. Ecol. Environ. Conserv. 27 (1), 382–393.

Pathak, N., Tran, V.H., Merenda, A., Johir, M.A.H., Phuntsho, S., Shon, H., 2020. Removal of organic micro-pollutants by conventional membrane bioreactors and high-retention membrane bioreactors. Appl. Sci. 2076-3417. 10 (8), 2969. doi:10.3390/app10082969.

Pegu, K., Samal, K.P., 2025. Navigating India's Legal Governance in Water Sector: A Scrutiny into National Performance under SDG Framework for Clean and Wastewater Management in India. J. Sustainable Dev. L. & Pol'y 16 (3), 210. doi:10.4314/jsdlp.v16i3.10.

Prasad, S., Saluja, R., Joshi, V., Garg, J.K., 2020. Heavy metal pollution in surface water of the Upper Ganga River, India: human health risk assessment. Environ. Monit. Assess. 15732959. 192 (11). Available from: https://link.springer.com/journal/10661. doi:10.1007/s10661-020-08701-8.

Qrenawi, L.I., Rabah, F.K.J., 2021. Sludge management in water treatment plants: literature review. Int. J. Environ. Waste Manage. 14789868. 27 (1), 93–125. Available from: http://www.inderscience.com/ijewm. doi:10.1504/IJEWM.2021.111909.

Ramzanipour, M.M., Mousavi, S.J., Vajargah, M.F., 2023. Overview of municipal wastewater and sludge treatment processes. Ecol. Conserv. Sci. 2, 555–596. doi:10.19080/ECOA.2023.02.555596.

Rosendo, J.C.M., da Paz, G.M., Rosendo, A., 2022. Constructed wetlands applied on domestic wastewater for decetralized systems: concepts, processes, modalities, combinations and enhancements; a review. Rev. Environ. Sci. Biotechnol. 15729826. 21 (2), 371–397. Available from: http://www.kluweronline.com/issn/1569-1705. doi:10.1007/s11157-022-09616-1.

Ruan, X., Li, K., Yang, M., Dong, Z., Wang, G., 2024. From China to the world: the implications of fiscal decentralization, government debt, and public health dynamics. Environ. Dev. Sustain. pp.1–25. doi:10.1007/s10668-024-05529-x.

Salano, L., Bozzini, M.M., Caspani, S., Bozzano, G., Manenti, F., 2024. Integration of an autothermal outer electrified reformer technology for methanol production from biogas: enhanced syngas quality production and CO_2 capture and utilization assessment. Processes. 22279717. 12 (8). Available from: http://www.mdpi.com/journal/processes. doi:10.3390/pr12081598.

Seow, T.W., Lim, C.K., Nor, M.H.M., Mubarak, M.F.M., Lam, C.Y., Yahya, A., Ibrahim, Z., 2016. Review on wastewater treatment technologies. Int. J. Appl. Environ. Sci. 11 (1), 111–126.

Sha, C., Shen, S., Zhang, J., Zhou, C., Lu, X., Zhang, H., 2024. A review of strategies and technologies for sustainable decentralized wastewater treatment. Water. 2073-4441. 16 (20), 3003. doi:10.3390/w16203003.

Sharma, R., Kumar, R., Satapathy, S.C., Al-Ansari, N., Singh, K.K., Mahapatra, R.P., Agarwal, A.K., Le, H.V., Pham, B.T., 2020. Analysis of water pollution using different physicochemical parameters: a study of Yamuna river. Front. Environ. Sci. 2296665X. 8. Available from: https://www.journal.frontiersin.org/journal/environmental-science. doi:10.3389/fenvs.2020.581591.

Sharma, T., Singh, A., Kumar, N., Chauhan, G., Singh, D.P., Singh, A., Rana, B.B., 2023. Emerging Pollutants in the Environment and Ecological Risks Management and Mitigation of Emerging Pollutants. Springer International Publishing, India, pp. 1–20. Available from: https://link.springer.com/book/10.1007/978-3-031-41005-5. doi:10.1007/978-3-031-41005-5_1.

Shen, Y., Linville, J.L., Urgun-Demirtas, M., Mintz, M.M., Snyder, S.W., 2015. An overview of biogas production and utilization at full-scale wastewater treatment plants (WWTPs) in the United States: challenges and opportunities towards energy-neutral WWTPs. Renew. Sustain. Energy Rev. 18790690. 50, 346–362. Available from: https://www.journals.elsevier.com/renewable-and-sustainable-energy-reviews. doi:10.1016/j.rser.2015.04.129.

Srivastava, R., Malhotra, R., Sachdeva, R., 2023. Wastewater pollution, declining river water quality, and socio-economic impacts in NCT of Delhi. Urban India 43, 179–204.

Tariq, A.M., 2023. Untreated wastewater reasons and causes: a review of most affected areas and cities. Int. J. Chem. Biochem. Sci. 22269614. 23 (1), 121–143. Available from: https://www.iscientific.org/wp-content/uploads/2023/05/15-IJCBS-23-23-22.pdf.

Tchobanoglous, G., Burton, F.L., Stensel, H.D. (2003). Wastewater Engineering: Treatment and Reuse, Boston: McGraw-Hill.

Tharasawatpipat, Choo-in, S., Kayee, P., Javadi, B., 2025. Evaluation of wastewater treatment and solar energy-based solutions for enhanced water quality improvement. J. Ecol. Eng. 22998993. 26 (4), 198–208. Available from: http://www.jeeng.net. doi:10.12911/22998993/199465.

Torres-Lozada, P., Amezquita-Marroquín, C.P., Agudelo-Martínez, K.D., Ortiz-Benítez, N., Martínez-Ducuara, D.S., 2018. Evaluation of turbidity and dissolved organic matter removal through double filtration technology with activated carbon. Dyna 85 (205), 234-239.

Varma, V.K.C., Rathinam, R., Suresh, V., Naveen, S., Satishkumar, P., Abdulrahman, I.S., Salman, H.M., Singh, P., Kumar, J.A., 2024. Urban waste water management paradigm evolution: decentralization, resource recovery, and water reclamation and reuse. Environ. Qual. Manage. 15206483. 33 (4), 523–540. Available from: http://onlinelibrary.wiley.com/journal/10.1002/(ISSN)1520-6483. doi:10.1002/tqem.22109.

Vásquez, L.A.H., García, F.P., Méndez, J.P., Lass-man, A.A., Sánchez, E.M.O., 2022. Artificial wetlands and floating islands: use of macrophytes. South Florida J. Dev. 2675-5459. 3 (1), 476–498. doi:10.46932/sfjdv3n1-036.

Wagner, E., Karthikeyan, K.G., 2022. Precipitating phosphorus as struvite from anaerobically-digested dairy manure. J. Clean. Prod. 09596526. 339, 130675. doi:10.1016/j.jclepro.2022.130675.

Wang, S., 2022. Innovative MBR–RO processes for reclamation of municipal wastewater to high-grade product water.

WHO Classification of Tumours Editorial Board,2022. WHO classification of endocrine and neuroendocrine tumours. Lyon, France: IARC.

Yang, X., Liu, L., Wang, Y., Lu, T., Wang, Z., Qiu, G., 2023. Sustainable and reagent-free cathodic precipitation for high-efficiency removal of heavy metals from soil leachate. Environ. Pollut. 18736424. 320. Available from: https://www.journals.elsevier.com/environmental-pollution. doi:10.1016/j.envpol.2023.121002.

Zhang, J., Li, G., Yuan, X., Li, P., Yu, Y., Yang, W., Zhao, S., 2023. Reduction of ultrafiltration membrane fouling by the pretreatment removal of emerging pollutants: a review. Membranes 13 (1), 77. doi:10.3390/membranes13010077.

Zhang, X., Zhang, Y., Shi, P., Bi, Z., Shan, Z., Ren, L., 2021. The deep challenge of nitrate pollution in river water of China. Sci. Total Environ. 18791026. 770. Available from: https://www.elsevier.com/locate/scitotenv. doi:10.1016/j.scitotenv.2020.144674.

Zhao, Y., Ma, D., Sun, W., Qian, G., 2024. Investigation of high-rate algal ponds for the treatment of domestic sewage with low carbon-to-nitrogen ratio. Chem. Eng. J. Adv. 26668211. 19. Available from: https://www.sciencedirect.com/science/journal/26668211. doi:10.1016/j.ceja.2024.100635.

Chapter 20

Physical, chemical, and bioremedial measures

Subhadwip Ghorai and Ankur Mukhopadhyay
Department of Agriculture, Brainware University, Kolkata, West Bengal, India

1 Introduction

Rivers are dynamic, self-regulating systems that bring together hydrological, geological, and biological processes along their catchments, from the headwaters to the delta. They support agriculture, industry, and urbanization, along with valuable biodiversity and sociocultural values. However, over the past few decades, rampant industrialization, growing urban settlements, and intensive agriculture have compromised their ecological foundation. Consequently, there is a widespread occurrence of intense degradation in many rivers, typified by high biochemical and chemical oxygen demand (BOD and COD), eutrophication, heavy metal pollution, and the occurrence of emerging pollutants like pharmaceuticals, personal care products, microplastics, and per- and polyfluoroalkyl substances (Md Anawar and Chowdhury, 2020). The varied sources and intricate fate of these contaminants present challenges in assessment and remedy, requiring science-based, integrated interventions.

The loss of river health is seldom linear or uniform; it is a result of the interaction between pollutant loads, hydrology, geomorphology, and biogeochemical cycling (Chakraborty, 2021a). Seasonal variations in discharge in monsoon-based basins, like those of the Ganga and Brahmaputra, control dilution, sediment transport, and residence time, resulting in intermittent phases of flushing and stagnation. On the contrary, temperate rivers experience spikes in pollution during low-flow or snowmelt events (Corriveau et al., 2013). Human activities, channel straightening, deforestation of riparian zones, and floodplain encroachment also dampen natural self-purification by limiting hyporheic exchange and biological buffering. The lack of effective wastewater treatment facilities, especially along periurban and industrial clusters, results in the direct release of untreated domestic and industrial effluents into rivers (Tariq and Mushtaq, 2023).

River Pollution. https://doi.org/10.1016/B978-0-443-45024-2.00009-9

Management of such multidimensional degradation requires an integrated river-basin management strategy that marries engineering ingenuity with nature restoration. Three main categories of remediation measures form the basis of such a strategy: physical, chemical, and biological (bioremediation) (Scott and Ollis, 1995; Grifoni et al., 2022). Physical measures entail mechanical or hydraulic treatments like dredging of sediments, aeration, filtration along riverbanks, and hydromorphological rehabilitation. These processes provide quick water quality enhancements and flow stabilization but tend to be energy-consuming, expensive, and can disturb aquatic ecosystems. Chemical processes utilize methods such as coagulation–flocculation, oxidation, and adsorption for the precipitation, transformation, or immobilization of contaminants (Rashid et al., 2021). Though useful for recalcitrant pollutants, they produce secondary wastes and have large operational footprints (Dai et al., 2023). Conversely, bioremediation leverages the metabolic activities of fungi, plants, and microorganisms to break down or immobilize pollutants in place. Xenobiotics are metabolized, hydrocarbons are oxidized, and nitrates are denitrified by microbial consortia, whereas macrophytes such as Typha, Phragmites, and Vetiveria species uptake heavy metals and nutrients (Jayaprada et al., 2023; Aryal, 2024). CWs couple these processes into low-cost, scalable systems that are compatible with circular bioeconomy frameworks (Soni and Soni, 2025).

Both global and regional experiences suggest that river restoration is sustainable only if it does not rely on a single technique. Examples from India, such as Namami Gange, have successfully combined sewer interception, aeration, and phytoremediation (Coates, 2013), while the Rhine, Danube, and Thames in Europe demonstrate the synergistic value of chemical polishing, habitat restoration, and strict policy implementation (Simon and Joshi, 2022). These initiatives are a testament to the doctrine of site-specific, adaptive remediation, prioritizing measures ranging from source reduction to in-stream treatment and ongoing ecological monitoring.

By moving beyond isolated interventions and adopting system-level integration, contemporary remediation regimes aim not only to achieve water quality goals but also to restore ecological resilience. These frameworks reconcile technological viability with economic and environmental sustainability, dovetailing with the United Nations' Sustainable Development Goal 6 (Clean Water and Sanitation) (Arora and Mishra, 2022) and India's current river rejuvenation efforts. They prioritize phased implementation, permanent monitoring with the aid of both physicochemical and biological markers, and direct stakeholder engagement (Kulwant and Yadav, 2024).

This chapter thus provides a comprehensive analysis of the prominent methods, physical, chemical, and bioremedial, employed in river pollution management. The chapter examines their mechanisms, design criteria, case studies, and synergistic attributes systematically. It then deliberates on hybrid and novel methodologies, such as nanotechnology-facilitated adsorbents,

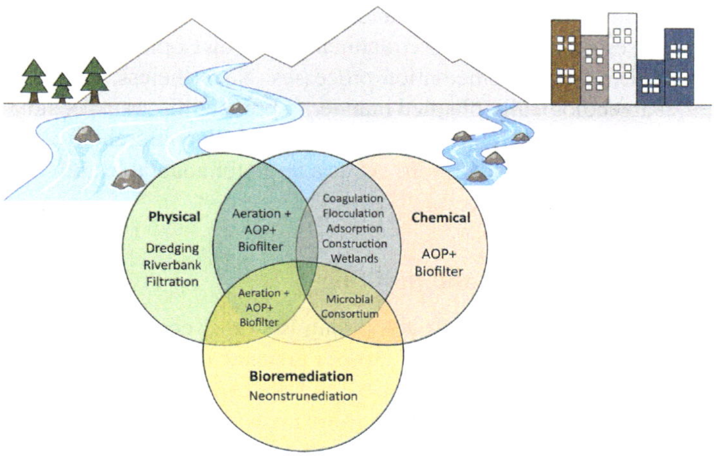

FIG. 20.1 Integrated remediation framework for river basins.

bioelectrochemical systems (BESs), and artificial intelligence (AI)-supported monitoring, followed by a cost, efficiency, and sustainability-based comparative analysis. The final sections cover challenges of implementation, policy implications, and needed future research to construct a self-sustaining and resilient riverine environment. The integration of physical, chemical, and bioremedial processes forms the cornerstone of sustainable river remediation. A holistic understanding of these approaches within a basin-scale framework is essential for developing efficient, low-carbon, and adaptive interventions (Fig. 20.1).

2 Physical remediation techniques

Physical remediation is the initial phase of river restoration, focusing on the mechanical and hydraulic processes that directly remove or redistribute the pollutants without modifying their chemistry. These interventions seek to restore the hydraulic balance and self-purification capability of the river, alleviating issues such as sedimentation, sludge deposition, and oxygen depletion due to untreated discharges, channel alteration, and excessive silt load (Md Anawar and Chowdhury, 2020). Through enhanced flow dynamics and oxygenation, they achieve instant visual and ecological enhancement, setting the stage for biological recovery (Geist and Hawkins, 2016).

The most significant physical measures are sediment dredging and desilting, aeration and reaeration systems, riverbank filtration (RBF), and hydromorphological restoration. Although dredging and aeration provide quick outcomes through contaminated sediment removal and dissolved oxygen (DO) addition (Yan and Li, 2023), RBF and channel reconstruction ensure long-term robustness through natural filtering and ecosystem restoration (Li et al., 2022). These processes also serve as pretreatment steps that optimize the effectiveness of chemical and bioremediation processes. Nevertheless, they need to be executed in an ecologically planned manner to prevent the resuspension of toxins or habitat destruction. When integrated into a more comprehensive management context, physical interventions act as a basis for adaptable and sustainable river rejuvenation.

2.1 Sediment dredging and desilting

Sediment dredging and desilting are among the most common physical interventions to eliminate contaminated bottom deposits that impinge on flow and accumulate toxic metals, hydrocarbons, and organic residues. Over time, the deposition of fine particles forms anaerobic areas from which toxic gases like methane and hydrogen sulfide escape, causing foul odors and poor water quality (Mutegoa and Sahini, 2023). Dredging mechanically removes these particles using hydraulic or mechanical means, enhancing channel depth, flow rate, and oxygen transfer (Soetan et al., 2023).

Controlled dredging under the Ganga Action Plan and comparable international efforts proved to dramatically increase river flow and decrease sediment-associated pollutants. On the contrary, uncontrolled dredging has the potential to resuspend contaminants and destroy benthic communities. Pretest assessment of sediment properties, grain size, and pollutant concentration is, therefore, critical. Dewatered sediments can be reused for land reclamation or construction purposes if stabilized or treated and reused safely, thus decreasing the disposal burden (Müller and Hiete, 2026).

2.2 Aeration and reaeration systems

Aeration is an essential physical process that replenishes lost DO levels caused by the overloading of organics (Singh, 2024). By improving oxygen transfer at the air–water interface, aeration aids microbial oxidation of organics and prevents the development of anaerobic conditions (Costa et al., 2025). Surface aerators are typical aeration systems.

Even though aeration quickly improves water quality, it should be optimized for flow conditions, energy consumption, and turbulence intensity. Excessive aeration can resuspend fine sediments or generate noise in urban settings. Natural reaeration features, including riffles, cascades, and stepped weirs, can enhance oxygenation further by inducing surface renewal. Various aeration devices differ in oxygen dispersion and turbulence patterns, influencing reaeration efficiency in streams and ponds. The schematic representation of commonly used aeration devices and their associated oxygen dispersion zones is shown in Fig. 20.2.

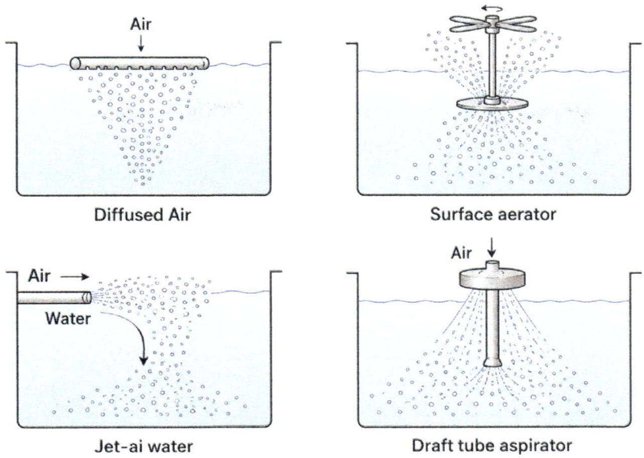

FIG. 20.2 Schematic illustration of typical aeration devices and associated oxygen dispersion patterns.

2.3 Riverbank filtration

RBF is a natural filtration and attenuation process where river water seeps through permeable alluvial strata prior to reaching nearby wells. During subsurface travel, contaminants are eliminated via filtration, adsorption, ion exchange, and biodegradation, yielding water of enhanced clarity and microbial quality (Al-Hashimi et al., 2021). The effectiveness of pollutant removal exists in categories and is given in Table 20.1, which shows average removal rates in field systems throughout India and Europe.

RBF systems at the Rhine (Germany), Danube (Austria), and Ganga (Haridwar, India) have shown 90%–99% removal of pathogens and 70%–85% removal of organic micropollutants (El-Fakharany, 2025). Major performance parameters include hydraulic conductivity, residence time, and sediment type. Though seasonal clogging and changing water tables can decrease infiltration, RBF is a low-cost, low-energy pretreatment method that is consistent with conventional treatment plants (Cui et al., 2021). The pollutant removal performance of RBF varies depending on aquifer properties, residence time, and hydraulic conductivity.

2.4 Hydromorphological and habitat restoration

Hydromorphological restoration aims at re-creating natural river form, flow, and ecological connectivity, reversing the effects of channelization, sand mining, and embankment (Mondal and Patel, 2022; Krapesch et al., 2024). Its measures, like re-meandering of channels, riparian buffer establishment, floodplain

TABLE 20.1 Representative pollutant removal efficiencies in riverbank filtration systems.

Pollutant type	Removal efficiency (%)	Dominant mechanism	Reference	Pollutant type
Pathogens (*E. coli*, coliforms)	95–99	Filtration and adsorption	(Pineda et al., 2021)	Pathogens (*E. coli*, coliforms)
Organic micropollutants	70–85	Sorption and biodegradation	(Fernandez-Fontaina et al., 2013; Ajaz et al., 2024)	Organic micropollutants
Nitrate	60–75	Denitrification	(Chen et al., 2024)	Nitrate

TABLE 20.2 Hydromorphological restoration measures and associated ecological functions.

Measure	Primary function	Ecological benefit	Reference
Re-meandering	Restores flow variability	Enhances oxygen and habitat diversity	
Riparian buffers	Nutrient and sediment trapping	Reduces eutrophication	(Wu et al., 2023)
Floodplain reconnection	Nutrient cycling and storage	Improves denitrification	(Wohl, 2021)
Floating treatment wetlands	Phytoremediation and biofilm support	Removes metals and nutrients	(Zeb et al., 2025)

reconnection, and floating treatment wetlands (FTWs), restore flow diversity and habitat complexity. The main ecological functions of these measures are presented in Table 20.2.

Riparian plants filter nutrients and sediments from runoff, stabilize banklines, and moderate temperature (Osborne and Kovacic, 1993). Reconnected floodplains increase carbon sequestration and denitrification, whereas FTWs couple physical and biological mechanisms for heavy-metal and nutrient removal. The Namami Gange Mission in India involved large-scale riparian plantation and wetland restoration, resulting in quantifiable enhancement of sediment stability and biodiversity (Mehta and Jade, 2024).

2.5 Integrated role and significance of physical measures

Together, physical measures both instantaneously alleviate overt pollution and create a base from which long-term recovery can take hold. By minimizing sediment-bound pollutants, enhancing aeration, and reviving flow variability, they make chemical and biological treatments more effective (Zhang et al., 2020; Sun et al., 2025).

Their sustainable use, however, demands site-specific planning, ongoing monitoring, and eco-friendly sediment management. If incorporated within adaptive river basin management structures, physical techniques not only restore degraded lengths but also enable natural recovery, building the foundation for sustainable river rejuvenation (Kumar et al., 2025).

3 Chemical remediation methods

Chemical remediation entails the use of reagents or reactive materials for the conversion, precipitation, or removal of pollutants from river water and sediments (Bartolomeu et al., 2018). It is specifically useful for persistent and recalcitrant pollutants that are resistant to biological degradation, including heavy metals, dyes, phenolics, and industrial organics. Chemical approaches are normally utilized as secondary or polishing processes following physical pretreatment to meet regulatory discharge criteria (Taylor et al., 2019; Lai et al., 2025).

The major chemical methods are coagulation, flocculation, oxidation and advanced oxidation processes (AOPs), as well as adsorption and ion exchange (Yu et al., 2020). These methods involve various physicochemical mechanisms but share the common aim of transforming toxic compounds into less toxic or more easily removable forms (Hama Aziz et al., 2024). A comparative synopsis of important chemical processes, their mechanisms, and main limitations is provided in Table 20.3.

3.1 Coagulation–flocculation

Coagulation and flocculation continue to be one of the most well-established processes for removing suspended and colloidal particles, organic material, and dyes from river water (Sonal and Mishra, 2021; Zaharia et al., 2024). The process involves the destabilization of negatively charged colloids by multivalent coagulants like alum ($Al_2(SO_4)_3 \cdot 18\,H_2O$), ferric chloride ($FeCl_3$), lime ($Ca(OH)_2$), or polymeric coagulants like polyaluminum chloride (PAC). The destabilized particles form larger flocs that can settle or be filtered out with ease. The over-all reaction mechanism can be written as:

$$Al^{3+} + 3H_2O \rightarrow Al(OH)_3 + 3H^+$$

Gelatinous $Al(OH)_3$ flocs remove suspended impurities and three organic molecules, clarifying turbid river reaches effectively.

TABLE 20.3 **Major chemical remediation methods and their fundamental characteristics.**

Process	Mechanism	Advantages	Limitations	Reference
Coagulation–flocculation	Destabilization of colloids and particle aggregation	Rapid clarification; simple operation	Sludge disposal required; high chemical usage	(Shammas et al., 2021)
Oxidation and AOPs	Generation of radicals for pollutant degradation	Effective for refractory organics	Energy- and reagent-intensive	(Mukherjee et al., 2023)
Adsorption and ion exchange	Sorption onto porous media or ion-exchange resins.	High selectivity; reusable adsorbents	Adsorbent regeneration needed	(Barman et al., 2023; Jatoi et al., 2023)

Field-scale treatment, for example, in the Yamuna River, has removed turbidity by 90% and achieved considerable reductions in total suspended solids with the use of PAC dosing at 30–40 mg L^{-1} (Obura et al., 2023; Verma et al., 2023). The process also produces massive metal-rich sludge that must be disposed of safely or stabilized. Overdosing will also raise residual aluminum or iron levels, which poses a health risk (Randhawa, 2024). Coagulation should, therefore, be combined with sedimentation and filtration for sustainable utilization.

3.2 Oxidation and AOPs

Oxidation processes aim to break down refractory organic substances and decontaminate polluted water through the formation of reactive species like hydroxyl radicals (•OH). Traditional oxidants include chlorine, ozone, and hydrogen peroxide, whereas AOPs (Farzanehsa et al., 2023), such as Fenton, photo-Fenton, ozonation, photocatalysis (e.g., TiO_2, ZnO), and electrochemical oxidation, can mineralize complex contaminants (Laghrib et al., 2021; Pandis et al., 2022). The core of the Fenton reaction is expressed as:

$$Fe_{2+} + H_2O_2 \rightarrow Fe_{3+} + \bullet OH + OH^-$$

These radicals have a high oxidation potential (E° = +2.8 V) and cleave aromatic rings, degrading toxic organics into CO_2 and H_2O. More than 80% reduction in COD has been achieved through TiO_2-based photocatalysis, but electrochemical oxidation systems have provided considerable color elimination

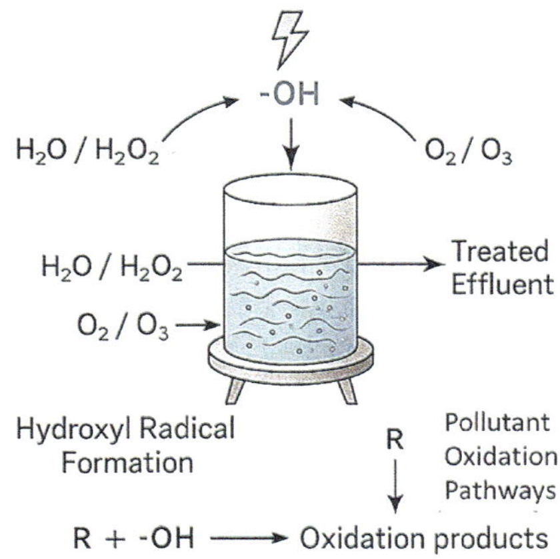

FIG. 20.3 Mechanistic diagram of an AOP system, showing hydroxyl radical formation and pollutant oxidation pathways.

(Sebastian et al., 2024). Still, considerable energy consumption, along with the possible formation of toxic intermediates, remains an important issue. Optimization via hybrid configurations that integrate AOPs with biological polishing has been proposed to enhance cost-effectiveness (Satyam and Patra, 2025; Gong et al., 2025). The mechanistic steps of hydroxyl radical generation and subsequent pollutant oxidation pathways are depicted in Fig. 20.3, illustrating the roles of UV, ozone, and catalytic systems in mineralization.

3.3 Adsorption and ion exchange

Adsorption and ion exchange are the most flexible and environmentally friendly chemical methods for removing soluble impurities, such as heavy metals, dyes, and organic micropollutants. The method is based on the adherence of contaminants to a solid surface (adsorbent) through physical forces or chemical bonding (Enaime et al., 2020; Raji et al., 2023). Activated carbon, biochar, zeolites, bentonite, and graphene oxide composites are typical adsorbents.

The adsorbent–pollutant interaction is described using adsorption isotherms such as Langmuir (monolayer) and Freundlich (multilayer) (Guo and Wang, 2019; Alafnan et al., 2021; Amrutha et al., 2023; Wang and Guo, 2023). Surface area, porosity, and functional groups on the adsorbent surface influence the adsorption efficiency and selectivity.

In various Indian river revitalization pilots, crop residue-derived biochar removed 85%–90% of phosphate and nitrate from eutrophic sections (Chamoli et al., 2025), and graphene oxide composites exhibited >95% removal of pharmaceutical residues (Brito et al., 2023). While being highly efficient, the drawback is the regeneration and disposal of the adsorbent. Designing reusable nanocomposite adsorbents and combining adsorption with bioreactors are the upcoming directions in sustainable water treatment (Kumar et al., 2023a).

3.4 Integrated view of chemical methods

Chemical technologies offer quick and effective elimination of contaminants and are particularly well-suited to industrial effluent-affected river reaches (Ibitoye et al., 2025). They are capable of nearly complete decolorization and organic load reduction in a matter of hours, surpassing biological technologies for recalcitrant substances. Nevertheless, their sustainability relies on the optimization of reagents, minimization of wastes, and combination with physical or biological processes (Selvasembian et al., 2022; Monira and Mostafa, 2023; Jagaba et al., 2024).

When used wisely, after sediment removal and followed by biological polishing, chemical remediation is an effective link between mechanical cleanup and ecological rehabilitation. Its effectiveness in contemporary river basin management is based on the pairing of chemical reactivity and environmental stewardship, ensuring that pollutant removal does not lead to secondary contamination (Chakraborty, 2021a; Xu et al., 2025; Akhtar et al., 2021).

4 Bioremediation techniques

Bioremediation employs the metabolic capacity of microorganisms, plants, and their interactions to convert or immobilize pollutants into harmless products, providing an environmentally friendly and sustainable option compared to physical and chemical processes (Xu et al., 2025). Bioremediation works through mechanisms like biodegradation, biosorption, bioaccumulation, enzymatic conversion, and phyto-uptake to address organic pollutants, nutrients, and heavy metals (Karnwal et al., 2024). In contrast to chemical oxidation or coagulation, bioremediation is energy-conserving and generates very little secondary waste, which aligns with the objectives of sustainable river management and the circular bioeconomy (Jayaswal et al., 2025).

This chapter reviews three major strategies: microbial bioremediation, phytoremediation, and CWs, and then discusses the integration of their roles in the restoration of rivers.

4.1 Microbial bioremediation

Microbial bioremediation utilizes bacteria, fungi, and actinomycetes for degrading or converting pollutants through metabolic processes. Microbes use contaminants as a carbon source, an energy source, or cometabolize them into less toxic compounds (Espinosa-Ortiz et al., 2022; Singh, 2025). The main processes include bioaccumulation (intracellular metal sequestration), biosorption (surface adsorption by cell wall polymers), and enzymatic degradation (redox or hydrolytic cleavage) (Xu and Wang, 2023; Suhaimi et al., 2025).

Laboratory and field experiments have reported effective degraders like *Pseudomonas putida* (phenols, hydrocarbons), *Bacillus subtilis* (metal ions), *Rhodococcus erythropolis* (chlorinated organics), and *Trichoderma harzianum* (synthetic dyes) (Adetunji et al., 2021; Kumari and Das, 2023). Performance is highly dependent on pH, temperature, oxygen, and nutrient ratio (C/N/P). The key strength of microbial methods is their potential for adaptability and in situ application, though sustaining biomass performance under changing conditions remains an issue (Elshobary et al., 2025).

4.2 Phytoremediation

Phytoremediation utilizes vascular plants and their associated rhizospheric microbes for pollutant absorption, accumulation, or transformation. It comprises various submechanisms: phytoextraction, phytostabilization, rhizofiltration, phytodegradation, and phytovolatilization, each controlled by plant physiology and contaminant chemistry (Ogundele et al., 2025). Vegetation such as *Eichhornia crassipes* (water hyacinth) (Singh et al., 2023), *Typha latifolia* (cattail) (Minkina et al., 2021), and *Vetiveria zizanioides* (vetiver grass) (Dorafshan et al., 2023) has demonstrated excellent performance in the removal of heavy metals and nutrients from eutrophic or industrially polluted rivers.

Phytoremediation efficiency is influenced by root surface area, transpiration rate, and metal tolerance. Though floating plants are superior in removing nutrients, rooted emergent plants are superior for sedimentation and immobilization of metals. *E. crassipes* may remove 85% of cadmium and 80% of chromium within 10 days in tropical environments (Gaballah et al., 2021). Genetic enhancement of hyperaccumulators and plant growth-promoting rhizobacteria are new frontier areas to increase uptake efficiency and multimetal stress tolerance (Mondal et al., 2023).

4.3 CWs and bioengineered systems

CWs are a synergistic amalgam of physical, chemical, and biological processes within a planned landscape environment (Spieles, 2022). Water travels above (free-water-surface type) or below (subsurface-flow type) a planted bed filled with gravel, sand, and organic material (Kumar et al., 2023a). Inside this matrix, microbial consortia and plant roots facilitate the removal of organic matter, nutrients, and metals through sedimentation, adsorption, nitrification–denitrification, and plant uptake (Moazzem et al., 2023).

CWs are able to eliminate BOD, nitrogen, and phosphorus, in addition to creating habitat and esthetic values (Ghangrekar, 2022). The East Kolkata Wetlands, a major functional natural wetland system in Asia, treats around

900 million L/day of municipal wastewater while sustaining fisheries and agriculture (Biswas and Prasad, 2023).

4.4 Integrated and applied perspective

Microbial, plant, and wetland systems perform optimally as complementary elements of a holistic bioremediation system. Microbes break down organic contaminants and enable nutrient cycling, whereas plants stabilize sediments and promote oxygen penetration through the rhizosphere (Gao et al., 2025). CWs, on the contrary, offer the physical matrix upon which both microbial and vegetation communities reside. Such synergistic integration improves resilience, efficacy, and scalability for various river conditions (Feng et al., 2021; Santos, 2025).

Bioremediation is also propitious for carbon sequestration, restoration of biodiversity, and the circular use of biomass in the form of compost, biochar, or biogas (Kowalska and Biczak, 2025). Nonetheless, its performance in the field may be temperature-, hydraulic loading-, and pollutant concentration-dependent. Hence, monitoring and adaptive management have to be ongoing.

5 Integrated and emerging remediation approaches

The drawbacks of single, standalone physical, chemical, or biological processes have triggered the development of integrated and hybrid remediation systems that leverage the strengths of more than one process. These strategies endeavor to optimize pollutant removal efficiency, reduce the amount of energy and reagents required, and increase ecological sustainability (Gong et al., 2025; Akhtar et al., 2024). Integration facilitates sequential or concurrent treatment, in which physical removal lowers particulate load, chemical oxidation focuses on recalcitrant compounds, and bioprocesses restore ecological function (Sravan et al., 2024).

Integrated systems are now more often designed with modular treatment trains that can be configured based on river hydrology, pollutant profiles, and land availability. For example, a typical hybrid sequence is sediment dredging → aeration → adsorption or oxidation → CW polishing. Multistage trains were successfully applied to the Ganga and Yamuna rejuvenation initiatives, achieving over 80% removal of COD and BOD, as well as enhancement of ecological indicators (Srivastava and Prathna, 2022).

5.1 Nanotechnology-enabled remediation

Recent developments in nanoscience have given rise to the formulation of nanomaterial-based adsorbents and catalysts for contaminant removal. Metal oxide nanoparticles (e.g., TiO_2, ZnO, Fe_3O_4), carbon nanotubes, and biochar–nanocomposites possess large surface areas, adjustable reactivity, and high sorption capacities (Chahar et al., 2023; Ahmaruzzaman et al., 2025). They are capable of degrading or adsorbing trace organics, dyes, and heavy metals at low concentrations where traditional treatments are ineffective.

Nanoadsorbents dissociate cationic metals through electrostatic forces and ion exchange, whereas photocatalytic nanoparticles enhance oxidative degradation under UV or visible light. For example, Fe_3O_4–biochar composites have recorded >95% removal of Pb^{2+} and Cd^{2+} in river water simulant (Tahir et al., 2021; Bedasa and Mishra, 2025).

Although promising, nanomaterials have stability, toxicity, and recoverability concerns. Immobilizing nanoparticles on biodegradable supports and using magnetic separation methods can minimize these concerns. Incorporation with CWs or AOPs can further increase pollutant degradation while avoiding nanoparticle leakage into aquatic environments (Arliyani et al., 2024; Cavalcante et al., 2024).

5.2 Bioelectrochemical and microbial fuel cell systems

BESs, such as microbial fuel cells (MFCs), represent novel technologies that integrate bioremediation and energy production. In BESs, electroactive microorganisms (e.g., *Geobacter*, *Shewanella*) reduce organic contaminants at the anode, mobilizing electrons that travel to the cathode to produce electricity (Garbini et al., 2023; Kumar et al., 2023a). Such processes simultaneously degrade pollutants and recover renewable energy.

Laboratory- and pilot-scale experiments have proved COD removal efficiencies of around 70%–85% and power yields of 1.2–1.5 W m^{-2} in household wastewater-fed MFCs (Kumar et al., 2023a). BES technologies need optimization with respect to electrode design, scaling, and maintenance costs (Herkendell, 2021). Interfacing BES units with wetlands or aeration modules can smoothen performance while producing additional energy for monitoring and operation (Jamaludin et al., 2024).

5.3 AI- and Internet of Things-assisted smart monitoring systems

Emerging digital technologies like AI, the Internet of Things (IoT), and machine learning (ML) are transforming river pollution monitoring and treatment optimization (Miller et al., 2025). Continuous in situ measurements of DO, turbidity, conductivity, and nutrient levels can be provided by real-time sensors and sent to cloud-based platforms for automatic control. AI algorithms forecast pollution episodes and optimize aeration or reagent dosing, reducing operating costs (Monday et al., 2024; Stojanovic et al., 2023).

For instance, adaptive control of aeration intensity by AI in the Yamuna pilot restoration project minimized energy consumption by about 25% while sustaining DO levels above 5 mg L^{-1} (Monday et al., 2024). Likewise, IoT-instrumented CWs with flow and redox sensors sustained stable BOD removal

TABLE 20.4 Applications of AI and IoT tools in river pollution control.

Application	Function	Outcome	Reference
AI-based aeration control	Adjusts blower operation using DO prediction	Energy saving	(Miller et al., 2025)
IoT-enabled wetland	Monitors flow, pH, and redox status	Enhanced nutrient stability	(Sandhwar et al., 2025)
ML-based pollution forecasting	Predicts contamination peaks	Improved preventive management	(Rajesh et al., 2025)
Smart biosensors	Real-time detection of metals and organics	Faster remediation response	(Anchidin-Norocel et al., 2025)

through predictive regulation of hydraulic loading (Chen et al., 2023). The major characteristics and advantages of digital integration are listed in Table 20.4.

5.4 Holistic integration and future vision

Integrating and emerging methodologies are the future of sustainable river restoration, synergizing engineering acumen with ecological wisdom. The synergistic integration of nanotechnology-assisted adsorption, bioelectrochemical energy harvesting, and AI-assisted adaptive management guarantees greater efficiency and durability (Islam et al., 2025).

Future research directions include the creation of biodegradable nanomaterials, the advancement of AI-based real-time decision models, and the facilitation of community-inclusive digital river monitoring platforms. Such synergy between ecological engineering and smart technology has the potential to bridge conventional remediation with self-optimizing, self-sustaining systems that restore water quality in addition to regenerating ecosystem functions (den Elzen et al., 2021).

6 Conclusion and future prospects

Rivers are dynamic ecosystems whose health reflects the equilibrium between natural processes and human interference. Restoration of their integrity implies a holistic concept unifying physical, chemical, and bioremedial methods and not depending on any one technique. Physical techniques like dredging, aeration, RBF, and hydromorphological restoration provide quick hydraulic improvement and oxygen enrichment. Chemical treatments such as coagulation–flocculation, advanced oxidation, and adsorption offer focused removal of recalcitrant contaminants, whereas bioremediation utilizes microorganisms and vegetation to catalyze natural detoxification and nutrient cycling.

Sequential and combinative deployment of these methods—mechanical cleanup with subsequent chemical polishing and biological stabilization—produces the most cost-efficient and sustainable outcomes. Hybrid approaches, such as aeration + CWs or dredging + AOP + phytoremediation, consistently remove over 90% of BOD, COD, and nutrient loads while restoring aquatic biodiversity. Their success reflects the need for adaptive, science-driven management, together with community involvement within national initiatives like Namami Gange and global programs like the EU Water Framework Directive.

Looking forward, the way of the future for river remediation is technological convergence and ecological intelligence. Nanocomposite adsorbents, BESs, and AI- or IoT-enabled smart monitoring can significantly enhance efficiency and facilitate real-time control. At the same time, nature-based solutions—riparian buffers, floodplain reconnection, and CWs—will bolster resilience to pollution and climate stress. Linking these innovations to Integrated Water-Resource Management and the UN Sustainable Development Goals (6, 13, 15) will improve water security, induce a circular bioeconomy, and enhance climate adaptation.

References

Adetunji, C.O., Olaniyan, O.T., Igere, B.E., Ekundayo, T.C., Anani, O.A., Bodunrinde, R.E., Olisaka, F.N., Inobeme, A., Uwadiae, E.O., Obayagbona, O.N., Dauda, W.P., Inamuddin, 2021. Microbial Degradation of Chlorophenolic Compounds. Springer Science and Business Media LLC, pp. 313–349. doi:10.1007/978-981-16-0518-5_12.

Ahmaruzzaman, M., Roy, S., Singha, A., Rtimi, S., Aminabhavi, T.M., 2025. Emerging nanotechnologies in adsorption of dyes: a comprehensive review of carbon and metal oxide-based nanomaterials. Adsorption. 15728757. 31 (2). Available from: https://www.springer.com/journal/10450. doi:10.1007/s10450-024-00588-y.

Ajaz, S., Hassan, A.A., Michael, R.N., Leusch, F.D., 2024. Removal of organic micropollutants in biologically active filters: a systematic quantitative review of key influencing factors. J. Environ. Manag. 353, 120203. doi:10.1016/j.jenvman.2024.120203.

Akhtar, M.S., Ali, S., Zaman, W., 2024. Innovative adsorbents for pollutant removal: exploring the latest research and applications. Molecules 29 (18), 4317. doi:10.3390/molecules29184317.

Akhtar, N., Syakir Ishak, M.I., Bhawani, S.A., Umar, K., 2021. Various natural and anthropogenic factors responsible for water quality degradation: a review. Water 13 (19), 2660. doi:10.3390/w13192660.

Al-Hashimi, O., Hashim, K., Loffill, E., Čebašek, T.M., Nakouti, I., Faisal, A.A.H., Al-Ansari, N., 2021. A comprehensive review for groundwater contamination and remediation: occurrence, migration and adsorption modelling. Molecules. 1420-3049. 26 (19), 5913. doi:10.3390/molecules26195913.

Alafnan, S., Awotunde, A., Glatz, G., Adjei, S., Alrumaih, I., Gowida, A., 2021. Langmuir adsorption isotherm in unconventional resources: applicability and limitations. J. Pet. Sci. Eng. 09204105. 207, 109172. doi:10.1016/j.petrol.2021.109172.

Amrutha, Jeppu, G., Girish, C.R., Prabhu, B., Mayer, K., 2023. Multi-component adsorption isotherms: review and modeling studies. Environ. Process. 21987505. 10. Available from: https://www.springer.com/journal/40710. doi:10.1007/s40710-023-00631-0.

Anchidin-Norocel, L., Bosancu, A., Iatcu, O.C., Lobiuc, A., Covasa, M., 2025. Real-time detection of heavy metals and other pollutants in wastewater using chemical sensors: a strategy to limit the spread of antibiotic-resistant bacteria. Chemosensors 13 (9), 352. doi:10.3390/chemosensors13090352.

Arliyani, I., Noori, M.T., Ammarullah, M.I., Tangahu, B.V., Mangkoedihardjo, S., Min, B., 2024. Constructed wetlands combined with microbial fuel cells (CW-MFCs) as a sustainable technology for leachate treatment and power generation. RSC Adv. 20462069. 14 (44), 32073–32100. Available from: http://pubs.rsc.org/en/journals/journal/ra. doi:10.1039/d4ra04658g.

Arora, N.K., Mishra, I., 2022. Sustainable development goal 6: global water security. Environ. Sustain. 2523-8922. 5 (3), 271–275. doi:10.1007/s42398-022-00246-5.

Aryal, M., 2024. Phytoremediation strategies for mitigating environmental toxicants. Heliyon. 24058440. 10 (19), e38683. doi:10.1016/j.heliyon.2024.e38683.

Barman, M.K., Bhattarai, A., Saha, B., 2023. Applications of ion exchange resins in environmental remediation. Vietnam J Chem 61 (5), 533–550. doi:10.1002/vjch.202300027.

Bartolomeu, M., Neves, M.G.P.M.S., Faustino, M.A.F., Almeida, A., 2018. Wastewater chemical contaminants: remediation by advanced oxidation processes. Photochem. Photobiol. Sci. 1474-905X. 17 (11), 1573–1598. doi:10.1039/c8pp00249e.

Bedasa, M.K., Mishra, D., 2025. Efficient adsorption techniques for heavy metal removal from east african paint industry wastewater. Future Perspect. Med. Pharm. Environ. Biotechnol. 3009-7762. 2 (2), 13–40. doi:10.21608/fpmpeb.2025.397177.1016.

Biswas, J.K., Prasad, M.N.V., 2023. Wastewater-based circular economy operations in east Kolkata wetlands (the Largest Ramasar Site in South Asia), India. In: Advances in Science, Technology and Innovation. Springer Nature, India, pp. 241–254. Available from: https://www.springer.com/series/15883. doi:10.1007/978-3-031-18165-8_19.

Brito, C.H.V., Gloria, D.C.S., Santos, E.B., Domingues, R.A., Valente, G.T., Vieira, N.C.S., Gonçalves, M., 2023. Porous activated carbon/graphene oxide composite for efficient adsorption of pharmaceutical contaminants. Chem. Eng. Res. Des. 02638762. 191, 387–400. Available from: http://www.elsevier.com/wps/find/journaldescription.cws_home/713871/description#description. doi:10.1016/j.cherd.2023.01.044.

Cavalcante, A.L., Dari, D.N., da Silva Aires, F.I., de Castro, E.C., Dos Santos, K.M., Dos Santos, J.C., 2024. Advancements in enzyme immobilization on magnetic nanomaterials: toward sustainable industrial applications. RSC Adv. 14 (25), 17946–17988. doi:10.1039/D4RA02939A.

Chahar, M., Khaturia, S., Singh, H.L., Solanki, V.S., Agarwal, N., Sahoo, D.K., Yadav, V.K., Patel, A., 2023. Recent advances in the effective removal of hazardous pollutants from wastewater by using nanomaterials—A review. Front. Environ. Sci. 2296665X. 11. Available from: https:/www.journal.frontiersin.org/journal/environmental-science. doi:10.3389/fenvs.2023.1226101.

Chakraborty, 2021a. Trophic Interactions and Biogeochemical Cycles in River Ecosystem. In: Riverine Ecology, Vol. 1. Springer, Cham. doi:10.1007/978-3-030-53897-2_4.

Chamoli, A., Karn, S.K., Kumari, M., Sivaramasamy, E., 2025. Biochar mediated fixation of nitrogen compounds (ammonia and nitrite) in soil: a review. Biodegradation. 15729729. 36 (2). Available from: https://www.springer.com/journal/10532. doi:10.1007/s10532-025-10116-6.

Chen, T., Wang, M., Su, J., Ikram, R.M., Li, J., 2023. Application of internet of things (IoT) technologies in green stormwater infrastructure (GSI): a bibliometric review. Sustainability 15 (18), 13317. doi:10.3390/su151813317.

Chen, X., Duan, F., Yu, X., Xie, Y., Wang, Z., Ni, S.Q., 2024. Uncovering pathway and mechanism of simultaneous thiocyanate detoxicity and nitrate removal through anammox and denitrification. Npj Clean Water. 20597037. 7 (1). Available from: https://www.nature.com/npjcleanwater/. doi:10.1038/s41545-024-00402-w.

Coates, P., 2013. A story of six rivers: history, culture and ecology. London: Reaktion Books.
Corriveau, J., Chambers, P.A., Culp, J.M., 2013. Seasonal variation in nutrient export along streams in the northern great plains. Water, Air, Soil Pollut. 15732932. 224 (7). Available from: http://www.kluweronline.com/issn/0049-6979/. doi:10.1007/s11270-013-1594-1.
Costa, L., Duarte, M.S., Magalhães, C.P., Pereira, M.A., Cavaleiro, A.J., 2025. Micro-aeration for improving anaerobic treatment and biogas production from organic pollutants. Appl. Microbiol. Biotechnol. 14320614. 109 (1). Available from: https://www.springer.com/journal/253. doi:10.1007/s00253-025-13519-w.
Cui, G., Xiaosi, S., Liu, Y., Zheng, S., 2021. Effect of riverbed sediment flushing and clogging on river-water infiltration rate: a case study in the Second Songhua River, Northeast China. Hydrogeol. J. 1431-2174. 29 (2), 551–565. doi:10.1007/s10040-020-02218-7.
Dai, W., Pang, J.W., Ding, J., Wang, Y.Q., Zhang, L.Y., Ren, N.Q., Yang, S.S., 2023. Study on the removal characteristics and degradation pathways of highly toxic and refractory organic pollutants in real pharmaceutical factory wastewater treated by a pilot-scale integrated process. Front. Microbiol. 1664302X. 14. Available from: https://www.frontiersin.org/journals/microbiology#. doi:10.3389/fmicb.2023.1128233.
den Elzen M. Portugal-Pereira J. Rogelj J. 2021 The emissions gap, in UNEP Emissions Gap Report 2021 Nairobi: United Nations Environment Programme (UNEP) 25–35 doi:10.18356/9789280738124c007
Dorafshan, M.M., Abedi-Koupai, J., Eslamian, S., Amiri, M.J., 2023. Vetiver grass (*Chrysopogon zizanoides* L.): a hyper-accumulator crop for bioremediation of unconventional water. Sustainability. 2071-1050. 15 (4), 3529. doi:10.3390/su15043529.
El-Fakharany, Z.M., 2025. Potential of using river bank filtration to augment water resources: a review. In: Springer Water, Vol. 9, pp. 97–126. Available from: https://www.springer.com/series/13419. doi:10.1007/978-3-031-79122-2_4.
Elshobary, M., Ismail, G.A., Allah, K.S., Keshta, A., El-Shenody, R., Saber, A.A., Huo, S., Liu, W., Zhu, F., El-Sheekh, M., 2025. Regulation of Phosphorus and Nitrogen Conversion by Cyanobacteria and Its Importance in Saline-Alkali Soil. In: Plant-Microbe Interactions for Environmental and Agricultural Sustainability. Springer Science and Business Media LLC, pp. 633–660. doi:10.1007/978-3-031-84939-8_24.
Enaime, G., Baçaoui, A., Yaacoubi, A., Lübken, M., 2020. Biochar for wastewater treatment—conversion technologies and applications. Appl. Sci. 2076-3417. 10 (10), 3492. doi:10.3390/app10103492.
Espinosa-Ortiz, E.J., Rene, E.R., Gerlach, R., 2022. Potential use of fungal-bacterial co-cultures for the removal of organic pollutants. Crit. Rev. Biotechnol. 15497801. 42 (3), 361–383. Available from: http://www.tandfonline.com/loi/ibty20. doi:10.1080/07388551.2021.1940831.
Farzanehsa, M., Vaughan, L.C., Zamyadi, A., Khan, S.J., 2023. Comparison of UV-Cl and UV-H2O2 advanced oxidation processes in the degradation of contaminants from water and wastewater: a review. Water Environ. J. 17476593. 37 (4), 633–643. Available from: http://onlinelibrary.wiley.com/journal/10.1111/(ISSN)1747-6593. doi:10.1111/wej.12868.
Feng, L., Shufei, H., Wei, L., Zhang, J., Haiming, W., 2021. Impacts of aeration and biochar on physiological characteristics of plants and microbial communities and metabolites in constructed wetland microcosms for treating swine wastewater. Environ. Res. 00139351. 200, 111415. doi:10.1016/j.envres.2021.111415.
Fernandez-Fontaina, E., Pinho, I., Carballa, M., Omil, F., Lema, J.M., 2013. Biodegradation kinetic constants and sorption coefficients of micropollutants in membrane bioreactors. Biodegradation 24 (2), 165–177. doi:10.1007/s10532-012-9568-3.
Gaballah, M.S., Ismail, K., Aboagye, D., Ismail, M.M., Sobhi, M., Stefanakis, A.I., 2021. Effect of design and operational parameters on nutrients and heavy metal removal in pilot floating

treatment wetlands with Eichhornia Crassipes treating polluted lake water. Environ. Sci. Pollut. Res. 16147499. 28 (20), 25664–25678. Available from: https://link.springer.com/journal/11356. doi:10.1007/s11356-021-12442-7.

Gao, T., Li, Y., Yang, N., Xiong, W., Liang, X., Wang, Y., 2025. Plant-rhizosphere microbe interactions and their roles in nitrogen cycles under periodic flooding: from cooperation mechanisms to ecological responses. Crit. Rev. Environ. Sci. Technol. 1064-3389. 55 (17), 1358–1382. doi:10.1080/10643389.2025.2530941.

Garbini, G.L., Caracciolo, A.B., Grenni, P., 2023. Electroactive bacteria in natural ecosystems and their applications in microbial fuel cells for bioremediation: a review. Microorganisms. 2076-2607. 11 (5), 1255. doi:10.3390/microorganisms11051255.

Geist, J., Hawkins, S.J., 2016. Habitat recovery and restoration in aquatic ecosystems: current progress and future challenges. Aquat. Conserv. 10990755. 26 (5), 942–962. Available from: http://onlinelibrary.wiley.com/journal/10.1002/(ISSN)1099-0755. doi:10.1002/aqc.2702.

Ghangrekar, M.M., 2022. Ponds and Wetlands for Treatment of Wastewater. Springer Science and Business Media LLC, pp. 555–592. doi:10.1007/978-981-19-4048-4_13.

Gong, H., Liu, Y., Yang, T., Liu, L., Xuechen, L., Chuantao, G., Jin, Z., Debao, L., Wang, W., 2025. Advancements in hybrid and combined biological technologies for treating polluted gases: a comprehensive review. Water, Air, & Soil Pollution. 0049-6979. 236 (3). doi:10.1007/s11270-025-07779-5.

Grifoni, M., Franchi, E., Fusini, D., Vocciante, M., Barbafieri, M., Pedron, F., Rosellini, I., Petruzzelli, G., 2022. Soil remediation: towards a resilient and adaptive approach to deal with the ever-changing environmental challenges. Environments. 2076-3298. 9 (2), 18. doi:10.3390/environments9020018.

Guo, X., Wang, J., 2019. Comparison of linearization methods for modeling the Langmuir adsorption isotherm. J. Mol. Liq. 01677322. 296, 111850. doi:10.1016/j.molliq.2019.111850.

Hama Aziz, K.H., Fatah, N.M., Muhammad, K.T., 2024. Advancements in application of modified biochar as a green and low-cost adsorbent for wastewater remediation from organic dyes. R. Soc. Open Sci. 20545703. 11 (5). Available from: https://www.rsos.royalsocietypublishing.org/. doi:10.1098/rsos.232033.

Herkendell, K., 2021. Status update on bioelectrochemical systems: prospects for carbon electrode design and scale-up. Catalysts. 2073-4344. 11 (2), 278. doi:10.3390/catal11020278.

Ibitoye, O.A., Agbaje, A.B., Anyanwu, C., Fasogbon, I.V., Dangana, R.S., Akinola, S.A., Reuben, M., 2025. Impact of industrial wastewater on the microbiome and resistome of waterbodies: a systematic review. F1000Research. 2046-1402. 14, 603. doi:10.12688/f1000research.164918.1.

Islam, F.S., Islam, M.N., 2025. AI-driven integration of nanotechnology and green nanotechnology for sustainable energy and environmental remediation. J. Eng. Res. Rep. 27 (7), 260–311. doi:10.9734/jerr/2025/v27i71574.

Jagaba, A.H., Abdulazeez, I., Lawal, D.U., Affam, A.C., Mu'azu, N.D., Soja, U.B., Usman, A.K., Noor, A., Lim, J.W., Aljundi, I.H., 2024. A review on the application of biochar as an innovative and sustainable biocarrier material in moving bed biofilm reactors for dye removal from environmental matrices. Environ. Geochem. Health. 15732983. 46 (9). Available from: https://www.springer.com/journal/10653. doi:10.1007/s10653-024-02122-z.

Jamaludin, M., Tsai, Y.C., Lin, H.T., Huang, C.Y., Choi, W., Chen, J.G., Sean, W.Y., 2024. Modeling and control strategies for energy management in a wastewater center: a review on aeration. Energies. 19961073. 17 (13). Available from: http://www.mdpi.com/journal/energies/. doi:10.3390/en17133162.

Jatoi, A.S., Baloch, H.A., Mazari, S.A., Mubarak, N.M., Sabzoi, N., Aziz, S., Soomro, S.A., Abro, R., Shah, S.F., 2023. Extractive fermentation via ion exchange adsorption resins: opportunities, challenges, and future prospects. Biomass Convers. Bioref. 13 (5), 3543–3554. doi:10.1007/s13399-021-01417-w.

Jayaprada, N.V.T., Wewalwela, J.J., Galahitigama, G.A.H., Pandipperuma, P.A.N.P., 2023. Microbial Interactions with Aquatic Plants Current Status of Fresh Water Microbiology. Springer International Publishing, Sri Lanka, pp. 135–160. Available from: doi:10.1007/978-981-99-5018-8. doi: 10.1007/978-981-99-5018-8_6.

Jayaswal, K., Siabbweka, M., Makwana, H., 2025. Heavy Metals Removal from Wastewater. Informa UK Limited, pp. 225–241. doi:10.1201/9781003663799-13.

Karnwal, A., Martolia, S., Dohroo, A., Al-Tawaha, A.R.M.S., Malik, T., 2024. Exploring bioremediation strategies for heavy metals and POPs pollution: the role of microbes, plants, and nanotechnology. Front. Environ. Sci. 2296-665X. 12. doi:10.3389/fenvs.2024.1397850.

Kowalska, A., Biczak, R., 2025. Phytoremediation and environmental law: harnessing biomass and microbes to restore soils and advance biofuel innovation. Energies. 1996-1073. 18 (7), 1860. doi:10.3390/en18071860.

Krapesch, M., Klösch, M., ten Brinke, W., Habersack, H., 2024. The rhine catchment: a review of sediment-related knowledge, monitoring, and a future research perspective. Water. 2073-4441. 16 (8), 1121. doi:10.3390/w16081121.

Kulwant, M., Yadav, A.K., 2024. Challenges and Opportunities for Water Quality Monitoring and Management in India. In: Water Science and Technology Library. Springer Science and Business Media LLC, pp. 121–137. doi:10.1007/978-3-031-62079-9_7.

Kumar, N., Gusain, R., Pandey, S., Ray, S.S., 2023a. Hydrogel nanocomposite adsorbents and photocatalysts for sustainable water purification. Adv. Mater. Interfaces. 21967350. 10 (2). Available from: http://onlinelibrary.wiley.com/journal/10.1002/(ISSN)2196-7350. doi:10.1002/admi.202201375.

Kumar, S.S., Wani, O.A., Kumar, S., Devi, K., 2025. Watershed management for soil restoration and environmental sustainability. In: Babu, S., Das, A., Rathore, S.S., Singh, R., Singh, R. (Eds.), Ecological Solutions to Agricultural Land Degradation, Springer, Singapore. doi:10.1007/978-981-96-3392-0_14.

Kumari, S., Das, S., 2023. Bacterial enzymatic degradation of recalcitrant organic pollutants: catabolic pathways and genetic regulations. Environ. Sci. Pollut. Res. 1614-7499. 30 (33), 79676–79705. doi:10.1007/s11356-023-28130-7.

Laghrib, F., Bakasse, M., Lahrich, S., Mhammedi, M.A.E., 2021. Advanced oxidation processes: photo-electro-Fenton remediation process for wastewater contaminated by organic azo dyes. Int. J. Environ. Anal. Chem. 10290397. 101 (15), 2947–2962. Available from: https://www.tandf.co.uk/journals/titles/03067319.asp. doi:10.1080/03067319.2020.1711892.

Lai, L.L., Wang, S.F., Qi, Z.Q., Zhang, Y.S., Wang, R., Yuan, L.J., 2025. A deep treatment process for chemical polishing wastewater towards resource recovery: optimization and performance. Sep. Purif. Technol. 18733794. 358. Available from: https://www.sciencedirect.com/science/journal/13835866. doi:10.1016/j.seppur.2024.130285.

Li, P., Li, D., Sun, X., Chu, Z., Xia, T., Zheng, B., 2022. Application of ecological restoration technologies for the improvement of biodiversity and ecosystem in the river. Water. 20734441. 14 (9). Available from: https://www.mdpi.com/2073-4441/14/9/1402/pdf. doi:10.3390/w14091402.

Md Anawar, H., Chowdhury, R., 2020. Remediation of polluted river water by biological, chemical, ecological and engineering processes. Sustainability. 2071-1050. 12 (17), 7017. doi:10.3390/su12177017.

Mehta, G., Jade, D., 2024. Nine steps towards rehabilitation and restoration of degraded urban rivers. In: Managing Urban Rivers. Elsevier BV, pp. 241–259. doi:10.1016/b978-0-323-85703-1.00004-3.

Miller, T., Durlik, I., Kostecka, E., Kozlovska, P., Łobodzińska, A., Sokołowska, S., Nowy, A., 2025. Integrating artificial intelligence agents with the internet of things for enhanced environmental monitoring: applications in water quality and climate data. Electronics. 2079-9292. 14 (4), 696. doi:10.3390/electronics14040696.

Minkina, T., Fedorenko, G., Nevidomskaya, D., Konstantinova, E., Pol'shina, T., Fedorenko, A., Chaplygin, V., Mandzhieva, S., Dudnikova, T., Hassan, T., 2021. The morphological and functional organization of cattails typha laxmannii lepech. And typha australis schum. and thonn. under soil pollution by potentially toxic elements. Water. 20734441. 13 (2). Available from: https://www.mdpi.com/2073-4441/13/2/227/pdf. doi:10.3390/w13020227.

Moazzem, S., Bhuiyan, M., Muthukumaran, S., Fagan, J., Jegatheesan, V., 2023. Microbiome wetlands in nutrient and contaminant removal. Curr. Pollut. Rep. 2198-6592. 9 (4), 694–709. doi:10.1007/s40726-023-00280-9.

Mondal, P.P.P., 2022. Incorporating hydromorphological assessments for restorative river science: context, concepts and criteria. In: Fluvial Systems in the Anthropocene. Available from: doi:10.1007/978-3-031-11181-5_4.

Mondal, S., Mukherjee, S.K., Hossain, S.T., 2023. Exploration of plant growth promoting rhizobacteria (PGPRs) for heavy metal bioremediation and environmental sustainability: recent advances and future prospects. In: Modern Approaches in Waste Bioremediation: Environmental Microbiology. Springer International Publishing, India, pp. 29–55. Available from: https://link.springer.com/book/10.1007/978-3-031-24086-7. doi:10.1007/978-3-031-24086-7_3.

Monday, C., Zaghloul, M.S., Krishnamurthy, D., Achari, G., 2024. A review of ai-driven control strategies in the activated sludge process with emphasis on aeration control. Water. 2073-4441. 16 (2), 305. doi:10.3390/w16020305.

Monira, U., Mostafa, M.G., 2023. Leather industrial effluent and environmental concerns: a review. Sustain. Water Resour. Manag. 23635045. 9 (6). Available from: https://www.springer.com/journal/40899. doi:10.1007/s40899-023-00969-1.

Mukherjee, J., Lodh, B.K., Sharma, R., Mahata, N., Shah, M.P., Mandal, S., Ghanta, S., Bhunia, B., 2023. Advanced oxidation process for industrial wastewater: strategies, mechanisms, bottlenecks and prospects. Chemosphere 345, 140473. doi:10.1016/j.chemosphere.2023.140473.

Müller, D.P., Hiete, M., 2026. Recycled content and recycling of electric vehicle batteries in the context of EU battery regulation: Methodological deficits and strategic approaches for sustainable battery materials. Resour. Conserv. Recycl. 231, p. 108903. doi:10.1016/j.resconrec.2026.108903.

Mutegoa, E., Sahini, M.G., 2023. Approaches to mitigation of hydrogen sulfide during anaerobic digestion process – a review. Heliyon. 24058440. 9 (9), e19768. doi:10.1016/j.heliyon.2023.e19768.

Obura, D., Akamushaba, O., Khaldi, A., Dadebo, D., 2023. Determination and implementation of an onsite management system for motor-vehicle wash effluents: a field-scale commercial application toward sector sustainability. J. Water Process Eng. 22147144. 56. Available from: http://www.journals.elsevier.com/journal-of-water-process-engineering/. doi:10.1016/j.jwpe.2023.104513.

Ogundele, O.D., Ogundele, D.T., Popoola, O.J., Adelekun, A.A., Olagunju, V.A., 2025. Phytoremediation in Wetlands. Springer Science and Business Media LLC, pp. 199–237. doi:10.1007/978-3-031-91982-4_10.

Osborne, L.L., Kovacic, D.A., 1993. Riparian vegetated buffer strips in water-quality restoration and stream management. Freshw. Biol. 29 (2), 243–258. doi:10.1111/j.1365-2427.1993.tb00761.x.

Pandis, P.K., Kalogirou, C., Kanellou, E., Vaitsis, C., Savvidou, M.G., Sourkouni, G., Zorpas, A.A., Argirusis, C., 2022. Key points of advanced oxidation processes (AOPs) for wastewater, organic pollutants and pharmaceutical waste treatment: a mini review. Chem. Eng. 23057084. 6 (1). Available from: https://www.mdpi.com/2305-7084/6/1/8/pdf. doi:10.3390/chemengineering6010008.

Pineda, E., García-Ruiz, M.J., Guaya, D., Manrique, J., Osorio, F., 2021. Removal of total coliforms and E. *coli* using filtration with natural clays and silica sand. Environ. Geochem. Health 43 (1), 195–207. doi:10.1007/s10653-020-00623-1.

Rajesh, M., Babu, R.G., Moorthy, U., Easwaramoorthy, S.V., 2025. Real-time air quality assessment and predictive environmental health risk mapping via machine learning. Sci. Rep. 15 (1), 28801. doi:10.1038/s41598-025-14214-6.

Raji, Z., Karim, A., Karam, A., Khalloufi, S., 2023. Adsorption of heavy metals: mechanisms, kinetics, and applications of various adsorbents in wastewater remediation—a review. Waste. 2813-0391. 1 (3), 775–805. doi:10.3390/waste1030046.

Randhawa, K.S., 2024. Microplastics removal by coagulation: cutting-edge coagulants and processes. Pigment Resin Technol. 10, 29–84. doi:10.1039/D3EW00119A.

Rashid, R., Shafiq, I., Akhter, P., Iqbal, M.J., Hussain, M., 2021. A state-of-the-art review on wastewater treatment techniques: the effectiveness of adsorption method. Environ. Sci. Pollut. Res. 16147499. 28 (8), 9050–9066. Available from: https://link.springer.com/journal/11356. doi:10.1007/s11356-021-12395-x.

Sandhwar, V.K., Saxena, S., Saxena, D., Tiwari, A., Parikh, S.M., 2025. Future trends and emerging technologies in water quality management. In: Dubey, A.K., Srivastav, A.L., Kumar, A., Marquez, F.P.G., Giannakoudakis, D.A. (Eds.), Computational Automation for Water Security. Elsevier. doi: 10.1016/B978-0-443-33321-7.00012-3

Santos, E., 2025. Nature-based solutions for water management in Europe: what works, what does not, and what's next? Water. 2073-4441. 17 (15), 2193. doi:10.3390/w17152193.

Satyam, S., Patra, S., 2025. The evolving landscape of advanced oxidation processes in wastewater treatment: challenges and recent innovations. Processes. 22279717. 13 (4). Available from: http://www.mdpi.com/journal/processes. doi:10.3390/pr13040987.

Scott, J.P., Ollis, D.F., 1995. Integration of chemical and biological oxidation processes for water treatment: review and recommendations. Environ. Prog. 15475921. 14 (2), 88–103. doi:10.1002/ep.670140212.

Sebastian, S.L., Kalivel, P., Subbiah, K., Murphy, M.S.A., David, J.J., Palanichamy, J., 2024. Assessing titanium vs. aluminium electrodes for wastewater remediation in the small-scale industries (SSI) textile sector. Environ. Nanotechnol. Monit. Manag. 22151532. 21. Available from: https://www.sciencedirect.com/science/journal/22151532. doi:10.1016/j.enmm.2024.100950.

Selvasembian, R., Mal, J., Rani, R., Sinha, R., Agrahari, R., Joshua, I., Santhiagu, A., Pradhan, N., 2022. Recent progress in microbial fuel cells for industrial effluent treatment and energy generation: fundamentals to scale-up application and challenges. Bioresour. Technol. 18732976. 346. Available from: https://www.elsevier.com/locate/biortech. doi:10.1016/j.biortech.2021.126462.

Shammas, N.K., Hahn, H.H., Wang, M.H., Wang, L.K., 2021. Fundamentals of chemical coagulation and precipitation. In: Wang, L.K., Wang, M.H.S., Shammas, N.K., Aulenbach, D.B. (Eds.), Environmental Flotation Engineering. Handbook of Environmental Engineering, Vol. 21. Springer, Cham. doi:10.1007/978-3-030-54642-7_3.

Simon, M., Joshi, H., 2022. Story of the Ganga River: Its Pollution and Rejuvenation. Springer Science and Business Media LLC, pp. 21–55. doi:10.1007/978-3-030-87067-6_2.

Singh, J., Kumar, P., Eid, E.M., Taher, M.A., El-Morsy, M.H.E., Osman, H.E.M., Al-Bakre, D.A., Kumar, V., 2023. Phytoremediation of nitrogen and phosphorus pollutants from glass industry effluent by using water hyacinth (*Eichhornia crassipes* (Mart.) Solms): application of RSM and ANN techniques for experimental optimization. Environ. Sci. Pollut. Res. 16147499. 30 (8), 20590–20600. Available from: https://www.springer.com/journal/11356. doi:10.1007/s11356-022-23601-9.

Singh, S., 2025. The role of environmental microbiota in bioremediation: harnessing bacteria for pollutant degradation. Clin. Med. Health Res. J. 1694-4305. 5 (02), 1228–1240. doi:10.18535/cmhrj.v5i02.464.

Singh, V., 2024. Textbook of Environment and Ecology Textbook of Environment and Ecology. Springer Nature, India, pp. 1–359. Available from: https://link.springer.com/book/9789819988464. doi:10.1007/978-981-99-8846-4.

Soetan, O., Nie, J., Viteritto, M., Feng, H., 2023. Evaluation of sediment dredging in remediating toxic metal contamination — a systematic review. Environ. Sci. Pollut. Res. 1614-7499. 30 (27), 69837–69856. doi:10.1007/s11356-023-27489-x.

Sonal, S., Mishra, B.K., 2021. Role of Coagulation/Flocculation Technology for the Treatment of Dye Wastewater: Trend and Future Aspects Water Pollution and Management Practices. Springer Singapore, India, pp. 303–331. Available from: https://link.springer.com/book/10.1007/978-981-15-8358-2. doi:10.1007/978-981-15-8358-2_13.

Soni, S.K., Soni, R., 2025. Techno-Economic and Environmental Assessment. Springer Science and Business Media LLC, pp. 291–314. doi:10.1007/978-981-96-9013-8_9.

Spieles, D.J., 2022. Wetland construction, restoration, and integration: a comparative review. Land. 2073-445X. 11 (4), 554. doi:10.3390/land11040554.

Sravan, J.S., Matsakas, L., Sarkar, O., 2024. Advances in biological wastewater treatment processes: focus on low-carbon energy and resource recovery in biorefinery context. Bioengineering. 23065354. 11 (3). Available from: https://www.mdpi.com/journal/bioengineering. doi:10.3390/bioengineering11030281.

Srivastava A. Prathna T.C. 2022. Yamuna action plan-III: Impact on water quality of river Yamuna, India Fine Chem. Eng. 1–10

Stojanovic M. Asujic D. Vuletic M. Kostic M. Savic M. 2023 Video-assisted thoracoscopic surgery for thymectomy in myasthenia gravis patients, one year single centre experience J. Cardiothorac. Vasc. Anesth. 37 46–47 doi:10.1053/j.jvca.2023.08.091

Suhaimi, S., Ramlee, N.N., Rahman, N.H.A., 2025. Enzymatic and Microbial-Based Systems for Bioremediation of Wastewater. Springer Science and Business Media LLC, pp. 341–380. doi:10.1007/978-981-96-6699-7_15.

Sun, X., Shang, J., Zhao, L., Li, H., Hao, R., 2025. Optimizing push-flow aeration in urban shallow lakes: threshold-based algal control and nutrient dynamics through field-validated numerical modeling. J. Water Process Eng. 22147144. 75. Available from: https://www.sciencedirect.com/science/journal/22147144. doi:10.1016/j.jwpe.2025.108041.

Tahir, M.B., Asiri, A.M., Malik, M.F., Khalid, N.R., Iqbal, T., Rafique, M., Kiran, H., Muhammad, S., Siddeeg, S.M., Shahzad, K., 2021. Role of nano-photocatalysts in detoxification of toxic heavy metals. Curr. Anal. Chem. 18756727. 17 (2), 126–137. Available from: https://www.eurekaselect.com/177910/article. doi:10.2174/1573411016666191230151455.

Tariq, A., Mushtaq, A., 2023. Untreated wastewater reasons and causes: a review of most affected areas and cities. Int. J. Chem. Biochem. Sci. 22269614. 23 (1), 121–143. Available from: https://www.iscientific.org/wp-content/uploads/2023/05/15-IJCBS-23-23-22.pdf.

Taylor, M., Alabdrabalameer, H., Skoulou, V., 2019. Choosing physical, physicochemical and chemical methods of pre-treating lignocellulosic wastes to repurpose into solid fuels. Sustainability. 2071-1050. 11 (13), 3604. doi:10.3390/su11133604.

Verma, M., Chakraborty, S., Kumari, S., Gupta, A., Kumar, D., Iqbal, J., Banu, J.R., Pugazhendi, A., Kumar, R.N., 2023. Co-treatment of stabilized landfill leachate and municipal wastewater in a granular activated carbon-sequencing batch reactor (GAC-SBR). Process Saf. Environ. Prot. 09575820. 174, 424–432. Available from: http://www.elsevier.com/wps/find/journaldescription.cws_home/713889/description#description. doi:10.1016/j.psep.2023.04.015.

Wang, J., Guo, X., 2023. Adsorption kinetics and isotherm models of heavy metals by various adsorbents: an overview. Crit. Rev. Environ. Sci. Technol. 1064-3389. 53 (21), 1837–1865. doi:10.1080/10643389.2023.2221157.

Wohl, E., 2021. An integrative conceptualization of floodplain storage. Rev. Geophys. 59 (2), e2020RG000724. doi:10.1029/2020RG000724.

Wu, S., Bashir, M.A., Raza, Q.U., Rehim, A., Geng, Y., Cao, L., 2023. Riparian buffer zones in agricultural non-point source pollution control—a review. Front. Sustain. Food Syst. 7, 985870.

Xu, F., Wang, D., 2023. Review on soil solidification and heavy metal stabilization by microbial-induced carbonate precipitation (MICP) technology. Geomicrobiol. J. 15210529. 40 (6), 503–518. Available from: http://www.tandf.co.uk/journals/titles/01490451.asp. doi:10.1080/01490451.2023.2208113.

Xu, S., Lin, J., Luo, H., Li, S., Qian, Y., Long, Y., Wu, Z., Zhu, G., 2025. Technologies for the remediation of nitrogen pollution and advances in the application of metal–phenolic networks. Processes. 22279717. 13 (7). Available from: http://www.mdpi.com/journal/processes. doi:10.3390/pr13072131.

Yan, J., Fang, L., 2023. Effects of sediment dredging on freshwater system: a comprehensive review. Environ. Sci. Pollut. Res. 1614-7499. 30 (57), 119612–119626. doi:10.1007/s11356-023-30851-8.

Yu, X., Tang, Y., Pan, J., Shen, L., Begum, A., Gong, Z., Xue, J., 2020. Physico-chemical processes. Water Environ. Res. 1061-4303. 92 (10), 1751–1769. doi:10.1002/wer.1430.

Zaharia, C., Musteret, C.P., Afrasinei, M.A., 2024. The use of coagulation–flocculation for industrial colored wastewater treatment—(i) the application of hybrid materials. Appl. Sci. 20763417. 14 (5). Available from: https://www.mdpi.com/journal/applsci/. doi:10.3390/app14052184.

Zeb, B.S., Mahmood, Q., Irshad, M., Zafar, H., Wang, R., 2025. Synergistic phytoremediation in biofilm wetlands treating combined industrial wastewater. Int. J. Phytoremediation 27 (1), 128–134. doi:10.1080/15226514.2024.2403037.

Zhang, Y., Luo, P., Zhao, S., Kang, S., Wang, P., Zhou, M., Lyu, J., 2020. Control and remediation methods for eutrophic lakes in the past 30 years. Water Sci. Technol. 0273-1223. 81 (6), 1099–1113. doi:10.2166/wst.2020.218.

Chapter 21

Remediation of polluted rivers using novel techniques

Sonali Kumari, Ravinsh Kumar, Ashutosh Singh, and Amrita Srivastava
Department of Life Science, Central University of South Bihar, Gaya, Bihar, India

1 Introduction

Rivers are essential to preserving both ecological harmony and human civilization. Along with population increase, the wastewater effluents from industries, agricultural runoff, and urban effluents have significantly contributed to river pollution. Due to the world's population growth and economic expansion, there are an increasing need for domestic, industrial, and agricultural water and a growing scarcity of freshwater. Freshwater resources are limited, and at the same time, the water quality is deteriorating, which leads to a shortage of water resources. Indeed, water shortages caused by poor quality are becoming a major global concern, particularly in emerging nations. Thus protecting river water and creating innovative, affordable methods to control river water pollution have become top global priorities (Azevedo et al., 2000).

The main sources of pollution in rivers are organic pollutants and heavy metals (Goldar and Banerjee, 2004). Surface water pollution is caused by anthropogenic activities such as urbanization, industrial development, and agricultural practices. The discharge of wastewater, which can carry pathogens, heavy metals, pharmaceutical waste, and other hazardous pollutants, has become increasingly prevalent in surface water sources. Farming methods, industrial development, urbanization, and other anthropogenic activities contribute to the pollution of surface water. Surface water sources are increasingly being used for the discharge of wastewater, which can contain heavy metals, pharmaceutical waste, pathogens, and other dangerous pollutants (Gayathri et al., 2021). The main causes of heavy metal contamination in river waters are various industries, including breweries, textile manufacturing, steel plants, mining industries, paint manufacturing, rubber industries, leather tanning, pulp and paper processing, dye production, synthetic fiber production, and metal industries. This contamination is also caused by the uncontrolled use of pesticides and fertilizers containing heavy metals in agricultural areas.

River Pollution. https://doi.org/10.1016/B978-0-443-45024-2.00015-4

In India, about 50% of total human waste is discharged into rivers and other water bodies without proper treatment (Jamwal, 2017). In many developing countries, including India, the remediation of polluted rivers is a burning issue that can be addressed by either in situ water treatment or pollution control at the source plant. Different methods can be applied for the remediation of polluted river water, including physical, chemical, biological, ecological, and hybrid techniques. Conventional methods of remediation include a combination of physical, chemical, and biological methods such as sedimentation, filtration, aeration, coagulation–flocculation, chemical oxidation/reduction, and activated sludge processes (Table 21.1). Although these conventional techniques have been effective in pollution control, they have some drawbacks, including high energy requirements, the generation of toxic substances, high operational costs, and incomplete removal of micropollutants. To address these challenges, novel remediation techniques for river pollution have emerged that combine advanced materials with innovative physical, chemical, and biological techniques and hybrid designs. These include bioremediation, phytoremediation, nanotechnology-based remediation, hybrid nanobioremediation, electrochemical oxidation processes, and membrane and absorptive systems (Table 21.2).

Novel approaches such as bioremediation, phytoremediation enhanced by nanomaterials, and nanobioremediation are more ecofriendly, ecologically feasible, and efficient in the complete degradation and transformation of pollutants like heavy metals, as well as organic and inorganic pollutants. Biological processes are less costly, more environmentally friendly, and self-sustaining compared to physical and chemical ones. However, these methods require a long period of time for microbial growth, ranging from months to years, and occasionally their efficiency is impacted by several environmental conditions, including temperature and rainfall (Bai et al., 2020).

These novel approaches have shown promising results in both laboratory and pilot scales. They aim not only at mere pollutant removal but also emphasize sustainable solutions to mitigate the global freshwater crisis. However, these techniques have their own advantages and limitations, and large-scale remediation processes of river water remain a complex and challenging endeavor. Therefore further research is required to improve these techniques and determine which individual or combined approaches are most effective, sustainable, eco-friendly, and economically viable.

2 Conventional vs novel approach

By assessing three parameters—physical, chemical, and biological—the status of water quality is reported. Through this approach, the health and sustainability of water bodies are evaluated. Physical parameters such as color, turbidity (cloudiness), odor, taste, and temperature are monitored to provide insights into the overall condition of water bodies. Any deviations from normal values indicate pollution sources, climate change impacts, or habitat alterations.

TABLE 21.1 Conventional methods of water remediation.

Method	Technique/Process	Description	Drawbacks	References
Physical methods	**Sedimentation**	Suspended solids settle by gravity in tanks or ponds; heavier particles sink, leaving clarified water above.	Ineffective for dissolved pollutants; requires large space and time; resuspension may occur in high flow.	(Tripathi and Verma, 2022)
	Filtration (sand/activated carbon/membrane)	Water passes through filter media (sand, gravel, carbon, or membranes) that trap particles and adsorb organic compounds.	Filters clog easily; high maintenance cost; membranes suffer from fouling and limited removal of micropollutants.	(Najid et al., 2022; Aziz et al., 2024)
	Aeration/Air Stripping	Enhances dissolved oxygen (DO) levels by diffusing air through water; promotes oxidation of organic matter.	Energy-intensive; limited for deep or stagnant waters; may cause CO_2 loss or volatilization of pollutants.	(Md Anawar and Chowdhury, 2020)
Chemical methods	**Coagulation–flocculation**	Addition of coagulants (alum, Fe salts) to destabilize colloids, forming flocs that settle easily.	Generates large sludge volumes; residual metals may remain; costly disposal.	(Ukiwe et al., 2014; Iwuozor, 2019)
	Ion exchange	Uses resins or zeolites to exchange ions in solution (e.g., removing Pb^{2+}, Ca^{2+}, NO_3^-).	Resin fouling; chemical regeneration produces saline waste; less effective in mixed-ion matrices.	(Ezugbe and Rathilal, 2020)
	Chemical precipitation	Addition of lime, alum, or ferric salts forms insoluble precipitates that settle as sludge.	Sludge handling issues; chemical cost; may not remove trace pollutants.	(Chen et al., 2009)
	Chemical oxidation/reduction	Uses oxidants (Cl_2, O_3, $KMnO_4$) or reductants ($FeSO_4$, $NaHSO_3$) to transform pollutants into less harmful forms.	May form toxic by-products, such as chlorinated organics; needs precise control.	(Ullah et al., 2020)

Continued

TABLE 21.1 Conventional methods of water remediation—cont'd

Method	Technique/Process	Description	Drawbacks	References
	Disinfection (chlorination/ozonation)	Chemical oxidation destroys pathogens; chlorine and ozone are the most common.	Produces carcinogenic disinfection by-products (THMs, HAAs); ozone costly to generate.	(World Health Organization, 2004; (Metcalf and Eddy, 2014)
Biological methods	**Activated sludge process**	Microorganisms suspended in aerated tanks degrade organic matter; biomass is separated and recycled.	High energy use (especially for aeration), large sludge production, sensitivity to fluctuations in influent, and the requirement for skilled operation.	(Sravan et al., 2024)
	Trickling filters/ biofilters	Wastewater trickles over a bed of media colonized by biofilm microbes, which degrade pollutants.	Inefficient for nitrogen/phosphorus removal; clogging and maintenance issues; limited capacity under high loads.	(Sravan et al., 2024)
	Stabilization ponds/ lagoons	Shallow ponds allow natural microbial, algal, and sunlight-driven degradation of pollutants over time.	Requires a very large land area, slow treatment rates, and limited performance in cold climates or under heavy pollutant loads.	(Sravan et al., 2024)

TABLE 21.2 Novel methods for water remediation.

Method	Technique	Description	Major advantages	Drawbacks	References
Biological methods	**Bioremediation and microbial technologies**	Microorganisms, such as bacteria, fungi, and algae, degrade or transform pollutants through enzymatic and metabolic pathways. Engineered microbial consortia and biofilms enhance pollutant breakdown and process stability.	Environmentally safe, low sludge generation, minimal energy requirement, and effective for diverse pollutants.	Sensitive to pH, temperature, and toxic shocks; slower degradation occurs under harsh conditions.	(Kulshreshtha et al., 2014; Sreedevi et al., 2022)
	Genetically modified and extremophilic microbes	Engineered and extremophilic microbes produce pollutant-degrading enzymes and metal-chelating proteins; function effectively under stress conditions.	High specificity; efficient degradation, even in extreme environments.	Biosafety concerns; limited large-scale application; costly containment.	(Bustamante-Torres et al., 2024; Jeong and Choi, 2020)
Plant-based systems	**Phytoremediation in constructed wetlands**	Plants and their associated microbes absorb, stabilize, or degrade metals, nutrients, and organic pollutants. Constructed wetlands mimic natural filtration by using plants such as *Typha latifolia* and *Eichhornia crassipes*.	Low-cost, self-sustaining, promotes biodiversity, and minimal chemical use.	Slow rate; large land area required; performance affected by season and climate.	(Md Anawar and Chowdhury, 2020; Zinicovscaia et al., 2025)
Adsorptive processes	**Adsorption and biosorption**	Use of adsorbents such as activated carbon, zeolites, biochar, or biological materials (e.g., algae, fungi, bacteria) to capture pollutants on surface sites.	Simple operation, reusable, cost-effective, and highly efficient.	Scalability of synthesis procedures, costly processes, and maintenance of adsorbent efficiency.	(Akhtar et al. 2024)

Continued

TABLE 21.2 Novel methods for water remediation—cont'd

Method	Technique	Description	Major advantages	Drawbacks	References
Membrane technologies	**Ultrafiltration (UF), nanofiltration (NF), reverse osmosis (RO)**	Semi-permeable membranes remove fine particles, salts, and microorganisms; new designs use nanomaterials or biopolymers for better fouling resistance.	Produces high-quality water; compact; suitable for decentralized treatment.	Prone to fouling, high maintenance costs, and energy required for pressure operation.	(Ezugbe and Rathilal, 2020; Aziz et al., 2024)
Chemical–physical hybrids	**Advanced oxidation processes (AOPs)**	Generates hydroxyl radicals (OH) through photocatalysis, ozonation, or Fenton's reaction to oxidize persistent pollutants.	Destroys recalcitrant organics; effective for pharmaceuticals, dyes, and pesticides; can be integrated with biological systems.	Not cost-effective, high energy and reagent demand; possible toxic by-products if uncontrolled.	(Deng and Zhao, 2015; Pandis et al., 2022)
Nanotechnology applications	**Nanoparticles and nanocomposites**	Metal or metal oxide nanoparticles (e.g., TiO_2, Fe_3O_4, ZnO) remove contaminants through adsorption, reduction, or photocatalysis; often combined with biological or polymer supports.	High surface reactivity; multipollutant removal; customizable for specific contaminants.	Nanoparticle aggregation; recovery difficulty; potential ecotoxicity.	(Zamora-Ledezma et al., 2021)

Similarly, alterations in chemical parameters such as pH, total dissolved solids, electrocoagulation (EC), dissolved oxygen (DO), hardness, and alkalinity are evaluated to check water pollution. Monitoring indicator species and biodiversity falls under biological parameters to assess the condition of water. Remediation technologies involve physical, chemical, and biological processes to remove contaminants from effluents. The selection of the treatment depends on the characteristics of the water to be processed. It focuses on cost-effectiveness, feasibility, efficiency, practicability, reliability, environmental impact, sludge production, operational difficulty, pretreatment requirements, and toxic by-products produced before opting for a particular treatment plan. Different methods required to remediate water pollution are categorized as physical, chemical, biological, ecological, and some engineering technologies. Conventional methods for removing pollution from river water rely on a combination of physical, chemical, and biological techniques that have been refined through decades of environmental engineering. These methods aim to restore water quality by removing suspended particles, organic matter, heavy metals, and pathogens, thereby protecting both human and ecological health (Tripathi and Verma, 2022).

Conventional methods are associated with several drawbacks, such as high energy consumption, the generation of toxic substances, high operational costs, and the requirement of complex infrastructure; there is a need to move toward a more effective approach.

Rapid population growth, growing industrialization, and increasing anthropogenic activities have led to a greater load of pollutants on water sources and have put immense stress on freshwater resources, leading to a global water crisis. A paradigm shift in water treatment strategy is necessary in the present scenario, as the United Nations Sustainable Development Goal 6 aims to ensure "availability and sustainable management of water and sanitation for all" by 2030.

The novel methods for remediation must aim to reduce dependence on harsh and toxic chemicals, cut down energy consumption, and minimize toxic sludge production. The focus should also be on low operational costs so that it can be accessible to both urban and rural areas. The use of nanotechnology-based filtration techniques, advanced oxidation processes, bioremediation processes, and low-cost natural adsorbents could play a promising role in the future (Fig. 21.1).

3 Bioremediation and microbial innovations

It is an eco-friendly and state-of-the-art technique that utilizes microbes or biological processes to completely remove toxic contaminants. It employs living organisms to degrade or detoxify hazardous substances to normal or below toxic concentration limits. On the basis of treating contaminated areas, it is generally classified as in situ or ex situ. When bioremediation involves treating the

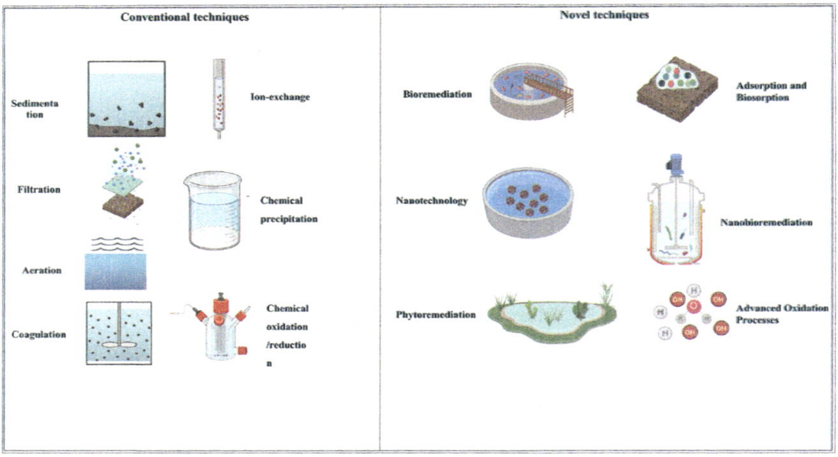

FIG. 21.1 Conventional and novel techniques. The figure shows a schematic representation of various conventional and novel remediation techniques that are used for addressing river pollution.

contaminated area at the site, it is referred to as in situ, while treating the contaminated site elsewhere is referred to as ex situ (Kensa, 2011). Bioremediation includes several techniques such as degradation, extraction, sequestration, and volatilization to address the technical components of the process. Through these strategies, microorganisms or plants are able to immobilize or modify the intricate components of contaminants, hence facilitating the cleanup of both land and water environments (Kaur and Sood, 2025). Optimization of bioremediation depends on several factors, which include the presence of a microbial population proficient in degrading the contaminants, exposure of pollutants to the microbial population, and environmental factors such as pH, types of soil, and temperature (Sharma, 2012). However, while bioremediation is a sustainable approach toward water and soil pollution remediation, it does have some limitations. It is a time-consuming process compared to other physiochemical remediation techniques. The scale-up of the process, limitation of nutrients, low population or absence of microbes with degradative capabilities, and pollutant bioavailability are among the major pitfalls that may hinder the success of bioremediation (Azubuike et al., 2016; Haque and Gazi-Khan, 2025).

3.1 Engineered microbial consortia

In our environment, the presence of complex pollutants made up of many different compounds, such as crude oil or petroleum, is not easily degraded by single species; rather, it requires the cooperation of more than one species. Assemblages of microorganisms cause the availability of diverse enzymes; they share nutrients and divide metabolic tasks (Ghazali et al., 2004). Microbial consortia have more advantages compared to isolated bacteria, as the former are found to be more compliant and stable within the growth environment. When a variety of cells with different functions are assembled, a dynamic balance is established and maintained through complex interactions between cells, leading to a more adaptable and stable system when facing environmental fluctuations (Cao et al., 2022). In nature, microbial communities generally compete for the available nutrients, but if microbial consortia are manipulated intentionally by designing and assembling a group of microorganisms, it leads to enhanced cooperation and stability. In such engineered microbial consortia, substances produced by one species are used as food or signaling molecules by another species. An engineered microbial consortium is different from a naturally available one because it is intentionally created in the laboratory by coculturing selected microorganisms according to specific needs (Li et al., 2021).

The main role of a synthetic microbial consortium in bioremediation is to break down and remove harmful pollutants from the environment. For this reason, microorganisms that can directly degrade toxic compounds are called degraders, which form the core component of such consortia. However, degraders alone are often not sufficient for efficient pollutant removal. Other types of microorganisms with supportive roles are also essential to improve the overall performance of bioremediation. These additional species perform various complementary functions that help create favorable conditions for the degraders to work effectively. For example, organic pollutants with low water solubility are difficult for microbes to access. In such cases, microbes that produce surfactants (biosurfactants) are valuable. Surfactants help increase the solubility and availability of pollutants, making them easier for degraders to break down. Similarly, in environments that are nutrient-poor, such as polluted soils with limited carbon or nitrogen, introducing microbes that can fix nitrogen or produce carbon compounds can provide essential nutrients. This supports the growth and activity of degraders, ensuring continuous and efficient bioremediation.

A good example of this cooperative mechanism is found in a pyrene-degrading microbial consortium, where a *Bacillus* strain—although unable to degrade pyrene itself—enhances the bioavailability of pyrene by producing biosurfactants (Wanapaisan et al., 2018). Similarly, a consortium containing Pseudomonas and Actinobacteria strains capable of degrading polycyclic aromatic hydrocarbons (PAHs) also shows emulsifying activity in the presence of PAHs. This emulsification process helps dissolve and disperse the pollutants, significantly improving the rate and extent of their biodegradation (Isaac et al., 2015).

3.2 Genetically modified organisms

In this ecosystem, a wide variety of microbial flora exists that has the ability to degrade complex pollutants. This is due to the fact that some toxic or complex pollutants contain elements that are not degraded by microbes due to a lack of catabolic pathways. This inadequacy is addressed using genetic engineering approaches. Through this approach, the bioremediation potential of microbes has been improved. The development of genetically modified microbes generally comprises four major approaches to improve microbial performance and monitoring during bioremediation (Kumar et al., 2017).

3.2.1 *Alteration of enzyme specificity and affinity*

It mainly focuses on modifying or engineering enzymes responsible for breaking down pollutants, so they can act on a broader range of substrates or work more efficiently on specific contaminants.

3.2.2 *Gene construction and regulation pathways*

In this approach, the focus is mainly on manipulating microbial genes and metabolic pathways to improve pollutant degradation capacity. This may include inserting new catabolic genes, overexpressing existing ones, or optimizing regulatory mechanisms that control enzyme production. By redesigning genetic circuits, scientists can ensure that microbes activate degradation pathways only in the presence of pollutants, thus saving energy and increasing stability. In addition, combining genes from different organisms can create novel biodegradation pathways capable of processing complex mixtures of pollutants.

3.2.3 *Process development, monitoring, and controlling of bioremediation*

This approach involves developing and refining the process engineering aspects of bioremediation—such as optimizing pH, temperature, oxygen, and nutrient levels—to support microbial activity.

3.2.4 *Application of sensor-based bioaffinity bioreporters*

This modern approach involves the use of biosensors, such as incorporating reporter genes (like lux, gfp, or lacZ) through genetic engineering, so that they respond to specific pollutants or toxic conditions by producing a measurable signal, such as fluorescence or luminescence.

Gong et al. (2018) used genetically altered *Pseudomonas putida* KT2440 for organophosphate and pyrethroid bioremediation. Enzymes secreted by white rot fungi are used to degrade 2,4,6-trinitrotoluene, polychlorinated biphenyls (PCBs), and PAHs (Sharma et al., 2024).

3.3 Extremophilic microorganisms

Microorganisms that are able to thrive in extreme conditions are known as extremophiles. The use of extremophiles in bioremediation is the most suitable approach, as they not only detoxify harmful and toxic pollutants but are also able to thrive in such extreme conditions (Singh et al., 2008). Since polluted areas contain toxic organic substances, heavy metals, and radioactive materials, this also creates extreme conditions. Conventional physicochemical methods—such as chemical oxidation, precipitation, or adsorption—often fail to completely eliminate contaminants. This failure usually occurs because pollutants may be deeply embedded in soils or sediments, making them inaccessible to chemical treatments. Moreover, these methods can sometimes cause secondary contamination, where the treatment itself introduces new toxic by-products or alters the environment in undesirable ways.

Extremophilic microorganisms are capable of thriving under extreme environmental stress and have evolved unique biological and biochemical mechanisms that enable them to survive and function where most organisms cannot. Along with this, they have developed the ability not only to tolerate high levels of toxic substances but also to transform or detoxify pollutants into stable, less harmful, or even useful compounds for their own metabolism. These remarkable adaptations have attracted significant scientific interest as natural models for sustainable bioremediation (Jeong and Choi, 2020).

According to Khemili-Talbi et al. (2015), *Natrialba* sp. C21—an archaebacterium that was able to grow in 25% saline conditions containing aromatic hydrocarbons and was isolated from oil-contaminated saline water in Ain Salah (Algeria)—was able to degrade naphthalene (3% v/v) and pyrene (3% v/v) under high salinity conditions with high growth, enzymatic activity, and biosurfactant production. Similarly, *Pseudomonas* sp. ST41 a cold-adapted bacterium collected from Antarctica was able to degrade hydrocarbons (Stallwood et al., 2005).

3.4 Biofilms for pollutant degradation

A cluster of microbial cells forming a community that is irreversibly adhered to a living or inert surface is known as a biofilm. Microbial cells are encased in a self-produced matrix of extracellular polymeric substances that primarily consists of polysaccharides, proteins, lipids, and nucleic acids and constitutes up to 90% of the biofilm's dry mass (Saini et al., 2023). The application of microbial communities in the form of biofilms, or biofilm-mediated remediation, has emerged as a promising and cost-effective alternative to conventional physical and chemical cleanup methods for various contaminants (Mani, 2020). Microbes in biofilms are more resistant and efficient than free-floating (planktonic) microbes because they cooperate, share nutrients, and exchange genetic information. Due to their genetic diversity, metabolic flexibility, and adaptability, these microbial communities can survive in contaminated environments and actively participate in removing pollutants. They detoxify harmful metals and other pollutants through bioaccumulation, biotransformation, biomineralization, and biosorption (Mishra et al., 2022). Because of these capabilities, biofilm-based remediation is considered safer, more effective, and more organized than using individual (planktonic) microorganisms. It represents a modern, sustainable, and innovative biological approach for degrading or neutralizing environmental contaminants such as heavy metals, organic toxins, and other pollutants. Although siderophores are known for their iron chelation ability, they also play an important role in the detoxification of heavy metal-contaminated sites.

3.5 Application of microbial products in bioremediation

3.5.1 Siderophores

It is a low-molecular-weight nonribosomal peptide secreted by microbes primarily to chelate iron from the surroundings under iron-deficient conditions (Khan et al., 2018). Siderophores have drawn attention due to their potential role in various fields, such as their role in bioremediation (Saha et al., 2016). Mohanty et al., (2018) reported that siderophores isolated from both bacteria and fungi are able to reduce asbestos fiber toxicity.

3.5.2 Biosurfactants

These are amphiphilic molecules secreted extracellularly by microbes. They increase the accessibility of hydrophobic, toxic organic pollutants to degrading microorganisms. Rhamnolipids and surfactin are biosurfactants that reduce the surface and interfacial tension between immiscible liquids, thereby emulsifying water-insoluble contaminants like petroleum hydrocarbons, PAHs, and certain pesticides (Parthipan et al., 2022). They help emulsify oil layers and make pollutants more bioavailable for microbial degradation.

3.5.3 Microbial enzymes

Microbes such as bacteria and fungi secrete various enzymes that act on pollutants and convert them into innocuous products. The use of enzymes in bioremediation is a simple, rapid, eco-friendly, and socially acceptable method. Enzymes, such as oxidases, reductases, transferases, as well as hydrolytic and degrading enzymes, disrupt major chemical bonds of toxic compounds and result in a reduction of toxicity (Karigar and Rao, 2011; Yagnik et al., 2023). The advantage of using enzymes for the biotransformation of toxic compounds is that they do not create any hazardous by-products, which usually occur when using nonbiological systems (Shome, 2020).

4 Nanotechnology-based remediation

Nanotechnology-based remediation provides a powerful tool for the removal of pollutants in past decades. Its unique physicochemical properties enable efficient degradation, adsorption, and transformation of environmental pollutants compared to their bulkier counterparts. In addition, nanomaterials are effective in both soil and water because they possess a high surface-to-volume ratio, which increases their reactivity and adsorption capability. Interestingly, nanomaterials' surface chemistry can be easily modified to target specific hazardous metals for remediation purposes. Nanomaterials have exceptional redox potential properties that help in binding, catalyzing, and transforming pollutants even at low concentrations (Guerra et al., 2018; Thangavelu, 2022).

4.1 Principles and mechanisms

The remediation potential of nanomaterials depends on their adsorption and catalytic properties. Nanoparticles interact with contaminants through electrostatic attraction and the generation of reactive oxygen species (ROS) in the presence of photocatalytic conditions (Takhar and Singh, 2025). For example, titanium dioxide (TiO_2) nanoparticles (NPs) act as photocatalysts under UV or visible light, generating hydroxyl radicals that oxidize organic metal pollutants and microbes. In a similar case, zero-valent iron (nZVI) NPs and bimetallic composites like Fe/Pd and Fe/Ni mediate electron transfer reactions that reduce chlorinated hydrocarbons, dyes, and heavy metals such as Cr(VI), Pb(II), and Cd(II) to less toxic or insoluble forms (Gao et al., 2016).

Nanoparticles can penetrate microbial membranes or root surfaces, which accelerates biotransformation reactions and promotes synergistic remediation.

4.2 Types of nanomaterials used

4.2.1 Metal and metal oxide NPs

In recent years, metal-based NPs such as Fe, Cu, Ag, ZnO, and TiO_2 have been extensively used for the degradation of organic and inorganic contaminants. Specifically, iron and iron oxide NPs have shown high efficiency in immobilizing heavy metals (Ni^{2+}, Cu^{2+}, Co^{2+}, Cr^{6+}). Their ability to reduce chlorinated solvents through electron transfer mechanisms is extensively reported. The Fe^0 (zero-valent nanoiron) core acts as a reducing agent and is stable in water, while its oxide shell has excellent adsorption capacity, which targets cationic metals (Guerra et al., 2018). Due to this property, researchers utilize Fe^0 in water treatment. Another metal NP, silver nanoparticles (AgNPs) are effective antimicrobial agents and are used in water disinfection. They exhibit antimicrobial properties against *E. coli* and *Staphylococcus aureus*. Titanium dioxide NPs, doped with transition metals like Ag or Fe, enhance visible-light absorption and improve the photocatalytic degradation activity of dyes and phenolic pollutants (Akhter et al., 2022).

4.2.2 Carbon-based nanomaterials and nanocomposites

Carbon nanotubes (CNTs), graphene oxide, and activated carbon nanocomposites have large surface areas and abundant functional groups that enhance their adsorption capacity for heavy metals and organic micropollutants. CNTs have excellent chemical stability and biocompatibility, making them suitable for use in different conjugates for bioremediation purposes. The incorporation of magnetic iron oxide into carbon frameworks facilitates easy magnetic recovery. Graphene–Fe_3O_4 nanocomposites, for example, have been successfully used for the removal of phenazopyridine, Cr (VI), and Pb (II) from wastewater, combining high adsorption capacity with excellent reusability (Thangavelu and Veeraragavan, 2022).

4.2.3 Green-synthesized NPs

Recent advances emphasize ecofriendly synthesis methods using plant extracts like polysaccharides or microbial metabolites as reducing agents. Such "green" NPs reduce secondary pollution risks to the environment and also enhance biocompatibility. For example, Fe and Cu NPs synthesized using *Azadirachta indica* and *Ocimum sanctum* extracts demonstrated high Cr (VI) reduction efficiency without generating toxic by-products (Fatima et al., 2025). El-Sheekh et al. (2021) reported that green-synthesized iron oxide nanoparticles (Fe_3O_4 NPs) produced from brown algae (*Petalonia fascia, Colpomenia sinuosa*, and *Padina pavonica*) are effective in removing nitrogen, phosphorus, and harmful algae from polluted seawater. This eco-friendly, low-cost nanobioremediation approach offers a sustainable solution for restoring marine water quality.

4.2.4 Nanobubbles and hybrid systems

Nanobubbles, sometimes also known as micronanobubble gas with diameters <200 nm, are increasingly used in hybrid remediation systems. When contaminants form a coating around nanobubbles, they create a diffusion barrier, and upon bursting, these nanobubbles generate hydroxyl radicals (·OH) that play a crucial role in the oxidative degradation of organic pollutants in wastewater (Wang et al., 2024). When nanobubbles coupled with nanozeolite matrices, nanobubbles elevate DO and promote microbial degradation. Patel et al. (2025) reported *Typha* and *Canna indica* systems showed >70% total phosphorus reduction, a 179% increase in DO, and enhanced contaminant breakdown and plant health.

4.3 Case studies of nano-based remediation

Field and laboratory reports have confirmed the efficiency of nanotechnology-based remediation approaches under a wide range of environmental conditions.

4.3.1 nZVI for groundwater dechlorination

Emulsified formulations of nZVI have demonstrated >90% efficiency in removing trichloroethylene and other chlorinated volatile organic compounds from contaminated aquifers, highlighting their potential for in situ groundwater remediation (Ebrahimbabaie & Pichtel, 2021; Guerra et al., 2018).

4.3.2 Graphene–Fe_3O_4 nanocomposites

Graphene–Fe_3O_4 nanocomposites efficiently adsorbed Pb^{2+} and Cd^{2+} ions, with adsorption capacities above 250 mg g^{-1}, surpassing the performance of traditional activated carbon adsorbents (Chowdhury et al., 2022; Thangavelu and Veeraragavan, 2022).

4.3.3 TiO_2 and metal-doped nanocatalysts

Ag-doped TiO_2 nanofibers exhibited enhanced degradation of 2-chlorophenol under UV irradiation and revealed improved reaction kinetics compared to pure TiO_2. This improvement is attributed to the synergistic effects of photoactivation and the formation of efficient metal–semiconductor interfaces (Onkani, 2019; Guerra et al., 2018).

4.3.4 Nanobiocomposites in wastewater treatment

Integration of NPs with Plant Growth Promoting Rhizobacteria (PGPR) improved heavy metal uptake in *Brassica juncea* and *Phragmites australis*, suggesting synergistic detoxification between nanomaterials and biological systems (Gomes, 2025; Saha et al., 2021).

4.4 Environmental implications and future prospects

Despite promising outcomes, large-scale deployment of nanotechnology in remediation faces challenges, including NP aggregation, ecotoxicity, and uncertain long-term stability. Research increasingly focuses on biodegradable, magnetic, and recoverable nanomaterials that minimize secondary pollution. Green synthesis and surface functionalization with natural ligands or biopolymers can mitigate toxicity and enhance reusability. Future advancements are expected in CRISPR-engineered nanobiosystems, smart responsive nanocomposites, and nanobubble–microbe coupling for dynamic pollutant removal (Wentzell, 2025; Dhanwal et al., 2017).

Overall, nanotechnology-based remediation represents a transformative step toward sustainable environmental management, merging physicochemical precision with biological adaptability to achieve efficient, eco-friendly, and scalable pollutant removal.

5 Phytoremediation and plant-based advances

Phytoremediation is a sustainable, cost-effective technology in which plants and their associated microbiomes are exploited to absorb, sequester, transform, or stabilize contaminants found in soil and water systems. In recent years, this method has been extensively studied for selecting suitable plants for phytoremediation purposes by characterizing their potential for heavy metal remediation. In comparison with conventional physicochemical treatments, it offers a sustainable alternative to reduce secondary pollution and also provides benefits for soil and biodiversity (Saha et al., 2021).

5.1 Hyperaccumulator plants

Various studies have reported that plants can be used as hyperaccumulators due to their specialized physiological functions. These plants tolerate and accumulate exceptionally high concentrations of heavy metals such as cadmium (Cd), lead (Pb), zinc (Zn), and nickel (Ni) without their physiology being affected by metal toxicity (Manara et al., 2020). Over 450 hyperaccumulator plants, such as *B. juncea*, *Amaranthus hypochondriacus*, *Thlaspi caerulescens*, *Helianthus annuus*, *Sesbania drummondii*, *Salix matsudana*, *Populus deltoides*, and *Pteris vittata*, are recognized for their efficiency in removing metals through phytoextraction and phytostabilization (Srivastava, 2020). These plants employ multiple detoxification strategies. They upregulate metal transporter proteins and synthesize chelating agents such as phytochelatins, metallothioneins, and organic acids. In addition, they compartmentalize metals into vacuoles to reduce cytoplasmic toxicity. The use of hyperaccumulator species has been highly successful in cleaning soils contaminated by industrial discharge, mining activities, and metallurgical waste.

In aquatic environments, macrophytes such as *Eichhornia crassipes*, *Lemna minor*, *Pistia stratiotes*, and *Typha latifolia* show strong potential for pollutant remediation. These plants remove about 97% of metals under optimal conditions. *E. crassipes* is particularly efficient in bioaccumulating Fe, Cu, Zn, Cr, Cd, and As. Such macrophytes are valuable tools for wastewater treatment and ecological restoration (Wentzell, 2025).

5.2 Genetically engineered plants

Recent advances in genetic engineering have significantly increased the scope of phytoremediation by inserting microbial and synthetic genes into plants to enhance metal tolerance, accumulation, and detoxification capacities. Through precise genetic modification, plants can be equipped with pathways that improve their ability for bioremediation of a wider range of pollutants.

Transgenic lines of *Arabidopsis thaliana* and *B. juncea* expressing bacterial *merA* and *merB* genes have demonstrated enhanced reduction and volatilization of mercury, effectively converting toxic Hg^{2+} to its less harmful elemental form. Similarly, overexpression of the *AtPCS1* gene, which encodes phytochelatin synthase, has been shown to increase cadmium accumulation and tolerance by improving metal chelation and vacuolar sequestration (Saha et al., 2021). In one study, transgenic *Medicago sativa* (alfalfa) plants coexpressing the human cytochrome P450 2E1 (CYP2E1) and glutathione S-transferase genes revealed excellent efficiency in removing mixed mercury–trichloroethylene (Hg–TCE) pollutants (Zhang et al., 2013). The coordinated expression of these genes produced a strong synergistic effect, enhanced pollutant accumulation, and increased their tolerance of heavy metal–organic complexes. In a similar study, Tammam et al., (2021) reported that *Glebionis coronaria*, a species of flowering plant, has the ability to accumulate lead (Pb) in contaminated soil.

Beyond transgenic methods of genome editing, CRISPR/Cas9 offers precise manipulation of metal transporter and antioxidant genes, allowing targeted enhancement of phytoremediation efficiency without introducing foreign DNA (Venegas-Rioseco et al., 2021). This recent advancement accelerates the development of plants capable of removing pollutants and specific heavy metals.

5.3 Rhizosphere engineering

Microorganisms associated with plant roots play a crucial role in degrading organic pollutants within the rhizosphere, which is the primary region of plant–microbe interaction. Engineering this microenvironment has become an effective strategy to enhance phytoremediation efficiency. Several studies have reported that microbial inoculation, biochar amendment, and organic acid supplementation can improve soil conditions. These approaches increase metal solubility, nutrient availability, and the degradation of organic pollutants, thereby enhancing the plant's metal accumulation ability (Dhanwal et al., 2017).

PGPR, such as *Bacillus cereus*, *Pseudomonas fluorescens*, and *Rhizobium* spp., play a critical role in rhizosphere engineering by producing siderophores, phytohormones, and organic acids that mobilize heavy metals and stimulate root growth (Hakim et al., 2021). These microbes also secrete enzymes capable of degrading hydrocarbons and other xenobiotic compounds, facilitating the simultaneous remediation of organic and inorganic pollutants.

Similarly, arbuscular mycorrhizal fungi (AMF), such as *Glomus versiforme* and *Claroideoglomus etunicatum,* form mutualistic associations with plant roots, increasing the absorptive surface area and enhancing nutrient and metal uptake. These fungi improve plant tolerance to metal-induced oxidative stress and contribute to soil aggregation and stability. When combined with PGPR, AMF establish a synergistic network that regulates rhizosphere chemistry, promotes metal immobilization, and enhances overall remediation efficiency.

Such integrated rhizosphere engineering approaches create a balanced and resilient micro-ecosystem that supports plant growth while optimizing pollutant transformation and stabilization. This strategy holds promise for the sustainable remediation of multicontaminant environments.

5.4 Combined plant–microbe approaches

Plant–microbe consortia frequently outperform plant-only systems by coupling plant uptake with microbial biosorption, redox transformation, and cometabolism. For instance, inoculation of *H. annuus* with *Pseudomonas libanensis* plus *Claroideoglomus claroideum* enhanced Ni and Na removal in saline soils;

partnerships of *Sinorhizobium meliloti* with *Medicago lupulina* improved Cu immobilization while fixing nitrogen (Saha et al., 2021). In saline/metal-rich settings, consortia with halophytes (e.g., *Salicornia ramosissima*) increased Zn, Cu, Pb, As, and Ni uptake and improved photosynthetic efficiency (Mesa-Marín et al., 2020). Such designs also reduce effluent biochemical oxygen demand (BOD) and chemical oxygen demand (COD) where textile dye loads coexist with metals, as shown with *E. crassipes* and *Pistia* systems (Wentzell, 2025).

5.5 Nanotechnology in phytoremediation

Nanointerventions can improve mass transfer, root respiration, and contaminant bioavailability. Nanobubbles have increased DO and boosted Cd removal by about 17% in aquatic phytoremediation with *Alternanthera philoxeroides*, while nanobubble–zeolite combinations with *Typha*/*Canna* reduced total phosphorus by >70% and raised DO by about 179% relative to controls (Guo et al., 2022). Engineered NPs (e.g., TiO_2, ZnO) may enhance Pb/As/Cu/Cd immobilization through photocatalytic ROS generation and vacuolar sequestration; carbon nanodots can strengthen antioxidant defenses and tolerance in macrophytes. Field translation, however, requires robust ecotoxicological and cost assessments.

5.6 Phytoremediation of emerging contaminants

Recent research emphasizes the capability of aquatic plants to address emerging contaminants, including pharmaceuticals, personal care products, and microplastics. *L. minor* efficiently removes triclosan and bisphenol A, while *E. crassipes* reduces organic micropollutants via enzymatic degradation pathways (Polińska et al., 2021). Microplastics can be physically trapped within the root networks of *L. minor* and *E. crassipes*, preventing their dispersion; bacterial biofilms of *Pseudomonas* and *Rhodococcus* in rhizospheres may further aid polymer degradation. These combined plant–microbe systems thus provide a feasible route to mitigating novel pollutants resistant to traditional treatment methods.

5.7 Microbial synergy and plant–microbe consortia

Plant–microbe systems show significantly higher efficiency than plant-only systems. Inoculation of *S. ramosissima* with *Vibrio neocaledonicus*, *Thalassospira australica*, and *Pseudoarthrobacter oxydans* increased Zn, Cu, Pb, As, and Ni uptake while boosting photosynthetic efficiency (Mesa-Marín et al., 2020). Similarly, *P. australis* inoculated with native bacteria doubled Cu accumulation compared with controls. A PGPR consortium with *Arundo donax* achieved about 75% arsenic volatilization with minimal tissue accumulation, aided by epigenetic stress adaptations. Microbial consortia also degrade textile effluents, reducing BOD/COD and promoting biomass production in *E. crassipes* and *P. stratiotes*. These systems illustrate how cometabolic activity and rhizosphere colonization optimize pollutant uptake and resilience under stress.

6 Electrochemical and oxidation techniques

6.1 Electrocoagulation (EC)

Electrocoagulation is a cost-effective, highly reliable, and broad technique for wastewater treatment. It is based on the principle of electrolysis. Emulsified oils, heavy metals, total suspended solids (TSS), and microbes can all be eliminated from water using the advanced EC process (Moussa et al., 2017). EC can now function on fuel cells, windmills, and solar cells because it can run on low currents (Shahedi et al., 2020). Since chemical reagents are not added and electrode materials are readily available in the ground, it is regarded as environmentally friendly.

The EC system consists of an electrolytic cell with a cathode and an anode that are externally connected to a power source and submerged in an electrolytic solution. As current flows through the electrolytic cell, the anode oxidizes, and the metal separates into di- or trivalent metallic ions, releasing an equivalent number of electrons. On the cathode, the water molecules are reduced, resulting in the formation of hydrogen gas and hydroxyl ions. The hydroxyl group combines with metallic ions to form metal hydroxide. The metallic hydroxide, M(OH)n, possesses a wide surface area and a strong ability to form bonds with contaminants, thereby forming flocs and making it an excellent adsorbent of pollutants. The dissociated H^+ ions combine with other H^+ ions present in water to produce H_2 gas (Kuokkanen et al., 2013). To remove flocs that do not sink due to gravity, H_2 and O_2 gases play a crucial role (Kobya et al., 2020). This process, known as electroflotation, occurs when H_2 gas evolves, lifting the flocs to the solution's surface, where they can be skimmed off (Nidheesh and Singh, 2017).

6.2 Electrochemical advanced oxidation process

Electrochemical advanced oxidation processes (EAOPs) are advanced and emerging technologies for contaminated water treatment that use anodic oxidation of water to produce extremely reactive hydroxyl radicals (·OH), which break down a variety of persistent organic contaminants, frequently at diffusion-limited rates. Common electrode materials with high oxygen evolution overpotentials that facilitate effective ·OH production include doped-SnO_2 (Comninellis and Pulgarin, 1993), PbO_2 (Chai et al., 2011), boron-doped diamond (Iniesta et al., 2001), and doped TiO_2 (Carter and Farrell, 2008). EAOPs aid in direct electron transfer (DET) from pollutants to the anode in addition to oxidation via ·OH. DET is a crucial process for breaking down substances that are resistant to radical attack, including fluorinated organics (Zhuo et al., 2011).

6.3 Electro-Fenton technique

The electro-Fenton processing technology is a component of the most well-known EAOPs, which represent the most recent generation of AOPs that have been extensively developed over the past 10 years. The EAOP process is a new technology that offers clean, efficient, and economical methods for removing pollutants from water (Oturan and Oturan, 2018). In general, there are two types of EF (Electro-Fenton) processes: the first type uses an external Fenton reagent and the second configuration uses H_2O_2 added from an external process, while Fe^{2+} is supplied as a result of anode reduction (Atmaca, 2009).

Over the past 10 years, EF has been applied to the removal of contaminants from water and wastewater. This EF process depends on the appropriate impacting elements, which have been observed and compared. These factors include electrode type, pH, current density, Fe^{2+} and H_2O_2 concentration, O_2 flow rate, electrode gap, temperature, and electrolyte. EF has processed pollutants such as color, leachate, medications, pesticides, fertilizers, phenol, and biological pollutants, and this process is advancing technologies for use in industrial wastewater (Marlina and Purwanto, 2019).

7 Membrane and adsorptive technologies

The membrane adsorption (MA) technique has been utilized increasingly in recent years to treat inorganic wastewater. Adsorption is a very economical and cost-effective method with excellent metal removal efficiency. It is used as a rapid technique for all kinds of wastewater treatment. It is a process of mass transfer in which a material moves from the liquid phase to the solid surface, where it is bound by chemical and/or physical interactions. Consequently, two forms of adsorption exist:

1. The adsorbate can form a monolayer in chemical adsorption, also known as active adsorption. In addition, it is employed in catalytic processes. The following are generally the primary processes in the adsorption of pollutants on solid adsorbents:
   a. Pollutant transport occurs from the bulk solution to the external surface of the adsorbent. Pore diffusion transports the internal mass from the outer surface of the adsorbent to the inner surface of the porous structure.
   b. Adsorbate adsorption occurs on the adsorbent's active sites within its pores. Since the final stage of adsorption is quicker than the first two, the overall rate of adsorption is determined by either film formation, intraparticle diffusion, or both.
2. Physical adsorption can occur in any solid–liquid or solid–gas system and is a generic phenomenon. In the process of physical adsorption, van der Waals forces on the adsorbent surface cause the adsorbate to bond.

Functional groups are attached to the pore walls and surfaces of polymer membranes to create membrane adsorbents; the target pollutants are adsorbed onto the functional groups in a selective manner. The filtering capabilities of the membrane are efficiently combined with the membrane adsorbents. Because the target pollutants and the adsorbed active binding sites of the membrane adsorbents have a very short contact distance at the submicron scale, the functional active binding sites will combine with the target pollutants when contaminated water passes through the membrane, thereby removing contaminants from drinking water with a high adsorption rate and capacity (Zheng et al., 2015).

Activated carbon, zeolites, and natural fibers are examples of common adsorbents that have weak adsorption capabilities, poor selective sorption, and inadequate regeneration capacity. To address these issues, a variety of sophisticated adsorbents have been created, including porous boron nitride nanosheets, porous graphene, nanostructured metal oxides, carbon, and nanotubes (Yang et al., 2015).

7.1 Polymeric membranes

Originally created in the 1960s, traditional polymeric adsorbents have evolved into effective materials for eliminating contaminants through MA (Kunin, 1977), wherein pollutants are absorbed by functional groups on the sorbent layer or membrane surface as wastewater passes through. Polymers and biopolymers with functional groups such as amine, carboxyl, and sulfonic acid exhibit strong affinity for heavy metal ions and can regenerate after saturation (Vieira and Beppu, 2005; Saber-Samandari et al., 2014). Due to their simplicity of production, biocompatibility, and redox activity, conducting polymer-based adsorbents—particularly those made from polypyrrole (PPy)—have demonstrated remarkable effectiveness in the removal of metals such as Cr(VI), Zn(II), and Pb(II). Numerous PPy composites exhibit excellent adsorption capacity and structural adaptability, including PPy/biosorbents, PPy/Fe_3O_4 nanocomposites, PPy–polyaniline nanofibers, PPy–graphene nanocomposites, organically modified clay–PPy systems, and hierarchical porous PPy nanoclusters (Mahmud et al., 2016). Polymeric membranes can adsorb hazardous pollutants, potentially enabling the development of portable and reusable water purifiers. Membranes for MA effectively remove trace levels of contaminants, including cationic heavy metals, nitrates, and anionic phosphates, by performing both filtration and adsorption.

8 Integrated and hybrid techniques

8.1 Combining nanotechnology with bioremediation

The development of novel, efficient, environmentally friendly, and economically viable methods for the remediation of environmental contaminants is still needed, despite the widespread use of many traditional physicochemical techniques, including adsorption and electrochemical treatment (Fomina and Gadd, 2014).

Bioremediation technology employs low-tech and low-cost methods to eliminate contaminants by utilizing microorganisms. It depends on the presence of specific microbes and a confluence of suitable ecological conditions (Adams et al., 2015).

Technologies that are able to monitor, identify, and treat the relatively small amounts of contaminants in air, water, and soil are needed since contaminants are typically found in combinations. Technology capable of detecting, minimizing, preventing, and treating environmental pollution is required in this context. Nanotechnology offers a long-term, atomic- and molecular-level solution to these problems. It is possible to create materials with specific features that allow them to identify a particular contaminant within a combination. Because of their high surface-to-volume ratio and compact size, NPs can be detected with extreme sensitivity (Asztemborska et al., 2015; Das et al., 2015).

Nanomaterials used for water quality monitoring, notably those used for the detection of trace pollutants and pathogens, include CNTs, magnetic NPs, noble metal nanomaterials, and quantum dots (Kumari and Singh, 2016; Xue et al., 2016, 2017). Moreover, the use of elemental or zero-valent metals in nanoscale form, such as iron, nickel, and palladium, has shown encouraging results toward contaminated sites with various types of toxic substances (Li et al., 2016).

Microbes can effectively be used as convenient nanofactories for synthesizing NPs through "green" methods (Mishra et al., 2015). Phytonanotechnology, an environmentally benign, straightforward, quick, stable, and economical process, has recently opened up new possibilities for the synthesis of NPs. Consequently, the high demand for NPs with applications in the healthcare and environmental fields can be met by plant-derived NPs made from readily available plant components and the nontoxic nature of plants. Recently, extracts from the leaves and roots of the medicinal herbal plant Panax ginseng were successfully used to incorporate gold and silver NPs (Hussain et al., 2016). Furthermore, metal NPs have been synthesized using a variety of plants.

Nanobioremediation is the process of removing pollutants by increasing microbial activity (e.g., organic and inorganic toxins, heavy metals) utilizing NPs /nanomaterials molded by bacteria, fungi, and plants with the aid of nanotechnology (Singh and Walker, 2006).

The productivity of phytoremediation is increased by nanotechnology, whereby NPs can be used to clean up water contaminated with heavy metals, as well as both natural and inorganic pollutants. In compound-based bioremediation, NPs can be used in combination with phytoremediation (Pillai and Kottekottil, 2016).

9 Future perspective and challenges

In the future, river pollution remediation using novel technologies will require further improvement through the integration of multiple approaches that combine biological sustainability with physical and chemical techniques. This strategy aims to determine the best methods. Future remediation strategies are expected to focus on self-sustaining systems that use renewable energy and eco-compatible materials, enhancing pollutant degradation and reducing the risk of secondary contamination.

For example, scaling advanced phytoremediation is constrained by remediation time, contaminant specificity, biomass handling, and regulatory oversight. Risks include nanomaterial persistence and transgene flow from engineered plants (Gressel and Al-Ahmad, 2005); compliance with genetically modified organisms/nanomaterial frameworks (e.g., Cartagena Protocol; REACH) is essential (Wentzell, 2025). Pragmatic near-term routes include pairing robust hyperaccumulators with site-adapted microbial consortia and low-cost amendments; medium-term opportunities lie in CRISPR-enabled trait stacking and omics-guided strain/host selection to address mixed, variable loads (Dhanwal et al., 2017; Saha et al., 2021; Wentzell, 2025).

Despite remarkable progress and the success of novel remediation techniques, several challenges remain in achieving large-scale and long-term success. A primary concern is the scalability and economic feasibility of advanced methods. Due to high production costs and high energy demands, many promising methods are still confined to laboratory scales.

10 Conclusion

The degradation of rivers due to anthropogenic activities has become a global concern. Various conventional and novel remediation techniques discussed in this chapter can be applied to control river pollution. Every technology has its advantages and limitations, so the development of an integrated system of novel remediation technologies such as nanotechnology, bioremediation, and phytoremediation has revolutionized the strategies available for river pollution management. The combination of biofilms, genetically engineered microorganisms, and plant-based systems with nanomaterials has shown significant improvements in the detoxification and restoration of polluted ecosystems. In addition, the use of membrane-based and nano-assisted bioprocesses demonstrates a growing trend toward renewable and environmentally friendly techniques.

References

Adams, G.O., Fufeyin, P.T., Okoro, S.E., Ehinomen, I., 2015. Bioremediation, biostimulation and bioaugmention: a review. Int. J. Environ. Bioremediation Biodegrad. 3 (1), 28–39.

Akhtar, M.S., Ali, S., Zaman, W., 2024. Innovative adsorbents for pollutant removal: exploring the latest research and applications. Molecules. 29 (18), 4317.

Akhter, P., Arshad, A., Saleem, A., Hussain, M., 2022. Recent development in non-metal-doped titanium dioxide photocatalysts for different dyes degradation and the study of their strategic factors: a review. Catalysts. 2073-4344. 12 (11), 1331. doi:10.3390/catal12111331.

Asztemborska, M., Steborowski, R., Kowalska, J., Bystrzejewska-Piotrowska, G., 2015. Accumulation of platinum nanoparticles by Sinapis alba and Lepidium sativum plants. Water Air Soil Pollut. 15732932. 226 (4). Available from: http://www.kluweronline.com/issn/0049-6979/. doi:10.1007/s11270-015-2381-y.

Atmaca, E., 2009. Treatment of landfill leachate by using electro-Fenton method. J. Hazard. Mater. 03043894. 163 (1), 109–114. doi:10.1016/j.jhazmat.2008.06.067.

Azevedo, L.G.T.D., Gates, T.K., Fontane, D.G., Labadie, J.W., Porto, R.L., 2000. Integration of water quantity and quality in strategic river basin planning. J. Water Resour. Plan. Manag. 126 (2), 85–97.

Aziz, S., Mazhar, A.R., Ubaid, A., Shah, S.M.H., Riaz, Y., Talha, T., Jung, D.W., 2024. A comprehensive review of membrane-based water filtration techniques. Appl. Water Sci. 21905495. 14 (8). Available from: https://www.springer.com/journal/13201. doi:10.1007/s13201-024-02226-y.

Azubuike, C.C., Chikere, C.B., Okpokwasili, G.C., 2016. Bioremediation techniques–classification based on site of application: principles, advantages, limitations and prospects. World J. Microbiol. Biotechnol. 32 (11), 180.

Bai, X., Zhu, X., Jiang, H., Wang, Z., He, C., Sheng, L., Zhuang, J., 2020. Purification effect of sequential constructed wetland for the polluted water in urban river. Water. 2073-4441. 12 (4), 1054. doi:10.3390/w12041054.

Bustamante-Torres, M., Torres, O., Abad-Sojos, S., Pardo, S., Bucio, E., 2024. Application of genetically modified microorganisms for bioremediation of polluted environments. In: Inamudin, Adetunji, C.,O., Ahmed, I., Altahi, T., (Eds.), Genetically Engineered Organisms in Bioremediation. CRC Press, Ecuador, pp. 18–51. Available from: https://www.routledge.com/Genetically-Engineered-Organisms-in-Bioremediation/Inamuddin-Adetunji-Ahamed-Altalhi/p/book/9781032036960. doi:10.1201/9781003188568-2.

Cao, Z., Yan, W., Ding, M., Yuan, Y., 2022. Construction of microbial consortia for microbial degradation of complex compounds. Front. Bioeng. Biotechnol. 22964185. 10. Available from: http://journal.frontiersin.org/journal/bioengineering-and-biotechnology#archive. doi:10.3389/fbioe.2022.1051233.

Carter, K.E., Farrell, J., 2008. Oxidative destruction of perfluorooctane sulfonate using boron-doped diamond film electrodes. Environ. Sci. Technol. 0013936X. 42 (16), 6111–6115. doi:10.1021/es703273s.

Chai, S., Zhao, G., Li, P., Lei, Y., Zhang, Y.N., Li, D., 2011. Novel sieve-like SnO_2/TiO_2 nanotubes with integrated photoelectrocatalysis: fabrication and application for efficient toxicity elimination of nitrophenol wastewater. J. Phys. Chem. C. 19327455. 115 (37), 18261–18269. doi:10.1021/jp205228h.

Chen, Q., Luo, Z., Hills, C., Xue, G., Tyrer, M., 2009. Precipitation of heavy metals from wastewater using simulated flue gas: sequent additions of fly ash, lime and carbon dioxide. Water Res. 00431354. 43 (10), 2605–2614. Available from: https://www.sciencedirect.com/science/journal/00431354. doi:10.1016/j.watres.2009.03.007.

Chowdhury, I.R., Chowdhury, S., Mazumder, M.A.J., Al-Ahmed, A., 2022. Removal of lead ions (Pb^{2+}) from water and wastewater: a review on the low-cost adsorbents. Appl. Water Sci. 21905495. 12 (8). Available from: http://www.springer.com/earth+sciences+and+geography/hydrogeology/journal/13201. doi:10.1007/s13201-022-01703-6.

Comninellis, C., Pulgarin, C., 1993. Electrochemical oxidation of phenol for wastewater treatment using SnO_2, anodes. J. Appl. Electrochem. 15728838. 23 (2), 108–112. doi:10.1007/BF00246946.

Das, S., Sen, B., Debnath, N., 2015. Recent trends in nanomaterials applications in environmental monitoring and remediation. Environ. Sci. Pollut. Res. 22 (23), 18333–18344.

Deng, Y., Zhao, R., 2015. Advanced oxidation processes (AOPs) in wastewater treatment. Curr. Pollut. Rep. 21986592. 1 (3), 167–176. Available from: http://springer.com/environment/pollution+and+remediation/journal/40726. doi:10.1007/s40726-015-0015-z.

Dhanwal, P., Kumar, A., Dudeja, S., Chhokar, V., Beniwal, V., 2017. Recent advances in phytoremediation technology. In: Kumar, R., Kumar, A., Dudeja, S. (Eds.), Advances in Environmental Biotechnology. Springer Nature, Singapore, pp. 227–242. doi:10.1007/978-981-10-4041-2_14.

Ebrahimbabaie, P., Pichtel, J., 2021. Biotechnology and nanotechnology for remediation of chlorinated volatile organic compounds: current perspectives. Environ. Sci. Pollut. Res. 0944-1344. 28 (7), 7710–7741. doi:10.1007/s11356-020-11598-y.

El-Sheekh, M.M., El-Kassas, H.Y., Shams El-Din, N.G., Eissa, D.I., El-Sherbiny, B.A., 2021. Green synthesis, characterization applications of iron oxide nanoparticles for antialgal and wastewater bioremediation using three brown algae. Int. J. Phytoremediation. 15497879. 23 (14), 1538–1552. Available from: https://www.tandfonline.com/loi/bijp20. doi:10.1080/15226514.2021.1915957.

Ezugbe, E.O., Rathilal, S., 2020. Membrane technologies in wastewater treatment: a review. Water Environ. Res. 92 (9), 144–158.

Fatima, S., Taj, M.B., Carabineiro, S.A.C., 2025. Ocimum species as multifunctional green adsorbents for the removal of dyes and heavy metals and corrosion inhibition. In: Singh, A.K., Singh, A.K., Susan, M.A.B.H., (Eds), Advances in Nanomaterials for Detection, Control, and Removal of Environmental Pollutants. Springer Nature Switzerland, Cham, pp. 357–391. doi:10.1007/978-3-031-87409-3_14.

Fomina, M., Gadd, G.M., 2014. Biosorption: current perspectives on concept, definition and application. Bioresour. Technol. 18732976. 160, 3–14. Available from: www.elsevier.com/locate/biortech. doi:10.1016/j.biortech.2013.12.102.

Gao, Y., Wang, F., Wu, Y., Naidu, R., Chen, Z., 2016. Comparison of degradation mechanisms of microcystin-LR using nanoscale zero-valent iron (nZVI) and bimetallic Fe/Ni and Fe/Pd nanoparticles. Chem. Eng. J. 13858947. 285, 459–466. Available from: www.elsevier.com/inca/publications/store/6/0/1/2/7/3/index.htt. doi:10.1016/j.cej.2015.09.078.

Gayathri, S., Krishnan, K.A., Krishnakumar, A., Maya, T.M.V., Dev, V.V., Antony, S., Arun, V., 2021. Monitoring of heavy metal contamination in Netravati river basin: overview of pollution indices and risk assessment. Sustain. Water Resour. Manag. 23635045. 7 (2). Available from: http://springer.com/journal/40899. doi:10.1007/s40899-021-00502-2.

Ghazali, F.M., Rahman, R.N.Z.A., Salleh, A.B., Basri, M., 2004. Biodegradation of hydrocarbons in soil by microbial consortium. Int. Biodeterior. Biodegrad. 54 (1), 61–67. doi:10.1016/j.ibiod.2004.02.002.

Goldar, B., Banerjee, N., 2004. Impact of informal regulation of pollution on water quality in rivers in India. J. Environ. Manag. 03014797. 73 (2), 117–130. Available from: https://www.sciencedirect.com/journal/journal-of-environmental-management. doi:10.1016/j.jenvman.2004.06.008.

Gomes, M.P., 2025. Nanophytoremediation: advancing phytoremediation efficiency through nanotechnology integration. Discover Plants. 3005-1207. 2 (1). doi:10.1007/s44372-025-00090-x.

Gong, T., Xu, X., Dang, Y., Kong, A., Wu, Y., Liang, P., Wang, S., Yu, H., Xu, P. and Yang, C., (2018). An engineered Pseudomonas putida can simultaneously degrade organophosphates, pyrethroids and carbamates. Sci. Total Environ. 628-629, 1258–1265. doi:10.1016/j.scitotenv.2018.02.143.

Gressel, J., Al-Ahmad, H., 2005. Assessing ecological risks of transgenic plants used in phytoremediation. Environ. Biosafety Res. 4 (3), 135–150. doi:10.1051/ebr:2005016.

Guerra, F.D., Attia, M.F., Whitehead, D.C. and Alexis, F., (2018). Nanotechnology for Environmental Remediation: Materials and Applications. Molecules, 23(7), 1760. doi:10.3390/molecules23071760.

Guo, C., Wang, H., Wei, Y., Li, J., Peng, B., Shu, X., 2022. Experimental study of al-modified zeolite with oxygen nanobubbles in repairing black odorous sediments in river channels. Water. 2073-4441. 14 (14), 2194. doi:10.3390/w14142194.

Hakim, S., Naqqash, T., Nawaz, M.S., Laraib, I., Siddique, M.J., Zia, R., Mirza, M.S., Imran, A., 2021. Rhizosphere engineering with plant growth-promoting microorganisms for agriculture and ecological sustainability. Front. Sustain. Food Syst. 2571581X. 5. Available from: www.frontiersin.org/journals/sustainable-food-systems#. doi:10.3389/fsufs.2021.617157.

Haque, S.E., Gazi-Khan, L., 2025. Limitations and challenges of bioremediation approach: alternative solutions. In: Srivastav, A.L., Zinicovscaia, I., Cepoi, L., (Eds.), Biotechnologies for Wastewater Treatment and Resource Recovery. Elsevier. pp. 147–157.

Hussain, I., Singh, N.B., Singh, A., Singh, H. and Singh, S.C., (2016). Green synthesis of nanoparticles and its potential application. Biotechnol Lett, 38(4), 545–560. doi:10.1007/s10529-015-2026-7.

Iniesta, J., Michaud, P.A., Panizza, M., Cerisola, G., Aldaz, A., Comninellis, C.H., 2001. Electrochemical oxidation of phenol at boron-doped diamond electrode. Electrochim. Acta. 00134686. 46 (23), 3573–3578. doi:10.1016/S0013-4686(01)00630-2.

Isaac, P., Martínez, F.L., Bourguignon, N., Sánchez, L.A., Ferrero, M.A., 2015. Improved PAHs removal performance by a defined bacterial consortium of indigenous Pseudomonas and actinobacteria from Patagonia, Argentina. Int. Biodeterior. Biodegrad. 101, 23–31.

Iwuozor, K.O., 2019. Prospects and challenges of using coagulation-flocculation method in the treatment of effluents. Adv. J. Chem.-Sect. A 2 (2), 105–127.

Jamwal, P., 2017. Remediation of contaminated urban streams: a decentralized ecological wastewater treatment approach. In: Bhattacharya, S., Gupta, A., Gupta, A., Pandey, A., (Eds.), Water Remediation. Springer Singapore, Singapore, pp. 29–41.

Jeong, S.-W., Choi, Y.J., 2020. Extremophilic microorganisms for the treatment of toxic pollutants in the environment. Molecules. 1420-3049. 25 (21), 4916. doi:10.3390/molecules25214916.

Karigar, C.S., Rao, S.S., 2011. Role of microbial enzymes in the bioremediation of pollutants: a review. Enzyme Res. 20900414. 2011 (1). doi:10.4061/2011/805187.

Kaur, G., Sood, P., 2025. Significance of biological approaches/bioremediation of wastewater treatment over physicochemical methods: a comparative analysis. In: Srivastav, A.L., Zinicovscaia, I., Cepoi, L., (Eds.), Biotechnologies for Wastewater Treatment and Resource Recovery. Elsevier. pp. 211–225.

Kensa, V.M., 2011. Bioremediation-an overview. J. Ind. Pollut. Control. 27 (2), 161–168.

Khan, A., Singh, P., Srivastava, A., 2018. Synthesis, nature and utility of universal iron chelator–Siderophore: a review. Microbiol. Res. 212, 103–111.

Khemili-Talbi, S., Kebbouche-Gana, S., Akmoussi-Toumi, S., Angar, Y., Gana, M.L., 2015. Isolation of an extremely halophilic arhaeon Natrialba sp. C21 able to degrade aromatic compounds and to produce stable biosurfactant at high salinity. Extremophiles. 14334909. 19 (6), 1109–1120. Available from: http://www.springerlink.com/app/home/journal.asp?wasp=a11a8a269a0040c1816cfffb60384506&referrer=parent&backto=linkingpublicationresults,1:100494,1. doi:10.1007/s00792-015-0783-9.

Kobya, M., Soltani, R.D.C., Omwene, P.I., Khataee, A., 2020. A review on decontamination of arsenic-contained water by electrocoagulation: reactor configurations and operating cost along with removal mechanisms. Environ. Technol. Innov. 23521864. 17. Available from: http://www.journals.elsevier.com/environmental-technology-and-innovation/. doi:10.1016/j.eti.2019.100519.

Kulshreshtha, A., Agrawal, R., Barar, M., Saxena, S., 2014. A review on bioremediation of heavy metals in contaminated water. IOSR J. Environ. Sci. Toxicol. Food Technol. 23192399. 8 (7), 44–50. doi:10.9790/2402-08714450.

Kumar, N.M., Muthukumaran, C., Sharmila, G., Gurunathan, B., 2017. Genetically modified organisms and its impact on the enhancement of bioremediation. In: Varjani, S., Agarwal, A., Gnansounou, E., Gurunathan, B., (Eds.), Bioremediation: Applications for Environmental Protection and Management. Springer Singapore, Singapore, pp. 53–76.

Kumari, B., Singh, D.P., 2016. A review on multifaceted application of nanoparticles in the field of bioremediation of petroleum hydrocarbons. Ecol. Eng. 09258574. 97, 98–105. Available from: www.elsevier.com/inca/publications/store/5/2/2/7/5/1. doi:10.1016/j.ecoleng.2016.08.006.

Kunin, R., 1977. The use of macroreticular polymeric adsorbents for the treatment of waste effluents. In: Eisenberg, H., (Ed.), Macromolecular Chemistry–11. Pergamon, pp. 205–211.

Kuokkanen, V., Kuokkanen, T., Rämö, J., Lassi, U., 2013. Recent applications of electrocoagulation in treatment of water and wastewater—a review. Green Sustain. Chem. 2160-6951. 03 (02), 89–121. doi:10.4236/gsc.2013.32013.

Li, L., Hu, J., Shi, X., Fan, M., Luo, J., Wei, X., 2016. Nanoscale zero-valent metals: a review of synthesis, characterization, and applications to environmental remediation. Environ. Sci. Pollut. Res. 16147499. 23 (18), 17880–17900. Available from: http://www.springerlink.com/content/0944-1344. doi:10.1007/s11356-016-6626-0.

Li, X., Wu, S., Dong, Y., Fan, H., Bai, Z., Zhuang, X., 2021. Engineering microbial consortia towards bioremediation. Water. 2073-4441. 13 (20), 2928. doi:10.3390/w13202928.

Mahmud, M.E., Nabi, H., Huq, A.K., Yahya, R.,, and (2016). The removal of heavy metal ions from wastewater/aqueous solution using polypyrrole-based adsorbents: a review. RSC Adv., 6(18), 14778–14791. doi:10.1039/C5RA24358K.

Manara, A., Fasani, E., Furini, A., DalCorso, G., 2020. Evolution of the metal hyperaccumulation and hypertolerance traits. Plant Cell Environ. 0140-7791. 43 (12), 2969–2986. doi:10.1111/pce.13821.

Mani, I., 2020. Biofilm in bioremediation. In: Pandey, V.C., Singh, V., (Eds.), Bioremediation of Pollutants. Elsevier. pp. 375–385.

Marlina, E., 2019. Electro-Fenton for industrial wastewater treatment: a review. In: E3S Web of Conferences, vol. 125. EDP Sciences, p. 03003.

Marlina, E. and Purwanto, 2019. October. Electro-Fenton for industrial wastewater treatment: A review. In E3S Web of Conferences (Vol. 125, p. 03003). EDP Sciences.

Md Anawar, H., Chowdhury, R., 2020. Remediation of polluted river water by biological, chemical, ecological and engineering processes. Sustainability. 12 (17), 7017.

Mesa-Marín, J., Pérez-Romero, J.A., Redondo-Gómez, S., Pajuelo, E., Rodríguez-Llorente, I.D., Mateos-Naranjo, E., 2020. Impact of plant growth promoting bacteria on salicornia ramosissima ecophysiology and heavy metal phytoremediation capacity in estuarine soils. Front. Microbiol. 1664302X. 11. Available from: https://www.frontiersin.org/journals/microbiology#. doi:10.3389/fmicb.2020.553018.

Metcalf & Eddy, A.E.C.O.M., 2014. Wastewater engineering treatment and resource recovery. McGraw-Hill Education.. Tchobanoglous, G., Stensel, H.D., Tsuchihashi, R. and Burton, F.L. (eds.)

Mishra, S., Dixit, S., Soni, S., 2015. Methods of nanoparticle biosynthesis for medical and commercial applications. In: Singh, O.V., (Ed.), Bio-Nanoparticles: Biosynthesis and Sustainable Biotechnological Implications. pp. 141–154.

Mishra, S., Huang, Y., Li, J., Wu, X., Zhou, Z., Lei, Q., Bhatt, P., Chen, S., 2022. Biofilm-mediated bioremediation is a powerful tool for the removal of environmental pollutants. Chemosphere. 00456535. 294, 133609. doi:10.1016/j.chemosphere.2022.133609.

Mohanty, S.K., Gonneau, C., Salamatipour, A., Pietrofesa, R.A., Casper, B., Christofidou-Solomidou, M. and Willenbring, J.K., (2018). Siderophore-mediated iron removal from chrysotile: Implications for asbestos toxicity reduction and bioremediation. J. Hazard. Mater. 341, 290–296. doi:10.1016/j.jhazmat.2017.07.033.

Moussa, D.T., El-Naas, M.H., Nasser, M., Al-Marri, M.J., 2017. A comprehensive review of electrocoagulation for water treatment: potentials and challenges. J. Environ. Manag. 10958630. 186, 24–41. Available from: https://www.sciencedirect.com/journal/journal-of-environmental-management. doi:10.1016/j.jenvman.2016.10.032.

Najid, N., Hakizimana, J.N., Kouzbour, S., Gourich, B., Ruiz-García, A., Vial, C., Stiriba, Y., Semiat, R., 2022. Fouling control and modeling in reverse osmosis for seawater desalination: a review. Comput. Chem. Eng. 00981354. 162, 107794. doi:10.1016/j.compchemeng.2022.107794.

Nidheesh, P.V., Singh, T.A., 2017. Arsenic removal by electrocoagulation process: recent trends and removal mechanism. Chemosphere. 181, 418–432.

Onkani, S.P., 2019. Synthesis and characterization of undoped and Ag doped TiO2, ZnO and ZnS nanoparticles for the photocatalytic degradation of 2-chlorophenol under UV irradiation (Master's thesis, Vaal University of Technology (South Africa)).

Oturan, N., Oturan, M.A., 2018. Electro-fenton process: background, new developments, and applications. In: Martínez-Huitle, C.A., Rodrigo, M.A., Scialdone, O., (Eds.), Electrochemical Water and Wastewater Treatment. Butterworth-Heinemann. pp. 193–221.

Pandis, P.K., Kalogirou, C., Kanellou, E., Vaitsis, C., Savvidou, M.G., Sourkouni, G., Zorpas, A.A., Argirusis, C., 2022. Key points of advanced oxidation processes (AOPs) for wastewater, organic pollutants and pharmaceutical waste treatment: a mini review. ChemEngineering. 23057084. 6 (1). Available from: https://www.mdpi.com/2305-7084/6/1/8/pdf. doi:10.3390/chemengineering6010008.

Parthipan, P., Cheng, L., Rajasekar, A., Parthiba Karthikeyan, O., Rahman, P.K., 2022. Biosurfactants—a next generation biomolecules for enhanced biodegradation of organic pollutants. Front. Microbiol. 13, 947801.

Patel, V., Sharma, S., Desai, C., Kikani, B., Madamwar, D., 2025. Synergistic removal of emerging contaminants using bacterial augmented floating treatment bed system (FTBs) of Typha latifolia and Canna indica for rejuvenation of polluted river water. Front. Microbiol. 1664302X. 16. Available from: https://www.frontiersin.org/journals/microbiology#. doi:10.3389/fmicb.2025.1512992.

Pillai, H.P.S., Kottekottil, J., 2016. Nano-phytotechnological remediation of endosulfan using zero valent iron nanoparticles. J. Environ. Prot. 2152-2197. 07 (05), 734–744. doi:10.4236/jep.2016.75066.

Polińska, W., Kotowska, U., Kiejza, D., Karpińska, J., 2021. Insights into the use of phytoremediation processes for the removal of organic micropollutants from water and wastewater; a review. Water. 2073-4441. 13 (15), 2065. doi:10.3390/w13152065.

Saber-Samandari, S., Gulcan, H., Ozan, S. and Gazi, M., (2014). Efficient Removal of Anionic and Cationic Dyes from an Aqueous Solution Using Pullulan-graft-Polyacrylamide Porous Hydrogel. Water Air Soil Pollut. 225 (11). doi:10.1007/s11270-014-2177-5.

Saha, L., Tiwari, J., Bauddh, K., Ma, Y., 2021. Recent developments in microbe–plant-based bioremediation for tackling heavy metal-polluted soils. Front. Microbiol. 1664302X. 12. Available from: https://www.frontiersin.org/journals/microbiology#. doi:10.3389/fmicb.2021.731723.

Saha, M., Sarkar, S., Sarkar, B., Sharma, B.K., Bhattacharjee, S., Tribedi, P., 2016. Microbial siderophores and their potential applications: a review. Environ. Sci. Pollut. Res. 16147499. 23 (5), 3984–3999. Available from: http://www.springerlink.com/content/0944-1344. doi:10.1007/s11356-015-4294-0.

Saini, S., Tewari, S., Dwivedi, J., Sharma, V., 2023. Biofilm-mediated wastewater treatment: a comprehensive review. Mater. Adv. 26335409. 4 (6), 1415–1443. Available from: https://www.rsc.org/journals-books-databases/about-journals/materials-advances/. doi:10.1039/d2ma00945e.

Shahedi, A., Darban, A.K., Taghipour, F., Jamshidi-Zanjani, A., 2020. A review on industrial wastewater treatment via electrocoagulation processes. Curr. Opin. Electrochem. 24519103. 22, 154–169. doi:10.1016/j.coelec.2020.05.009.

Sharma, S., 2012. Bioremediation: features, strategies and applications. Asian J. Pharm. Life Sci. 2231, 4423.

Sharma, S., Pathania, S., Bhagta, S., Kaushal, N., Bhardwaj, S., Bhatia, R.K., Walia, A., 2024. Microbial remediation of polluted environment by using recombinant *E. coli*: a review. Biotechnol. Environ. 2948-2356. 1 (1). doi:10.1186/s44314-024-00008-z.

Shome, R., 2020. Role of microbial enzymes in bioremediation. ELifePress. 1 (1), 15–20.

Singh, B.K., Walker, A., 2006. Microbial degradation of organophosphorus compounds. FEMS Microbiol. Rev. 15746976. 30 (3), 428–471. doi:10.1111/j.1574-6976.2006.00018.x.

Singh, P., Kim, Y.J., Yang, D.C., 2016. A strategic approach for rapid synthesis of gold and silver nanoparticles by Panax ginseng leaves. Artif. Cells Nanomed. Biotechnol. 44 (8), 1949–1957.

Singh, S., Kang, S.H., Mulchandani, A., Chen, W., 2008. Bioremediation: environmental cleanup through pathway engineering. Curr. Opin. Biotechnol. 09581669. 19 (5), 437–444. doi:10.1016/j.copbio.2008.07.012.

Sravan, J.S., Matsakas, L., Sarkar, O., 2024. Advances in biological wastewater treatment processes: focus on low-carbon energy and resource recovery in biorefinery context. Bioengineering. 11 (3), 281.

Sreedevi, P.R., Suresh, K., Jiang, G., 2022. Bacterial bioremediation of heavy metals in wastewater: a review of processes and applications. J. Water Process Eng. 48, 102884.

Srivastava, N., 2020. Phytoremediation of toxic metals/metalloids and pollutants by Brassicaceae plants. In: Hasanuzzaman, M. (Ed.), The Plant Family Brassicaceae: Biology and Physiological Responses to Environmental Stresses. Springer Singapore, Singapore, pp. 409–435. doi:10.1007/978-981-15-6345-4_14.

Stallwood, B., Shears, J., Williams, P.A., Hughes, K.A., 2005. Low temperature bioremediation of oil-contaminated soil using biostimulation and bioaugmentation with a Pseudomonas sp. from maritime Antarctica. J. Appl. Microbiol. 1364-5072. 99 (4), 794–802. doi:10.1111/j.1365-2672.2005.02678.x.

Takhar, V., Singh, S., 2025. Nanomaterials ROS: a comprehensive review for environmental applications. Environ. Sci. Nano 20518161. 12 (5), 2516–2550. Available from: http://pubs.rsc.org/en/journals/journal/en. doi:10.1039/d5en00049a.

Tammam, A., El-Aggan, W., Abou-Shanab, R. and Mubarak, M., 2021. Improved of growth and phytostabilization potential of lead (Pb) in Glebionis coronaria L. under the effect of IAA and GA3 alone and in combination with EDTA by altering biochemical attributes of stressed plants. Int. J. Phytoremediation, 23 (9), pp. 958–968.

Thangavelu, L., et al ., (2022). [Retracted] Role of Nanoparticles in Environmental Remediation: An Insight into Heavy Metal Pollution from Dentistry. Bioinorg. Chem. Appl. 2022 (1). doi:10.1155/2022/1946724.

Tripathi, M., Verma, A., 2022. Conventional methods for removal of emerging water pollutants emerging water pollutants: concerns and remediation technologies. In: Mazari, S.A., Mujawar, M.N., Sabzoi, N., (Eds.), Emerging Water Pollutants: Concerns and Remediation Technologies. Bentham Science Publishers, India, pp. 204–228. Available from: https://benthambooks.com/book/9789815040739/. doi:10.2174/9789781504073912201001l.

Ukiwe, L.N., Ibeneme, S.I., Duru, C.E., Okolue, B.N., Onyedika, G.O., Nweze, C.A., 2014. Chemical and electro-coagulation techniques in coagulation-flocculation in water and wastewater treatment-a review. J. Adv. Chem. 9 (3), 2321–2807.

Ullah, S., Guo, X., Luo, X., Zhang, X., Leng, S., Ma, N., Faiz, P., 2020. Rapid and long-effective removal of broad-spectrum pollutants from aqueous system by ZVI/oxidants. Front. Environ.

Sci. Eng. 2095221X. 14 (5). Available from: http://www.springerlink.com/content/2095-2201/. doi:10.1007/s11783-020-1268-3.

Venegas-Rioseco, J., Ginocchio, R., Ortiz-Calderón, C., 2021. Increase in phytoextraction potential by genome editing and transformation: a review. Plants. 11 (1), 86. doi:10.3390/plants11010086.

Vieira, R.S., Beppu, M.M., 2005. Mercury ion recovery using natural and crosslinked chitosan membranes. Adsorption. 11 (Suppl 1), 731–736.

Wanapaisan, P., Laothamteep, N., Vejarano, F., Chakraborty, J., Shintani, M., Muangchinda, C., Morita, T., Suzuki-Minakuchi, C., Inoue, K., Nojiri, H., Pinyakong, O., 2018. Synergistic degradation of pyrene by five culturable bacteria in a mangrove sediment-derived bacterial consortium. J. Hazard. Mater. 18733336. 342, 561–570. Available from: www.elsevier.com/locate/jhazmat. doi:10.1016/j.jhazmat.2017.08.062.

Wang, T., Yang, C., Sun, P., Wang, M., Lin, F., Fiallos, M., Khu, S.-T., 2024. Generation mechanism of hydroxyl free radicals in micro–nanobubbles water and its prospect in drinking water. Processes. 2227-9717. 12 (4), 683. doi:10.3390/pr12040683.

Wentzell, B.M., 2025. Phytoremediation for water quality improvement: current advances and future prospects. Biotechnol. Environ. 2948-2356. 2 (1). doi:10.1186/s44314-025-00022-9.

World Health Organization, 2004. Guidelines for Drinking-water Quality, vol. 1. World Health Organization.

Xue, X.Y., Cheng, R., Shi, L., Ma, Z., Zheng, X., 2016. Nanomaterials for monitoring and remediation of water pollution. In: Ranjan, S., Dasgupta, N., Lichtfouse, E., (Eds.), Nanoscience in Food and Agriculture 2. Springer International Publishing, Cham, pp. 207–233.

Xue, X.Y., Cheng, R., Shi, L., Ma, Z., Zheng, X., 2017. Nanomaterials for water pollution monitoring and remediation. Environ. Chem. Lett. 15 (1), 23–27.

Yagnik, S.M., Arya, P.S., Raval, V.H., 2023. Microbial enzymes in bioremediation. In: Brahmachari, G., (Ed.), Biotechnology of Microbial Enzymes. Academic Press. pp. 685–708.

Yang, R.X., Wang, T.T., Deng, W.Q., 2015. Extraordinary capability for water treatment achieved by a perfluorous conjugated microporous polymer. Sci. Rep. 5(1), 10155.

Zamora-Ledezma, C., Negrete-Bolagay, D., Figueroa, F., Zamora-Ledezma, E., Ni, M., Alexis, F., Guerrero, V.H., 2021. Heavy metal water pollution: a fresh look about hazards, novel and conventional remediation methods. Environ. Technol. Innov. 23521864. 22, 101504. doi:10.1016/j.eti.2021.101504.

Zhang, Y., Liu, J., Zhou, Y., Gong, T., Wang, J. and Ge, Y., (2013). Enhanced phytoremediation of mixed heavy metal (mercury)–organic pollutants (trichloroethylene) with transgenic alfalfa co-expressing glutathione S-transferase and human P450 2E1. J. Hazard. Mater. 260, 1100–1107. doi:10.1016/j.jhazmat.2013.06.065.

Zheng, G., Ye, H., Zhang, Y., Li, H., Lin, L., Ding, X., 2015. Removal of heavy metal in drinking water resource with cation-exchange resins (Type 110-H) mixed pes membrane adsorbents. J. Hazard. Toxic Radioact. Waste. 21535515. 19 (2). Available from: http://ascelibrary.org/hzo/. doi:10.1061/(ASCE)HZ.2153-5515.0000229.

Zhuo, Q., Deng, S., Yang, B., Huang, J., Yu, G., 2011. Efficient electrochemical oxidation of perfluorooctanoate using a Ti/SnO_2-Sb-Bi anode. Environ. Sci. Technol. 15205851. 45 (7), 2973–2979. doi:10.1021/es1024542.

Zinicovscaia, I., Svozilíková Krakovská, A., Yushin, N., Peshkova, A., Grozdov, D., 2025. Phytoremediation of zinc-contaminated industrial effluents with phragmites australis and typha latifolia in constructed wetlands. Water. 2073-4441. 17 (16), 2358. doi:10.3390/w17162358.

Index

Note: Page numbers followed by "*f*" indicate figures, "*t*" indicate tables.

F

G

H

I

N

O

P

S

Y

Z

9780443450242